Vitamin C

Vitamin C

Function and biochemistry in animals and plants

Edited by

Han Asard
Department of Biochemistry, Beadle Center,
University of Nebraska-Lincoln,
1901 Vine Street, Lincoln,
NE 68588, USA

James M. May
715 Preston Research Building,
Vanderbilt University School of Medicine,
2220 Pierce Ave., Nashville,
TN 37232-6303, USA

Nicholas Smirnoff
School of Biological Sciences,
University of Exeter,
Washington Singer Laboratories,
Perry Road, Exeter EX4 4QG, UK

BIOS Scientific Publishers
Taylor & Francis Group
LONDON AND NEW YORK

First published 2004

A CIP catalogue record for this book is available from the British Library.

ISBN 1 85996 293 9

Garland Science/BIOS Scientific Publishers
4 Park Square, Milton Park, Abingdon, Oxon, OX14 4RN, UK and
29 West 35th Street, New York, NY 10001–2299, USA
World Wide Web home page: www.bios.co.uk

Garland Science/BIOS Scientific Publishers is a member of the Taylor & Francis Group.

Distributed in the USA by
Fulfilment Center
Taylor & Francis
10650 Toebben Drive
Independence, KY 41051, USA
Toll Free Tel.: +1 800 634 7064; E-mail: taylorandfrancis@thomsonlearning.com

Distributed in Canada by
Taylor & Francis
74 Rolark Drive
Scarborough, Ontario M1R 4G2, Canada
Toll Free Tel.: +1 877 226 2237; E-mail: tal_fran@istar.ca

Distributed in the rest of the world by
Thomson Publishing Services
Cheriton House
North Way
Andover, Hampshire SP10 5BE, UK
Tel.: +44 (0)1264 332424; E-mail: salesorder.tandf@thomsonpublishingservices.co.uk

Library of Congress Cataloging-in-Publication Data
Vitamin C : its function and biochemistry in animals and plants / edited by Han Asard, James May, Nicholas Smirnoff.

p. cm.
Includes bibliographical references and index.
ISBN 1-85996-293-9 (hardcover : alk. paper)
1. Vitamin C. I. Asard, Han. II. May, James. III. Smirnoff, N.

QP772.A8V567 2003
572′.58--dc22
2003017985

Production Editor: Andrew Watts
Typeset by Saxon Graphics Ltd, Derby, UK
Printed and bound by MPG Books, Bodmin, UK

Table of contents

Contributors

Asard, H. Department of Biochemistry, Beadle Center, University of Nebraska-Lincoln, 1901 Vine Street, Lincoln, NE 68588, USA

Bánhegyi, G. Department of Medical Chemistry, Molecular Biology and Pathobiology, Semmelweis University, 9 Puskin Street, 1088 Budapest, Hungary

Buettner, G.R. University of Iowa, Free Radical Research Group, EMRB 68 Iowa City, IA 52242–1101, USA

Córdoba-Pedregosa, M. del Carmen Departamento de Biologia Celular, Fisiologia e Immunologia, Universidad de Córdoba, Campus de Rabanales, Edificio C-6, Planta 3a, 14014 Córdoba, Spain

Corpe, C. Molecular and Clinical Nutrition Section, Bld 10, Rm 4D52, MSC 1372 National Institute of Diabetes, Digestive, and Kidney Diseases, National Institutes of Health, Bethesda, MD 20892–1372, USA

De Gara, L. Dipartimento di Biologia e Patologia Vegetale, V. E Orabona, 4 I-70126 Bari, Italy

De Tullio, M. Dipartimento di Biologia e Patologia Vegetale, V. E Orabona, 4 I-70126 Bari, Italy

Eck, P. Molecular and Clinical Nutrition Section, Bld 10, Rm 4D52, MSC 1372 National Institute of Diabetes, Digestive, and Kidney Diseases, National Institutes of Health, Bethesda, MD 20892–1372, USA

Foyer, C.H. Crop Performance and Improvement Division, Rothamsted Research, Harpenden, Herts AL5 2JQ, United Kingdom

Frei, B. Linus Pauling Institute, 571 Weniger Hall, Oregon State University, Corvallis, OR 97331–6512, USA

Gatzek, S. Max Planck Institute of Molecular Plant Physiology, Am Mühlenberg 1, D-14476 Golm, Germany

González-Reyes, J.A. Departamento de Biologia Celular, Fisiologia e Immunologia, Universidad de Córdoba, Campus de Rabanales, Edificio C-6, Planta 3a, 14014 Córdoba, Spain

Hagen, T.M. Linus Pauling Institute, Oregon State University, Corvallis, OR 97331, USA

Heinecke, J. Medicine/Metabolism, Box 356426, University of Washington, 1959 NE Pacific Street, HSB BB512, Seattle, WA 98195–6426, USA

Higdon, J.V. Linus Pauling Institute, Oregon State University, Corvallis, OR 97331, USA

Katz, A. Molecular and Clinical Nutrition Section, Bld 10, Rm 4D52, MSC 1372 National Institute of Diabetes, Digestive, and Kidney Diseases, National Institutes of Health, Bethesda, MD 20892–1372, USA

Kwon, U. Molecular and Clinical Nutrition Section, Bld 10, Rm 4D52, MSC 1372 National Institute of Diabetes, Digestive, and Kidney Diseases, National Institutes of Health, Bethesda, MD 20892–1372, USA

Lawson, S. Linus Pauling Institute, Oregon State University, Corvallis, OR 97331, USA

Lee, J.-H. Molecular and Clinical Nutrition Section, Bld 10, Rm 4D52, MSC 1372 National Institute of Diabetes, Digestive, and Kidney Diseases, National Institutes of Health, Bethesda, MD 20892–1372, USA

Levine, M. Molecular and Clinical Nutrition Section, Bld 10, Rm 4D52, MSC 1372 National Institute of Diabetes, Digestive, and Kidney Diseases, National Institutes of Health, Bethesda, MD 20892–1372, USA

Loewus, F.A. 1700 NE Upper Drive, Pullman, WA 99163–4624, USA

Loft, S. Department of Clinical Pharmacology, Q-7642, Rigshopitalet, Copenhagen University Hospital, 20 Tagensvej, DK-2200 Copenhagen N, Denmark

Lykkesfeldt, J. Department of Clinical Pharmacology, Q-7642, Rigshopitalet, Copenhagen University Hospital, 20 Tagensvej, DK-2200 Copenhagen N, Denmark

May, J.M. 715 Preston Research Building, Vanderbilt University School of Medicine, 2220 Pierce Ave., Nashville, TN 37232–6303, USA

Michels, A.J. 571 Weniger Hall, Linus Pauling Institute, Oregon State University Corvallis, OR 97331, USA

Møller, P. Department of Clinical Pharmacology, Q-7642, Rigshopitalet, Copenhagen University Hospital, 20 Tagensvej, DK-2200 Copenhagen N, Denmark

Padayatty, S. Molecular and Clinical Nutrition Section, Bld 10, Rm 4D52, MSC 1372 National Institute of Diabetes, Digestive, and Kidney Diseases, National Institutes of Health, Bethesda, MD 20892–1372, USA

Peng, S. Bio-Technical Resources, 1035 South 7th Street, Manitowoc, WI 54220, USA

Poulsen, H.E. Department of Clinical Pharmacology, Q-7642, Rigshopitalet, Copenhagen University Hospital, 20 Tagensvej, DK-2200 Copenhagen N, Denmark

Reiter, R.J. Department of Cellular and Structural Biology, The University of Texas Health Science Center, San Antonio, TX 78284–7762, USA

Rosson, R.A. Bio-Technical Resources, 1035 South 7th Street, Manitowoc, WI 54220, USA

Running, J.A. Bio-Technical Resources, 1035 South 7th Street, Manitowoc, WI 54220, USA

Schafer, F.Q. University of Iowa, Free Radical & Radiation Biology, Dept of Radiation Oncology, EMRB 68, Iowa City, IA 52242–1101, USA

Smirnoff, N. School of Biological Sciences, University of Exeter, Washington Singer Laboratories, Perry Road, Exeter EX4 4QG, United Kingdom

Stocker, R. Centre for Thrombosis and Vascular Research, School of Medical Sciences, University of New South Wales, Sydney, UNSW 2052, Australia

Villalba, J.-M. Departamento de Biologia Celular, Fisiologia e Immunologia, Universidad de Córdoba, Campus de Rabanales, Edificio C-6, Planta 3a, 14014 Córdoba, Spain

Wang, Y. Molecular and Clinical Nutrition Section, Bld 10, Rm 4D52, MSC 1372 National Institute of Diabetes, Digestive, and Kidney Diseases, National Institutes of Health, Bethesda, MD 20892–1372, USA

Weimann, A. Department of Clinical Pharmacology, Q-7642, Rigshopitalet, Copenhagen University Hospital, 20 Tagensvej, DK-2200 Copenhagen N, Denmark

Wilson, J.X. Department of Physiology, University of Western Ontario, London, ON Canada N6A 5C1, Canada

Witting, P.K. Biochemistry Group, The Heart Research Institute, 145 Missenden Road, Camperdown, Sydney, NSW, 2050, Australia

Acknowledgments

The authors gratefully acknowledge the financial support to their laboratories from the Biotechnology and Biological Sciences Research Council (UK) to N.S., from the Bio-Technical resources (Wisconsin, USA) to N.S., from the National Institutes of Health to J.M. (RO1 grant DK 50435) and to H.A. (COBRE grant 1P20RR17675). N.S. also acknowledges the contribution of Dr. Glen L. Wheeler to the work on ascorbate metabolism.

Abbreviations

α-TOH	α-tocopherol
2,3-DHBA	2,3-dihydroxybenzoate
2,5-DKG	2,5-diketo-D-gluconate
2-KLG	2-keto-L-gluonic acid
2-ODD	2-oxoglutarate-dependent dioxygenase
2-OG	2-oxoglutarate
5-KGA	5-keto-D-gluconic acid
6-OHDA	6-hydroxydopamine
Aβ	amyloid β-peptide
AA	ascorbic acid
ABA	abscisic acid
AD	Alzheimer's disease
AI	adequate intake
ALS	amyotrophic lateral sclerosis
AO	ASC oxidase
AP-1	activator protein-1
APX	ASC peroxidase
Ara	D-arabinose
AraL	D-arabinono-1,4-lactone
ASAP	Antioxidant Supplementation in Atherosclerosis Prevention (study)
ASC	ascorbate
BTR	Bio-Technical Resources
CAD	coronary artery disease
CaOx	calcium oxalate
CMS	cytoplasmic male sterile
CNS	central nervous system
CP	chloroplast protein
CSF	cerebrospinal fluid
Cu,Zn-SOD	copper, zinc superoxide dismutase
CuAO	multicopper ascorbate oxidase
CVD	cardiovascular disease
Cytb_5	cytochrome b_5
Cytb_5R	NADH-cytochrome b_5 reductase
DA	dopamine
D-GalUA	D-galacturonic acid
D-GlcUA	D-glucuronate
DHA	dehydroascorbate
DHAR	dehydroascorbate reductase
DOPA	3,4-hydroxyphenyl alanine
DRI	Dietary Reference Intakes

EAA	D-*glycero*-pent-2-enono-1,4-lactone
EAR	Estimated Average Requirement
EASC	D-erythroascorbate
ECF	extracellular fluid
EC-SOD	extracellular superoxide dismutase
EMDHA	D-erythromonodehydroascorbate
EMDHAR	EMDHA reductase
ENDOIII	endonuclease III
eNOS	endothelial nitric oxide synthase
EPR	electron paramagnetic resonance spectroscopy
ER	endoplasmic reticulum
ESCODD	European Standards Committee on Oxidative DNA Damage
EST	expressed sequence tag
FBPase	fructose-1, 6-bisphosphatase
FMD	flow-mediated dilation
FPG	formamidopyrimidine DNA glycosylase
FPLC	fast protein liquid chromatography
G6PDH	glucose 6-phosphate dehydrogenase
GA	gibberellic acid
gAPX	glyoxysome-bound APX
Glc	D-glucose
GLUT	glucose transporter
GMP	GDP-Man pyrophosphorylase
GPX	glutathione peroxidases
GR	glutathione reductase
GSH	reduced glutathione
GSSG	glutathione disulphide
GST-peroxidase	selenium independent peroxidase
HD	Huntington disease
HIF 1α	hypoxia-inducible factor 1α
HPS	Heart Protection Study
I/R	ischemia/reperfusion
IMT	intima-to-media thickness
IPTG	isopropyl-δ-D-thiogalactopyranoside
L-Ascorbate	vitamin C; L -*threo*-hex-2-enono-1,4-lactone
LDL	low density lipoprotein
L-Gal	L-galactose
L-GalDH	L-galactose dehydrogenase
L-GalL	L-galactono-1,4-lactone
L-GalLDH	L-galactono-1,4-lactone dehydrogenase
L-GulL	L-gulono-1,4-lactone
L-GulLO	L-gulono-1,4-lactone oxidase
LH	hydrogen-containing lipid
LHRH	luteinizing hormone releasing hormone
LOAEL	lowest observed adverse effect level
Man	mannose

MCAO	middle cerebral artery occlusion
MDHA	monodehydroascorbate
MDHAR	monodehydroascorbate reductase
mGPDH	mitochondrial glycerol-3-phosphate dehydrogenase
Mn-SOD	manganese superoxide dismutase
MPO	myeloperoxidase
n.d.	not determined
ND	not detectable
NHANES II	National Institutes of Health and Nutrition Examination
$Ni(en)_3^{2+}$	tris-(ethylenediamine)-nickel (II) chloride 2-hydrate
NO	nitric oxide
OM	outer mitochondrial
OxA	oxalic acid
PAF	platelet-activating factor
PAM	peptidyl-α-amidating monooxygenase
pAPX	peroxisome-bound APX
PCD	programmed cell death
*p*CMBS	*para*-(chloromercuri)benzenesulfonic acid
PCR	polymerase chain reaction
PDI	protein disulphide isomerase
PKC	protein kinase C
PM	plasma membrane
PMA	phorbol 12-myristate 13-acetate
POX	peroxidases
PR1	pathogenesis-related protein 1
PRX	2-cys peroxiredoxin
PS	photosystem
pVHL	von Hippel–Lindau tumor suppression protein
QTL	quantitative trait loci
RDA	Recommended Dietary Allowance
RNS	reactive nitrogen species
ROS	reactive oxygen species
RT-PCR	reverse transcriptase polymerase chain reaction
SBPase	sedoheptulose-1,7-bisphosphatase
SDR	short-chain dehydrogenase/reductase
SDS/PAGE	sodium dodecyl sulfate polyacrylamide gel electrophoresis
SMR	standardized mortality ratio
SNP	single nucleotide polymorphisms
SOD	superoxide dismutase
SVCT	sodium-dependent vitamin C transport
TA	L-(+)-tartaric acid
tAPX	thylakoid-bound APX
t-BHP	*t*-butyl hydroperoxide
TCA	tricarboxylic acid
ThA	L-threonic acid
TMP	tocopherol-mediated peroxidation
TRH	thyrotropin-releasing hormone

TTFA	2-thenoyl trifluoroacetone
UDP-Glc	UDP-glucose
UDP-GlcUA	UDP-glucuronic acid
UL	tolerable upper intake level
VDE	violaxanthin de-epoxidase
WAVE	Women's Angiographic Vitamin and Estrogen (study)

The optimum intake of vitamin C: history and controversy

Stephen Lawson, Jane V. Higdon and Balz Frei

Diseases likely to be scurvy have been reported throughout written history. Known as the 'calamity of sailors,' scurvy has also been recorded during famines, sieges, imprisonment and long expeditions over land. James Lind reported the benefits of citrus fruits in treating scurvy in his *Treatise on Scurvy* in 1753, but it was not until 1795 that the British admiralty mandated a daily dose of citrus juice for British seamen, the origin of the term, 'limey' (Carpenter, 1986). Although it was acknowledged that citrus fruits could prevent and cure scurvy, the concept that the disease was caused by the lack of an essential nutrient in the diet was not generally accepted at the beginning of the twentieth century. Reports by Axel Holst and Theodor Frölich in 1907 that scurvy could be produced experimentally in guinea pigs by feeding a diet lacking fresh fruits or vegetables, and the proposal by Casmir Funk in 1912 that scurvy, pellagra, rickets and beriberi were due to dietary deficiencies of factors he called 'vitamines' led to 20 years of intensive efforts toward isolating the anti-scorbutic factor (Sauberlich, 1997). The complementary findings of the research groups of Charles King at the University of Pittsburgh in the U.S. and Albert Szent-Györgyi at the University of Szeged in Hungary (*Figure 1*) led to the discovery of vitamin C as the anti-scorbutic factor in 1932 (King and Waugh, 1932; Svirbely and Szent-Györgyi, 1932). Using material isolated by Szent-Györgyi, a group led by Walter Norman Haworth at the University of Birmingham (*Figure 2*) elucidated the structure of vitamin C in 1933 (Haworth and Hirst, 1933), and vitamin C was first synthesized in the laboratory the same year. Szent-Györgyi was awarded the 1937 Nobel Prize for Physiology or Medicine, in part, for his work in isolating vitamin C as the anti-scorbutic factor, and Haworth was awarded the Nobel Prize for Chemistry the same year. The isolation, identification, and synthesis of ascorbate (ASC) laid the foundation for research into the role of vitamin C in health and disease that continues today.

In the 1950s, the free radical theory of aging hypothesized that free radicals arising from enzymatic and nonenzymatic reactions inside and outside cells

Vitamin C, edited by Han Asard, James M May and Nicholas Smirnoff

Figure 1: Albert Szent-Györgyi. Nobel Prize for Physiology or Medicine, 1937. © The Nobel Foundation.

Figure 2: Walter Norman Haworth. Nobel Prize for Chemistry, 1937. © The Nobel Foundation.

contribute to the aging process (Harman, 1956). However, the presence of free radicals in biological systems was not generally considered likely until the discovery of superoxide dismutase in 1969 (McCord and Fridovich, 1969). Since then, a great deal of evidence has accumulated implicating reactive oxygen and nitrogen species in the pathology of a number of chronic diseases and age-associated functional declines (Knight, 1998; Halliwell and Gutteridge, 1999). Ascorbate readily scavenges many physiologically relevant reactive oxygen and nitrogen species, and is the most effective endogenous aqueous-phase antioxidant in human plasma under many different oxidizing conditions (Frei *et al.*, 1989; Frei, 1991). Although other endogenous antioxidants are able to decrease the rate of lipid and protein oxidation in plasma, only ASC is reactive enough to intercept oxidants before they can cause detectable oxidative damage. These experimental data are in agreement with a thermodynamic hierarchy or 'pecking order' indicating that vitamin C is the terminal small-molecule antioxidant in biological systems (Buettner, 1993).

Nobel Laureate Linus Pauling (*Figure 3*), perhaps the most renowned advocate of vitamin C, became interested in a possible biochemical basis for mental illness in the late 1950s, resulting in the publication in 1968 of his seminal paper *Orthomolecular Psychiatry* in *Science* (Pauling, 1968). Bringing together work in microbial genetics on the pyridoxine-requiring mutant of *Neurospora sitophila* (Beadle and Tatum, 1941) and the relationships between substrate concentrations and reaction rates in enzyme-catalyzed reactions, Pauling hypothesized that some types of mental illness may be effectively treated by providing the optimum

Figure 3: Linus Pauling. Nobel Prize for Peace, 1962. © The Nobel Foundation.

concentration of micronutrients, especially B vitamins and vitamin C, in the brain. He later codified this concept as *orthomolecular medicine*, i.e., 'the preservation of good health and the treatment of disease by varying the concentration in the human body of substances that are normally present in the body'. In support of Pauling's hypothesis, a recent review identified approximately 50 genetic diseases involving defective enzymes that can be remedied by high concentrations of the vitamin component of their coenzymes (Ames *et al.*, 2002).

In the mid-1960s Pauling focused his attention on vitamin C, mainly because of its manifold biochemical functions and provocative but neglected clinical literature from the 1940s and 1950s. Pauling used cross-species comparisons, evolutionary arguments, and the vitamin C content of diets based on raw plant foods to argue for an optimal intake of vitamin C greater than the current recommended dietary allowance (Pauling, 1970). After reviewing the clinical literature on the prevention and treatment of colds with vitamin C, Pauling published *Vitamin C and the Common Cold* in 1970 (Pauling, 1970). He believed that adequate amounts of supplemental vitamin C would help prevent colds, shorten their duration, and ameliorate symptoms. Subsequent clinical trials have provided some support for the latter two effects (Hemila, 1994).

Pauling turned to the possible therapeutic application of vitamin C in cancer in the early 1970s. His clinical collaboration in this field with Ewan Cameron, a Scottish surgeon, spanned two decades and produced a popular book *Cancer and Vitamin C* (Cameron and Pauling, 1979). Cameron and Pauling argued for a role of vitamin C in host resistance to cancer by enhancing immunocompetence, optimizing collagen synthesis for tumor encapsulation, and stabilizing glycosaminoglycans through the stimulation of a hyaluronidase inhibitor (Cameron and Pauling, 1973). Cameron administered at least 10 g/day of vitamin C intravenously to terminal cancer patients for about 10 days, followed by an equivalent oral dose maintained indefinitely. Cameron and Pauling reported that supplemented patients survived more than four times longer than matched controls not treated with vitamin C (Cameron and Pauling, 1976). Decreased use of narcotic analgesics and 'increased feelings of well-being' were also reported in supplemented patients. However, two randomized, placebo-controlled clinical trials of the same dose of oral vitamin C in patients with advanced cancer conducted at the Mayo Clinic in the 1970s and 1980s failed to confirm the benefits of vitamin C observed by Cameron and Pauling (Creagan *et al.*, 1979; Moertel *et al.*, 1985). More recent pharmacokinetic data indicate that intravenous but not oral administration of vitamin C can produce plasma levels of ASC that are toxic to many cancer cell lines in culture, suggesting that the use of intravenous ASC in cancer treatment deserves further study (Padayatty and Levine, 2000).

In his tenth decade, Pauling returned to an earlier theme, that of a possible prophylactic role for vitamin C in heart disease. He hypothesized that lipoprotein(a), a pro-atherogenic lipoprotein, accumulated in arterial walls when vitamin C intake was insufficient, and published three case reports on the amelioration of angina pectoris with vitamin C and L-lysine (Pauling, 1991; McBeath and Pauling, 1993; Pauling, 1993). Angina pectoris is a result of insufficient coronary blood flow to meet the demands of the myocardium. Endothelium-derived nitric oxide (NO) induces vasodilation by stimulating vascular smooth muscle relaxation, in addition to inhibiting the potentially atherogenic processes of smooth muscle proliferation, platelet aggregation and leukocyte–endothelial cell interactions (Carr and Frei, 2000). In the vascular endothelium, NO is synthesized by endothelial nitric oxide synthase (eNOS). Treatment with vitamin C has consistently resulted in improved endothelium-dependent vasodilation in individuals with coronary artery disease, angina pectoris, hypercholesterolemia, hypertension or diabetes (Carr and Frei, 1999). Vitamin C has been found to enhance the activity of eNOS by maintaining its cofactor tetrahydrobiopterin in the reduced, and thus active, form (Huang *et al.*, 2000). The finding that vitamin C enhances eNOS activity through its activity as a reducing agent supports Pauling's idea that sufficient vitamin C intake may contribute to the prevention of cardiovascular diseases.

The consequences of insufficient vitamin C intake have been recorded throughout history, yet ASC was not isolated and recognized as the anti-scorbutic factor until 1932. Although the need for small amounts of vitamin C to prevent scurvy is now widely recognized, optimal intakes of vitamin C with respect to disease prevention and treatment remain controversial. Data from epidemiologic, biochemical and clinical studies indicate that vitamin C intakes between 100 and 200 mg/day are associated with tissue saturation in young healthy adults (Levine *et al.*, 1996, 2001) and reduced risk of chronic disease (Carr and Frei, 1999). However, pharmacokinetic and physiologic responses to vitamin C are known to vary considerably between individuals, and optimal intakes for children, older adults, and those suffering from acute and chronic diseases remain to be determined.

References

Ames BN, Elson-Schwab I, Silver EA (2002) High-dose vitamin therapy stimulates variant enzymes with decreased coenzyme binding affinity (increased Km): relevance to genetic disease and polymorphisms. *Am. J. Clin. Nutr.* **75**, 616–658.

Beadle GW, Tatum EL. (1941) Genetic control of biochemical reactions in Neurospora. *Proc. Natl Acad. Sci. USA* **27**, 499–506.

Buettner GR (1993) The pecking order of free radicals and antioxidants: lipid peroxidation, alpha-tocopherol, and ascorbate. *Arch. Biochem. Biophys.* **300**, 535–543.

Cameron E, Pauling L (1973) Ascorbic acid and the glycosaminoglycans. An orthomolecular approach to cancer and other diseases. *Oncology* **27**, 181–192.

Cameron E, Pauling L (1976) Supplemental ascorbate in the supportive treatment of cancer: Prolongation of survival times in terminal human cancer. *Proc. Natl Acad. Sci. USA* **73**, 3685–3689.

Cameron E, Pauling L (1979) *Cancer and Vitamin C.* The Linus Pauling Institute of Science and Medicine, Menlo Park, CA.

Carpenter KJ (1986) *The History of Scurvy and Vitamin C.* Cambridge University Press, Cambridge.

Carr A, Frei B (2000) The role of natural antioxidants in preserving the biological activity of endothelium-derived nitric oxide. *Free Radic. Biol. Med.* **28**, 1806–1814.

Carr AC, Frei B (1999) Toward a new recommended dietary allowance for vitamin C based on antioxidant and health effects in humans. *Am. J. Clin. Nutr.* **69**, 1086–1107.

Creagan ET, Moertel CG, O'Fallon JR, Schutt AJ, O'Connell MJ, Rubin J, Frytak S (1979) Failure of high-dose vitamin C (ascorbic acid) therapy to benefit patients with advanced cancer. A controlled trial. *N Engl. J. Med.* **301**, 687–690.

Frei B (1991) Ascorbic acid protects lipids in human plasma and low-density lipoprotein against oxidative damage. *Am. J. Clin. Nutr.* **54**, 1113S–1118S.

Frei B, England L, Ames BN (1989) Ascorbate is an outstanding antioxidant in human blood plasma. *Proc. Natl Acad. Sci. USA* **86**, 6377–6381.

Halliwell B, Gutteridge JMC (1999) *Free Radicals in Biology and Medicine,* 3rd Edn. Oxford University Press, New York, NY.

Harman D (1956) Aging: a theory based on free radical and radiation chemistry. *J. Gerontol.* **11** 298–230.

Haworth WN, Hirst EL (1933) Synthesis of ascorbic acid. *J. Soc. Chem. Ind.* **52**, 645–647.

Hemila H (1994) Does vitamin C alleviate the symptoms of the common cold? – a review of current evidence. *Scand. J. Infect Dis.* **26**, 1–6.

Huang A, Vita JA, Venema RC, Keaney JF, Jr. (2000) Ascorbic acid enhances endothelial nitric-oxide synthase activity by increasing intracellular tetrahydrobiopterin. *J. Biol. Chem.* **275**, 17399–17406.

King CG, Waugh WA (1932) The chemical nature of vitamin C. *Science* **75**, 357–358.

Knight JA (1998) Free radicals: their history and current status in aging and disease. *Ann. Clin. Lab. Sci.* **28**, 331–346.

Levine M, Conry-Cantilena C, Wang Y, Welch RW, Washko PW, Dhariwal KR, Park JB, Lazarev A, Graumlich JF, King J, Cantilena LR (1996) Vitamin C pharmacokinetics in healthy volunteers: evidence for a recommended dietary allowance. *Proc. Natl Acad. Sci. USA* **93**, 3704–3709.

Levine M, Wang Y, Padayatty SJ, Morrow J (2001) A new recommended dietary allowance of vitamin C for healthy young women. *Proc. Natl Acad. Sci. USA* **98**, 9842–9846.

McBeath M, Pauling L (1993) A case history: lysine/ascorbate-related amelioration of angina pectoris. *J. Orthomolecular Med.* **8**, 77–78.

McCord JM, Fridovich I (1969) Superoxide dismutase. An enzymic function for erythrocuprein (hemocuprein). *J. Biol. Chem.* **244**, 6049–6055.

Moertel CG, Fleming TR, Creagan ET, Rubin J, O'Connell MJ, Ames MM (1985) High-dose vitamin C versus placebo in the treatment of patients with advanced cancer who have had no prior chemotherapy. A randomized double-blind comparison. *N. Engl. J. Med.* **312**, 137–141.

Padayatty SJ, Levine M (2000) Reevaluation of ascorbate in cancer treatment: emerging evidence, open minds and serendipity. *J. Am. Coll. Nutr.* **19**, 423–425.

Pauling L (1968) Orthomolecular psychiatry. Varying the concentrations of substances normally present in the human body may control mental disease. *Science* **160**, 265–271.

Pauling L (1970) Evolution and the need for ascorbic acid. *Proc. Natl Acad. Sci. USA* **67**, 1643–1648.

Pauling L (1970) *Vitamin C and the Common Cold.* W. H. Freeman, San Francisco, CA.

Pauling L (1991) Case report: lysine/ascorbate-related amelioration of angina pectoris. *J. Orthomolecular Med.* **6**, 144–146.

Pauling L (1993) Third case report on lysine-ascorbate amelioration of angina pectoris. *J. Orthomolecular Med.* **8** 137–138.

Sauberlich HE (1997) A history of scurvy and vitamin C. In: *Vitamin C in Health and Disease* (eds Packer L, Fuchs J). Marcel Decker Inc., New York, NY, pp. 1–24.

Svirbely JL, Szent-Györgyi A (1932) Hexuronic acid as the antiscorbutic factor. *Nature* **129**, 576.

1

Ascorbate biosynthesis: a diversity of pathways

Nicholas Smirnoff, Jeffrey A. Running and Stephan Gatzek

1.1 Introduction

1.1.1 The chemistry of ascorbate

In this chapter we review the pathways of ascorbate (ASC) biosynthesis, consider the factors controlling its rate of synthesis and accumulation and review progress in manipulating ASC synthesis. L-Ascorbate (vitamin C; L-*threo*-hex-2-enono-1,4-lactone) is a hexose sugar derivative (*Figure 1.1*). The key part of the molecule in relation to its biological activity is the *ene*-diol group at carbon atoms 2 and 3. This group gives ASC its acidic and reducing (antioxidant) properties, as it can both ionize (pK_a = 4.17 and 11.17) and readily donate electrons. The *ene*-diol group provides the characteristic UV absorption maximum of ASC (undissociated form 245 nm; monodissociated form 265 nm). It is soluble in water (0.33 g·ml^{-1}), much less so in ethanol (0.033 g·ml^{-1}) and insoluble in lipophilic solvents (Jaffe, 1984). Further information on the chemistry of ASC can be found in the chapters by Buettner and Schafer (Chapter 10), and De Tullio (Chapter 9) in this volume and in Davies *et al.* (1991).

The oxidation of ASC is catalyzed by transition metal ions such as copper (II) or iron (III) that act as electron acceptors from ASC. ASC also reacts with a range of reactive oxygen species (ROS) such as hydrogen peroxide, superoxide and singlet oxygen, which is the basis of its antioxidant action. Reduction of hydrogen peroxide is catalyzed by ASC peroxidase (APX), an enzyme that is particularly prominent in plants (Shigeoka *et al.*, 2002; see Foyer, Chapter 4). The role of ASC as a cofactor for a range of oxygenase and hydroxylase enzymes is also dependent on its reducing activity (see De Tullio, Chapter 9). Upon oxidation, by loss of one electron, the monodehydroascorbate (MDHA) radical is formed and this is usually the initial product of ASC oxidation in biological systems (Smirnoff, 1996; Noctor and Foyer, 1998). It is sometimes referred to as the ascorbate free radical (AFR). The key property of MHDA is that it is a relatively stable radical, which means that it can scavenge ROS and other free radicals without forming an equally reactive and destructive product (Sturgeon *et al.*,

Vitamin C, edited by Han Asard, James M May and Nicholas Smirnoff

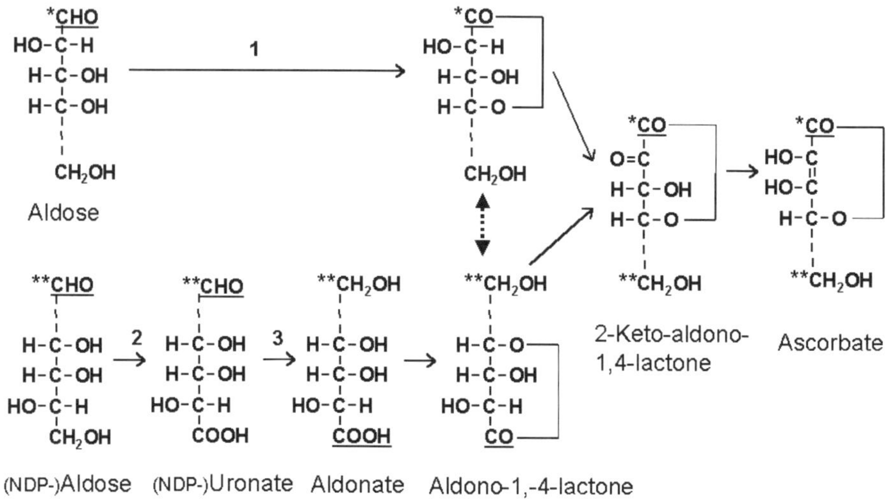

Figure 1.1: Two routes for the biosynthesis of ascorbate and its analogs *via* aldono-1,4-lactones.
The sugars and their derivatives are represented by straight chains rather than their actual ring structures for clarity. Carbon atom 1 (C1) is underlined. In the upper pathway (direct/retention/noninversion pathway) C1, labeled by a single asterisk, is oxidized and becomes C1 of ASC (step 1). In the lower pathway (inversion pathway), sequential oxidation at C6 (step 2) and reduction at C1 (step 3) results in C1, labeled by a double asterisk, becoming C6 of the aldono-1,4-lactone. The double-headed arrow indicates that the aldono-1,4-lactones produced from each pathway are identical. In the final step the aldono-1,4-lactone is oxidized at C2 forming a 2-keto intermediate that spontaneously enolizes to ASC. The dotted lines in the carbon skeletons indicate a variable number of C atoms.

1998). MDHA is recycled back to ASC in two ways. Firstly two MDHA molecules can disproportionate to form ASC and dehydroascorbate (DHA). DHA is relatively unstable, as it is subject to irreversible opening of the lactone ring to form 2,3-diketogulonic acid. DHA can be reduced back to ASC by thiols. The reaction is rapid at high pH but is also catalyzed by a glutathione-dependent DHA reductase and at a lower rate by thioredoxins and trypsin inhibitors (Morell *et al.*, 1997), thioredoxin reductase (May *et al.*, 1997), thiol transferase and protein disulfide isomerase (Wells *et al.*, 1990). MDHA can also be reduced directly back to ASC by a range of mechanisms, including NAD(P)H-dependent monodehydroascorbate reductases, cytochrome b_5 (Lee *et al.*, 2001, see May and Asard, Chapter 8, and Villalba *et al.*, Chapter 7) and by photosystem I in chloroplasts (Asada, 1999). The reactions that regenerate ASC from its oxidized forms are key to maintaining the ASC concentration of plant and animal tissues and are dealt with by May and Asard. ASC also interacts with the lipophilic antioxidant α-tocopherol (vitamin E) by regenerating it from its oxidized radical form (Asada, 1999; Smirnoff, 2000; see Buettner and Schaffer, Chapter 10).

1.1.2 Extraction and assay of ascorbate from biological samples

There is a huge literature on methods for extracting and assaying ASC and DHA from biological samples and foodstuffs (Davey *et al.*, 2000). The key issue is to avoid oxidation of ASC and degradation of DHA. This is achieved by using acidic extractants (e.g., metaphosphoric acid, perchloric acid or trichloroacetic acid) and an agent that binds transition metal ions (e.g., EDTA, oxalic acid or metaphosphoric acid). Assays for ASC include UV/colorimetric methods and chromatography. The change in UV absorbance after specifically oxidizing ASC with ASC oxidase (AO) is commonly used (Hewitt and Dickes, 1961). The ability of ASC to reduce Fe^{3+} in acidic solution with subsequent colorimetric determination of Fe^{2+}, for example by the red complex formed with bipyridyl, is a convenient but less specific assay (Kampfenkel *et al.*, 1995). In both cases ASC plus DHA is determined by first reducing DHA to ASC with a thiol reagent (e.g., dithiothreitol or homocysteine). DHA can be detected by its coloured 2,4-dinitrophenylhydrazine derivative of its de-lactonized product 2,3-diketogulonate (Roe and Kuether, 1943). Chromatography, recently mostly high pressure liquid chromatography (HPLC), can be used to increase the sensitivity and specificity of ASC determination. Ion exchange, reverse phase and ion pairing separations along with UV or electrochemical detection (which increases specificity when interfering compounds are present) have been used (Kimoto *et al.*, 1997; Kall and Andersen, 1999). DHA is determined by prior reduction to ASC or as the 2,4-dinitrophenylhydrazine derivative of its de-lactonized product 2,3-diketogulonate. ASC can also be oxidized before chromatography and determined as DHA. A number of chromatographic methods (e.g., thin layer chromatography on cellulose and borate impregnated silica plates) can distinguish between ASC and its 5-epimer D-araboascorbate (Roomi and Tsao, 1998). Given the existence of ASC analogs and conjugates, it is advisable to check for these by chromatography when investigating an organism from a previously uninvestigated taxonomic group. MDHA, being a radical, can be detected *in vivo* by electron paramagnetic resonance spectroscopy (Heber *et al.*, 1996; Hideg *et al.*, 1997) although this approach is insensitive because oxidative stress must be applied to see the signals clearly. The development of specific fluorescent probes for ASC or DHA that could be used for *in vivo* localization and quantification, as can be done for glutathione, would greatly advance our understanding of ASC function.

1.1.3 The occurrence of ascorbate and ascorbate analogs

ASC is synthesized by animals, green plants and by members of those protistan phyla that have been investigated. However in a few groups, notably guinea pigs, humans and other primates, loss of a functional form of the final enzyme of the biosynthetic pathway (L-gulonolactone oxidase) results in a dietary requirement for ASC. In contrast to the previous groups, most fungi synthesize a C_5 analog, D-erythroascorbate, although some species may also synthesize 6-deoxyascorbate (Sections 1.5.2 and 1.5.3). The 5-epimer of ASC, D-*arabo*ascorbate, is found in one fungal genus, *Penicillium* (Section 1.5.4). The occurrence of ASC, or its analogs, in prokaryotes has not been systematically studied: it appears to be generally absent but has been reported in some cyanobacteria (Wahal *et al.*, 1973; Aaronson *et al.*,

1977). ASC can form conjugates at C2 with various anions (e.g., sulfate, phosphate and fatty acids; Jaffe, 1984). These are more stable than ASC because the *ene*-diol group is protected and they are used in food supplements. There are few reports of these in nature other than in the brine shrimp, which contains ascorbate-2-hydrogen sulphate (Tolbert *et al.*, 1975). Its function is unknown though an ascorbate-2-sulfate sulfohydrolase has been described (Dabrowski *et al.*, 1993). The 5-O-glycosides of the ASC analogs D-erythroascorbate and 6-deoxyascorbate occur in fungi (Section 1.5). In addition to this diversity of ASC analogs, the biosynthetic pathway of L-ASC itself also differs between Kingdoms.

1.2 Ascorbate biosynthesis in mammals

1.2.1 Aldonolactones are the precursors of ascorbate in all known cases

Although the pathways of ASC biosynthesis differ between the major groups of organisms, the immediate precursor is always an aldono-1,4-lactone. Aldonolactones are derived from the corresponding aldose sugar, which is oxidized at C1 to produce an aldonic acid. Rearrangement of the ring structure results in the carboxylate group of the aldonic acid forming a lactone ring with C4 (*Figure 1.1*). ASC is formed by oxidation of the aldonolactone at C2, followed by spontaneous formation of an *ene*-diol group.

1.2.2 The biosynthetic pathway of ascorbate in mammals

ASC biosynthesis is apparently localized in the liver in mammals. In egg-laying mammals, reptiles and amphibians the kidney is probably the site of synthesis, whereas in birds it may be synthesized in either or both organs (Chatterjee, 1973). The pathway is illustrated in *Figure 1.2*. The detailed evidence for this pathway, and the properties of the enzymes involved, are reviewed in more detail by Burns and Conney (1960) and Smirnoff (2001). The aldonolactone precursor of ASC in mammals is L-gulono-1,4-lactone (L-GulL). It is readily converted to ASC when supplied to rats (Isherwood *et al.*, 1954). The enzyme catalyzing this conversion, L-gulono-1,4-lactone oxidase (L-GulLO), is membrane bound and associated with the endoplasmic reticulum. ASC accumulates inside microsomal vesicles supplied with L-GulL, so the enzyme may be localized on the lumenal side (Puskas *et al.*, 1998). It has a covalently bound FAD cofactor and paradoxically the other reaction product is hydrogen peroxide, so a high rate of ASC synthesis also generates an oxidant. L-GulLO is the only mammalian ASC biosynthesis enzyme to have been studied in any detail. Presumably the reason for this focus is that loss of a functional L-GulLO gene accounts for their inability to synthesize ASC (see Section 1.2.3).

L-GulL is derived from uridine diphosphate glucose (UDP-Glc). UDP-Glc is oxidized at C6 by UDP-Glc dehydrogenase to produce UDP-glucuronic acid (UDP-GlcUA). Free GlcUA is then formed, although nothing is known about the enzymes involved (Smirnoff, 2001). UDP-GlcUA is used to detoxify xenobiotics and UDP-Glc dehydrogenase activity is induced by various drugs (e.g., chloretone and barbital). Interestingly, these treatments also increase ASC synthesis in rat

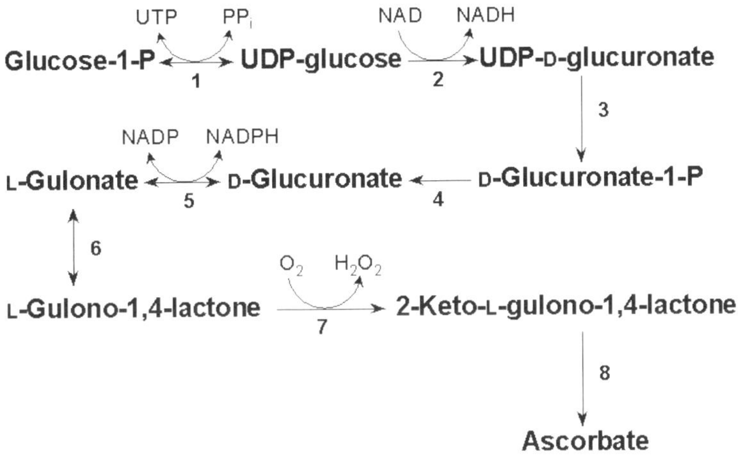

Figure 1.2: The biosynthetic pathway of ascorbate in mammals.
(1) UDP-glucose pyrophosphorylase. (2) UDP-glucose dehydrogenase. (3) UDP-glucuronate pyrophosphatase (speculative). (4) Glucuronate-1-P phosphatase (speculative). (5) Glucuronate reductase. (6) L-Gulono-1,4-lactone hydrolase (aldonolactonase, lactonase I). (7) L-Gulono-1,4-lactone oxidase. (8) Spontaneous.

liver, providing evidence that GlcUA is a precursor of ASC (Hollmann and Touster, 1962). GlcUA is converted to L-GulL by reduction of the C1 aldehyde. There are two possible routes: either glucuronic acid or its 3,6-lactone form could be the substrate. The balance of evidence suggests that glucuronic acid is the substrate and the product L-gulonic acid is then lactonized to L-GulL by aldonolactonase (*Figure 1.2*; Smirnoff, 2001). The enzyme catalyzing glucuronate reduction, L-gulonic acid: NADP oxidoreductase (glucuronic acid reductase), has been detected (York, 1961; Smirnoff, 2001). In this reaction scheme, the reduction of C1 of a uronic acid results in a renumbering of the carbon skeleton because the rule of carbohydrate nomenclature requires that the resulting hydroxymethyl group is C6. This is the reason that in mammals, labeling experiments show that 6-[^{14}C]Glc produces ASC primarily labeled at C1 (*Figure 1.1*). The pathway has therefore been termed an 'inversion' pathway in distinction to the different pathway in green plants (Section 1.3.1). Detailed investigations of the ASC biosynthesis pathway, with the exception of the cloning of L-GulLO (presumably because this was the only aspect relevant to human biology), more or less ceased 30 years ago. Therefore we have little knowledge about how the pathway is controlled and integrated with other aspects of metabolism, as well as how the rate of ASC synthesis is controlled (Banhegyi *et al.*, 1997).

1.2.3 Loss of L-GulLO activity is the reason for inability to synthesis ascorbate in a wide range of animals

While ASC synthesis is the norm for vertebrates, that capacity has been lost during the evolution of several groups (e.g., primates, teleost fish) and species (e.g., guinea pigs, fruit bats, passeriform birds; Chatterjee, 1973). The situation in invertebrates is less clear, with suggestions that insects cannot synthesize ASC. However, *Drosophila melanogaster* (fruit fly) can survive on an ASC-free diet (Massie *et al.*, 1991). It is striking that in all the cases where ability to synthesize ASC is lacking, it is associated with lack of L-GulLO activity. This may indicate that L-GulLO is the only enzyme which is dedicated to ASC synthesis in vertebrates, while all upstream enzymes have other essential functions. Where detailed studies have been carried out, L-GulLO protein cannot be detected by antibodies and its transcript is not detectable (Nishikimi and Udenfriend, 1976; Nishikimi *et al.*, 1988). In humans and guinea pigs, a L-GulLO gene can be detected, but there is considerable sequence divergence from the functional rat gene (Nishikimi *et al.*, 1992 and 1994). The human gene contains retrovirus-like sequences that could be associated with its lack of expression (Challem and Taylor, 1998). It is not known if other enzymes in the pathway, for example those that convert UDP-GlcUA to L-GulL also have reduced activity in ASC-deficient animals. It is clear that whatever the mutational event was that caused loss of biosynthetic capacity, it was not deleterious because of a sufficient dietary supply of ASC. Possibly, loss of ASC biosynthesis could even have selective advantage because hydrogen peroxide formation in the liver or kidney, as a product of L-GulLO activity, is eliminated (Banhegyi *et al.*, 1997).

1.2.4 Metabolic engineering of ascorbate synthesis in animals

On the assumption that loss of functional L-GulLO is the only cause of ASC deficiency in primates and various other animals, introduction of a functional version of the gene is an obvious target for gene therapy (irrespective of the ethical issues). As some teleost fish are unable to synthesize ASC for the same reason as primates, they have been a target for a transgenic approach. The potential commercial value is that the need for ASC supplements for farmed fish would be eliminated. Expression of a L-GulLO gene in transgenic rainbow trout (a teleost fish; Krasnov *et al.*, 1998; 1999), medaka (a teleost fish; Toyohara *et al.*, 1996) and guinea pig (Krasnov *et al.*, 1998; 1999) has been carried out. However this was unsuccessful in introducing ASC synthesis into the fish: rainbow trout did not express active enzyme while medaka did. The functional expression of rat L-GulLO in monkey cells was reported by Yagi *et al.* (1991), although without information on ASC synthesis in these cells. In these approaches, it is assumed that species lacking L-GulLO will have retained expression of the genes needed to convert UDP-GlcUA to L-GulL. In the case of rainbow trout, kidney and liver homogenates could slowly convert UDP-GlcUA to L-GulL, suggesting the other biosynthetic enzymes are present in low amounts (Krasnov *et al.*, 1998). These steps were not examined in medaka (Toyohara *et al.*, 1996).

1.3 Ascorbate biosynthesis in green plants

1.3.1 Synthesis of ascorbate by the mannose/L-galactose pathway

The biosynthetic pathway of ASC in plants has been a source of conjecture since the first investigations in the late 1950s (Smirnoff *et al.*, 2001). The early evidence indicated a distinct pathway from that in mammals in which carbon atom 1 (C1) of a hexose precursor such as Glc, is incorporated into C1 of the resulting ASC (Loewus *et al.*, 1956). This so-called 'noninversion' or 'direct' pathway is different from the 'inversion' pattern in mammals where C1 of Glc is incorporated into C6 of ASC (Section 1.2.2; *Figure 1.1*). A pathway for plants that was consistent with the noninversion pattern of hexose incorporation was proposed by Wheeler *et al.* (1998) and for chlorophyte algae by Berry *et al.* (1999; Section 1.4.2). The pathway is shown in *Figure 1.3*. It is proposed that ASC is synthesized by a two-step oxidation of L-galactose (L-Gal). It is oxidized first by an NAD-dependent L-galactose dehydrogenase (L-GalDH) at C1 to form L-galactono-1,4-lactone (Wheeler *et al.*, 1998; Gatzek *et al.*, 2002) and then by L-galactono-1,4-lactone dehydrogenase (L-GalLDH) at C2 or C3. This enzyme utilizes cytochrome *c* as electron acceptor and is located on the inner mitochondrial membrane (Siendones *et al.*, 1999; Bartoli *et al.*, 2000). L-GalDH appears to be cytosolic (Gatzek *et al.*, 2002). L-Gal is a relatively rare sugar but is known to occur in small quantities in plant cell wall polysaccharides (Baydoun and Fry, 1988; Roberts and Harrer, 1973) and is also a component of polysaccharides in some algae and animals (Feingold, 1982). Antisense suppression of L-GalDH (Gatzek *et al.*, 2002) and L-GalLDH (Tabata *et al.*, 2001) expression in transgenic plants reduces ASC concentrations, providing direct evidence that these enzymes are involved in ASC biosynthesis. Various strands of evidence suggest that L-Gal is derived from guanosine phosphate-D-mannose (GDP-Man). This nucleotide sugar is the mannose donor for cell wall polysaccharides that contain Man and also of Man for protein glycosylation. Radiolabeling studies in peas showed that GDP-Man can be converted to GDP-L-Gal by a double epimerization reaction catalyzed by a GDP-Man-3,5-epimerase (Barber *et al.*, 1979; Hebda *et al.*, 1979; Wheeler *et al.*, 1998). This epimerase has recently been purified and cloned from *Arabidopsis thaliana* (Wolucka *et al.*, 2001). It has high specificity and affinity for GDP-Man. The pathway for GDP-Man synthesis from Man-1-P and GTP, catalyzed by GDP-Man pyrophosphorylase (GMP) is relatively well-established (*Figure 1.3*). That GMP is involved in ASC synthesis is shown by a low ASC *A. thaliana* mutant (*vtc1*) that has low GMP activity and a point mutation in the gene encoding GMP (Conklin *et al.*, 1999). Also, antisense suppression of GMP activity in potato plants reduced their ASC concentrations (Keller *et al.*, 1999). The biochemical and genetic evidence for synthesis of ASC by a Man/L-Gal pathway (the so-called Smirnoff–Wheeler–Running pathway) is strong. The only part of the pathway for which genes have not been cloned is the conversion of GDP- L-Gal to L-Gal. However, specific enzymes catalyzing this conversion have been identified and purified (S. Gatzek, S. Rolinski and N. Smirnoff, unpublished results). The pathway is reviewed in more detail by Smirnoff *et al.* (2001).

1.3.2 Do plants have other biosynthetic routes to ascorbate?

While the evidence described above combined with the patterns of radiolabeling deduced from early experiments by Loewus and coworkers (Loewus *et al.*, 1956, 1958; Loewus and Kelly, 1961; Loewus, 1963) suggest that the Man/L-Gal pathway is the major source of ASC in plants, there is the intriguing possibility

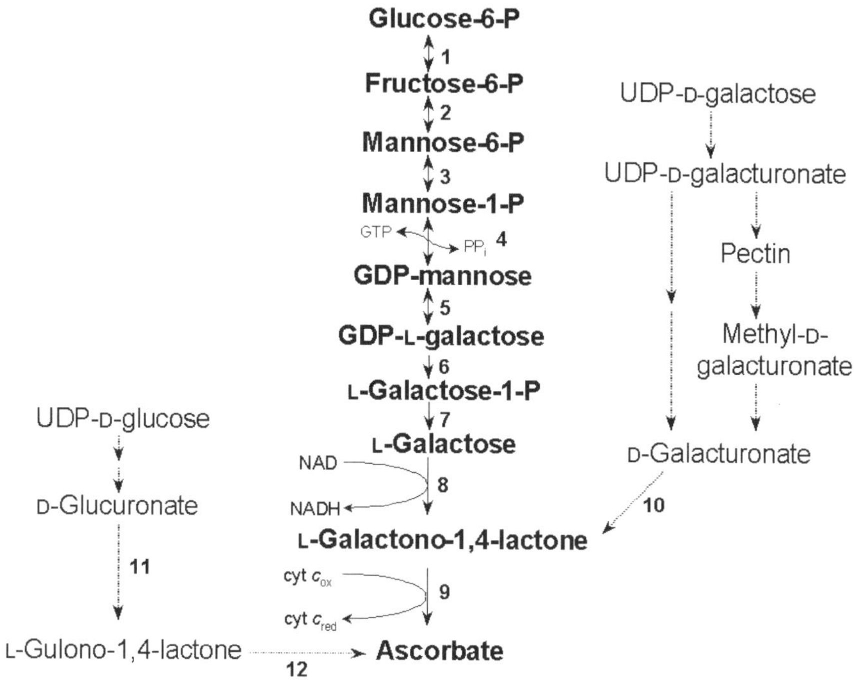

Figure 1.3: The biosynthetic pathway of ascorbate in green plants.
The mannose/L-galactose pathway is shown in bold font. Possible alternative pathways are shown with dotted arrows. Conversion of D-galacturonate to ASC has been engineered by transgenic expression of D-galacturonate reductase (step 10). D-Glucuronate can be converted to ASC but the enzymes (steps 11 and 12) are not characterized. The contribution of these pathways to ASC synthesis may usually be small. (1) Glucose phosphate isomerase. (2) Phosphomannose isomerase. (3) Phosphomannose mutase. (4) GDP-mannose pyrophosphorylase. (5) GDP-mannose-3,5-epimerase. (6,7) Uncharacterized steps. (8) L-Galactose dehydrogenase. (9) L-Galactono-1,4-lactone dehydrogenase. (10) D-Galacturonate reductase. (11) D-Glucuronate reductase. (12) L-Gulono-1,4-lactone oxidase/dehydrogenase.

that alternative minor pathways also operate. These are shown with dotted arrows in *Figure 1.3*. Radioactive D-galacturonic acid (D-GalUA) is incorporated into ASC (Loewus and Kelly, 1961) and administration of the methyl ester of D-GalUA increases ASC concentration (Davey *et al.*, 1999). This suggests that plants can convert D-GalUA to L-GalL in an inversion type pathway analogous to the mammalian route (*Figure 1.2*). Recently a D-GalUA reductase that specifically reduces D-GalUA to L-GalL has been cloned from strawberry fruit (Agius *et al.*, 2003), explaining this result. In addition to a route *via* D-GalUA, plants can also convert exogenously-supplied D-glucuronate (D-GlcUA) to ASC (Davey *et al.*, 1999; *Figure 1.3*). This is presumably reduced to L-GulL as in the mammalian pathway. Purified L-GalLDH from plants is specific for L-GalL (Imai *et al.*, 1998; Mutsuda *et al.*, 1995; Østergaard *et al.*, 1997), but most plants can convert L-GulL to ASC to a greater or lesser extent (Baig *et al.*, 1970; Isherwood *et al.*, 1954). This may imply the existence of an additional unidentified aldonolactone dehydrogenase/oxidase. Detailed analysis of the mass spectra of metabolites separated by gas chromatography from a variety of plant species has identified traces of gulonic acid (the de-lactonized form of gulonolactone), gluconic acid and D-*arabo*ascorbate (Wagner *et al.*, 2003). The presence of gulonate (assuming it to be the L-form) suggests that the pathway from D-glucuronic acid (*Figure 1.3*) could occur *in vivo* to some extent. Also, the presence of traces of gluconic acid (which could be derived from the common pentose phosphate pathway intermediate 6-phosphogluconolactone) and D-*arabo*ascorbate, provide the possibility that D-*arabo*ascorbate could be synthesized in a similar way to that reported in the ascomycete fungus *Penicillium* (Section 1.5.4 and *Figure 1.5*). Another pathway *via* glucosone and sorbosone (Loewus *et al.*, 1990; Saito *et al.*, 1990; Saito, 1996) is possible, but there is little evidence that it is physiologically important (Pallanca and Smirnoff, 1999; Smirnoff *et al.*, 2001). These alternative pathways provide a useful target for engineering increased ASC biosynthesis in plants, as described below.

1.3.3 Engineering ascorbate biosynthesis in plants

Engineering ASC biosynthesis is an attractive target. Plants with increased ASC would have improved nutritional value and might also have improved generic resistance to various environmental stresses as a result of improved antioxidant defense (see Foyer, Chapter 4). Also, plants with increased or decreased ASC would be useful tools to investigate its proposed functions. There are a number of alternative approaches that can be used for metabolic engineering. Firstly, transgenic approaches in which potentially useful genes are overexpressed to increase enzyme activity or expressed as antisense and RNAi constructs to reduce expression can be used (Ecker and Davis, 1986; Smith *et al.*, 2000; Weising *et al.*, 1988). Secondly, where genetic variation in ASC content exists, marker-assisted breeding techniques can be employed in which quantitative trait loci (QTLs) are used to identify and sequence genes involved in determining ASC content (Asins, 2002). Thirdly, an alternative transgenic approach, insertional mutagenesis, can be used to identify genes involved in controlling ASC synthesis and accumulation. In this case plants are transformed with a DNA construct containing a strong promoter that drives high expression of nearby genes (Kardailsky *et al.*, 1999; Weigel *et al.*, 2000). ASC content will be increased if the activated gene is a key biosynthetic enzyme or, more interestingly, a global regulator of the pathway. In plants such as *A. thaliana*, in which the genome sequence is known, the activated gene can be identified by PCR approaches. Several laboratories are actively pursuing the second and third routes at the time of writing.

Transgenic approaches to engineering increased ASC content can take a number of routes. The endogenous D-Man/L-Gal pathway could be targeted, alternative pathways could be introduced or recycling of ASC from DHA and MDHA could be enhanced. As the enzymatic basis for ASC catabolism has not been identified (see Chapter 2 by Banhegyi and Loewus), this route is not open for direct manipulation. Overexpression of the genes encoding the last two enzymes in the pathway, L-GalDH and L-GalLDH has little effect on ASC concentration of plants (Bauw *et al.*, 1998; Gatzek *et al.*, 2002). It is probably not surprising that manipulating the last enzymes of the pathway is generally ineffective, as substrate supply to the pathway may be limited further back. However, overexpression of phosphomannose isomerase and phosphomannose mutase, enzymes at the 'top' of the pathway also had little effect on ASC content (J. Dowdle, S. Gatzek, S. Rolinski and N. Smirnoff, unpublished results).

Manipulation of the enzymes that synthesize and break down GDP-L-Gal may be more profitable, since these control the entry of carbon skeletons into the part of the pathway dedicated to ASC synthesis. Indeed, increases in ASC content of mutants of the unicellular alga *Prototheca moriformis* are strongly linked to increases in the activity of GDP-Man-3,5-epimerase (see Section 1.3.1). Wolucka *et al.* (2001) have also demonstrated a strong correlation between epimerase activity and ASC content of *A. thaliana* cells.

Attempts to introduce alternative pathways have met with more success. Expression of the rat L-GulLO in tobacco and lettuce increased ASC content (Jain and Nessler, 2000). Although the relative increases were four- to sevenfold, the unusually low ASC content of the wild-type plants means that these represent small absolute increases. It is not clear if the enzyme utilized L-GalL (for which it has an affinity) synthesized via the Man/L-Gal pathway or if plants can also synthesize L-GulL. More substantial increases in ASC have been achieved by expressing a strawberry gene encoding D-galacturonic acid reductase in *A. thaliana* (Agius *et al.*, 2003; Smirnoff, 2003). This enzyme reduces D-GalUA acid to L-GalL, and thus facilitates one of the alternative pathways discussed in Section 1.3.2 (*Figure 1.3*). Further investigations will be needed to determine the origin of the D-GalUA acid, which could come from pectin breakdown (pectin, or polygalacturonic acid, is a prominent cell wall polysaccharide) or more directly from UDP-Gal (*Figure 1.3*). This result is important because it indicates that increases in ASC can be engineered by increasing synthesis without activating excessive catabolism. Early work on overexpression of glutathione reductase resulted in small increases in ASC (Foyer *et al.*, 1995), suggesting that improved recycling *via* the ASC-glutathione cycle (see May and Asard, Chapter 8) could stabilize ASC. This approach has now been taken further, and overexpression of a wheat glutathione-dependent dehydroascorbate reductase (DHAR) in tobacco and maize increases ASC concentration by two- to fourfold as well as reducing the proportion of the pool present as DHA (Chen *et al.*, 2003). A key role for DHAR in controlling the redox state of ASC and stabilizing the ASC pool in plants is apparent. It is an exciting possibility that combining increased recycling capacity with increased biosynthesis could result in even more substantial increases in ASC. Manipulation of ASC and DHA carriers, which so far have not been cloned from plants, might also be useful in affecting ASC concentration by altering intracellular compartmentalization and transport between tissues (Franceschi and Tarlyn, 2002).

1.3.4 The control of ascorbate biosynthesis in plants

ASC concentration varies greatly between different species and between tissues of the same species. For example, some fruits contain exceptionally high concentrations (e.g., more than 70 μmol·g^{-1} in the west indian cherry *Acerola*: Davey *et al.*, 2000). Concentrations also tend to be higher in meristematic tissue (i.e., regions of cell division; Conklin, 2001). ASC plus DHA concentrations tend to be low in mature dry seeds, but rapidly increase in the early stages of germination (Pallanca and Smirnoff, 1999). For this reason, germinating pea seedlings have been a useful model system for elucidating the biosynthetic pathway (Wheeler *et al.*, 1998). ASC concentrations are also influenced by the nutritional status of plants (Mozafar, 1993), and markedly by temperature and light intensity. Plants grown at low temperature tend to have more ASC in their leaves (Schöner and Krause, 1990). Also, leaves grown under high light intensity tend to have higher ASC concentration than those grown at lower light intensity (Smirnoff and Pallanca, 1996; Smirnoff, 2000). Low temperature and high light both increase the possibility of photo-oxidative stress from photosynthesis (Asada, 1999), so the responsiveness of ASC to these conditions may be associated with its antioxidant and photoprotective roles (Noctor and Foyer, 1998; Smirnoff, 2000). Now that the pathway of ASC biosynthesis is understood, at least in outline, more can be learnt about how ASC biosynthesis and accumulation is controlled. However, these studies are in their infancy (Smirnoff *et al.*, 2001). With respect to light, activities of several D-Man/L-Gal pathway enzymes, and their corresponding mRNA levels, change rapidly upon transferring plants from low to high light intensity (J. Dowdle and N. Smirnoff, unpublished results; Tabata *et al.*, 2002). The success of engineering increased ASC by overexpressing DHAR described in the previous section indicates that maintaining the ASC pool in a reduced state has a large effect on pool size. Therefore it will be important to take both biosynthesis and recycling into consideration to understand the control of ASC concentration. If there is a coordinate increase in expression of a number of biosynthetic genes when ASC synthesis increases, for example under high light intensity, this implies that they are responsive to light signaling.

1.4 Ascorbate biosynthesis in algae

1.4.1 Occurrence of ascorbate in algae

It has been known for quite some time that algae are capable of synthesizing ASC. References in the literature report the presence of ASC in the brown algae *Ascophyllum* and *Fucus* (Jensen, 1961) the red alga *Pterocladia* (Liso and Calabrese, 1974, 1975), the golden-brown alga *Poteriochromonas* (Aaronson *et al.*, 1971; Helsper *et al.*, 1982), the diatom *Cyclotella* (Grun and Loewus, 1984), the yellow-green alga *Chlorocloster* (Running *et al.*, 1994), the flagellated photosynthetic protist *Euglena* (Shigeoka *et al.*, 1979) and many green algae (Aaronson *et al.*, 1977; Baker *et al.*, 1981; Running *et al.*, 1994). Many of the early reports of ASC in algae were questionable due to assay methods that could not distinguish between the L- and D-isomers. The advent of methods such as HPLC and the AO assay now allows for definitive isomer determinations. These methods have been used to confirm the presence of the L-isomer in algae (Grun and Loewus, 1984; Running *et al.*, 1994).

1.4.2 Mannose/L-galactose pathway in chlorophytes

One of the first clues as to the pathway of ASC synthesis in green algae (chlorophytes) was the demonstration by Loewus' group (Renstrom *et al.*, 1983) that *Chlorella pyrenoidosa* synthesized ASC from ^{14}C-labeled Glc by retaining the carbon chain configuration (i.e., carbon-1 of Glc remained carbon-1 of ASC). Similar results were obtained by Running *et al.* (2003) in ^{13}C labeling studies of the chlorophyte *Prototheca moriformis*. During a program of classical strain improvement for increased L-ASC production (see Running, Peng and Rosson, Chapter 3), they created mutants of *P. moriformis* that varied in their abilities to synthesize ASC. This battery of mutants provided a means by which to elucidate steps in the biosynthetic pathway. Some mutants which made little or no ASC from Glc could still produce ASC when fed either L-Gal or L-GalL. Those that could still synthesize ASC from Glc converted mannose to ASC at a rate greater than the conversion rate from Glc. Various mutants were assayed for the activity of enzymes in the biosynthetic pathway from Glc to GDP-Man. They were also assayed for the activity of an enzyme originally described by Barber (Barber, 1971, 1975) in *C. pyrenoidosa*, GDP-Man-3,5-epimerase, which converts GDP-Man to GDP-L-Gal. None of the enzyme activities linking fructose-6-phosphate to GDP-Man showed any correlation to the mutants' ASC synthesizing abilities. There was a strong correlation between ASC production and epimerase activity (*Figure 1.4*). These data were consistent with a

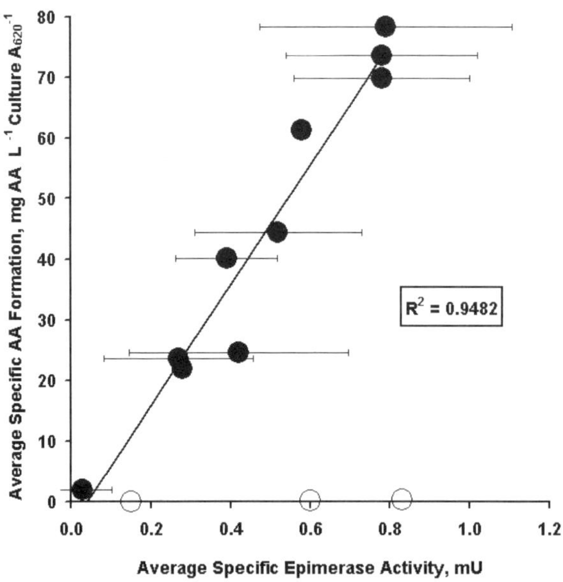

Figure 1.4: Ascorbate productivity vs. epimerase activities in 15 strains of *P. moriformis*. Correlation of average specific epimerase activities with average whole broth ASC specific formations of *P. moriformis* ATCC 75669 and 14 mutants ultimately derived from it. The ASC nonproducing strains represented by open symbols probably are affected in steps between GDP- L-Gal and L-Gal.

biosynthetic pathway (Berry *et al.*, 1999; Running *et al.*, 2003) that is virtually identical to that proposed separately by Wheeler *et al.* (Wheeler *et al.*, 1998), based on their work in *A. thaliana* and peas (Section 1.3.1).

1.4.3 Inversion pathway in photosynthetic protists

One of the first proposals for a biosynthetic pathway from Glc to ASC in an alga was that of Shigeoka *et al.* (1979), studying *Euglena gracilis*. Their results indicated a pathway whereby the carbon chain of Glc was 'inverted' (carbon-1 of Glc became carbon-6 of ASC). They found significant conversions into ASC of uronic acids; Running *et al.* (2003) found no such conversion in *Prototheca*. Loewus' group (Helsper *et al.*, 1982; Grun and Loewus, 1984) subsequently showed that the chrysophyte *Poteriochromonas (Ochromonas) danica* and the diatom *Cyclotella cryptica* also inverted the Glc carbon chain in ASC. There is little more biochemical detail available on the pathway intermediates in these protists. However, it is possible that they use D-GalUA and D-GalL, as in the minor plant 'inversion' type pathway described in Section 1.3.2 (*Figure 1.3*; Shigeoka *et al.*, 1979; Helsper *et al.*, 1982; Grun and Loewus, 1984; Smirnoff *et al.*, 2001).

1.5 Biosynthesis of ascorbate analogs in fungi

1.5.1 Introduction

There are numerous reports in the early literature concerning the presence of ASC in filamentous fungi and yeasts (Sastry and Sarma, 1957; Takahashi *et al.*, 1960; Bleeg, 1966; Heick *et al.*, 1969). However the assay methods employed (absorbance peak of the *ene*-diol group, absorbance at 540 nm of 2,4-dinitrophenylhydrazine derivatives, decolorization of 2,6-dichlorophenolindophenol, paper chromatography) could not distinguish between ASC and other compounds with the same chemical properties. More recent investigations using improved analytical techniques suggest that fungi do not generally contain ASC but instead contain the C_5 analog erythroascorbate (D-*glycero*-pent-2-enono-1,4-lactone, EAA) (*Figure 1.5*). Some fungi also contain 6-deoxyascorbate. The occurrence of the 5-epimer of ASC, D-*arabo*ascorbic acid (D-erythorbic acid; D-isoascorbic acid) is restricted to one genus (*Penicillium*). Another apparently unique feature of fungi is that glycosides of EAA and 6-deoxyascorbate are also present. In this context the true fungi are considered to be members of the Ascomycetes, Basidiomycetes and Zygomycetes.

1.5.2 D-Erythroascorbate

As early as 1967, Yasuda reported the presence in a yeast (*Candida*) of EAA, a five-carbon analog of ASC with very similar reducing properties. Murakawa *et al.* (1977) tested nearly 100 strains of *Candida* and found EAA as the only ASC analog produced amongst them. Using an HPLC method that could detect 0.1 mg·l^{-1} of either ASC or EAA, Running *et al.* (1994) were unable to detect either in any extract of 20 yeast species from nine genera. Leung and Loewus (1985) have also questioned

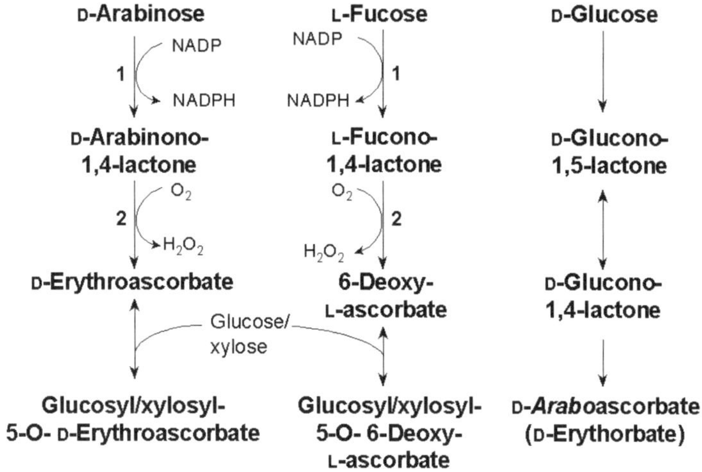

Figure 1.5: Synthesis of ascorbate analogs in fungi.
Ascomycete, basidiomycete and zygomycete fungi synthesize D-erythroascorbate and its glycosides. Basidiomycete fungi also synthesize 6-deoxyascorbate and its glycosides. D-*Arabo*ascorbate has only been found in some *Penicillium* species (Ascomycetes). The enzymes of glycoside synthesis and breakdown are unknown. (1) Arabinose dehydrogenase. (2) Arabinono-1,4-lactone oxidase.

the early reports of ASC in fungi and re-tested the yeasts *Saccharomyces cerevisiae* and *Lipomyces starkeyi* using HPLC, finding not ASC, but a similar unknown reducing compound subsequently identified as EAA (Nick *et al.*, 1986). EAA has also been reported in the filamentous fungi *Neurospora* (Dumbrava and Pall, 1987) and *Sclerotinia* (Loewus *et al.*, 1995). All the preceding species are Ascomycete fungi. The Basidiomycetes (which includes mushrooms) also appear to lack ASC but contain EAA and additionally have D-*arabo*ascorbate (Okamura, 1998). The only group of true fungi in which ASC metabolism has not been investigated in detail is the Zygomycetes. Examination of a representative of this group (*Phycomyces blakesleeanus*) shows that it lacks ASC but contains EAA (A. Baroja and N. Smirnoff, unpublished results). This suggests that EAA synthesis is characteristic of the true fungi. EAA has the same reactivity as ASC and presumably plays the same roles in fungi as ASC does in other organisms. For example disrupting EAA biosynthesis or recycling in yeast increases sensitivity to oxidative stress (Huh *et al.*, 1998; Lee *et al.*, 2001), although the concentration of EAA in yeast cells is very much lower than in plant cells (Spickett *et al.*, 2000). A substantial proportion of EAA in both Ascomycetes (e.g., *Sclerotinia sclerotiorum*, Keates *et al.*, 1998) and Basidiomycetes (Okamura, 1998) is present as glycosides in which C1 of galactose or Glc is attached to C5 of EAA. The glycosides retain the reducing activity of EAA but the significance of this conjugation to sugars is unknown, as are the enzymes involved in formation and hydrolysis of the glycosides.

The biosynthetic pathway of EAA in yeasts (e.g., *S. cerevisiae* and *Candida*) is shown in *Figure 1.5*. Feeding D-arabinose (Ara) and D-arabinono-1,4-lactone

(AraL) to yeast results in increased EAA pool size (Spickett *et al.*, 2000). The pathway is therefore similar to the direct plant pathway (*Figure 1.2*) and an analogous aldose-1-dehydrogenase to plant L-GalDH, AraDH, catalyzes oxidation of Ara to AraL. The enzyme uses NADP as a cofactor and has a rather low affinity (~ 10 mM) for Ara (Kim *et al.*, 1996). It is also active with L-Gal and L-fucose (L-Fuc, 6-deoxy-L-Gal). The genes encoding AraDH have been identified from *Candida albicans* (Kim *et al.*, 1996) and *S. cerevisiae* (Kim *et al.*, 1998). In the final step of the pathway AraL is oxidized to EAA by AraL oxidase (AraLO; Huh *et al.*, 1994). This enzyme is localized in mitochondria, like the plant L-GalLDH, but uses oxygen as electron acceptor, like mammalian L-GulLO. As well as oxidizing AraL, it can form ASC from L-GalL (Huh *et al.*, 1994). The gene encoding AraLO has been cloned from yeast (Huh *et al.*, 1998). Homologues can be identified in other fungi, which have about 26% amino acid sequence identity to the plant and mammalian orthologues (*Figure 1.6*). Proof that the proposed pathway operates in

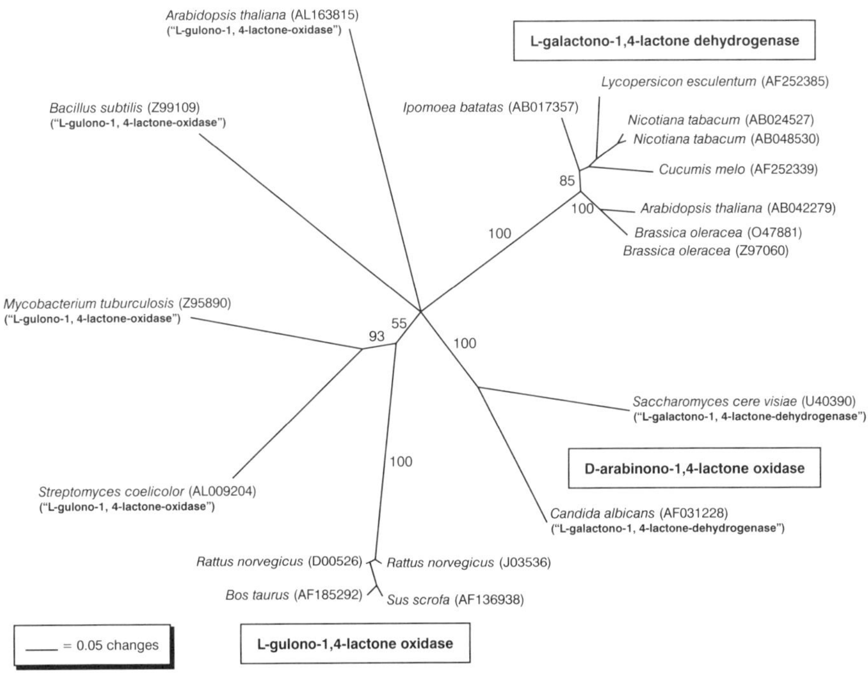

Figure 1.6: A phylogenetic tree of aldonolactone/oxidase and dehydrogenases. Members of the aldonolactone/oxidase and dehydrogenase family were identified by sequence similarity and a phylogenetic tree was constructed on the basis of predicted amino acid sequences. The plant enzymes (L-galactolactone dehydrogenases), mammalian enzymes (L-gulonolactone oxidases) and fungal enzymes (D-arabinonolactone oxidases) form distinct clusters and are labeled. There is currently no proof that the bacterial sequences and one plant sequence (*A. thaliana*) annotated as L-gulonolactone oxidases have enzyme activity. The phylogram was constructed by bootstrapped (100 replicates) neighbor-joining analysis. Bootstrap values > 50% are shown for all major nodes. Sequences were aligned using Clustal X and phylogenetic analysis was performed using PAUP*4 (Swofford, 1998; J.R. Stevens and N. Smirnoff, unpublished results).

yeast was provided by knocking out expression of AralLO in yeast: the resulting cells were EAA deficient (Huh *et al.*, 1998). In the earlier literature, before identification of EAA in yeast, this enzyme was assumed to be a L-GalL oxidase (Bleeg, 1966; Nishikimi *et al.*, 1978). It seems that not all fungi can convert L-GalL into ASC. Onofri *et al.* (1997) demonstrated this conversion in *S. cerevisiae, Clavispora* and *Pichia*, but could not detect it in *Cryptococcus*.

1.5.3 6-Deoxyascorbate

In addition to EAA and its glycosides, Basidiomycete fungi contain 6-deoxyascorbate and its 5-xylosyl and 5-glucosyl glycosides (Okamura, 1994; 1998). 6-Deoxyascorbate could be derived from L-Fuc. Yeast AraDH has the same affinity for L-Fuc as Ara, and could therefore form L-fucono-1,4-lactone. AraL oxidase could then oxidize the lactone to 6-deoxyascorbate. In support of this pathway, the zygomycete *P. blakesleeanus* can convert exogenous L-Fuc to 6-deoxyascorbate (A. Baroja and N. Smirnoff, unpublished results). The same pathway can operate using the plant ASC pathway because L-Fuc-fed *Chlorella* cells produce a reducing compound that gives a strong positive reaction in the bipyridyl colorimetric assay for ASC, but which was shown by HPLC to not be ASC (J. A. Running, unpublished results). GDP-L-Fuc is formed in *Escherichia coli* from GDP-D-Man in a three-step reaction catalyzed by two enzymes, GDP-4, 6-dehydratase and a dual-function 3,5-epimerase/reductase (Somers *et al.*, 1998). Plants are known to have these activities (Bonin *et al.*, 1997; Bonin and Reiter, 2000), and thus GDP-D-Man can be converted by one 3,5-epimerase to GDP-L-Gal (as in the plant ASC biosynthesis pathway) and *via* two enzymes that also result in 3,5-epimerase to the 6-deoxy-derivative of GDP-L-Gal, GDP-L-Fuc.

1.5.4 D-Araboascorbate

Takahashi *et al.* (1960) performed infra-red analysis of 2,4-dinitrophenylhydrazone derivatives of a reducing compound produced by *Penicillium* species, and concluded it was D-*arabo*ascorbate. They proposed a three-step synthesis route from Glc to D-gluconate to D-glucono-1,4-lactone to D-*arabo*ascorbate. This proposal was later modified when evidence showed that D-gluconate was not converted to D-glucono-1,4-lactone directly, but rather was first converted enzymatically to D-glucono-1,5-lactone, which could be non-enzymatically converted to the 1,4-lactone (Takahashi and Mitsumoto, 1961; Fig 1.5). The sequence is thus Glc, glucono-1,5-lactone, glucono-1,4-lactone, D-*arabo*ascorbate, the last reaction catalyzed by D-gluconolactone dehydrogenase (Takahashi *et al.*, 1976; Murakawa and Takahashi, 1977). Researchers at Fujizawa Co. screened 4800 fungi and found only 20 strains that produced D-*arabo*ascorbate (Yagi *et al.*, 1967), all of them *Penicillium* species. This group developed a fermentation process with *Penicillium* to produce 80 $g \cdot l^{-1}$ D-*arabo*ascorbate in laboratory-scale fermentors (Shimizu *et al.*, 1967; Yagi *et al.*, 1967). *Penicillium* is an imperfect (anamorphic) genus of fungus, representing morphologically similar strains in which a sexual stage has not been observed. When sexuality is observed in an anamorphic fungus, it is re-classified into one of the three individual classes of fungi for which sexual reproduction is known. Many 'penicillia' have thus been re-classified as members of the perfect (teleomorphic) genera *Eupenicillium, Hamigera, Talaromyces* (LoBuglio *et al.*, 1993). Since *Penicillium* is the only fungal genus in which D-*arabo*ascorbate has been detected, it is possible that those strains are the anamorphic forms of a single teleomorphic genus, that alone among the fungi can produce D-*arabo*ascorbate.

1.6 Phylogenetic considerations and conclusions

It is evident from our review of ASC biosynthesis that much remains to be learnt. Most of the significant information on ASC biosynthesis in mammals was gained over 30 years ago, with very little knowledge being based on more recent molecular approaches. In relation to improving our understanding of the functions of ASC, it is very likely that a deeper understanding of the mechanisms that control its synthesis in mammals will be revealing. The occurrence of ASC analogs and the biosynthetic pathways in invertebrates and protists has hardly been explored, although the evidence reviewed in this chapter suggests that at least some protists differ from mammals and green plants. The ASC biosynthesis pathway in plants has only recently been elucidated and research is now underway on the factors controlling ASC synthesis and engineering increased production. This could improve the nutritional value of food plants and possibly improve their stress resistance. Also, plants and algae with altered ASC content will provide useful tools for understanding its functions.

The diversity of ASC biosynthesis pathways prompts speculation on the origin of ASC. It is generally thought that prokaryotes lack ASC. However, it has been reported to occur in some cyanobacteria (Tel-Or *et al.*, 1985) along with the enzyme APX (Miyake *et al.*, 1991). If correct, this possibility would be interesting and requires further investigation. Ancestors of the cyanobacteria were the first cells to acquire oxygenic photosynthesis, causing the large increase in atmospheric oxygen that resulted in aerobic metabolism. However, it seems clear that many components of the antioxidant system (e.g., catalase and superoxide dismutase) already existed in early 'anaerobic' cells to deal with ROS and other free radicals derived from traces of oxygen and generated by UV radiation. Therefore did ASC biosynthesis, which requires oxygen, 'evolve' before or after oxygenic photosynthesis? Green plants (chlorophyte algae, bryophytes and vascular plants) share the Man/L-Gal pathway: could this have been derived from the endosymbiotic cyanobacteria that became chloroplasts? Vertebrates (and possibly invertebrates) share the L-GulL pathway. The true fungi all synthesize EAA and this confirms them as a very distinct group of organisms. The diverse groups of protists (unicellular or simple colonial organisms) have hardly been studied: some have a pathway similar to mammals but use L-GalL as the precursor instead of L-GulL. Further studies of this group may contribute evidence for their relationships. The aldonolactone oxidases/dehydrogenases that have been sequenced so far have regions of high sequence similarity with overall amino acid identity of 26–31% (Smirnoff *et al.*, 2001). A phylogenetic tree of amino acid sequences related to aldonolactone oxidases/dehydrogenases is shown in *Figure 1.6*. Plant, animal and fungal sequences form their own clusters. Interestingly, a sequence more similar to the mammalian enzyme also occurs in a plant (*A. thaliana*) and related sequences of unknown function occur in bacteria. What is not clear is if ASC synthesis appeared in an early eukaryotic cell and the pathways then diverged (along with gene sequences) as the animals, plants, protists and fungi evolved or if ASC (and EAA and 6-deoxyascorbate) synthesis evolved independently in each group.

References

Aaronson S, DeAngelis B, Frank O, Baker H (1971) Secretion of vitamins and amino acids into the environment by *Ochromonas danica*. *J. Phycol.* **7**, 215–218.

Aaronson S, Dhawale SW, Patni NJ, DeAngelis B, Frank O, Baker H (1977) The cell content and secretion of water-soluble vitamins by several freshwater algae. *Arch. Microbiol.* **112**, 57–59.

Agius F, González-Lamothe R, Caballero JL, Munoz-Blanco J, Botella MA, Valpuesta V (2003) Engineering increased vitamin C levels in plants by overexpression of a D-galacturonic acid reductase. *Nature Biotechnol.* **21**, 177–181.

Asada K (1999) The water-water cycle in chloroplasts: Scavenging of active oxygens and dissipation of excess photons. *Ann. Rev. Plant Phys. Plant Mol. Biol.* **50**, 601–639.

Asins MJ (2002) Present and future of quantitative trait locus analysis in plant breeding. *Plant Breeding* **121**, 281–291.

Baig MM, Kelly S, Loewus FA (1970) L-Ascorbic acid biosynthesis in higher plants from L-gulono-1,4-lactone and L-galactono-1,4-lactone. *Plant Physiol.* **46**, 277–280.

Baker ER, McLaughlin JJA, Hutner SH, DeAngelis B, Feingold S, Frank O, Baker, H (1981) Water-soluble vitamins in cells and spent culture supernatants of *Poteriochromonas stipitata*, *Euglena gracilis*, and *Tetrahymena thermophila*. *Arch. Microbiol.* **129**, 310–313.

Bánhegyi G, Braun L, Csala M, Puskas F, Mandl J (1997) Ascorbate metabolism and its regulation in animals. *Free Rad. Biol. Med.* **23**, 793–803.

Barber GA (1971) The synthesis of L-glucose* by plant enzyme systems. *Arch. Biochem. Biophys.* **147**, 619–623. *This is a typographical error in the published paper: 'L-glucose' should read 'L-galactose'.

Barber GA (1975) The synthesis of guanosine 5′-diphosphate-L-galactose by extracts of *Chlorella pyrenoidosa*. *Arch. Biochem. Biophys.* **167**, 718–722.

Barber GA (1979) Observations on the mechanism of the reversible epimerization of GDP-D-mannose to GDP-L-galactose by an enzyme from *Chlorella pyrenoidosa*. *J. Biol. Chem.* **254**, 7600–7603.

Bartoli CG, Pastori GM, Foyer CH (2000) Ascorbate biosynthesis in mitochondria is linked to the electron transport chain between complexes III and IV. *Plant Physiol.* **123**, 335–343.

Bauw GJC, Davey MW, Van Montagu M, Østergaard J (1998) Production of ascorbic acid in plants. World Intellectual Property Organization (WIPO) patent WO9850558.

Baydoun EAH, Fry SC (1988) [2-H-3] Mannose incorporation in cultured plant-cells – investigation of L-galactose residues of the primary-cell wall. *J. Plant Physiol.* **132**, 484–490.

Berry A, Running JA, Severson DK, Burlingame RP (1999) Vitamin C production in microorganisms and plants. World Intellectual Property Organization (WIPO) patent WO9964618.

Bleeg HS (1966) L-ascorbic acid in yeast and isolation of L-galactono-γ-lactone oxidase from the mitochondria. *Enzymologia* **31**, 105–112.

Bonin C, Potter I, Vanzin GF, Reiter W-D (1997) The MUR1 gene of *Arabidopsis thaliana* encodes an isoform of GDP-D-mannose-4,6-dehydratase, catalyzing the first step in the *de novo* synthesis of GDP-L-fucose. *Proc. Natl Acad. Sci. USA* **94**, 2085–2090.

Bonin CP, Reiter WD (2000) A bifunctional epimerase-reductase acts downstream of the MUR1 gene product and completes the de novo synthesis of GDP-L-fucose in *Arabidopsis thaliana*. *Plant J.* **21**, 445–454.

Burns JJ, Conney AH (1960) Water-soluble vitamins. 1. (ascorbic acid, nicotinic acid, vitamin-B6, biotin, inositol). *Annu. Rev. Biochem.* **29**, 413–436.

Challem JJ, Taylor EW (1998) Retroviruses, ascorbate, and mutations, in the evolution of *Homo sapiens*. *Free Rad. Biol. Med.* 25, 130–132.

Chatterjee IB (1973) Evolution and biosynthesis of ascorbic-acid. *Science* **182**, 1271–1272.

Chen Z, Young E, Ling J, Chang SC, Gallie DR (2003) Increasing vitamin C content of plants through enhanced ascorbate recycling. *Proc. Natl Acad. Sci. USA* **100**, 3525–3530.

Conklin PL (2001) Recent advances in the role and biosynthesis of ascorbic acid in plants. *Plant Cell Env.* **24**, 383–394.

Conklin PL, Norris SR, Wheeler GL, Williams EH, Smirnoff N, Last RL (1999) Genetic evidence for the role of GDP-mannose in plant ascorbic acid (vitamin C) biosynthesis. *Proc. Natl Acad. Sci. USA* **96**, 4198–4203.

Dabrowski K, Lackner R, Doblander C (1993) Ascorbate-2-sulfate sulfohydrolase in fish and mammal – comparative characterization and possible involvement in ascorbate metabolism. *Comp. Biochem. Phys. B* **104**, 717–722.

Davies MB, Austin J, Partridge DA (1991) *Vitamin C: Its Chemistry and Biochemistry*. Royal Society of Chemistry, Cambridge, UK.

Davey MW, Gilot C, Persiau G, Ostergaard J, Han Y, Bauw GC, Van Montagu MC (1999) Ascorbate biosynthesis in *Arabidopsis* cell suspension culture. *Plant Physiol.* **121**, 535–543.

Davey MW, Van Montagu M, Inze D, Sanmartin M, Kanellis A, Smirnoff N, Benzie IJJ, Strain JJ, Flavell D, Fletcher J (2000) Plant L-ascorbic acid: chemistry, function, metabolism, bioavailability and effects of processing. *J. Sci. Food Agric.* **80**, 825–860.

Dumbrava V-A, Pall ML (1987) Control of nucleotide and erythroascorbic acid pools by cyclic AMP in *Neurospora crassa*. *Biochim. Biophys. Acta* **926**, 331–338.

Ecker JR, Davis RW (1986) Inhibition of gene expression in plant cells by expression of antisense RNA. *Proc. Natl Acad. Sci. USA* **83**, 5372–5376.

Feingold DS (1982) Aldo (and keto) hexoses and uronic acids. In: *Encyclopedia of Plant Physiology. Plant Carbohydrates I: Intracellular Carbohydrates* (eds Loewus FA, Tanner W). Springer, New York, pp. 3–76.

Foyer CH, Souriau N, Perret S, Lelandais M, Kunert KJ, Pruvost C, Jouanin L (1995) Overexpression of glutathione reductase but not glutathione synthetase leads to increases in antioxidant capacity and resistance to photoinhibition in poplar trees. *Plant Physiol.* **109**, 1047–1057.

Franceschi VR, Tarlyn NM (2002) L-Ascorbic acid is accumulated in source leaf phloem and transported to sink tissues in plants. *Plant Physiol.* **130**, 649–656.

Gatzek S, Wheeler GL, Smirnoff N (2002) Antisense suppression of L-galactose dehydrogenase in *Arabidopsis thaliana* provides evidence for its role in ascorbate synthesis and reveals light modulated L-galactose synthesis. *Plant J.* **30**, 541–553.

Grun M, Loewus FA (1984) L-ascorbic acid biosynthesis in the euryhaline diatom *Cyclotella cryptica*. *Planta* **160**, 6–11.

Hebda PA, Behrman EJ, Barber GA (1979) Guanosine 5′-diphosphate D-mannose: guanosine 5′-diphosphate L-galactose epimerase of *Chlorella pyrenoidosa*. Chemical synthesis of guanosine 5′-diphosphate L-galactose and further studies of the enzyme and the reaction it catalyzes. *Arch. Biochem. Biophys.* **194**, 496–502.

Heber U, Miyake C, Mano J, Ohno C, Asada K (1996) Monodehydroascorbate radical detected by electron paramagnetic resonance spectrometry is a sensitive probe of oxidative stress in intact leaves. *Plant Cell Physiol.* **37**, 1066–1072.

Heick HM, Stewart HB, Graff GL, Humpers JEC (1969) Occurrence of ascorbic acid in the yeast *Lipomyces starkeyi*. *Can. J. Microbiol.* **47**, 752–753.

Helsper JP, Kagan L, Hilby CL, Maynard TM, Loewus FA (1982) L-ascorbic acid biosynthesis in *Ochromonas danica*. *Plant Physiol.* **69**, 465–468.

Hewitt EJ, Dickes GJ (1961) Spectrophotometric measurements on ascorbic acid and their use for estimation of ascorbic acid and dehydroascorbic acid in plant tissues. *Biochem. J.* **78**, 384–391.

Hideg E, Mano J, Ohno C, Asada K (1997) Increased levels of monodehydroascorbate radical in UV-B-irradiated broad bean leaves. *Plant Cell Physiol.* **38**, 684–690.

Hollmann S, Touster O (1962) Alterations in tissue levels of uridine diphosphate glucose dehydrogenase, uridine diphosphate glucuronic acid pyrophosphatase and glucuronyl transferase induced by substances influencing production of ascorbic acid. *Biochim. Biophys. Acta* **62**, 338–352.

Huh W-K, Kim S-T, Yang K-S, Seok Y-J, Hah Y-C, Kang S-O (1994) Characterization of D-arabinono-1,4-lactone oxidase from *Candida albicans* ATCC 10231. *Eur. J. Biochem.* **223**, 1073–1079.

Huh W-K, Lee B-Y, Kim S-T, Kim Y-R, Rhie G-E, Baek Y-W, Hwang C-S, Lee J-S, Kang S-O (1998) D-Erythorbic acid is an important antioxidant molecule in *Saccharomyces cerevisiae*. *Mol. Micrbiol.* **30**, 895–903.

Imai T, Karita S, Shiratori G, Hattori M, Nunome T, Oba K, Hirai M (1998) L-Galactono-γ-lactone dehydrogenase from sweet potato: Purification and cDNA sequence analysis. *Plant Cell Physiol.* **39**, 1350–1358.

Isherwood FA, Chen YT, Mapson LW (1954) Synthesis of L-ascorbic acid in plants and animals. *Biochem. J.* **56**, 1–15.

Jaffe GM (1984) Ascorbic acid. In: *Encyclopedia of Chemical Technology* (eds Kirk RE, Othmer DF). Wiley, New York, NY, pp. 8–40.

Jain AK, Nessler CL (2000) Metabolic engineering of an alternative pathway for ascorbic acid biosynthesis in plants. *Mol. Breeding* **6**, 73–78.

Jensen A (1961) Ascorbic acid in *Ascophyllum nodosum*, *Fucus serratus* and *Fucus vesiculosus*. *Int. Seaweed Symp.* **4**, 319–325.

Kall MA, Andersen C (1999) Improved method for simultaneous determination of ascorbic acid and dehydroascorbic acid, isoascorbic acid and dehydroisoascorbic acid in food and biological samples. *J. Chrom. B* **730**, 101–111.

Kampfenkel K, Van Montagu M, Inzé D (1995) Extraction and determination of ascorbate and dehydroascorbate from plant tissue. *Anal. Biochem.* **225**, 165–167.

Kardailsky I, Shukla VK, Ahn JH, Dagenais N, Christensen SK, Nguyen JT, Chory J, Harrison MJ, Weigel D (1999) Activation tagging of the floral inducer FT. *Science* **286**, 1962–1965.

Keates SE, Loewus FA, Helms GL, Zink DL (1998) 5-O-(α-D-galactopyranosyl)-D-glycero-pent-2-enono-1,4-lactone: characterization in the oxalate-producing fungus, *Sclerotinia sclerotiorum*. *Phytochemistry* **49**, 2397–2401.

Keller R, Springer F, Renz A, Kossmann J (1999) Antisense inhibition of the GDP-mannose pyrophosphorylase reduces the ascorbate content in transgenic plants leading to developmental changes during senescence. *Plant J.* **19**, 131–141.

Kim S-T, Huh W-K, Lee B-H, Kang S-O (1998) D-arabinose dehydrogenase and its gene from *Saccharomyces cerevisiae*. *Biochim. Biophys. Acta – Protein Struct. M* **1429**, 29–39.

Kim S-T, Huh W-K, Kim J-Y, Hwang S-W, Kang S-O (1996) D-Arabinose dehydrogenase and biosynthesis of erythorbic acid in *Candida albicans*. *Biochim. Biophys. Acta* **1297**, 1–8.

Kimoto E, Terada S, Yamaguchi T (1997) Analysis of ascorbic acid, dehydroascorbic acid, and transformation products by ion-pairing high-performance liquid chromatography with multiwavelength ultraviolet and electrochemical detection. *Vitamins and Coenzymes, PT I Meth. Enzymol.* **279**, 3–12.

Krasnov A, Reinisalo M, Pitkanen TI, Nishikimi M, Molsa H (1998) Expression of rat gene for L-gulono-γ-lactone oxidase, the key enzyme of L-ascorbic acid biosynthesis, in guinea pig cells and in teleost fish rainbow trout (*Oncorhynchus mykiss*). *Biochim. Biophys. Acta* **1381**, 241–248.

Krasnov A, Pitkanen TI, Molsa H (1999) Gene transfer for targeted modification of salmonid fish metabolism. *Genet. Anal. Biomol.* **15**, 115–119.

Lee J-S, Huh W-K, Lee B-H, Baek Y-U, Hwang C-S, Kim S-T, Kim Y-R, Kang SO (2001) Mitochondrial NADH-cytochrome b5 reductase plays a crucial role in the reduction of D-erythroascorbyl free radical in *Saccharomyces cerevisiae*. *Biochim. Biophys. Acta* **1527**, 31–38.

Leung CT, Loewus FA (1985) Concerning the presence and formation of ascorbic acid in yeasts. *Plant Sci.* **38**, 65–69.

Liso R, Calabrese G (1974) Research on ascorbic acid physiology in red algae. 2. Dehydroascorbic acid compartmentation in the cell. *Phycologia* **13**, 205–208.

Liso R, Calabrese G (1975) Research on ascorbic acid physiology in red algae. 3. Lycorine-ascorbic acid interaction in growth control. *Phycologia* **14**, 9–11.

LoBuglio KF, Pitt JI, Taylor JW (1993) Phylogenetic analysis of two ribosomal DNA regions indicates multiple independent losses of sexual Talaromyces state among asexual Penicillium species in subgenus Biverticillium. *Mycologia* **85**, 592–604.

Loewus FA (1963) Tracer studies on ascorbic acid formation in plants. *Phytochemistry* **2**, 109–128.

Loewus FA, Jang R, Seegmiller CG (1956) Conversion of C-14 labeled sugars to L-ascorbic acid in ripening strawberries. *J. Biol. Chem.* **222**, 649–664.

Loewus FA, Kelly S (1961) Metabolism of D-galacturonic acid and its methyl ester in detached ripening strawberry. *Arch. Biochem. Biophys.* **95**, 483–493.

Loewus FA, Saito K, Suto RK, Maring E (1995) Conversion of D-arabinose to D-erythroascorbic acid and oxalic acid in *Sclerotinia sclerotiorum*. *Biochem. Biophys. Res. Comm.* **212**, 196–203.

Loewus MW, Bedgar DL, Saito K, Loewus FA (1990) Conversion of L-sorbosone to L-ascorbic acid by a NADP-dependent dehydrogenase in bean and spinach leaf. *Plant Physiol.* **94**, 1492–1495.

Massie HR, Shumway ME, Whitney SJP, Sternick SM, Aiello VR (1991) Ascorbic acid in Drosophila and changes during aging. *Exp. Gerontol.* **26**, 487–494.

May JM, Mendiratta S, Hill KE, Burk RF (1997) Reduction of dehydroascorbate to ascorbate by the selenoenzyme thioredoxin reductase. *J. Biol. Chem.* **272**, 22607–22610.

Miyake C, Michihata F, Asada K. (1991) Scavenging of hydrogen peroxide in prokaryotic and eukaryotic algae – acquisition of ascorbate peroxidase during the evolution of cyanobacteria. *Plant Cell Physiol.* **32**, 33–43.

Morell S, Follmann H, DeTullio M, Häberlein I (1997) Dehydroascorbate and dehydroascorbate reductase are phantom indicators of oxidative stress in plants. *FEBS Lett.* **414**, 567–570.

Mozafar A (1993) Nitrogen fertilizers and the amount of vitamins in plants – a review. *J. Plant Nutr.* **16**, 2479–2506.

Murakawa S, Sano S, Yamashita H, Takahashi T (1977) Biosynthesis of D-erythroascorbic acid by *Candida*. *Agric. Biol. Chem.* **41**, 1799–1800.

Murakawa S, Takahashi T (1977) Biosynthesis of a new ascorbic acid analog by D-gluconolactone dehydrogenase of *Penicillium cyaneo-fulvum*. *Agric. Biol. Chem.* **41**, 2103–2104.

Mutsuda M, Ishikawa T, Takeda T, Shigeoka S (1995) Subcellular localization and properties of L-galactono-gamma-lactone dehydrogenase in spinach leaves. *Biosci. Biotech. Biochem.* **59**, 1983–1984.

Nick JA, Leung CT, Loewus FA (1986) Isolation and identification of erythroascorbic acid in *Saccharomyces cerevisiae* and *Lipomyces starkeyi*. *Plant Sci.* **46**, 181–187.

Nishikimi M, Noguchi E, Yagi K. (1978) Occurrence in yeast of L-galactonolactone oxidase which is similar to a key enzyme for ascorbic acid biosynthesis in animals, L-gulonolactone oxidase. *Arch. Biochem. Biophys.* **191**, 479–486.

Nishikimi M, Udenfriend S (1976) Immunological evidence that gene for L-gulono-γ-lactone oxidase is not expressed in animals subject to scurvy. *Proc. Natl Acad. Sci. USA* **73**, 2066–2068.

Nishikimi M, Koshizaka T, Ozawa T, Yagi K (1988) Occurrence in humans and guinea-pigs of the gene related to their missing enzyme L-gulono-γ-lactone oxidase. *Arch. Biochem. Biophys.* **267**, 842–846.

Nishikimi M, Kawai T, Yagi K (1992) Guinea-pigs possess a highly mutated gene for L-gulono-γ-lactone oxidase, the key enzyme for L-ascorbic-acid biosynthesis missing in this species. *J. Biol. Chem.* **267**, 21967–21972.

Nishikimi M, Fukuyama R, Minoshima S, Shimizu N, Yagi K (1994) Cloning and chromosomal mapping of the human nonfunctional gene for L-gulono-γ-lactone oxidase, the enzyme for L-ascorbic-acid biosynthesis missing in man. *J. Biol. Chem.* **269**, 13685–13688.

Noctor G, Foyer CH (1998) Ascorbate and glutathione: Keeping active oxygen under control. *Ann. Rev. Plant Phys. Plant Mol. Biol.* **49**, 249–279.

Okamura, M. (1994) Distribution of ascorbic acid analogs and associated glycosides in mushrooms. *J. Nutr. Sci. Vitaminol.* **40**, 81–94.

Okamura M (1998) Separative determination of ascorbic acid analogs contained in mushrooms by high-performance liquid chromatography. *J. Nutr. Sci. Vitaminol.* **44**, 25–35.

Onofri S, Poerio E, Serangeli P, Tosi S, Garuccio I, Arrigoni O (1997) Influence of L-galactonic acid γ-lactone on ascorbate production in some yeasts. *Ant. van Leeuw.* **71**, 277–280.

Østergaard J, Persiau G, Davey MW, Bauw G, Van Montagu M (1997) Isolation of a cDNA coding for L-galactono-γ-lactone dehydrogenase, an enzyme involved in the biosynthesis of ascorbic acid in plants – Purification, characterization, cDNA cloning, and expression in yeast. *J. Biol. Chem.* **272**, 30009–30016.

Pallanca JE, Smirnoff N (1999) Ascorbic acid metabolism in pea seedlings. A comparison of D-glucosone, L-sorbosone, and L-galactono-1,4-lactone as ascorbate precursors. *Plant Physiol.* **120**, 453–461.

Puskas F, Braun L, Csala M, Kardon T, Marcolongo P, Benedetti A, Mandl J, Bánhegyi G (1998) Gulonolactone oxidase activity-dependent intravesicular glutathione oxidation in rat liver microsomes. *FEBS Lett.* **430**, 293–296.

Renstrom B, Grun M, Loewus FA (1983) Biosynthesis of L-ascorbic acid in *Chlorella pyrenoidosa*. *Plant Sci. Lett.* **28**, 299–305.

Roberts RM, Harrer E (1973) Determination of L-galactose in polysaccharide material. *Phytochemistry* **12**, 2679–2682.

Roe JH, Kuether CA (1943) The determination of ascorbic acid in whole blood and urine through the 2,4-dinitrophenylhydrazine derivative of dehydroascorbic acid. *J. Biol. Chem.* **147**, 399–407.

Roomi MW, Tsao CS (1998) Thin-layer chromatographic separation of isomers of ascorbic acid and dehydroascorbic acid as sodium borate complexes on silica gel and cellulose plates. *J. Agric. Food Chem.* **46**, 1406–1409.

Running JA, Huss RJ, Olson PT (1994) Heterotrophic production of ascorbic acid by microalgae. *J. Appl. Phycol.* **6**, 99–104.

Running JA, Burlingame RP, Berry A (2003) The pathway of L-ascorbic acid biosynthesis in the colorless microalga *Prototheca moriformis*. *J. Exp. Bot.* **54**, 1841–1849.

Saito K (1996) Formation of L-ascorbic acid and oxalic acid from D-glucosone in *Lemna minor*. *Phytochemistry* **41**, 145–149.

Saito K, Nick JA, Loewus FA (1990) D-Glucosone and L-sorbosone, putative intermediates of L-ascorbic acid biosynthesis in detached bean and spinach leaves. *Plant Physiol.* **94**, 1496–1500.

Sastry K, Sarma P (1957) Glucuronic acid, a precursor of ascorbic acid in *Aspergillus niger*. *Nature (Lond.)* **179**, 44–45.

Schöner S, Krause GH (1990) Protective systems against active oxygen species in spinach – response to cold-acclimation in excess light. *Planta* **180**, 383–389.

Shigeoka S, Nakano Y, Kitaoka S (1979) The biosynthetic pathway of L-ascorbic acid in *Euglena gracilis Z*. *J. Nutr. Sci. Vitaminol.* **25**, 299–307.

Shigeoka S, Ishikawa T, Tamoi M, Miyagawa Y, Takeda T, Yabuta Y, Yoshimura K (2002) Regulation and function of ascorbate peroxidase isoenzymes. *J. Exp. Bot.* **53**, 1305–1319.

Shimizu K, Nishiyama K, Inoue T, Takano N, Mikata M, Yamazaki M, Azuma T, Osawa S (1967) Studies on erythorbic acid production by fermentation. Part II. Erythorbic acid production by jar fermentor. *Agric. Biol. Chem.* **31**, 346–352.

Siendones E, González-Reyes JA, Santos-Ocana C, Navas P, Córdoba F (1999) Biosynthesis of ascorbic acid in kidney bean. L-Galactono-γ-lactone dehydrogenase is an intrinsic protein located at the mitochondrial inner membrane. *Plant Physiol.* **120**, 907–912.

Smirnoff N (2000) Ascorbate biosynthesis and function in photoprotection. *Phil. Trans. Roy. Soc. B* **355**, 1455–1464.

Smirnoff N (2001) L-Ascorbic acid biosynthesis. *Vitam. Horm.* **61**, 241–266.

Smirnoff N (2003) Vitamin C booster. *Nature Biotechnol.* **21**, 134–136.

Smirnoff N, Conklin PL, Loewus FA (2001) Biosynthesis of ascorbic acid in plants: A renaissance. *Annu. Rev. Plant Physiol. Plant Mol. Biol.* **52**, 437–467.

Smirnoff N, Pallanca JE (1996) Ascorbate metabolism in relation to oxidative stress. *Biochem. Soc. Trans.* **24**, 472–478.

Smith NA, Singh SP, Wang MB, Stoutjesdijk PA, Green AG, Waterhouse PM (2000) Gene expression – Total silencing by intron-spliced hairpin RNAs. *Nature* **407**, 319–320.

Somers WS, Stahl ML, Sullivan FX (1998) GDP-fucose synthetase from *Escherichia coli*: structure of a unique member of the short-chain dehydrogenase/reductase family that catalyzes two distinct reactions at the same active site. *Structure* **6**, 1601–1612.

Spickett CM, Smirnoff N, Pitt AR (2000) The biosynthesis of erythroascorbate in *Saccharomyces cerevisiae* and its role as an antioxidant. *Free Rad. Biol. Med.* **28**, 183–192.

Sturgeon BE, Sipe HJ, Barr DP, Corbett JT, Martinez JG, Mason RP (1998) The fate of the oxidizing tyrosyl radical in the presence of glutathione and ascorbate – Implications for the radical sink hypothesis. *J. Biol. Chem.* **273**, 30116–30121.

Swofford DL (1998) PAUP*. Version 4, Sinauer Associates, Sunderland, Massachusetts.

Tabata K, Oba K, Suzuki K, Esaka M (2001) Generation and properties of ascorbic acid-deficient transgenic tobacco cells expressing antisense RNA for L-galactono-1,4-lactone dehydrogenase. *Plant J.* **27**, 139–148.

Tabata K, Takaoka T, Esaka M (2002) Gene expression of ascorbic acid-related enzymes in tobacco. *Phytochemistry* **61**, 631–635.

Takahashi T, Mitsumoto M (1961) Studies on the pathway of D-araboascorbic acid synthesis by *Penicillium*. *Biochim. Biophys. Acta* **51**, 410–412.

Takahashi T, Mitsumoto M, Kayamori H (1960) Production of D-araboascorbic acid by *Penicillium*. *Nature (Lond.)* **188**, 411–412.

Takahashi T, Yamashita H, Kato E, Mitsumoto M, Murakawa S (1976) Purification and some properties of D-glucono-γ-lactone dehydrogenase, D-erythorbic acid producing enzyme of *Penicillium cyaneo-fulvum*. *Agric. Biol. Chem.* **40**, 121–129.

Tel-Or E, Huflejt M, Packer L (1985) The role of glutathione and ascorbate in hydroperoxide removal in cyanobacteria. *Biochem. Biophys. Res. Commun.* **132**, 533–539.

Tolbert BM, Downing M, Carlson RW, Knight MK, Baker EM (1975) Chemistry and metabolism of ascorbic-acid and ascorbate sulfate. *Ann. N.Y. Acad. Sci.* **258**, 48–69.

Toyohara H, Nakata T, Touhata K, Hashimoto M, Kinishita M, Sakaguchi M, Nishikimi K, Yagi Y, Ozato K (1996) Transgenic expression of L-gulono-γ-lactone oxidase in medaka (*Oryzias latipes*), a teleost fish that lacks this enzyme necessary for L-ascorbic acid biosynthesis. *Biochem. Biophys. Res. Commun.* **223**, 650–653.

Wagner C, Sefkow M, Kopka J (2003). Construction and application of a mass spectral and retention time index database generated from plant GC/EI-TOF-MS metabolite profiles. *Phytochemistry* **62**, 887–900.

Wahal CK, Bhattacharya NC, Talpasayi ERS (1973) Ascorbic acid and heterocyst development in the blue-green alga *Anabaena ambigua*. *Physiol. Plant* **28**, 424–429.

Weigel D, Ahn JH, Blazquez MA, Borevitz JO, Christensen SK, Fankhauser C, *et al.* (2000) Activation tagging in *Arabidopsis*. *Plant Physiol.* **122**, 1003–1013.

Weising K, Schell J, Kahl G. (1988) Foreign genes in plants – transfer, structure, expression, and applications. *Annu. Rev. Genet.* **22**, 421–477.

Wells WW, Xu DP, Yang Y, Rocque PA (1990) Mammalian thioltransferase (glutaredoxin) and protein disulfide isomerase have dehydroascorbate reductase activity. *J. Biol. Chem.* **265**, 15361–15364.

Wheeler GL, Jones MA, Smirnoff N (1998) The biosynthetic pathway of vitamin C in higher plants. *Nature (Lond.)* **393**, 365–369.

Wolucka BA, Persiau G, Van Doorsselaere J, Davey MW, Demol H, Vandekerckhove J, Van Montagu M, Zabeau M, Boerjan W (2001) Partial purification and identification of GDP-mannose 3′,5′-epimerase of *Arabidopsis thaliana*, a key enzyme of the plant vitamin C pathway. *Proc. Natl Acad. Sci. USA* **98**, 14843–14848.

Yagi J, Yamashita T, Kato K, Takagi Y, Sakai H (1967) Studies on erythorbic acid production by fermentation. Part 1. Erythorbic acid-producing strain and cultural condition. *Agric. Biol. Chem.* **31**, 340–345.

Yagi K, Koshizaka T, Kito M, Ozawa T, Nishikimi M (1991) Expression in monkey cells of the missing enzyme in L-ascorbic acid biosynthesis, L-gulono-γ-lactone oxidase. *Biochem. Biophys. Res. Commun.* **177**, 659–663.

Yasuda N. (1967) Japan patent Sho 42–11990.

York JL, Grollman AP, Bublitz C (1961) TPN-L-gulonate dehydrogenase. *Biochim. Biophys. Acta* **47**, 298–306.

2

Ascorbic acid catabolism: breakdown pathways in animals and plants

Gábor Bánhegyi and Frank A. Loewus

2.1 Introduction

Isolation, characterization, and synthesis of Vitamin C (L-ascorbic acid, ASC L-threo-hex-2-enono-1,4-lactone) were accomplished during a relatively brief period of 7 years (1928–1934). These advances emerged from Albert Szent-Györgyi's discovery of a rich source of Vitamin C (Szent-Györgyi, 1937; Carpenter, 1986; Moss, 1988; Davies *et al.*, 1991) in Hungarian paprika (*Capsicum* sp.; Csonka and Váradi, 1907) and led to production of a large quantity of highly purified crystalline product which enabled a select group of carbohydrate chemists to deduce its structure (Herbert *et al.*, 1933). Within a year, several syntheses of ASC were described (reviewed by Crawford, 1982). These studies provided a wealth of chemical detail applicable to subsequent research on ASC biosynthesis and metabolism (Bánhegyi *et al.*, 1997; Wheeler *et al.*, 1998; Conklin, 2001; Davey *et al.*, 2000, 2002). In this chapter, attention is given to catabolic processes that lead to ASC breakdown.

2.2 Catabolism of ascorbate in animals

2.2.1 Introduction

Reactions and enzymes participating in ascorbate (ASC) breakdown in animals have not been explored in full. Experimental difficulties in these investigations are due to the instability of several intermediates of the catabolic pathways under physiological conditions, the existence of numerous nonenzymatic reactions, and

Vitamin C, edited by Han Asard, James M May and Nicholas Smirnoff

finally, but most importantly, the relatively low substrate flow. Effective ASC sparing mechanisms prove the regeneration of the vitamin; despite the many ASC oxidizing reactions, the half-time of ASC disappearance is several days in those animals which are unable to synthesize it. Humans, for example, catabolize as little as 100 mg ASC per day. Therefore, ASC catabolism can be regarded as a metabolic pathway of minor capacity and importance.

However, in the light of recent observations the pathway deserves attention: namely, several intermediates are suspected to participate in pathophysiological processes. Moreover, other findings suggest that the contribution of ASC breakdown to substrate supply for the intermediary (carbohydrate) metabolism is not negligible in certain cell types and conditions. This chapter summarizes the known steps of ASC catabolism and emphasizes the implications of certain intermediates in pathology.

2.2.2 Reactions and enzymes in ascorbate catabolism

Characteristics of ASC catabolism depend on the investigated species, the quantity of the ingested ASC and the nutritional status. The fate of excess ASC varies with the species. In humans, the superfluous ASC is excreted into the urine, while in guinea pig it is rapidly metabolized to CO_2. The metabolic path in animals can be didactically divided into three parts. The first one is dominated by overwhelmingly oxidative – not necessarily enzymatic – reactions. The five- and four-carbon products are transformed to intermediates of the pentose phosphate pathway in the second phase. The third phase uses the known enzymes of the pentose phosphate pathway, resulting in the formation of glycolytic/gluconeogenetic precursors.

ASC, as an important water-soluble antioxidant and cofactor for several enzymes, first undergoes one-or two-electron oxidation yielding monodehydroascorbate (MDHA or ascorbyl free radical) or dehydroascorbic acid (DHA), respectively (*Figure 2.1*). The nonenzymatic disproportionation of MDHA finally leads to DHA formation. Chemical reactions include ASC oxidation by the ions of transition metals or by oxygen- and other radicals. ASC-dependent enzymes (prolyl- and lysyl-hydroxylases, etc.) also convert ASC to DHA. It has been thought that animals lack ASC oxidases (AO) and peroxidases (APX), but some findings indicate (Sun *et al.*, 1984; Nakamura and Ohtaki, 1993; Szarka *et al.*, 2002) or even prove (Wilkinson *et al.*, 2002) otherwise in the case of APX.

DHA can be reduced back to ASC efficiently both enzymatically and nonenzymatically. As a result of the balance between oxidation and reduction, a few percent of the ASC pool is present as DHA, which may rise upon oxidative challenges. DHA is unstable even in the absence of oxygen, and rapidly hydrolyzes at physiological pH to 2,3-L-diketogulonate (Bode *et al.*, 1990). The half-life of DHA *in vivo* may be as short as 2 min (Mitton *et al.*, 1997). The cleavage of the lactone ring is promoted by bicarbonate (Koshiishi *et al.*, 1998a). In fact, DHA is converted to 2,3-diketogulonate in the blood circulation of rats, from which aldonolactonase activity is absent (Koshiishi *et al.*, 1998b). Aldonolactonases, which are present in the liver of almost all mammals, also hydrolyze DHA to 2,3-diketogulonate (Winkelman and Lehninger, 1958; Yamada *et al.*, 1959; Kagawa *et al.*, 1961). Human liver seems to

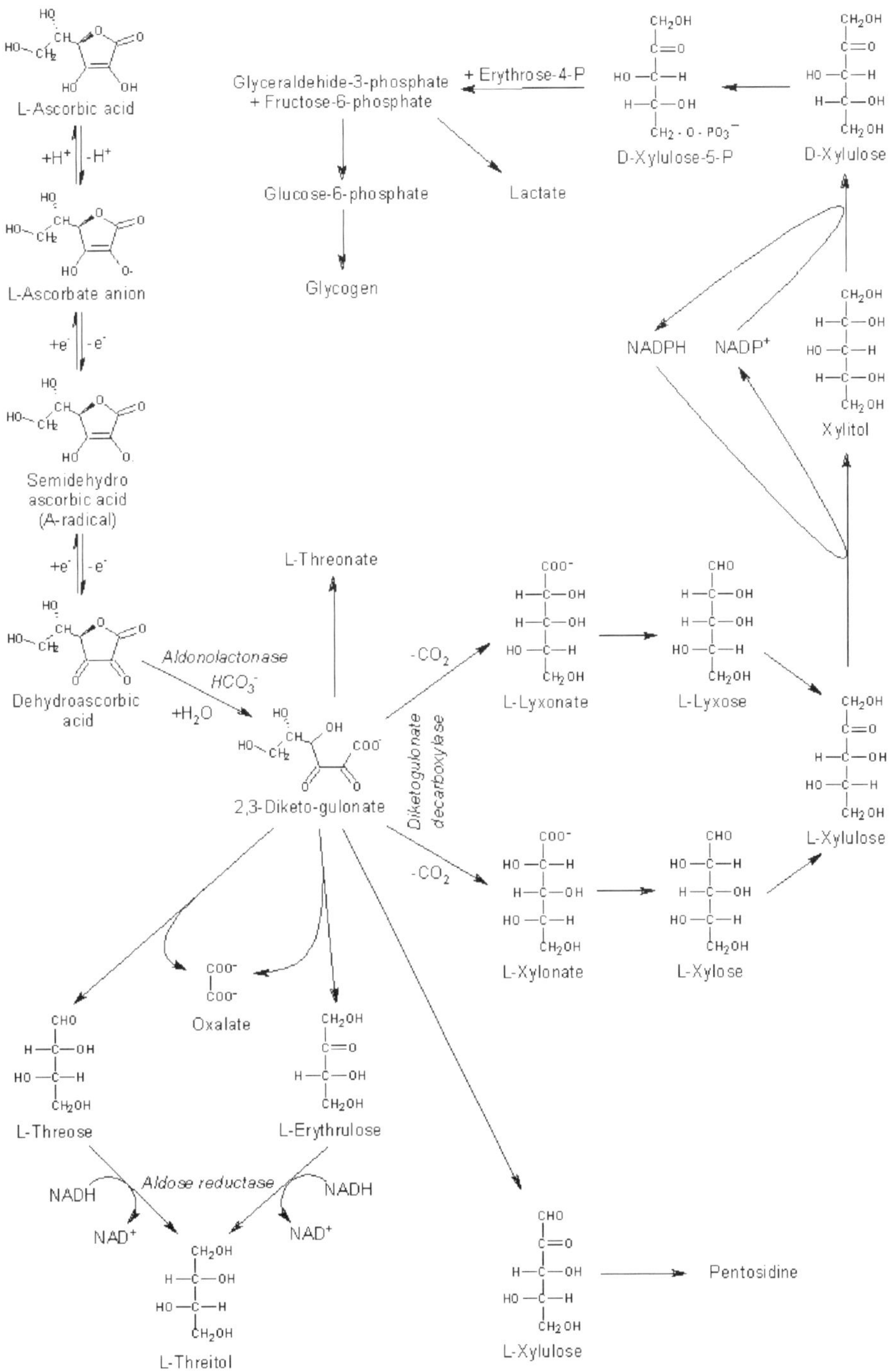

Figure 2.1: Ascorbate catabolizing pathways in animals

lack these enzymes (Cox and Wichelow, 1975). Diketogulonate, being also very unstable, degrades further. Despite its fast degradation, elevated diketogulonate levels could be detected in human erythrocytes under oxidative stress *in vitro* (Himmelreich *et al.*, 1998), and in lens of diabetic rats (Saxena *et al.*, 1996).

Divergent results were published by different researchers investigating the further fate of diketogulonate. L-Xylose and CO_2 (Chan *et al.*, 1958), xylosone and CO_2 (Kang *et al.*, 1982), L-xylonate, L-lyxonate and CO_2 (Kanfer *et al.*, 1960; Kagawa *et al.*, 1961; Kagawa, 1962) oxalate and L-threonate (Kagawa, 1962; Kang *et al.*, 1982), and L-threose (López and Feather, 1992) were all reported as the products of degradation (*Figure 2.1*). Several other products have been identified in *in vitro* systems after oxidative or nonoxidative alkali-catalyzed degradation of ASC, which can be regarded as a model for the normal metabolism (Niemelä, 1987; Deutsch, 1998). A recent paper identifies oxalate and L-erythrulose as the nonoxidative degradation products at physiological conditions. Human lens (containing aldose reductase) can reduce L-erythrulose to L-threitol. On the other hand, in oxidative conditions, i.e., in the presence of hydrogen peroxide, decomposition of diketogulonate results in the formation of L-threonate, oxalate and CO_2 (Simpson and Ortwerth, 2000). The same products are derived from the reaction of DHA and H_2O_2 at pH 7 (Isbell and Frush, 1979). Beside the nonenzymatic processes, enzymes (diketogulonate decarboxylases) can also transform 2,3-diketogulonate (Kanfer *et al.*, 1960; Kagawa, 1962). In this case five-carbon products (L-xylonate and L-lyxonate) are formed (Kanfer *et al.*, 1960), which might enter the pentose phosphate pathway through less well characterized steps (Tolbert *et al.*, 1975).

2.2.3 Final products of ascorbate catabolism

The metabolism of ASC in experimental animals (rat, guinea pig) leads to carbon dioxide production (Chan *et al.*, 1958; Hellman and Burns, 1958). Early human studies revealed that only a very minor fraction of labeled carbon atoms of ASC was expired in the breath as CO_2 (Hellman and Burns, 1958; Baker *et al.*, 1963). It was suggested that carbon dioxide formation is merely due to a bacterial or chemical degradation of ASC in the intestine (Kallner *et al.*, 1985). Approximately 10% of the ASC metabolites in the urine was oxalate; as well as two major and two minor unidentified, chemically stable products (Baker *et al.*, 1969, 1971b). It should be noted that in certain species a remarkable ASC-fueled oxalate production can be observed. For example, in the marine sponge *Chondrosia reniformis*, ASC-dependent calcium oxalate deposition was reported (Cerrano *et al.*, 1999).

Experiments on guinea pig liver preparations showed that the ^{14}C atoms of labeled ASC appeared in glucose (Glc) and glycogen, indicating that ASC catabolism reached the central pathways of carbohydrate metabolism (Chan *et al.*, 1958).

Physiological ASC concentration increased the flux of glycolysis in human red blood cells. In mild oxidative stress the flux through the hexose monophosphate shunt was also elevated (Sullivan and Stern, 1983). In agreement with these findings, addition of ASC or DHA to human erythrocytes or murine blood resulted in higher lactate formation. Oxidative challenge increased the effect (Braun *et al.*, 1997). However, the inability of human erythrocytes to metabolize ASC or DHA to lactate was also reported (Himmelreich *et al.*, 2000).

ASC or DHA addition to isolated hepatocytes prepared from fasted mice or to HepG2 cells resulted in a concentration-dependent Glc production and in the elevation of intracellular xylulose-5-phosphate level. Stimulation of ASC oxidation or the inhibition of DHA reduction increased, while the inhibition of transketolases by oxythiamine decreased Glc production (Braun *et al.*, 1996).

To summarize the different findings, it is highly probable that carbon atoms from ASC reach the main highways of carbohydrate metabolism (*Figure 2.1*). Therefore, ASC metabolism in ASC-synthesizing species can be regarded as a cyclic pathway with inter-organ cooperation (Bánhegyi *et al.*, 1997).

2.2.4 Conjugation reactions

Besides its degradation, ASC can also be converted by several enzymes of the second phase of biotransformation. It can be conjugated with sulphate (Baker *et al.*, 1971a; Bond *et al.*, 1972; Tolbert *et al.*, 1975; Pillai *et al.*, 1991), phosphate (Miyasaki *et al.*, 1991), glucuronic acid (Gallice *et al.*, 1994), Glc (Muto *et al.*, 1991) or methyl groups (Blaschke and Hertting, 1971) at the second carbon. The enzymes participating in the synthesis or in the hydrolysis (Tolbert *et al.*, 1975; Wakamiya *et al.*, 1992; Dabrowski *et al.*, 1993) of these compounds are present both in ASC-synthesizing and -nonsynthesizing species. The exact metabolic role of these derivatives is not known, but they could serve as an easily accessible ASC source enlarging the available endogenous pool.

2.2.5 Regulation

The regulation of ASC catabolism is an almost completely unexplored field, in spite of its putative importance. The higher rate of ASC catabolism in oxidative conditions is probably due to the diminished capacity of the recycling mechanisms and not to the increased activity and/or expression of the enzymes of catabolism (Bánhegyi *et al.*, 1998).

Diminished ASC levels have been found in plasma and tissues of diabetic patients and animals (Will and Byers, 1996). The diabetes-related changes are due to the altered gene expression and cofactor supply of enzymes in ASC metabolism. Enhanced ASC breakdown has been reported in animal models for both type II (Kashiba *et al.*, 2000) and type I (Kashiba *et al.*, 2002) diabetes. Although the higher activity of ASC catabolizing enzymes cannot be excluded, the phenomenon is mainly attributed to the impaired recycling of DHA. In any case, ASC supplementation seems to be beneficial in the prevention of diabetic complications (Szaleczky *et al.*, 1998; Price *et al.*, 2001). The higher ASC-synthesizing capacity of male rats is accompanied by a faster catabolism of ASC, indicated by higher DHA and 2,3-diketogulonate metabolism in liver and kidney. Similar differences between sexes were found in guinea pig (Muddeshwar *et al.*, 1992).

2.2.6 Ascorbate catabolism and human pathology

Pathological aspects of ASC breakdown can be grouped into two important topics. First, oxalate deriving from ASC catabolism was suspected to cause urolithiasis.

Second, ASC itself, but more importantly the reactive intermediates of its catabolism, can form protein adducts. These adverse effects are evident in cells containing ASC in high concentration and exposed to oxidative challenges (e.g., erythrocytes, eye lens etc.). These (putative or proven) pathological effects indicate the possible risk of ASC overdose.

The formation of oxalic acid upon excessive ASC intake has been used as an argument against pharmacological doses, especially in patients who are sensitive to urolithiasis. However, daily supply of ASC in high doses (several grams) resulted in a negligible increase of the normal oxalate excretion (Moser and Hornig, 1982). Long-term ASC supplementation did not increase the incidence of nephrolithiasis (Gerster, 1997; Auer *et al.*, 1998; Curhan *et al.*, 1996). Apparently, normal persons without metabolic disturbances are not adversely affected by ASC overdose.

Protein ascorbylation, due to the ASC-derived aldehydes is a well-known phenomenon. ASC oxidation in the presence of transition metals leads to the generation of reactive oxygen species and reactive aldehydes. Moreover, the generation of these reactive compounds may be higher in pathophysiological states characterized by enhanced ASC oxidation, such as diabetes mellitus. The process has been investigated most intensively in human lens. Lens crystallins incubated for a long time in the presence of physiological concentration of ASC developed a brownish color (Bensch *et al.*, 1985). It was supposed that a Maillard reaction between ASC-derived compounds and lens proteins was responsible for the coloration and deterioration of the lens (Ortwerth and Olesen, 1988a). Indeed, at an early stage of cataractogenesis an increased ascorbyl radical signal can be detected in the lens by measuring electron spin resonance spectra (Lohmann *et al.*, 1986). Other products of catabolism, i.e., DHA and 2,3-diketogulonate (Slight *et al.*, 1990; Saxena *et al.*, 1996), oxalate (Bron and Brown, 1987; Nagaraj *et al.*, 1999), threose (Ortwerth *et al.*, 1994; Nagaraj and Monnier 1995), threitol (Ortwerth *et al.*, 1994) and L-erythrulose/oxalate (Simpson and Ortwerth, 2000) could be shown in the lens. Proteins glycated by the ASC degradation products were also detected. Formyl threosyl pyrrole (Nagaraj and Monnier, 1995), pentosidine (Nagaraj *et al.*, 1991) and oxalate monoalkylamide (Nagaraj *et al.*, 1999) were described as advanced glycation end-products formed in the Maillard reaction of ASC catabolites with lens proteins. The adverse effects of ASC appear when a shortage of reducing equivalents, reflected in a decrease of GSH concentration, is present in the lens (Ortwerth and Olesen, 1988b). In these circumstances both ASC salvage mechanisms (DHA reduction) and the reduction of reactive aldehydes are deficient.

2.3 Catabolism of ascorbic acid in plants

2.3.1 Introduction

Apart from dormant tissues, for example seeds, L-ascorbic acid is probably a ubiquitous constituent in the growth cycle of plants. Generally speaking, highest levels of ASC are found in actively growing regions (sprouting seeds, root tips, bud development, expanding leaves, flower formation and fruit growth). In many instances ASC is retained in tissues of dried or senescing organs; examples are rose hips and walnut hulls. Whether such occurrences reflect continued biosynthesis, diminished catabolism, or storage has yet to be explored.

ASC biosynthesis in plants (see Smirnoff *et al.*, Chapter 1) involves a cascade of eight enzymic reactions beginning with D-glucose-6-P (Smirnoff *et al.*, 2001; Conklin, 2001). The final step, oxidation of L-galactono-1,4-lactone to ASC, occurs in mitochondria and ASC must be released or transported into cytosol prior to intracellular involvement and/or further intercellular movement (Potters *et al.*, 2000, Horemans *et al.*, 2000a, 2000b; Davey *et al.*, 2000, 2002; Gatzek *et al.*, 2002; Franceschi and Tarlyn, 2002) where it participates in a multitude of functional roles including enzymatic cofactor, antioxidant, and less definable activities involving cell growth and development (Horemans *et al.*, 2000a, 2000b; Davey *et al.*, 2000, 2002; Tabata *et al.*, 2000). Most of these roles preserve ASC either in its reduced form, or as MDHA or DHA; the former a free radical that disproportionates to produce ASC + DHA and the latter which is readily re-reduced back to ASC enzymatically (Noctor and Foyer, 1998; Chen *et al.*, 2003). Catabolic processes transform ASC or DHA irreversibly to products no longer linked to functional ASC activities.

A subtle difference, not generally appreciated in the case of ASC, distinguishes oxidoreduction from catabolism of ASC. Oxido-reductive processes only alter the balance of reduced/oxidized ASC whereas catabolic processes require ASC biosynthesis to replenish ASC requirements. Among studies on ASC turnover are reports that as much as 50% of the ASC pool must be replaced in about 24 h (Conklin *et al.*, 1997; Davey *et al.*, 2000; Pallanca and Smirnoff, 2000; Franceschi and Tarlyn, 2002; Chen *et al.*, 2003).

In the following sections, attention is given to significant catabolic processes involving cleavage of the carbon chain of ascorbic acid (abbreviated as ASC regardless of its disassociated or undisassociated state as reduced ascorbic acid).

2.3.2 Cleavage of the carbon chain of ASC

In their historic paper on the chemical structure of ASC, Herbert *et al.* (1933) identified two distinct stages of oxidation. The first involved a hydrated DHA followed by opening of the 1,4-lactone ring to form 2,3-diketo-L-gulonic acid. The second stage was cleavage between C2 and C3 to produce oxalic acid (OxA) and L-threonic acid (ThA) in equivalent amounts. Further oxidation of the latter yielded L-(+)-tartaric acid (TA). These observations clearly demonstrated the absence of a free carboxyl group. The acidic properties of ASC stemmed from the presence of an activated –CH.OH adjacent to a carbonyl group.

A stereochemical correspondence of C4–C5 in ASC to C2–C3 in TA prompted Hough and Jones (1956) to suggest C3–C6 of ASC as a potential precursor of TA. When this idea was tested by labeling a detached grape leaf (Vitaceae) with [6-^{14}C]ASC, only an insignificant amount of ^{14}C appeared in TA (Loewus and Stafford, 1958). Nearly 25% of the soluble label was recovered as ASC. Labeled sucrose, glucose, fructose and xylose accounted for about 45% of the neutral fraction and the remaining label was found in higher molecular carbohydrate constituents. In 1969, Saito and Kasai re-examined the question of ASC as a precursor of TA by infusing [1-^{14}C]ASC through peduncles into immature grape berries (Saito and Kasai, 1969). They found 72% of the ^{14}C accumulated in TA during the first 24 h of feeding. Analysis of TA revealed that virtually all of this ^{14}C was localized in carboxyl carbon. Clearly, ASC was a precursor of TA in the grape berry, presumably derived from C1–C4 of ASC.

2.3.3 Cleavage of ascorbate between C4 and C5 in vitaceous plants

Initial findings on ASC-related formation of TA in grape (Saito and Kasai, 1969) were supported by further studies of ^{14}C distribution into products of [1-^{14}C]-, [4-^{3}H,1-^{14}C]-, [4-^{14}C]-, [5-^{14}C]-, [6-^{3}H]-, [6-^{14}C]-or [U-^{14}C]ASC-labeled grape or Virginia creeper leaves (Williams and Loewus, 1978; Williams *et al.*, 1979; Saito and Loewus, 1979, 1989b; Helsper *et al.*, 1981; Saito and Kasai, 1982, 1984; Helsper and Loewus 1985; Saito *et al.*, 1997). Virginia creeper leaves provided evidence that the C2 fragment, possibly glycolaldehyde, recycled into respiratory CO_2, sugars, polysaccharides, and malic acid (Helsper *et al.*, 1981; Saito and Kasai, 1984). The C4 fragment leading to TA has yet to be isolated and identified (Saito and Loewus, 1989a).

In a remarkable study on C6 intermediates between ASC and its C4/C5 cleavage in grape, Saito and Kasai identified three compounds, L-xylo-2-hexulosonate (2-keto-L-idonate or 2-keto-L-gulonate), L-idonate, and D-xylo-5-hexulosonate (5-keto D-gluconate or 5-keto-L-idonate) (Saito and Kasai, 1982, 1984; Saito *et al.*, 1997; Saito and Loewus, 1989b). A comparison of ^{14}C-labeled products from [1-^{14}C]-versus [6-^{14}C]-D-xylo-5-hexulosonate-labeled grape apices found the former with 92% of its soluble label in TA whereas the latter had less that 2%, the bulk of the ^{14}C appearing in respired CO_2, sugars and polysaccharidic products (Saito and Kasai, 1984). A pathway of TA biosynthesis for vitaceous plants consistent with these findings is seen in *Figure 2.2*.

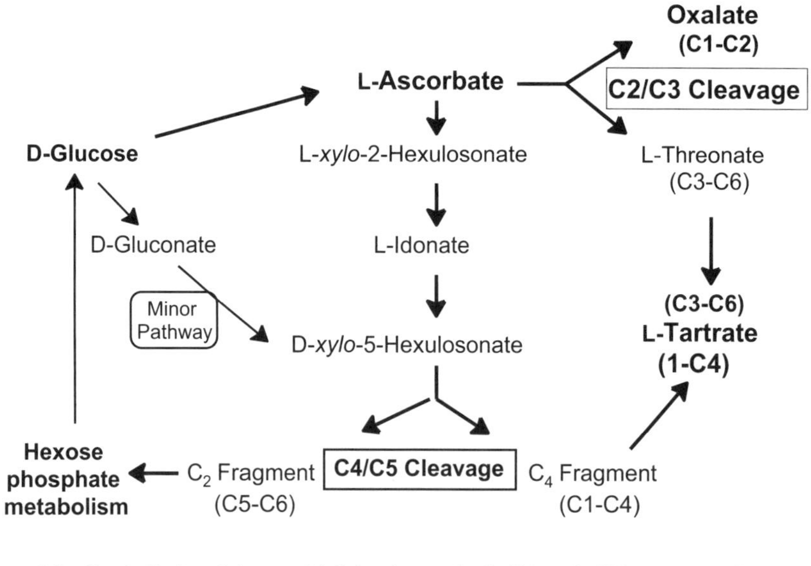

Figure 2.2: Catabolic breakdown of AA in plants via C2/C3 or C4/C5 cleavage of the carbon chain. A minor pathway from D-glucose to L-tartrate that bypasses AA is also shown.

2.3.4 Cleavage of ascorbate between C2 and C3 in tartaric acid accumulating plants

In contrast to grape, when comparable studies were undertaken with detached leaf apices of a TA-accumulating geranium, *Pelargonium crispum* (Geraniaceae), supplied with [6-^{14}C]ASC, nearly one-third of the soluble ^{14}C was recovered in TA (corresponding to C3–C6 of ASC) labeled exclusively in the carboxyl carbon. TA from [1-^{14}C]ASC-labeled apices contained only 0.4% of the soluble ^{14}C but OxA (corresponding to C1–C2 of ASC) contained 12% of the soluble label. Additional ^{14}C was recovered as $^{14}CO_2$ due to further oxidation of OxA (Loewus *et al.*, 1975; Saito *et al.*, 1984). There is experimental evidence that geranium leaves also have a minor pathway to TA that resembles C4/C5 cleavage in grape (Saito *et al.*, 1984).

OxA is a common plant constituent (Hodgkinson, 1977; Franceschi and Loewus, 1995; Loewus, 1999) whereas occurrence of TA is limited largely to plants in Vitaceae, Geraniaceae and Leguminosae families (Saito and Loewus, 1989a, 1989b). This prompted a closer search for a possible metabolic connection between ASC catabolism and OxA accumulation in oxalate-accumulating plants where C2/C3 cleavage of ASC seldom leads to TA accumulation. Instead of oxidation, ThA or its decarboxylated C3 product, L-glyceric acid, undergoes further catabolism to products which re-enter triose/hexose phosphate metabolism (see *Figure 2.2*).

Although both cleavage sites involved in ASC catabolism have been confirmed through use of radioisotopic markers (Loewus, 1999), enzymatic processes involved have yet to be identified. Important information in this regard was provided by tracing uptake and incorporation of ^{18}O from $^{18}O_2$ or $H_2{}^{18}O$ into young leaves of geranium, parsley, bean and grape that were administered D-xylo-5-[1-^{14}C]hexulosonic acid as well as [1,6-^{14}C]ASC in the case of geranium (Saito *et al.*, 1997). Data from these ^{18}O experiments strongly suggest that cleavage of ASC at C4/C5 involves both oxygenase and hydrolase reactions whereas cleavage of 5-keto-D-gluconic acid at C4/C5 involves only hydrolysis. This study also found that [1,6-^{14}C]ASC-labeled geranium leaves produced ^{14}C-labeled OxA, ThA and TA, confirmation of earlier studies made separately with [1-^{14}C]- or [6-^{14}C]ASC. In other studies involving L-[U-^{14}C]ThA-fed geranium leaves, young leaves converted up to 75% of the labeled compound to TA compared to only 5% in mature leaves (Helsper and Loewus, 1982; Saito *et al.*, 1984).

2.3.5 Overall conversion of D-glucose or D-glucosone to oxalic acid and tartaric acid in plants

ASC is a product of D-glucose metabolism in plants (Wheeler *et al.*, 1998). D-Glucosone (D-arabino-hexos-2-ulose), the C2 oxidation product of D-glucose, is also readily converted to ASC (Saito *et al.*, 1990; Saito and Loewus, 1992; Saito, 1996; Pallanca and Smirnoff, 1999). In geranium, a TA-accumulator, more ^{14}C was found in OxA when D-[5-^{3}H, 1-^{14}C]glucosone was the source of label than when D-[5-^{3}H, 6-^{14}C]glucosone was supplied. The converse applied to TA (Saito and Loewus, 1992). Only traces of ^{3}H remained in TA recovered from tissues treated with either D-[5-^{3}H, 1-^{14}C]- or D-[5-^{3}H, 6-^{14}C]glucosone, a result to be anticipated if epimerization at C5 of the carbon chain involves an exchange of ^{3}H with water

(Grün *et al.*, 1982). When plantlets of duckweed (*Lemna minor*), a typical oxalate accumulator, were pulse-labeled with D-[1-^{14}C]glucosone for 24 h and floated on inorganic medium over a 6-day period, the amount of labeled ASC produced during the pulsed period slowly declined while soluble and insoluble labeled OxA increased with conversion of soluble OxA to insoluble calcium oxalate over time, indicative of the metabolic relationship between ASC, OxA and calcium oxalate (Saito, 1996).

2.3.6 Biosynthesis of oxalic acid from ascorbate in oxalate-accumulating plants

Numerous metabolic intermediates, including glycolate, glyoxylate, isocitrate or oxaloacetate, have been proposed as precursors of OxA but the physiological significance of these compounds has yet to be established (Hodgkinson, 1977; Fujii *et al.*, 1993). The role of ASC role in OxA formation by plants was first observed in the TA accumulator *Pelargonium* (Wagner and Loewus, 1973). Studies involving administration of [1-^{14}C]ASC to oxalate-accumulating plants such as spinach, sorrel, beet, and several weeds provided further evidence that ASC's role as a precursor of OxA was widespread and, possibly, a major path of OxA biosynthesis (Yang and Loewus, 1975; Nuss and Loewus, 1978). Subsequent reports have strengthened this view (Franceschi and Horner, 1979; Wagner, 1981; Helsper and Loewus, 1982; Franceschi, 1987; Saito, 1996; Horner *et al.*, 2000; Keates *et al.*, 2000; Kostman *et al.*, 2001; Monje and Baran, 2002).

Keates *et al.* (2000) floated axenic *Pistia stratiotes* plants on aqueous media containing selected radiolabeled carbon sources (L-[1-^{14}C]galactose, [1-^{14}C]ASC, [6-^{14}C]ASC, D-[1-^{14}C]erythorbic acid, [1-^{14}C]glycolate, Ca salt, or [^{14}C]OxA) to test their potential as substrate for CaOx formation in the specialized leaf cells that accumulate calcium oxalate (CaOx) crystals (idioblasts). Plants were floated on the labeled medium for 12 h followed by a 12-h chase with unlabeled substrate. Leaves were examined by confocal microscopy to obtain reflected and transmitted images of ^{14}C-labeled tissues. In plants pulsed with [^{14}C]OxA, essentially a control, some ^{14}C was found in crystal idioblasts but substantial ^{14}C also appeared in meristematic tissue, possibly due to oxalate oxidase-generated release of $^{14}CO_2$. Carbon-14 from [1-^{14}C]ASC, D-[1-^{14}C]erythorbic acid, and L-[1-^{14}C]galactose was heavily incorporated into CaOx crystals within idioblasts. Carbon-14 from [6-^{14}C]ASC appeared in mesophyll cells but not in crystal idioblasts. Carbon-14 from [1-^{14}C]glycolate incorporated into insoluble residues of developing mesophyll cells but not into idioblasts.

Effective labeling of idioblasts by [1-^{14}C]ASC but not by [6-^{14}C]ASC confirmed the specific nature of C2/C3 cleavage in ASC with the C_2 fragment as OxA. L-[1-^{14}C]Galactose was as effective as [1-^{14}C]ASC in producing labeled CaOx crystals, a clear demonstration of the role played by this ASC precursor during OxA and CaOx production. Pending new studies, the only comment one can make regarding effective labeling of idioblasts with D-[1-^{14}C]erythorbic acid is a suggestion that oxidative cleavage of ASC and D-erythorbic acid at C2/C3 in *P. stratiotes* is nonspecific as regards C5 of these diastereomers. Failure of [1-^{14}C]glycolate to function as a labeling source of CaOx bears out earlier observations which found that idioblasts failed to oxidize glycolate to OxA (Li and Franceschi, 1990).

Recently, Kostman *et al.* (2001) succeeded in isolating viable crystal idioblasts from *P. stratiotes*. When these idioblasts were suspended for 4 h in media containing selected labeled substrates (D-[1-^{14}C]mannose, L-[1-^{14}C]galactose, L-[1-^{14}C]ASC, L-[6-^{14}C]ASC, L-[1-^{14}C]erythorbic acid, [1-^{14}C]glycolic acid, [1-^{14}C]glyoxylic acid or [^{14}C]oxalic acid), and then examined by confocal microscopy to obtain reflected and transmitted images of ^{14}C-labeled cells and their crystalline contents, isolated idioblasts exposed to [^{14}C]OxA accumulated ^{14}C in crystal bundles, evidence that isolated idioblasts were functionally capable of incorporating labeled oxalate into crystal bundles. Individual crystals of CaOx idioblasts released during post-label processing were heavily labeled with ^{14}C.

Incubation of isolated idioblasts with [1-^{14}C]ASC produced labeled crystal bundles whereas [6-^{14}C]ASC was ineffective. These results show that isolated crystal idioblasts contain the entire metabolic machinery needed to synthesize OxA from C1 + C2 of ASC and that this OxA is incorporated into crystal bundles as CaOx. Addition of 1 mM OxA to the medium during incubation of isolated idioblasts with [1-^{14}C]ASC noticeably reduced the amount of ^{14}C in idioblasts as compared with labeled ASC alone, an indication of feedback regulation in the pathway from ASC to OxA that limits OxA biosynthesis to amounts required for crystal formation.

Isolated idioblasts supplied with D-[1-^{14}C]mannose or L-[1-^{14}C]galactose, precursors of ASC (Wheeler *et al.*, 1998), also produced labeled crystal bundles. Qualitatively, L-galactose was much more effective than D-mannose as one might expect on the basis of its location in the biosynthetic pathway (Wheeler *et al.*, 1998, see Smirnoff *et al.*, Chapter 1).

Clearly, crystal idioblasts of *P. stratiotes* contain all enzymes necessary for conversion of hexose phosphate to ASC (and OxA) by the Smirnoff–Wheeler pathway.

2.3.7 Catabolism of myo-inositol and its possible temporal involvement in the formation and catabolism of ascorbate

A gene (GalUR) from strawberry that encodes an NADPH-dependent D-galacturonate reductase has recently been identified (Agius *et al.*, 2003). Their finding provides fresh evidence for an alternative biosynthetic pathway from D-galacturonic acid to ASC in plants (see Smirnoff *et al.*, Chapter 1). These authors revisited earlier studies in which attempts were made to find a link between uronic acid production during [^{3}H]- or [^{14}C]-labeled myo-inositol catabolism in ripening strawberries and ASC production (Loewus *et al.*, 1962; Loewus, 1963, 1965; Loewus and Murthy, 2000). The earlier studies used strawberries at green or green-white stages of ripening and failed to find labeled ASC although D-galacturonosyl residues of pectin were strongly labeled. Agius *et al.* (2003) examined strawberry fruit over the full range of development and found expression of GalUR to be positively correlated to ASC content with highest levels of expression in the fully ripe berry where hydrolytic pectic products rich in D-galacturonate occur. They speculate that the Smirnoff–Wheeler pathway of ASC biosynthesis prevails during the photosynthetic phase of growth but is either supplemented or supplanted by the D-galacturonate→L-galactonate→L-galactono-1,4-lactone→ASC pathway during ripening. It would be of interest to extend Agius *et al.*'s findings to plants like the grape that accumulate catabolic products of ASC such as oxalate (leaves and stems) or TA (berries) in this regard.

2.4 Overview and future prospects

2.4.1 Catabolism of ascorbic acid in animals

Although the first steps and the final products of ASC breakdown are known, further work is needed to clarify the middle part of the pathway and to identify the enzymes catalyzing that process. Earlier findings indicate the astounding diversity of ASC catabolism in various species and cell types; it would be particularly interesting to assess the differences of the pathway in ASC-synthesizing and nonsynthesizing species, from both qualitative and quantitative aspects.

Perhaps the most peculiar feature of ASC catabolism is that from an antioxidant molecule (ASC) another antioxidant compound (i.e., oxalate; Kayashima and Katayama, 2002) can be synthesized through highly reactive, electrophile, pro-oxidant intermediates. These intermediates may have a causal role in human pathology, e.g., in cataractogenesis and in the development of diabetic complications (Delamere, 1996; Hunt, 1996). Nevertheless, ASC catabolism seems to join with important physiological processes, such as oxygen sensing (Jaakkola *et al.*, 2001), oxidative protein folding (Bánhegyi *et al.*, 2002) and redox regulation (Bijur *et al.*, 1999). The prospective progression in these fields may contribute to the better understanding of the pathomechanism of scurvy.

2.4.2 Catabolism of ascorbic acid in plants

Isotopic markers provide convincing evidence for C2/C3 and C4/C5 cleavage of ASC but enzymatic activities associated with these processes have yet to be identified. In vitaceous plants where C4/C5 cleavage is related to accumulation of TA in the fruit (Saito and Kasai, 1969), CaOx accumulates in crystal idioblasts in the leaves (Webb *et al.*, 1995), presumably a product of ASC cleavage at C2/C3, but still to be tested experimentally. Although OxA and/or its salts often accumulate in plants, this product also undergoes further oxidation to CO_2 and hydrogen peroxide, catalyzed by oxalate oxidase (Franceschi and Loewus, 1995). The possibility that CO_2 generated through breakdown of ASC can be recycled by photosynthesis has been discussed (Loewus, 1999). Finally, little attention has been given to the catabolic fate of the C_4 fragment stemming from C2/C3 cleavage other than its oxidation to TA and even here, the enzymic nature of this oxidation remains unknown. A significant number of plants can decarboxylate ThA but the catabolic fate of its C_3 product needs further study. Hopefully, rapid progress in the area of molecular biology will contribute fresh insight.

References

Agius F, González-Lamonthe R, Caballero JL, Muñoz-Blanco J, Botella MA, Valpuesta V (2003) Engineering increased vitamin C levels in plants by overexpression of a D-galacturonic acid reductase. *Nature Biotechnol.* **21**, 177–181.

Auer BL, Auer D, Rodgers AL (1998) The effect of ascorbic acid ingestion on the biochemical and physicochemical risk factors associated with calcium oxalate kidney stone formation. *Clin. Chem. Lab. Med.* **36**, 143–147.

Baker EM, Levandovski NG, Sauberlich HE (1963) Respiratory catabolism in man of the degradative intermediates of L-ascorbic-1-^{14}C acid. *Proc. Soc. Exp. Biol.* **113**, 379–383.

Baker EM, Hodges RE, Hood J, Sauberlich HE, March SC (1969) Metabolism of ascorbic-1-^{14}C acid in experimental human scurvy. *Am. J. Clin. Nutr.* **22**, 549–558.

Baker EM, Hammer DC, March SC, Tolbert BM, Canham JE (1971a) Ascorbate sulfate: a urinary metabolite of ascorbic acid in man. *Science* **173**, 826–827.

Baker EM, Hodges RE, Hood J, Sauberlich HE, March SC, Canham JE (1971b) Metabolism of ^{14}C- and ^{3}H-labeled L-ascorbic acid in human scurvy. *Am. J. Clin. Nutr.* **24**, 444–454.

Bánhegyi G, Braun L, Csala M, Puskás F, Mandl J (1997) Ascorbate metabolism and its regulation in animals. *Free Radic. Biol. Med.* **23**, 793–803.

Bánhegyi G, Braun L, Csala M, Puskás F, Somogyi A, Kardon T, Mandl J (1998) Ascorbate and environmental stress. *Ann. N. Y. Acad. Sci.* **851**, 292–303.

Bánhegyi G, Csala M, Benedetti A, Mandl J (2002) Role of ascorbate in oxidative protein folding. In: *Thiol Metabolism and Redox Regulation of Cellular Functions* (eds A Pompella, G Bánhegyi, M Wellmann-Rousseau). IOS Press, Amsterdam. NATO Science Series, Life and Behavioural Sciences, vol. 347.

Bensch KG, Fleming JE, Lohmann W (1985) The role of ascorbic acid in senile cataract. *Proc. Natl Acad. Sci. USA* **82**, 7193–7196.

Bijur GN, Briggs B, Hitchcock CL, Williams MV (1999) Ascorbic acid-DHA induces cell cycle arrest at G2/M DNA damage checkpoint during oxidative stress. *Environ. Mol. Mutagen.* **33**, 144–152.

Blaschke E, Hertting G (1971) Enzymatic methylation of L-ascorbic acid by catechol-*O*-methyltransferase. *Biochem. Pharmacol.* **20**, 1363–1370.

Bode AM, Cunningham L, R (1990) Spontaneous decay of oxidized ascorbic acid (dehydro-L-ascorbic acid) evaluated by high-pressure liquid chromatography. *Clin. Chem.* **36**, 1807–1819.

Bond AD, McClelland BW, Einstein JR, Finamore FJ (1972) Ascorbic acid-2-sulfate of the brine shrimp, *Artemia salina*. *Arch. Biochem. Biophys.* **153**, 207–214.

Braun L, Puskás F, Csala M, Gyorffy E, Garzó T, Mandl J, Bánhegyi G (1996) Gluconeogenesis from ascorbic acid: ascorbate recycling in isolated murine hepatocytes. *FEBS Lett.* **390**, 183–186.

Braun L, Puskás F, Csala M, Mészáros Gy, Mandl J, Bánhegyi G (1997) Ascorbate as a substrate for glycolysis or gluconeogenesis. Evidence for an interorgan ascorbate cycle. *Free Radic. Biol. Med.* **23**, 804–808.

Bron AJ, Brown NAP (1987) Perinuclear lens retrodots: a role for ascorbate in cataractogenesis. *Br. J. Ophthalmol.* **71**, 86–95.

Carpenter, K.J. (1986) *The History of Vitamin C and Scurvy*. Cambridge University Press, Cambridge, UK.

Cerrano C, Bavestrello G, Arillo A, Benatti U, Bonpadre S, Cattaneo-Vietti R, *et al* (1999) Calcium oxalate production in the marine sponge *Chondrosia reniformis*. *Mar. Ecol. Prog. Ser.* **179**, 297–300.

Chan PC, Becker RR, King CG (1958) Metabolic products of L-ascorbic acid. *J. Biol. Chem.* **231**, 231–240.

Chen Z, Young TE, Ling J, Chang S-C, Gallie DE (2003) Increasing vitamin C content of plants through enhanced ascorbate recycling. *Proc. Natl Acad. Sci. USA* **100**, 3525–3530.

Conklin PL (2001) Recent advances in the role and biosynthesis of ascorbic acid in plants. *Plant Cell Environ.* **24**, 383–394.

Conklin PL, Pallanca JE, Last RL, Smirnoff N (1997) L-Ascorbic acid metabolism in the ascorbic acid-deficient *Arabidopsis* mutant *vtc1*. *Plant Physiol.* **115**, 1277–1285.

Cox BD, Whichelow MJ (1975) The measurement of dehydroascorbic acid and diketogulonic acid in normal and diabetic plasma. *Biochem. Med.* **12**, 183–193.

Crawford TC (1982) Synthesis of L-ascorbic acid. In: *Ascorbic Acid: Chemistry, Metabolism, and Uses* (eds PA Seib, BM Tolbert). American Chemistry Society, Washington, DC, Advances in Chemistry Series, No. 200, pp. 1–36.

Csonka F, Váradi G (1907) *Der Szegeder paprika und der Szegeder paprikahandel*. Buchdruckerei L. Engel, Szeged.

Curhan GC, Willett WC, Rimm EB, Stampfer MJ (1996) A prospective study of the intake of vitamins C and B6, and the risk of kidney stones in men. *J. Urol.* **155**, 1847–1851.

Dabrowski K, Lackner R, Doblander C (1993) Ascorbate-2-sulfate sulfohydrolase in fish and mammal. Comparative characterization and possible involvement in ascorbate metabolism. *Comp. Biochem. Physiol.* **104**, 717–722.

Davey, MW, Van Montagu M, Inzé D (2002) Ascorbate metabolism and stress. In: *Oxidative Stress in Plants* (eds M Van Montagu, D Inzé). Taylor & Francis, London and NY, pp. 271–296.

Davey MW, Van Montagu M, Inzé D, Sanmartin M, Kanellis A, Smirnoff N, Benzie IJJ, Strain JJ, Favell D, Fletcher J (2000) Plant L-ascorbic acid: chemistry, function, metabolism, bioavailability and effects of processing. *J. Sci. Agric.* **80**, 825–860.

Davies MB, Austin J, Partridge DA (1991) *Vitamin C. Its Chemistry and Biochemistry.* The Royal Society of Chemistry, Letchworth, UK.

Delamere NA (1996) Ascorbic acid and the eye. In: *Subcellular Biochemistry, Vol. 25. Ascorbic Acid: Biochemistry and Biomedical Cell Biology*. (ed. JR Harris JR). Plenum Press, New York, pp. 313–330.

Deutsch J (1998) Oxygen-accepting antioxidants which arise during ascorbate oxidation. *Anal. Biochem.* **265**, 238–245.

Franceschi VR (1987) Oxalic acid metabolism and calcium oxalate formation in *Lemna minor* L. *Plant Cell Environ.* **10**, 397–406.

Franceschi VR, Horner HT Jr. (1979) use of *Psychotria punctata* callus in study of calcium oxalate crystal idioblast formation. *Z. Pflanzenphysiol.* **92**, 61–75.

Franceschi VR, Loewus FA (1995) Oxalate biosynthesis and function in plants and fungi. In: *Calcium Oxalate in Biological Systems* (ed. S Khan). CRC Press, Boca Raton, FL, pp.113–130.

Franceschi VR, Tarlyn NM (2002) L-Ascorbic acid is accumulated in source leaf phloem and transported to sink tissues in plants. *Plant Physiol.* **130**, 649–656.

Fujii N, Watanabe M, Watanabe Y, Shimada N (1993) Rate of oxalate biosynthesis from glycolate and ascorbic acid in spinach leaves. *Soil Sci. Plant Nutr.* **39**, 627–643.

Gallice P, Sarrazin F, Polverelli M, Cadet J, Berland Y, Crevat A. (1994) Ascorbic acid-2-*O*-β-glucuronide, a new metabolite of vitamin C identified in human urine and uremic plasma. *Biochim. Biophys. Acta* **1199**, 305–310.

Gatzek S, Wheeler GL, Smirnoff N (2002) Antisense suppression of L-galactose dehydrogenase in *Arabidopsis thaliana* provides evidence for its role in ascorbic acid synthesis and reveals light modulated L-galactose synthesis. *The Plant J.* **30**, 541–553.

Gerster H (1997) No contribution of ascorbic acid to renal calcium oxalate stones. *Ann. Nutr. Metab.* **41**, 269–282.

Grün M, Renstrøm B, Loewus FA (1982) Loss of hydrogen from carbon 5 of D-glucose during conversion of D-[5-^{3}H, 6-^{14}C]glucose to L-ascorbic acid in *Pelargonium crispum* (L.) Hér. *Plant Physiol.* **70**, 1233–1235.

Hellman L, Burns JJ (1958) Metabolism of L-ascorbic acid-1-^{14}C in man. *J. Biol. Chem.* **230**, 923–930.

Helsper JP, Loewus FA (1982) Metabolism of L-threonic acid in *Rumex × acutus* L, *Pelargonium crispum* (L.) L'Hér. *Plant Physiol.* **69**, 1365–1368.

Helsper JP, Loewus FA (1985) Studies on L-ascorbic acid biosynthesis and metabolism in *Parthenocissus quinquefolia* L. (Vitaceae). *Plant Sci.* **40**, 105–109.

Helsper JP, Saito K, Loewus FA (1981) Biosynthesis and metabolism of L-ascorbic acid in virginia creeper (*Parthenocissus quinquefolia* L.). *Planta* **152**, 171–176.

Herbert RW, Hirst EL, Percival EGV, Reynolds RJW, Smith F (1933) The constitution of ascorbic acid. *J. Chem. Soc.* 1270–1290.

Himmelreich U, Drew KN, Serianni AS, Kuchel PW (1998) ^{13}C NMR studies of vitamin C transport and its redox cycling in human erythrocytes. *Biochemistry* **37**, 7578–7588.

Himmelreich U, Emling M, Drew KN, Serianni AS, Kuchel PW (2000) ^{13}C NMR evidence of the failure of human erythrocytes to metabolize ascorbate and dehydroascorbate to lactate. *Free Radic. Biol. Med.* **28**, 1607–1610.

Hodgkinson A (1977) *Oxalic Acid in Biology and Medicine*. Academic Press, New York.

Horemans N, Foyer CH, Asard H (2000a) Transport and action of ascorbate at the plant plasma membrane. *Trends Plant Sci.* 5, 263–267.

Horemans, N, Foyer CH, Potter G, Asard H (2000b) Ascorbate function and associated transport systems in plants. *Plant Physiol. Biochem.* **38**, 531–540.

Horner, HT, Kausch AP, Wagner BL (2000) Ascorbic acid: a precursor of oxalate in crystal idioblasts of *Yucca torreyi* in liquid root culture. *Intern. J. Plant Sci.* **161**, 861–868.

Hough L, Jones JKN (1956) The biosynthesis of monosaccharides. In: *Advances in Carbohydrate Chemistry II.* Academic Press, New York, pp. 185–262 (see also p. 240).

Hunt JV (1996) Ascorbic acid and diabetes mellitus. In: *Subcellular Biochemistry, Vol. 25. Ascorbic Acid: Biochemistry and Biomedical Cell Biology*. (ed. JR Harris). Plenum Press, New York, pp. 369–406.

Isbell HS, Frush HL (1979) Oxidation of L-ascorbic acid by hydrogen peroxide: preparation of L-threonic acid. *Carbohydr. Res.* **72**, 301–304.

Jaakkola P, Mole DR, Tian YM, Wilson MI, Gielbert J, Gaskell SJ ***et al.*** (2001) Targeting of HIF-α to the von Hippel–Lindau ubiquitylation complex by O_2-regulated prolyl hydroxylation. *Science* **292**, 468–472.

Kagawa Y (1962) Enzymatic studies on ascorbic acid catabolism in animals. I. Catabolism of 2,3-diketogulonic acid. *J. Biochem.* **51**, 134–144.

Kagawa Y, Takiguchi H, Shimazono N (1961) Enzymic delactonization of dehydro-L-ascorbate in animal tissues. *Biochim. Biophys. Acta* **51**, 413–415.

Kallner A, Hornig D, Pellikka R (1985) Formation of carbon dioxide from ascorbate in man. *Am. J. Clin. Nutr.* **41**, 609–613.

Kanfer J, Ashwell G, Burns JJ (1960) Formation of L-lyxonic and L-xylonic acids from L-ascorbic acid in rat kidney. *J. Biol. Chem.* **235**, 2518–2521.

Kang SO, Sapper H, Lohmann W (1982) The oxidative degradation of L-ascorbic acid via an α-ketoaldehyde. *Z. Naturforsch. [C]* **37**, 1064–1069.

Kashiba M, Oka J, Ichikawa R, Kageyama A, Inayama T, Kageyama H, Ishikawa T, Nishikimi M, Inoue M, Inoue S (2000) Impaired reductive regeneration of ascorbic acid in the Goto–Kakizaki diabetic rat. *Biochem. J.* **351**, 313–318.

Kashiba, M, Oka J, Ichikawa R, Kasahara, E, Inayama T, Kageyama A, et al. (2002) Impaired ascorbic acid metabolism in streptozotocin-induced diabetic rats. *Free Radic. Biol. Med.* **33**, 1221–1230.

Kayashima T, Katayama T (2002) Oxalic acid is available as a natural antioxidant in some systems. *Biochim. Biophys. Acta* **1573**, 1–3.

Keates SE, Tarlyn NM, Loewus FA, Franceschi VR (2000) L-Ascorbic acid and L-galactose are sources for oxalic acid and calcium oxalate in *Pistia stratiotes*. *Phytochemistry* **53**, 433–440.

Koshiishi I, Mamura Y, Imanari T (1998a) Bicarbonate promotes a cleavage of lactone ring of dehydroascorbate. *Biochim. Biophys. Acta* **1379**, 257–263.

Koshiishi I, Mamura Y, Liu J, Imanari T (1998b) Degradation of dehydroascorbate to 2,3-diketogulonate in blood circulation. *Biochim. Biophys. Acta* **1425**, 209–214.

Kostman TA, Tarlyn NM, Loewus FA, Franceschi VR (2001) Biosynthesis of L-ascorbic acid and conversion of carbons 1 and 2 of L-ascorbic acid to oxalic acid occurs within individual calcium oxalate crystal idioblasts. *Plant Physiol.* **125**, 634–640.

Li X, Franceschi VR (1990) Distribution of peroxisomes and glycolate metabolism in relation to calcium oxalate formation in *Lemna minor* L. *Eur. J. Cell Biol.* **51**, 9–16.

Loewus FA (1963) Tracer studies of ascorbic acid formation in plants. *Phytochemistry* **2**, 109–128.

Loewus F (1965) Inositol metabolism and cell wall formation in plants. *Fed. Proc, Fed. Am. Soc. Expt. Biol.* **24**, 855–862.

Loewus FA (1999) Biosynthesis and metabolism of ascorbic acid in plants and of analogs of ascorbic acid in fungi. *Phytochemistry* **52**, 193–210. Corrigendum *ibid.* (2000). **54**, 449.

Loewus FA, Kelly S, Neufeld EF (1962) Metabolism of *myo*-inositol in plants: Conversion to pectin, hemicellulose, D-xylose and sugar acids. *Proc. Natl Acad. Sci. USA* **48**, 421–425.

Loewus FA, Murthy PPN (2000) *myo*-Inositol metabolism in plants. *Plant Sci.* **150**, 1–19.

Loewus FA, Stafford HA (1958) Observations on the incorporation of ^{14}C into tartaric acid and the labeling pattern of D-glucose from an excised grape leaf administered L-ascorbic acid-6-^{14}C. *Plant Physiol.* **33**, 155–156.

Loewus FA, Wagner G, Yang JC (1975) Biosynthesis and metabolism of ascorbic acid in plants. *Ann. New York Acad. Sci.* **258**, 7–23.

Lohmann W, Schmehl W, Strobel J (1986) Nuclear cataract: oxidative damage to the lens. *Exp. Eye Res.* **43**, 859–862.

López MG, Feather MS (1992) The production of threose as a degradation product from L-ascorbic acid. *J. Carbohydr. Chem.* **11**, 799–806.

Mitton KP, Dzialoszynski T, Sanford SE, Trevithick JR (1997) Cysteine and ascorbate loss in the diabetic rat lens prior to hydration changes. *Curr. Eye Res.* **16**, 564–571.

Miyasaki T, Sato M, Yoshinaka R, Sakaguchi M (1991) Synthesis of ascorbyl-2-phosphate by liver enzyme of rainbow trout *Oncorhynchus mykiss*. *Comp. Biochem. Physiol. B.* **100**, 711–716.

Monje PV, Baran EJ (2002) Characterization of calcium oxalates generated as biominerals in cacti. *Plant Physiol.* **128**, 707–713.

Moser U, Hornig D (1982) High intakes of vitamin C: a contributor to oxalate formation in man? *Trends Pharmacol. Soc.* **3**, 480–483.
Moss RW (1988) *Free Radical: Albert Szent-Györgyi and the battle over Vitamin C.* Paragon House, New York, NY.
Muddeshwar MG, Nath N, Chari SN (1992) Sex variation in ascorbic acid catabolism. *Indian J. Physiol. Pharmacol.* **36**, 263–266.
Muto N, Ban Y, Akiba M, Yamamoto I (1991) Evidence for the in vivo formation of ascorbic acid 2-*O*-β-glucoside in guinea pigs and rats. *Biochem. Pharmacol.* **42**, 625–631.
Nagaraj RH, Monnier VM (1995) Protein modification by the degradation products of ascorbate: formation of a novel pyrrole from the Maillard reaction of L-threose with proteins. *Biochim. Biophys. Acta* **1253**, 75–84.
Nagaraj RH, Sell DR, Prabhakaram M, Ortwerth BJ, Monnier VM (1991) High correlation between pentosidine protein crosslinks and pigmentation implicates ascorbate oxidation in human lens senescence and cataractogenesis. *Proc. Natl Acad. Sci. USA* **88**, 10257–10261.
Nagaraj RH, Shamsi FA, Huber B, Pischetsrieder M (1999) Immunochemical detection of oxalate monoalkylamide, an ascorbate-derived Maillard reaction product in the human lens. *FEBS Lett.* **453**, 327–330.
Nakamura M, Ohtaki S (1993) Formation and reduction of ascorbate radicals by hog thyroid microsomes. *Arch. Biochem. Biophys.* **305**, 84–90.
Niemelä K (1987) Oxidative and non-oxidative alkali-catalysed degradation of L-ascorbic acid. *J. Chromatogr.* **399**, 235–243.
Noctor G, Foyer CH (1998) Ascorbate and glutathione: keeping active oxygen under control. *Annu. Rev. Plant Physiol. Plant Mol. Biol.* **49**, 249–279.
Nuss RF, Loewus FA (1978) Further studies on oxalate-accumulating plants. *Plant Physiol.* **61**, 590–593.
Ortwerth BJ, Olesen PR (1988a) Ascorbic acid-induced crosslinking of lens proteins. Evidence supporting a Maillard reaction. *Biochim. Biophys. Acta* **956**, 10–22.
Ortwerth BJ, Olesen PR (1988b) Glutathione inhibits the glycation and crosslinking of lens proteins by ascorbic acid. *Exp. Eye Res.* **47**, 737–750.
Ortwerth BJ, Speaker JA, Prabhakaram M, López MG, Li EY, Feather MS (1994) Ascorbic acid glycation: the reactions of L-threose in lens tissue. *Exp. Eye Res.* **58**, 665–674.
Pallanca JE, Smirnoff N (1999) Ascorbic acid metabolism in pea seedlings. A comparison of D-glucosone, L-sorbosone and L-galactono-1,4-lactone as ascorbate precursors. *Plant Physiol.* **120**, 453–461.
Pallanca JE, Smirnoff N (2000). The control of ascorbic acid synthesis and turnover in pea seedlings. *J. Exp. Bot.* **51**, 669–674.
Pillai GR, Indira M, Vijayammal PL (1991) Role of exogenous ascorbic acid in tissue status of ascorbic acid-2–sulphate in guinea pigs. *Indian J. Exp. Biol.* **29**, 1127–1130.
Potters G, Horemans N, Caubergs RJ, Asard H (2000) Ascorbate and dehydroascorbate influence cell cycle progression in a tobacco cell suspension. *Plant Physiol.* **124**, 17–20.
Price KD, Price CS, Reynolds RD (2001) Hyperglycemia-induced ascorbic acid deficiency promotes endothelial dysfunction and the development of atherosclerosis. *Atherosclerosis* **158**, 1–12.
Saito K (1996) Formation of L-ascorbic acid and oxalic acid from D-glucosone in *Lemna minor. Phytochemistry* **41**, 145–149.
Saito K, Kasai, Z. (1969) Tartaric acid synthesis from L-ascorbic acid-1–^{14}C in grape berries. *Phytochemistry*, **8**, 2177–2182.
Saito K, Kasai Z (1982) Conversion of L-ascorbic acid to L-idonic acid, L-idono-γ-lactone and 2-keto–L-idonic acid in slices of immature grapes. *Plant Cell Physiol.* **23**, 499–507.
Saito K, Kasai Z (1984) Synthesis of L-(+)-tartaric acid from L-ascorbic acid via 5-keto-D-gluconic acid in grapes. *Plant Physiol.* **76**, 170–174.
Saito K, Loewus FA (1979) The metabolism of L-[6-^{14}C]ascorbic acid in detached grape leaves. *Plant Cell Physiol.* **20**, 1481–1488.
Saito K, Loewus FA (1989a) Formation of L-(+)-tartaric acid in leaves of the bean, *Phaseolus vulgaris* L.: Radioisotopic studies with putative percursors. *Plant Cell Physiol.* **30**, 629–636.
Saito K, Loewus FA (1989b) Formation of tartaric acid in vitaceous plants: Relative contributions of L-ascorbic acid-inclusive and -noninclusive pathways. *Plant Cell Physiol.* **30**, 905–910.
Saito K, Loewus FA (1992) Conversion of D-glucosone to oxalic acid and L-(+)-tartaric acid in detached leaves of *Pelargonium. Phytochemistry* **31**, 3341–3344.

Saito K, Morita S, Kasai Z (1984) Synthesis of L-(+)-tartaric acid from 5-keto-D-gluconic acid in Pelargonium. *Plant Cell Physiol.* **25**, 1223–1232.

Saito K, Nick JA, Loewus FA (1990) D-Glucosone and L-sorbosone, putative intermediates of L-ascorbic acid biosynthesis in detached bean and spinach leaves. *Plant Physiol.* **94**, 1496–1500.

Saito K, Ohmoto J, Kuriha N (1997) Incorporation of ^{18}O into oxalic, L-threonic and L-tartaric acids during cleavage of L-ascorbic acid and 5-keto-D-glucunic acid in plants. *Phytochemistry* **44**, 805–809.

Saxena P, Saxena AK, Monnier VM (1996) High galactose levels in vitro and in vivo impair ascorbate regeneration and increase ascorbate-mediated glycation in cultured rat lens. *Exp. Eye Res.* **63**, 535–545.

Simpson GL, Ortwerth BJ (2000) The non-oxidative degradation of ascorbic acid at physiological conditions. *Biochim. Biophys. Acta* **1501**, 12–24.

Slight SH, Feather MS, Ortwerth BJ (1990) Glycation of lens proteins by the oxidation products of ascorbic acid. *Biochim. Biophys. Acta* **1038**, 367–374.

Smirnoff N, Conklin PL, Loewus FA (2001) Biosynthesis of ascorbic acid in plants: A renaissance. *Ann. Rev. Plant Physiol. Plant Mol. Biol.* **52**, 437–467.

Sullivan SP, Stern A (1983) Effects of physiologic concentrations of lactate, pyruvate and ascorbate on glucose metabolism in unstressed and oxidatively stressed human red blood cells. *Biochem. Pharmacol.* **32**, 2891–2902.

Sun I, Morre DJ, Crane FL, Safranski K, Croze EM (1984) Monodehydroascorbate as an electron acceptor for NADH reduction by coated vesicle and Golgi apparatus fractions of rat liver. *Biochim. Biophys. Acta* **797**, 266–275.

Szaleczky E, Prechl J, Ruzicska É, Fehér J, Braun L, Bánhegyi G, Csala M, Mandl J, Somogyi A. (1998) Reduction of glycated hemoglobin levels by long term, high dose ascorbic acid supplementation in healthy and diabetic patients. *Med. Sci. Monit.* **4**, 241–244.

Szarka A, Stadler K, Jenei V, Margittai É, Csala M, Jakus J, Mandl J, Bánhegyi G (2002) Ascorbyl free radical and dehydroascorbate formation in rat liver endoplasmic reticulum. *J. Bioenerg. Biomembr.* **34**, 317–323.

Szent-Györgyi A (1937) Oxidation, energy transfer, and vitamins. Nobel Lecture. http: //www.nobel.se/medicine/laureates/1937/szent–gyorgyi–lecture.pdf.

Tabata K, Takaoka T, Esaka M (2002) Gene expression of ascorbic acid-related enzymes in tobacco. *Phytochemistry* **61**, 631–635.

Tolbert BM, Downing M, Carlson RW, Knight MK, Baker EM (1975) Chemistry and metabolism of ascorbic acid and ascorbate sulfate. *Ann. N. Y. Acad. Sci.* **258**, 48–69.

Wagner G, Loewus F (1973) The biosynthesis of (+)-tartaric acid in *Pelargonium crispum*. *Plant Physiol.* **52**, 651–654.

Wagner GJ (1981) Vacuolar deposition of ascorbate–derived oxalic acid in barley. *Plant Physiol.* **67**, 591–593.

Wakamiya H, Suzuki E, Yamamoto I, Akiba M, Otsuka M, Arakawa N (1992) Vitamin C activity of 2-*O*-β-D-glucopyranosyl-L-ascorbic acid in guinea pigs. *J. Nutr. Sci. Vitaminol. (Tokyo)* **38**, 235–245.

Webb MA, Cavaletto JM, Carpita NC, Lopez LE, Arnett HJ (1995) The intravacuolar organic matrix associated with calcium oxalate crystals in leaves of *Vitis*. *The Plant J.* **7**, 633–648.

Wheeler GL, Jones MA, Smirnoff N (1998) The biosynthetic pathway of vitamin C in higher plants. *Nature* **303**, 365–369.

Wilkinson SR, Obado SO, Mauricio IL, Kelly JM (2002) *Trypanosoma cruzi* expresses a plant-like ascorbate-dependent hemoperoxidase localized to the endoplasmic reticulum. *Proc. Natl Acad. Sci. USA* **99**, 13453–13458.

Will JC, Byers T (1996) Does diabetes mellitus increase the requirement for vitamin C? *Nutr. Rev.* **54**, 193–202.

Williams M, Loewus FA (1978) Biosynthesis of (+)-tartaric acid from L-[4-^{14}C]ascorbic acid in grape and geranium. *Plant Physiol.* **61**, 672–674.

Williams M, Saito K, Loewus FA (1979) Ascorbic acid metabolism in geranium and grape. *Phytochemistry* **18**, 953–956.

Winkelman J, Lehninger AL (1958) Aldono- and uronolactonases of animal tissues. *J. Biol. Chem.* **233**, 794–799.

Yamada K, Ishikawa S, Shimazono N (1959) On the microsomal and soluble lactonases. *Biochim. Biophys. Acta* **32**, 253–255.

Yang JC, Loewus FA (1975) Metabolic conversion of L-ascorbic acid to oxalic acid in oxalate-accumulating plants. *Plant Physiol.* **56**, 283–285.

3

The biotechnology of ascorbic acid manufacture

Jeffrey A. Running, Susan Peng and Reinhardt A. Rosson

3.1 Introduction

The large-scale industrial production of L-ascorbic acid (ascorbate; ASC) dates from the commercialization of the Reichstein synthesis in the 1930s (Delic *et al.*, 1989). With the rise of the modern fermentation industry, attention was given to the idea of replacing some of the steps in this almost-entirely chemical synthesis with biological ones. Fifty years ago, Yamazaki (1953) reported on attempts to synthesize ASC biologically by first producing Reichstein intermediates through the oxidation of L-idonate and D-gluconate in fermentations of a *Pseudomonas* sp. Most current alternatives to the Reichstein synthesis are based on similar methods, especially the biological production of the last Reichstein intermediate, 2-keto-L-gulonic acid (2-KLG). This chapter will briefly review the microbial production of 2-KLG; a thorough treatment of the details can be found in four reviews (Delic *et al.*, 1989; Chotani *et al.*, 2000; Hancock and Viola, 2001, 2002). Only recently have biotechnology-based processes advanced to the point where they seriously compete with the Reichstein process on the commercial scale. In addition, future generations of fermentation processes are being developed. One of these will be discussed in some detail – an attempt to produce ASC in a one-step fermentation process from Glc by yeast genetically engineered with genes from the known Smirnoff–Wheeler–Running biosynthetic pathway in plants.

3.2 Multi-step ascorbic acid synthesis processes

3.2.1 Processes to produce ascorbic acid from 2-keto-L-gulonate

Reichstein process

The Reichstein process is a seven-step mixed fermentation/chemical process. Glc is chemically hydrogenated to form sorbitol, which is used as a fermentation

Vitamin C, edited by Han Asard, James M May and Nicholas Smirnoff

substrate to form D-sorbose. Reaction with sulfuric acid and acetone forms a tricylic ring structure, which serves to protect critical reactive groups during the next permanganate oxidation step (Jaffe, 1984). Subsequent hydrolysis yields 2-KLG, which is esterified and lactonized to form ASC. The efficiency of these steps in the Reichstein process have been improved to the point where each has a greater than 81% efficiency and the overall process has about a 50% yield of ASC from Glc (Hancock and Viola, 2001). The process requires large energy inputs and uses considerable quantities of organic and inorganic solvents and reagents, resulting in significant disposal costs. The difficulty of achieving further improvements in this mature process has been a motivating factor in the development of alternative production methods.

Biological routes to 2-keto-L-gulonate

Biotechnology-based alternatives to the Reichstein process have focused on the biological production of 2-KLG, one of the last intermediates of the Reichstein process. The conversion of 2-KLG to ASC, in all cases, uses the chemistry of the last steps of the Reichstein process.

Development of microbial 2-KLG production has proceeded along four routes

The first converts Glc to gluconate, 5-keto-D-gluconic acid (5-KGA) (Stewart, 1959), L-idonate/L-gulonate and finally 2-KLG (Chotani *et al.*, 2000). This route is not considered to be of major importance on an industrial scale (Hancock and Viola, 2001). The second and third, the 'sorbitol' and '2,5-diketo-D-gluconate' (2,5-DKG) routes, respectively, have achieved yields that make them promising candidates to replace the Reichstein process. A fourth route utilizes a genetically engineered bacterium for the one-step conversion of Glc to 2-KLG.

Sorbitol pathway

Tengerty (1961) and Huang (1962) first described the bacterial formation of 2-KLG from sorbose. Later work described microbial 2-KLG production from sorbose using L-idose and L-idonic acid as intermediates (Isono *et al.*, 1968; Okazaki *et al.*, 1969; Kanzaki and Okazaki, 1970). Zizheng *et al.* (1981) extended this idea by converting sorbose to 2-KLG in mixed cultures of *Gluconobacter oxydans* and *Pseudomonas striata*. Sugisawa *et al.* (1990) used classical strain improvement methods to increase yields of 2-KLG from sorbose to 60 $g{\cdot}l^{-1}$. Tsukada and Perlman (1972a, 1972b, 1972c) achieved conversion of sorbose to 2-KLG via a route that passed through sorbosone. Makover *et al.* (1975) extended these results, and Shinjoh *et al.* (1990, 1994, 1995) engineered *G. oxydans* with sorbosone dehydrogenase for increased 2-KLG yields. Saito *et al.* (1997), at Fujisawa Co., mutagenized recombinant *G. oxydans* to reduce loss of yield from L-idonate formation. This work achieved 130 $g{\cdot}l^{-1}$ 2-KLG in fermentors.

2,5-Diketo-D-gluconate pathway

Microorganisms capable of converting Glc to 2-KLG were unknown in the early 1980s. Sonoyama *et al.* (1982), at Shionogi Co., developed a promising two-step conversion of Glc to 2-KLG. Glc was first converted to 2,5-DKG by an *Erwinia* species, the resulting fermentation medium was treated with detergent to kill most of the producing cells and this detergent-treated broth was used as a feed for a second fermentative conversion of the 2,5-DKG to 2-KLG by a *Corynebacterium* species. This process is actively being developed in China. The process appears to be a clear improvement that should achieve superior economics relative to the equivalent Reichstein process steps to 2-KLG (see section 5 below).

A microbe capable of producing 2-KLG from Glc was engineered by Anderson *et al.* (1985) at Genentech. They succeeded in engineering the 2,5-DKG reductase from the *Corynebacterium* spp. into the *Erwinia* spp. described above. Initial yields were about 1 $g{\cdot}l^{-1}$. Grindley *et al.* (1988), at Biogen, cloned an improved reductase into a different *Corynebacterium* strain and achieved nearly 20 $g{\cdot}l^{-1}$ 2-KLG in 72 h. More recent improvements at Genencor have yielded more than 120 $g{\cdot}l^{-1}$ 2-KLG in about 5 days (Fowler and Causey, 1998).

Genencor subsequently developed, in conjunction with other technology partners, an *in vitro* Glc to 2-KLG process that produces 42 $g{\cdot}l^{-1}$ 2-KLG, with an overall productivity of more than 2 $g{\cdot}l^{-1}\,h^{-1}$ (Chotani *et al.*, 2000). The high cost of cofactor was a significant barrier to its commercial implementation. Productivity was also limited by the high K_ms and the thermal instability of the *Corynebacterium* enzyme(s) used. 2,5-DKG reductases with significant improvements in both parameters have been recently isolated by Eschenfeldt *et al.* (2001). Incorporation of these improvements could result in a new, viable ASC production process.

Conversion of 2-keto-L-gulonate to ascorbic acid

The conversion of 2-KLG to ASC entails first an esterification to 2-KLG methyl ester, formation of a metal ascorbate salt and lactonization with acid into ASC (Hancock and Viola, 2001). These reactions are integral to the Reichstein process. Scientists at Genencor have recently reported the direct conversion of 2-KLG to ASC by strains of *Candida* and *Cryptococcus* (Kumar, 2002). The overall conversion, however, was very poor: with 5 $g{\cdot}l^{-1}$ 2-KLG and 300 $g{\cdot}l^{-1}$ wet cells, 6.6 $mg{\cdot}l^{-1}$ ASC was made in 20 h.

3.2.2 Process to produce ascorbic acid by non-2-keto-l-gulonate biotransformations

Besides processes to produce 2-KLG, other attempts have been made for the industrial production of ASC. Danehy (1981) detailed a process to produce ASC from the lactose in whey after hydrolysis, oxidation to uronic acids, reduction to aldonic acids, dehydration to lactones and conversion to ASC with crude pea extracts. No yields were given. Cayle *et al.* (1986) described a similar multi-step process in which L-galactono-1,4-lactone (L-GalL) is produced from whey and fed to yeast for conversion into ASC. Classical strain improvement methods increased the ASC yields from the lactone sixfold. With 38 $g{\cdot}l^{-1}$ cells and 14 $g{\cdot}l^{-1}$ substrate they achieved nearly 8 $g{\cdot}l^{-1}$ ASC in 34 h (0.22 $g{\cdot}l^{-1}{\cdot}h^{-1}$). The fact that this process has not been commercialized attests to the high costs of producing the lactone, and the need to significantly increase the product concentration.

3.3 One-step ascorbic acid fermentation processes from microalgae

The one-step heterotrophic production of ASC by fermentation requires, among other things, a microbe that makes the L-isomer of ASC. Heterotrophic microalgae satisfy this condition. The development of such a process has been hindered by ASC's oxygen lability, resulting in degradation of any ASC made and excreted by cells in an aerobic fermentor. Stable ASC production should result from fermentation at low dissolved oxygen and/or low pH.

3.3.1 Classical strain improvement of heterotrophic algae

Photosynthetic microalgae

Many organisms referred to as 'algae' produce ASC directly from Glc (see Smirnoff *et al.*, Chapter 1). Disregarding the red and brown macroalgae, most of the remaining microalgae have characteristics that preclude them from being useful for producing ASC on large scales. Virtually all microalgae are fundamentally photosynthetic organisms, although some microalgae such as *Poteriochomonas* and *Euglena* have modest heterotrophic capabilities and some microalgae, such as *Chlorella*, have robust heterotrophic capabilities. The extremely high costs of illuminating large production-scale vessels for industrial production of ASC would alone preclude the development of any such process. Ignoring that, the technological requirements to develop and operate a large-scale, axenic, economical photosynthetic production process would be insuperable. Two additional characteristics of most microalgae that argue against their use as production organisms are slow growth rates and the inability to be grown to high cell densities. Finally, many of the microalgae that synthesize ASC are not amenable to genetic modification, either classical mutagenesis techniques or metabolic engineering methods.

Heterotrophic microalgae

Chlorella. For decades, many researchers studying plant biochemistry have employed species of the unicellular green microalga, *Chlorella* (Fogg, 1988). Renstrøm *et al.* (1982/1983) demonstrated the presence of ASC in *Chlorella pyrenoidosa* Chick UTEX 343. This strain has limited heterotrophic growth capabilities. Under ideal conditions, its growth rate on Glc is very slow, measured in days rather than hours. In the mid-1980s, Skatrud and Huss (1991) at Bio-Technical Resources (BTR) focused on ASC production by *C. sorokiniana* (*pyrenoidosa*) UTEX 1230 (Tx 7–11–05, ATCC 22521, CCAP 211/8k), a strain used in research for decades (Miller *et al.*, 1971; Starr, 1971; Heath, 1979; Krampitz and Yarris, 1983; Prunkard *et al.*, 1986; Komov *et al.*, 1988; Brand *et al.*, 1989; Kessler, 1995). This organism can grow heterotrophically in the dark on Glc minimal media at temperatures as high as 42°C, doubling about every 3 h. These researchers confirmed that this strain grew well in closed fermentors, and that it produced the L-isomer of ASC. This organism served as the parent in a strain improvement program utilizing classical mutagenesis. The best mutant strains produced 70 times the wild-type levels of ASC (Running *et al.*, 1994). The researchers also determined that ASC in aerobic solution was stabilized by acidification. However, when the pH of the *Chlorella* fermentation process was lowered below pH 5.5, the point where ASC began to accumulate extracellularly, cell viability was drastically reduced. The *Chlorella* process was ultimately abandoned because growth and ASC production under conditions that favored extracellular accumulation of ASC, low pH and/or lowered oxygen concentration, led to loss of cell viability (Running, unpublished results).

Prototheca. Running *et al.* (1999, 2002) detected ASC in another closely related microalga, *Prototheca*. *Prototheca* species were once thought of as yeast-like fungi, but they have since been shown to be green algae, albeit colorless ones lacking photosynthetic pigments (Pore, 1993). Like *Chlorella* species, *Prototheca* species make the L-isomer of ASC.

Researchers at Fujizawa Co. (Shimizu *et al.*, 1967; Yagi *et al.*, 1967) reported that the fungus *Penicillium* could accumulate extracellularly a related antioxidant, D-erythorbic acid (the D-isomer of ASC, also called D-*arabo*ascorbate, see Smirnoff *et al.*, Chapter 1), to high levels under aerobic conditions. A key parameter of that process was its operation at acidic pH, which allowed the extracellular accumulation of erythorbate. By analogy, it should be possible to produce high levels of ASC by fermentation with other acidophilic or acidotolerant microorganisms. Running *et al.* (1999) selected acidophilic ASC-producing *Prototheca* species as candidate organisms for development of a one-step fermentation process to produce ASC.

Prototheca strains were identified that grew faster than *C. sorokiniana* and produced ASC at pH values as low as 3.0 (Running *et al.*, 2002). Under such conditions, *P. moriformis* accumulated ASC extracellularly with no negative effects on growth rate. A strain improvement program, based on random mutagenesis coupled with various differential and selection methods, produced mutant strains that accumulated increased levels of ASC, as well as some mutants that made no ASC. ASC production was increased by limiting the cells for magnesium. The battery of mutants created during the program was later integral in the elucidation of an algal ASC biosynthesis pathway that has since become recognized as the main ASC synthesis route in plants (see Chapter 1).

The development program with *P. moriformis* demonstrated the ability to produce ASC in aerated fermentors and have the ASC accumulate extracellularly to levels more than 250-fold higher (4–5 $g \cdot l^{-1}$) than in published accounts of microalgal culture. To achieve ASC levels high enough for an economical industrial process, the level and rate of production would have to be increased at least another 10-fold. The *in vitro* reaction rates of some of the enzymatic steps in the ASC biosynthetic pathway suggest that many of the steps are highly regulated (Running *et al.*, 2003). The lack of progress beyond this level and the lack of molecular genetic systems to directly overexpress pathway genes in *Prototheca* caused the BTR researchers to discontinue development of a *Prototheca* based fermentation process (J. Running, personal communication).

3.3.2 Production of ascorbic acid

Intracellular

The literature describes the detection of ASC in numerous microalgae. Concentrations on a cell basis were typically 0.1–2.0 mg $ASC \cdot g^{-1}$ dry cells

(Aaronson *et al.*, 1977; Baker *et al.*, 1981; Renstrøm *et al.*, 1982/1983a; Abalde and Fabregas, 1991; Running *et al.*, 1994). Takeyama *et al.* (1997) detected more than 4 mg ASC·g^{-1} in *Euglena*. Aaronson *et al.* (1977) found 15 mg ASC·g^{-1} in *C. vulgaris*. In the strain improvement program for *C. sorokiniana* described herein (Running *et al.*, 1994), ASC was detected in the parent strain in highly aerated cultures at a concentration of 0.77 mg ASC·g^{-1} dry cells. Overproductive mutant isolates had internal concentrations of more than 54μg ASC·g^{-1} cells (Doncheck *et al.*, 1996), but no extracellular ASC was detected.

Extracellular

In the rare cases where ASC has been detected in the supernates of microalgal cultures, the cultures had been grown photosynthetically without aeration in stationary culture. Baker *et al.* (1981) found 1.6–3.0 mg·l^{-1} ASC in supernatants of the photosynthetically-grown protists *Poteriochomonas stipitata* and *E. gracilis*. Aaronson *et al.* (1971) found 12–17 mg·l^{-1} in supernatants of *Poteriochomonas (Ochomonas) danica*.

In the *Chlorella* process described above, fermentation conditions were modified to incorporate an initial growth phase, followed by a 'synthesis' phase during which the aeration rate was reduced to ameliorate ASC degradation, while still providing enough oxygen for cellular metabolism. Under these conditions, cell growth was greatly reduced and extracellular ASC titers of more than 5 g·l^{-1} were attained. A subsequent strain improvement and fermentation development program with the acidophile *Prototheca moriformis* obviated the need for lowered dissolved oxygen concentrations during fermentation. The ASC was excreted by the cells and accumulated extracellularly, in the presence of dissolved oxygen, to about 5 g·l^{-1} (Running *et al.*, 2002).

3.4 New one-step fermentation process for the production of ascorbic acid

3.4.1 Novel approach for production of L-ascorbic acid: potential of yeast

A logical next step in the development of a one-step industrial process from Glc to ASC would involve the cloning and overexpression of one or more pathway genes into an organism that is appropriate for culture in fermentors on an industrial scale and that can be manipulated using recombinant DNA methods. Yeasts, such as *Saccharomyces cerevisiae*, are strong candidates, in that most of them can be grown under the acidic culture conditions necessary to stabilize ASC in an aerobic environment. Yeast do not produce ASC (see Chapter 1); genes necessary to produce ASC would have to be introduced. A number of different strategies, some based on plant ASC pathway genes, have been proposed (Cayle *et al.*, 1986; Berry *et al.*, 1999; Hancock and Viola, 2001, 2002; Sauer and Porro, 2002). This section will focus on establishing the plant ASC pathway in yeast.

3.4.2 Ascorbic acid synthesis in plants and green algae: controlling the key step of biosynthesis

Although the ASC pathway in animals was established in the 1950s, the pathway in plants and algae remained unresolved until recently. In 1998, Smirnoff's group proposed an ASC biosynthetic pathway based on their discovery of an enzyme that oxidizes the rare sugar, L-Gal (Wheeler *et al.*, 1998). That same year, BTR's patent application outlined a virtually identical pathway based on an analysis of enzyme activities in mutants of the microalga *Prototheca* (Berry *et al.*, 1999), particularly GDP-D-Man-3,5-epimerase (epimerase; see Chapter 1 for more details). The level of ASC accumulated in the different *Prototheca* mutants showed a correlation with epimerase activity. The enzyme catalyzes the formation of GDP-L-Gal from GDP-D-Man and has been proposed as the rate-limiting step for the biosynthesis of ASC in algae (Running *et al.*, 2003). It is anticipated that plants or algae with an existing ASC pathway, modified to overexpress the epimerase activity, will have an increased potential to synthesize and accumulate ASC. This section will focus on ASC production in yeast that have been modified by introducing and over-expressing key plant ASC pathway genes.

3.4.3 Identification, characterization, cloning and expression of GDP-D-Man-3,5-epimerases

GDP-D-Man-3,5-epimerase from Arabidopsis

GDP-D-Man-3,5-epimerase (EC 5.1.3.18) catalyzes the conversion of GDP-D-Man into GDP-L-Gal. The enzyme activity was first detected in a snail (Goudsmit and Neufeld, 1967), later in the microalgae *C. pyrenoidosa* (Barber, 1971), *P. moriformis* (Berry *et al.*, 1999) and *Chlamydomonas reinhardtii* (unpublished data, S. Peng, BTR), and in *A. thaliana* and pea (Wheeler *et al.*, 1998). For decades, attempts to characterize the epimerase have been frustrated by the difficulty of purifying the enzyme to homogeneity (Barber, 1971; Hebda *et al.*, 1979; Wolucka *et al.*, 2001a, 2001b). By combining biochemical and mass spectrometry techniques, the epimerase gene has now been isolated from a highly purified protein prepared from *A. thaliana* (Wolucka *et al.*, 2001b). This enzyme is a member of the short-chain dehydrogenase/reductase (SDR) family and shares 36% homology with the BmlG 5′-epimerase protein from *Streptomyces*, and 21% homology with the GDP-4-keto-6-deoxy-D-Man 3,5-epimerase-4-reductase from *A. thaliana* and GDP-L-Fuc synthetase (EC 1.1.1.271) from *Escherichia coli* (Wolucka *et al.*, 2001b).

Purification and properties of GDP-D-Man-3,5-epimerase from Prototheca

The epimerase has been partially purified from *C. pyrenoidosa* (Hebda *et al.*, 1979) and highly-purified from *A. thaliana* (Wolucka *et al.*, 2001b), but there have been no reports of purification of the enzyme to homogeneity. Susan Peng (BTR, unpublished results) has attempted purification of algal epimerases. An overproducing *P. moriformis* mutant, PTA-111 (Running *et al.*, 2002) was selected for enzyme purification, as it has an epimerase activity threefold greater than that of the wild-type strain ATCC 75669 (Running *et al.*, 2003). Fast protein liquid chromatography (FPLC; Pharmacia, Piscataway, New Jersey, USA) was used to partially purify the epimerase from cell extracts. *Table 3.1* presents a summary of enzyme recovery and purification at each step.

As seen by sodium dodecyl sulfate polyacrylamide gel electrophoresis (SDS/PAGE) (*Figure 3.1*), the final epimerase protein purity was significantly increased after successive chromatography with hydrophobic interaction, anion

Table 3.1: Purification of *Prototheca* GDP-D-Man-3,5-epimerase.

Fold purification	Total activity (nmol)[a]	Total protein (mg)	Specific activity (nmol·mg^{-1})	Fold purification	Yield (%)	
					Activity	Protein
Crude extract	217	1087	0.2	1	100	100
Ammonium sulfate[b]	109	364	0.3	2	60	33
Phenyl sepharose	63	211	0.3	2	67	58
MonoQ	34	8	4.2	21	17	0.7
Gel filtration	6	1	5.5	28	1	0.1

[a]As L-Gal formation from GDP-D-Man. [b]50% (w/v).

exchange and gel filtration columns. Considerable difficulty, however, was encountered because of instability of the enzyme; every successive purification step after ammonium sulfate resulted in a dramatic loss of enzyme activity. The enzyme was stable in crude extracts or ammonium sulfate solutions, and could be stored at –20°C for up to 3 weeks without significant loss of activity. The epimerase activity pooled from the phenyl sepharose column (*Table 3.1*; *Figure 3.1*, lane 3) had a half-life of 3 days at 4°C.

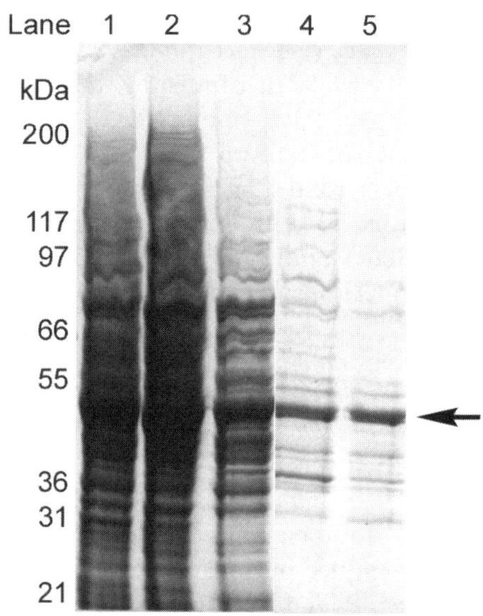

Figure 3.1: SDS/PAGE analysis of protein samples from unique purification steps of the *Prototheca* GDP-D-Man-3,5-epimerase.
Gel stained with Coomassie brilliant blue. Lane 1, 50 000 *g* supernatant fraction after French press disruption; lane 2, 50% ammonium sulfate precipitated fraction; lanes 3, 4 and 5, pooled active fractions from chromatography on hydrophobic interaction (phenyl sepharose), anion exchange (Mono Q) and gel filtration, respectively (all from Pharmacia, Piscataway, NJ). The arrow shows the protein band corresponding to the *Prototheca* GDP-D-Man-3,5-epimerase.

The properties of the *Prototheca* enzyme were determined with the partially purified enzyme collected from the Mono-Q column (*Table 3.1*). The pH optimum for epimerase activity was 7.2. The divalent cation, Ca^{2+}, was a very effective inhibitor; 1 mM Ca^{2+} reduced activity to 43%. Mg^{2+} had no effect on enzyme activity. Although EDTA, NADH and NADPH were activators of the epimerase from ammonium sulfate-precipitated fractions, these did not affect the partially purified epimerase (data not shown).

The kinetic properties of epimerase activity were analyzed by measuring initial velocities over a range of reaction times and concentrations of GDP-D-Man at several fixed enzyme concentrations. The formation of GDP-L-Gal was linear with incubation time and substrate and enzyme concentration. The Michaelis constant (K_m) value for GDP-D-Man was determined using a Lineweaver–Burk plot (*Figure 3.2*). The K_m for GDP-D-Man, calculated from the plot, was 30 μM. The K_m value was significantly lower for *Prototheca*, as compared to that determined by Hebda *et al.* (1979) for the *Chlorella* epimerase (97 μM).

Cloning of the GDP-D-Man-3,5-epimerase gene from C. reinhardtii *and comparison to homologous sequences in the genomes of* A. thaliana*, rice and maize*

Deng, McMullin and Peng (BTR) identified and aligned putative epimerase genes from genomic and expressed sequence tag (EST) databases (unpublished results). These sequences were aligned with the published *A. thaliana* epimerase sequence (Wolucka *et al.*, 2001b). Variable and conserved gene sequence regions were identified.

Cloning of C. reinhardtii, Prototheca *and* Chlorella *epimerases is in progress at BTR*

The cloning of the *C. reinhardtii* epimerase is reported here. PCR primers (degenerate and nondegenerate) were designed from conserved regions apparent in the

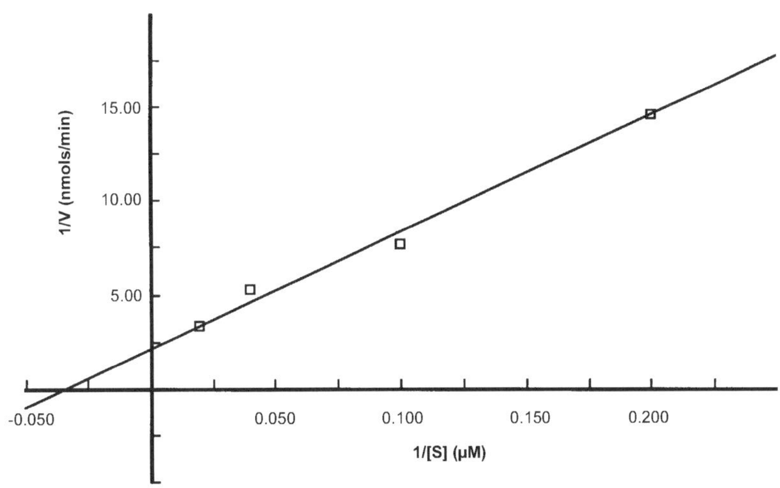

Figure 3.2: Lineweaver–Burk plot of partially purified *Prototheca* epimerase.

epimerase gene alignments described above. A PCR product of approximately 1200 bp was amplified from a cDNA library of *C. reinhardtii* (CC-1690, Chlamydomonas Genetic Center, Duke University, Durham, NC) and from genomic DNA prepared from *C. reinhardtii* (ATCC 18798). The PCR products were confirmed by sequencing. Products were cloned into the expression vector pET23-d(+) (Novagen, Madison, WI).

To identify additional related genes in other species, nucleotide and amino acid sequences of the cloned *C. reinhardtii* and *A. thaliana* epimerases were compared with public database sequences using standard nucleotide BLAST (Blastn) and standard protein BLAST (Blastp; NCBI, http: //www.ncbi.nlm.nih.gov/BLAST). Four sequences that showed homology to the cloned *C. reinhardtii* epimerase gene were identified in genomic sequences of rice, maize and *C. reinhardtii* (*Table 3.2*).

Functional expression of cloned GDP-d-Man-3,5-epimerases from A. thaliana *and* C. reinhardtii *in* E. coli *and* S. cerevisiae

E. coli. Overexpression of cloned epimerase genes from *A. thaliana* and *C. reinhardtii* was initially attempted in *E. coli* BL2 (DE3). An abundant protein band with a mass of 43 kDa was observed by SDS/PAGE with cells transformed with either gene and induced by isopropyl-δ-D-thiogalactopyranoside (IPTG) at either 25°C or 37°C (data not shown). The molecular mass of the overexpressed protein was consistent with the mass predicted from the cloned genes. To confirm that the cloned cDNA encoded epimerase activity, enzyme assays were carried out with soluble protein extracts isolated from cells induced with IPTG. GDP-D-Man was converted to GDP-L-Gal by both recombinant epimerases (data not shown).

S. cerevisiae. Constructs were made in a BTR derivative of the yeast expression vector yEp352 (Hill *et al.*, 1986) in order to express the recombinant *A. thaliana* epimerase. The epimerase genes were functionally expressed in three *S. cerevisiae* strains, including SWY5-ΔH1 (ATCC 28383, *ura3 his3*). As has been discussed, various studies showed that yeast cells are capable of synthesizing ASC when fed L-Gal (Hancock *et al.*, 2000) or L-GalL (Smirnoff *et al.*, 2001). The present studies confirmed that *S. cerevisiae* does not have necessary biochemical steps to convert

Table 3.2: Sequence similarities of GDP-D-Man-3,5-epimerases.

GDP-D-Man-3,5-epimerase (Public Database Records)	Coding sequence		Similarity to *A. thaliana* GDP-D-Man-3,5-epimerase (%)[a]	
	cDNA (bp)	Amino acids (#)	cDNA	Amino acids
A. thaliana	1134	377	100	100
C. reinhardtii[b]	1156	384	56	70
C. reinhardtii[c]	1185	394	58	71
Rice (NCBI)	1137	378	79	90
Maize (NCBI)	1143	380	78	89

[a]Conserved regions of aligned sequences were used to design PCR primers. [b]*C. reinhardtii* cloned epimerase cDNA (S. Peng, BTR). [c]*C. reinhardtii* consensus sequence from five EST sequences.

GDP-D-Man to L-Gal, as *S. cerevisiae* strains overexpressing the epimerase did not accumulate ASC in culture medium or cells.

3.4.4 Metabolic engineering of ascorbic acid synthesis in yeast

The use of microorganisms for ASC production has been intensively studied. Over the years, BTR has investigated several concepts of single-step ASC production in microalgae. Combining random mutagenesis with fermentation techniques, one-step fermentation processes were developed in *C. pyrenoidosa* and *P. moriformis*. These processes achieved yields up to 5 g·l^{-1} ASC (Running, 1998). Although a significant yield, it is not competitive with current manufacturing processes.

As has already been presented, recombinant technology will play an important role in the introduction of a new, highly efficient, one-step ASC fermentation process. Although *Prototheca* would seem to be a natural host strain in which to engineer an improved pathway for ASC synthesis on a commercial scale, the tools necessary to modify the existing pathway are unavailable – to date; no usable genetic system has been developed for this organism.

In contrast, fungi, and in particular, yeast, provide an attractive alternative as they can be manipulated genetically and have a proven track record in industrial fermentation processes. Particularly attractive characteristics include the observations that growing yeast have a high carbon flux through GDP-D-Man, the substrate for the epimerase (Hashimoto *et al.*, 1997). Yeast have native enzyme activities, counterparts to the last two steps of the plant ASC pathway: yeast D-arabinono dehydrogenase (counterpart to plant L-Gal dehydrogenase; L-GalDH) and D-arabinono-1,4-lactone oxidase (counterpart to plant L-GalL dehydrogenase; L-GalLDH) convert L-Gal and L-GalL, respectively, to ASC (Huh *et al.*, 1994; Kim *et al.*, 1996).

Missing enzymatic steps to complete the plant pathway in yeast include three enzyme activities: the GDP-D-Man-3,5-epimerase (described in this section), GDP-L-Gal phosphorylase and L-Gal-1-phosphatase. To complete an ASC biosynthetic pathway in yeast, therefore, at least two (epimerase plus GDP-L-Gal phosphorylase), potentially three (epimerase plus GDP-L-Gal phosphorylase plus L-Gal-1-phosphatase) and possibly four to five genes (additionally replace one or both of the last two native yeast counterpart genes for plant genes) would need to be cloned into the host cell.

The gene for the first of these key pathway steps, GDP-D-Man-3,5-epimerase, has been cloned from *A. thaliana* (Wolucka *et al.*, 2001b; Wolucka, 2002) and from a green microalga, *C. reinhardtii* (S. Peng, reported here). The plant gene encoding L-GalDH has also been cloned (Gatzek *et al.*, 2002), as has the gene for the last enzyme in the native plant pathway, L-GalLDH (Ostergaard *et al.*, 1997; Bauw *et al.*, 1998). What remains is the cloning of the two proposed genes involved in the conversion of GDP-L-Gal into free L-Gal. These have yet to be isolated. BTR and its collaborators are pursuing cloning of these genes and expect that these will be purified, cloned and functionally expressed in the near future. Once all of the genes are available, parallel expression of the enzymes in yeast should offer a novel opportunity for development of a low cost, single-step, ASC production process in yeast.

3.5 Commercial production as of 2003

3.5.1 Manufacturers

Roche, BASF and Chinese state pharmaceutical companies manufacture the bulk of the ASC produced commercially today, with Roche and BASF dominating product sales (Muth *et al.*, 1999; Competition Commission, 2001; Financial Times, 2002b). Current world production is estimated at about 85 000 metric tons per year (SRI Consulting, 1999; Financial Times, 2002b). The most widely used method of production continues to be the traditional Reichstein process (SRI Consulting, 1999; Competition Commission, 2001; Financial Times, 2002b). Roche and BASF both practice this process commercially.

All producers in China use the two-stage 2-KLG fermentation semi-synthetic processes to produce ASC. It has been reported to have lower fixed and capital costs. This results in an estimated overall production cost saving of about a third compared with the Reichstein process (Competition Commission, 2001). Roche has also licensed this technology, but there are no reports that Roche is commercially practicing it. Roche is said to be seeking partners in China to expand its ASC production using the 2-KLG process (Financial Times, 2002b; Dow Jones and Company, 2003). A joint venture between BASF and Cargill/Cerestar produces ASC in Germany using the 2-KLG process. Very recently, BASF announced a joint venture with a Chinese manufacture, Northeast General Pharmaceutical Factory (Chemical Week, 2002; Financial Times, 2003). This joint venture will reportedly expand its ASC capacity by building a new 30 000-ton ASC production line.

In practice, a variety of final products of differing purity are produced, each with its own pricing. These final products have different additives and are suitable for particular uses, depending on the needs of the customer.

3.5.2 World capacity and demand

Capacity to produce ASC has greatly exceeded demand in recent years, resulting in extremely low cost material. Total global demand has been estimated to be 85 000 tons and is expected to increase by 2000 tons per year (Financial Times, 2002b). Oversupply has led to frequent price wars in the world ASC market. Prices hovered around US $\$2.13 \cdot lb^{-1}$ at the beginning of 2000 and dropped to US $\$1.27 \cdot lb^{-1}$ by the end of 2000 (Chemical Marketing Reporter, 2001; Financial Times, 2002a; Sinopolis.com, 2003). By the end of 2002, the price had further eroded to about US $\$1.00 \cdot lb^{-1}$. Roche and BASF reduced production during 2002, resulting in a global price rebound in early 2003 to about US $\$4.50 \cdot lb^{-1}$. It is likely that ASC pricing will remain low and will, at times, again approach US $\$1.00 \cdot lb^{-1}$.

Barriers to introduction of new technology

Any new ASC process must be competitive with existing technology. Today, ASC is a commodity. To be competitive, synthetic-, semi-synthetic- or fermentation-based processes must be very low cost and must be practiced at very large scale (in order to benefit from economies of scale). Both Archer Daniels Midland and a Genencor International joint venture have announced 2-KLG technologies ready for commercial production, but neither have, so far, commercialized their processes (SRI Consulting, 1999). Other companies with new processes, even those that are low-cost, will find it difficult to justify the capital investment necessary to bring on new production capacity for the production of ASC in light of current ASC economics.

3.6 Summary

New methods for the industrial-scale production of ASC are now challenging the traditional Reichstein process. These new methods involve the biological or enzymatic production of the Reichstein intermediate 2-KLG, and the chemical conversion of it into ASC. Intriguing alternative methods would take advantage of enzymes of the recently-discovered Smirnoff–Wheeler–Running biochemical pathway of ASC synthesis in plants. Isolation of the remaining two genes in the pathway seems imminent. Even if such genes are used to engineer microbial (yeast) strains for ASC production, the full implementation of such technology may only be possible when the price of ASC would rise through increased demand.

References

Aaronson S, DeAngelis B, Frank O, Baker H (1971) Secretion of vitamins and amino acids into the environment by *Ochromonas danica*. *J. Phycol.* **7**, 215–218.

Aaronson S, Dhawale SW, Patni NJ, DeAngelis B, Frank O, Baker, H (1977) The cell content and secretion of water-soluble vitamins by several freshwater algae. *Arch. Microbiol.* **112**, 57–59.

Abalde J, Fabregas J (1991) β-Carotene, vitamin C and vitamin E content of the marine microalga *Dunaliella tertiolecta* cultured with different nitrogen sources. *Bioresource Technol.* **38**, 121–125.

Anderson S, Marks CB, Lazarus R, Miller J, Stafford K, Seymour J, Light D, Rastetter W, Estell D (1985) Production of 2-keto-L-gulonate, an intermediate in L-ascorbate synthesis, by a genetically modified *Erwinia herbicola*. *Science* **230**, 144–149.

Baker ER, McLaughlin JJA, Hutner SH, DeAngelis B, Feingold S, Frank O, Baker H (1981) Water-soluble vitamins in cells and spent culture supernatants of *Poteriochromonas stipitata*, *Euglena gracilis*, and *Tetrahymena thermophila*. *Arch. Microbiol.* **129**, 310–313.

Barber GA (1971) The synthesis of L-glucose by plant enzyme systems. *Arch. Biochem. Biophys.* **147**, 619–623.

Bauw GJC, Davey MW, Van Montagu MCE, Ostergaard J (1998) Production of ascorbic acid in plants. World Intellectual Property Organization (WIPO) patent WO9850558.

Berry A, Running JA, Severson DK, Burlingame RP (1999) Vitamin C production in microorganisms and plants. World Intellectual Property Organization (WIPO) patent WO9964618.

Brand JJ, Wright JN, Lien S (1989) Hydrogen production by eukaryotic algae. *Biotechnol. Bioeng.* **33**, 1482–1488.

Cayle T, Roland J, Mehnert D, Dinwoodie R, Larson R, Mathers J, *et al.* (1986) Production of L-ascorbic acid from whey. In: *Biotechnology in Food Processing* (eds SK Harlander, TP Labuza). Noyes, Park Ridge, NJ, pp. 157–169.

Chemical Marketing Reporter Dec., (2001) Vitamin C continues downward slide.

Chemical Week Nov., (2002) BASF plans joint venture with Chinese manufacturer.

Chotani G, Dodge T, Hsu A, Kumar M, LaDuca R, Trimbur D, Weyler W, Sanford K (2000) The commercial production of chemicals using pathway engineering. *Biochim. Biophys. Acta* **1543**, 434–455.

Competition Commission (2001) A report on the acquisition by BASF AG of certain assets of Takeda Chemicals Industries, Ltd. BASF AG and Takeda Industries, Ltd. London, UK.

Danehy JP (1981) Synthesis of ascorbic acid from lactose. US patent 4,259,443.

Delic V, Sunic D, Vlasic D (1989) Microbial reactions for the synthesis of vitamin C (L-ascorbic acid). In: *Biotechnology of Vitamins, Pigments and Growth Factors* (ed. EJ Vandamme). Elsevier, London, pp. 299–334.

Doncheck JA, Huss RJ, Running JA, Skatrud TJ (1996) L-ascorbic acid containing biomass of *Chlorella pyrenoidosa*. US patent 5,521,090.

Dow Jones, Company, Inc Apr., (2003) China pharmaceutical mulls alliance with DSM.

Eschenfeldt WH, Stols L, Rosenbaum H, Khambatta ZS, Quaite-Randall E, Wu S, Kilgore DC, Trent JD, Donnelly MI (2001) DNA from uncultured organisms as a source of 2,5-diketo-D-gluconic acid reductases. *Appl. Environ. Microbiol.* **67**, 4206–4214.

Financial Times Ltd Apr., (2002a) JVs may be vitamin-C industry's last hope.

Financial Times Ltd May, (2002b) Vitamin-C makers at crossroads.

Financial Times Ltd Feb., (2003) Domestic VC manufacturers facing hiding crisis.

Fogg GE (1988) Flexibility and variety of algal metabolism. In: *Biochemistry of the Algae and Cyanobacteria.* (eds LJ Rogers, JR Gallon). Clarendon Press, Oxford, pp. 4–12.

Fowler T, Causey SC (1998) Improved *Enterobacteriaceae* fermentation strains. World Intellectual Property Organization (WIPO) patent 9859054.

Gatzek S, Wheeler GL, Smirnoff N (2002) Antisense suppression of L-galactose dehydrogenase in *Arabidopsis thaliana* provides evidence for its role in ascorbate synthesis and reveals light modulated L-galactose synthesis. *Plant J.* **30**, 541–553.

Goudsmit EM, Neufeld EF (1967) Formation of GDP-L-galactose from GDP-D-mannose. *Biochem. Biophys. Res. Commun.* **26**, 730–735.

Grindley JF, Payton MA, van de Pol H, Hardy KG (1988) Conversion of glucose to 2-keto-L-gulonate, an intermediate in L-ascorbate synthesis by a recombinant strain of *Erwinia citreus*. *Appl. Environ. Microbiol.* **54**, 1770–1775.

Hancock RD, Galpin JR, Viola R (2000) Biosynthesis of L-ascorbic acid (vitamin C) by *Saccharomyces cerevisiae*. *FEMS Microbiol. Lett.* **186**, 245–250.

Hancock RD, Viola R (2001) The use of micro-organisms for L-ascorbic acid production: current status and future perspectives. *Appl. Microbiol. Biotechnol.* **56**, 567–576.

Hancock RD, Viola R (2002) Biotechnological approaches for L-ascorbic acid production. *Trends Biotechnol.* **20**, 299–305.

Hashimoto H, Sakakibara A, Yamasaki M, Yoda K (1997) *Saccharomyces cerevisiae* VIG9 encodes GDP-mannose pyrophosphorylase, which is essential for protein glycosylation. *J. Biol. Chem.* **272**, 16308–16314.

Heath RE (1979) A constitutive enzyme system for glucose transport by *Chlorella sorokiniana*. *Plant Physiol.* **64**, 224–227.

Hebda PA, Behrman EJ, Barber GA (1979) The guanosine 5′-diphosphate D-mannose: guanosine-diphosphate L-galactose epimerase of *Chlorella pyrenoidosa*. *Arch. Biochem. Biophys.* **194**, 496–502.

Hill JE, Myers AM, Koerner TJ, Tzagoloff A (1986) Yeast/*E. coli* shuttle vectors with multiple unique restriction sites. *Yeast* **2**, 163–167.

Huang HT (1962) Preparation of 2-keto-L-gulonic acid. US patent 3,043,749.

Huh W-K, Kim S-T, Yang K-S, Seok Y-J, Hah Y-C, Kang S-O (1994) Characterization of D-arabinono-1,4-lactone oxidase from *Candida albicans* ATCC 10231. *Eur. J. Biochem.* **223**, 1073–1079.

Isono M, Nakanishi K, Sasajima K, Motisuki K, Kanzaki T, Okazaki H, Yoshino H (1968) 2-keto-L-gulonic acid fermentation. I. Paper chromatographic characterization of metabolic products from sorbitol and L-sorbose by various bacteria. *Agric. Biol. Chem.* **32**, 414–431.

Jaffe GM (1984) Ascorbic acid. In: *Kirk–Othmer Encyclopedia of Chemical Technology, 3rd edn.* (ed. M Grayson). John Wiley and Sons, New York, pp. 8–40.

Kanzaki T, Okazaki H (1970) 2-keto-L-gulonic acid fermentation. IV. L-sorbose metabolism in *Pseudomonas aeruginosa*. *Agric. Biol. Chem.* **34**, 432–436.

Kessler E (1995) Comparative physiology and biochemistry of *Chlorella* species as the basis for their taxonomy and for their utilization in research and biotechnology. *Phycotalk* **1**, 141–153.

Kim S-T, Huh W-K, Kim J-Y, Hwang S-W, Kang S-O (1996) D-arabinose dehydrogenase and biosynthesis of erythorbic acid in *Candida albicans*. *Biochim. Biophys. Acta* **1297**, 1–8.

Komov E, Cho B–H, Kraus M (1988) The occurrence of the glucose-inducible transport systems for glucose, proline and arginine in different species of *Chlorella*. *Botanica Acta* **101**, 321–326.

Krampitz LO, Yarris CE (1983) Glycolate formation and excretion by *Chlorella pyrenoidosa* and *Netrium digitus*. *Plant Physiol.* **72**, 1084–1087.

Kumar M (2002) Production of ascorbic acid. US patent 6,358,715.

Makover S, Ramsey GB, Vane FM, Witt CG, Wright RB (1975) New mechanism for the biosynthesis and metabolism of 2-keto-L-gulonic acid in bacteria. *Biotechnol. Bioeng.* **17**, 1485–1514.

Miller R L, Wickline HE, Richardson B (1971) Effects of heterotrophic and autotrophic growth conditions on the composition of *Chlorella sorokiniana*. *J. Food Sci.* **36**, 774–777.

Muth MK, Anderson DW, Domanico JL, Smith JB, Wendling B (1999) Economic characterization of the dietary supplement industry. Contract No. 223–96–2290, RTI Project No. 6673–03. Research Triangle Institute. Research Triangle Park, NC.

Okazaki H, Kanzaki K, Sasajima K, Terada Y (1969) 2-keto-L-gulonic acid fermentation. III. Evaluation of the pathway of sorbitol metabolism in *Gluconobacter melanogenes*. *Agric. Biol. Chem.* **33**, 207–211.

Ostergaard J, Persiau G, Davey MW, Bauw G, van Montagu M (1997) Isolation of a cDNA coding for L-galactono-γ-lactone dehydrogenase, an enzyme in the biosynthesis of ascorbic acid in plants. *J. Biol. Chem.* **272**, 30009–30016.

Pore RS (1993) *Prototheca Kruger*. In: *The Yeasts* (eds CP Kurtzman, JW Fell). Elsevier, Amsterdam, pp. 883–888.

Prunkard DE, Bascomb NF, Robinson RW, Schmidt RR (1986) Evidence for chloroplastic localization of an ammonium-inducible glutamate dehydrogenase and synthesis of its subunit from a cytosolic precursor-protein in *Chlorella sorokiniana*. *Plant Physiol.* **81**, 349–355.

Renstrøm B, Grün M, Loewus FA (1982/1983) Biosynthesis of L-ascorbic acid in *Chlorella pyrenoidosa*. *Plant Sci. Lett.* **28**, 299–305.

Running JA (1998) Microbial process for the production of ascorbic acid using *Chlorella protothecoides*. US patent 5,792,631.

Running JA (1999) Process for the production of ascorbic acid with *Prototheca*. Patent 5,900,370.

Running JA, Burlingame RP, Berry A (2003) The pathway of L-ascorbic acid biosynthesis in the colorless microalga *Prototheca moriformis*. *J. Exp. Bot.* **54**, 1841–1849.

Running JA, Huss RJ, Olson PT (1994) Heterotrophic production of ascorbic acid by microalgae. *J. Appl. Phycol.* **6**, 99–104.

Running JA, Severson DK, Schneider KJ (2002) Extracellular production of L-ascorbic acid by *Chlorella protothecoides*, *Prototheca* species and mutants of *P. moriformis* during aerobic culturing at low pH. *J. Ind. Microbiol. Biotechnol.* **29**, 93–98.

Saito Y, Ishii Y, Hayashi H, Imao Y, Akashi T, Yoshikawa K, *et al.* (1997) Cloning of genes coding for L-sorbose and L-sorbosone dehydrogenases from *Gluconobacter oxydans* and microbial production of 2-keto-L-gulonate, a precursor of L-ascorbic acid, in a recombinant *G. oxydans* strain. *Appl. Environ. Microbiol.* **63**, 454–460.

Sauer M, Porro D (2002) Ascorbic acid production from yeasts. World Intellectual Property Organization (WIPO) patent WO0210425.

Shimizu K, Nishiyama K, Inoue T, Takano N, Mikata M, Yamazaki M, Azuma T, Osawa S (1967) Studies on erythorbic acid production by fermentation. Part II. Erythorbic acid production by jar fermentor. *Agric. Biol. Chem.* **31**, 346–352.

Shinjoh M, Setoguchi Y, Hoshino T, Fujiwara A (1990) L-sorbose dissimilation in 2-keto-L-gulonic acid-producing mutant UV10 derived from *Gluconobacter oxydans* IFO 3293. *Agric. Biol. Chem.* **54**, 2257–2263.

Shinjoh M, Sugisawa T, Masuda S, Hoshino T (1994) Efficient conversion of L-sorbosone to 2-keto-L-gulonic acid by *Acetobacter liquifaciens* strains. *J. Ferment. Bioeng.* **78**, 476–478.

Shinjoh M, Tomiyama N, Asakura A, Hoshino T (1995) Cloning and nucleotide sequencing of the membrane-bound L-sorbosone dehydrogenase gene of *Acetobacter liquefaciens* IFO 12258 and its expression in *Gluconobacter oxydans*. *Appl. Environ. Microbiol.* **61**, 413–420.

Sinopolis.com (2003) Domestic manufacturers plan to expand production; multinational companies do not want Chinese rivals to be successful.

Skatrud TJ, Huss RJ (1991) L-ascorbic acid production in microorganisms. US patent 5,001,059.

Smirnoff N, Conklin PL, Loewus FA (2001) Biosynthesis of ascorbic acid in plants: A renaissance. *Ann. Rev. Plant Physiol. Plant Mol. Biol.* **52**, 437–467.

Sonoyama T, Tani H, Matsuda K, Kageyama B, Tanimoto M, Kobayashi K, Yagi S, Kyotani H, Mitsushima K (1982) Production of 2-keto-L-gulonic acid from D-glucose by two-stage fermentation. *Appl. Environ. Microbiol.* **43**, 1064–1069.

SRI Consulting (1999) Ascorbic Acid. *PEP Review* 99–100.

(1971) Algal cultures – sources and methods of cultivation. *Methods Enzymol* **23**, 29–53.

Stewart DJ (1959) Production of 5-keto-gluconic acid by a species of *Pseudomonas*. *Nature* **183**, 1133–1134.

Sugisawa T, Hoshino T, Masuda S, Nomura S, Setoguchi Y, Tazoe M, Shinjoh M, Someha S, Fujiwara A (1990) Microbial production of 2-keto-L-gulonic acid from L-sorbose and D-sorbitol by *Gluconobacter oxydans*. *Agric. Biol. Chem.* **54**, 1201–1209.

Takeyama H, Kanamaru A, Yoshino Y, Kakuta H, Kawamura Y, Matsunaga T (1997) Production of antioxidant vitamins, β-carotene, vitamin C, and vitamin E, by two-step culture of *Euglena gracilis* Z. *Biotechnol. Bioeng.* **53**, 185–190.

Tengerty RP (1961) Redox potential changes in the 2-keto-L-gulonic acid fermentation – I. correlation between redox potential and dissolved-oxygen concentration. *J. Biochem. Microbiol. Technol. Eng.* **3**, 241–253.

Tsukada Y, Perlman D (1972a) The fermentation of L-sorbose by *Gluconobacter melanogenus*. I. General characteristics of the fermentation. *Biotehnol. Bioeng.* **14**, 799–810.

Tsukada Y, Perlman D (1972b) The fermentation of L-sorbose by *Gluconobacter melanogenus*. II. Inducible formation of enzyme catalyzing conversion of sorbose to 2-keto-L-gulonic acid. *Biotechnol. Bioeng.* **14**, 811–818.

Tsukada Y, Perlman D (1972c) The fermentation of L-sorbose by *Gluconobacter melanogenus*. III. Investigation of the metabolic pathway from sorbose to 2-keto-L-gulonic acid. *Biotechnol. Bioeng.* **14**, 1035–1038.

Wheeler GL, Jones MA, Smirnoff N (1998) The biosynthetic pathway of vitamin C in higher plants. *Nature (Lond.)* **393**, 365–369.

Wolucka BA (2002) GDP-mannose-3′-5′-epimerase and methods of use thereof. World Intellectual Property Organizatin (WIPO) patent WO02103001.

Wolucka BA, Davey MW, Boerjan W (2001a) A high-performance liquid chromatography radio method for determination of L-ascorbic acid and guanosine-5′-diphosphate-L-galactose, key metabolites of the plant vitamin C pathway. *Anal. Biochem.* **294**, 161–168.

Wolucka BA, Persiau G, Doorsselaere JV, Davey MW, Demol H, Vandekerckhove J, Montagu MV, Zabeau M, Boerjan W (2001b) Partial purification and identification of GDP-mannose-3′-5′-epimerase of *Arabidopsis thaliana*, a key enzyme of the plant vitamin C pathway. *Proc. Natl Acad. Sci. USA* **1998**, 14843–14848.

Yagi J, Yamashita T, Kato K, Takagi Y, Sakai H (1967) Studies on erythorbic acid production by fermentation. Part 1. Erythorbic acid-producing strain and cultural condition. *Agric. Biol. Chem.* **31**, 340–345.

Yamazaki M (1953) Production of vitamin C by fermentation. II. Oxidation of Ca-L-idonate and Ca-D-gluconate by *Pseudomonas* sp, isolation of the products. *J. Ferment. Technol.* **31**, 86–90.

Zizheng Y, Zengxin T, Longhua Y, Guanglin Y, Wenzhu N, Changhui W, *et al.* (1981) Studies on production of vitamin C precursor – 2-keto-L-gulonic acid from L-sorbose by fermentation. II. Conditions for submerged fermentation of 2-keto-L-gulonic acid. *Acta Microbiol. Sin.* **21**, 185–191.

4

The role of ascorbic acid in defense networks and signaling in plants

Christine H. Foyer

4.1 Introduction

Redox coupling between ascorbate (ASC) and glutathione is a universal phenomenon in cells that contain both compounds. It represents an instance of a biochemical trait whose discovery in plants preceded its recognition in animals (Szent-Gyorgyi, 1931; Mapson, 1958; Foyer and Halliwell, 1976) though this is rarely acknowledged explicitly (Meister, 1994). The possible existence of a type of ASC–glutathione cycle was first considered early in the twentieth century (Mapson, 1958). At this time these antioxidants were thought to have roles as intermediates in respiratory electron transport to oxygen, with ASC oxidase (AO) as the terminal enzyme. It was also considered possible that peroxidation of ASC *via* flavonoid peroxidases might be linked to glutathione cycling (Mapson, 1958). However, it was not until the 1970s that it first suggested that glutathione redox cycling in the chloroplast could be a necessary part of a H_2O_2 detoxification pathway, linked to ASC peroxidation (Foyer and Halliwell, 1976). By the early 1970s the concept of oxygen toxicity had become widely accepted. With the demonstration of superoxide generation and its universal removal by superoxide dismutase by Fridovich and colleagues, it was generally agreed that oxygen toxicity is caused by its propensity to form free radicals and other reactive intermediates (Halliwell, 1981). Throughout the 1970s the damaging effects of reactive oxygen species (ROS) were explored and documented (Halliwell and Gutteridge, 1989). It was much later that the crucial role of ROS, particularly H_2O_2, in controlling gene expression leading to stress tolerance was demonstrated (Levine *et al.*, 1994; Foyer *et al.*; 1997; Neill *et al.*, 2002b). Plant cells produce very large amounts of H_2O_2 as a result of photosynthesis particularly *via* pseudocyclic electron flow (Egneus *et al.*, 1975) and photorespiration (Brisson *et al.*, 1998; Azevedo *et al.*, 1998; Foyer and Noctor 2000; Novitskaya *et al.*, 2002). This is something of a paradox as some of the enzymes of CO_2 fixation in the chloroplasts

Vitamin C, edited by Han Asard, James M May and Nicholas Smirnoff

are inherently sensitive to inhibition by even very low amounts H_2O_2 (Kaiser, 1979). Since chloroplasts contain no catalase and very little nonspecific peroxidase activity it was clear that another chloroplast-located system must operate to detoxify H_2O. During my PhD studies, supervised by Barry Halliwell, I carried out a series of experiments that culminated in the hypothesis that ASC and glutathione function in ROS detoxification in chloroplasts (Foyer, 1977). To date no evidence has been found to discredit this hypothesis.

A great deal of attention was focused on ASC in the early 1970s as Pauling (1971) claimed that large doses were beneficial against the common cold and other infections and diseases. ASC was known to be more or less ubiquitous in the plant kingdom, where it was almost always present in the reduced state (Mapson, 1958). Observations that green leaves could contain as much ASC as chlorophyll prompted the analysis of why plants needed so much of this antioxidant and redox buffer (see Foyer *et al.*, 1983 and references therein). The early experiments of Arrigoni and his co-workers (1975) had indicated that ASC was an essential component of plant metabolism and that inhibition of ASC synthesis prevented growth. Several observations led us to consider that one of the cellular functions of ASC might be the inactivation of free radicals and other oxidants that might otherwise cause oxidative damage. Chloroplasts had been found to accumulate ASC at concentrations of up to 50 mM (Gerhardt, 1964). At these concentrations ASC effectively scavenges hydroxyl radicals as well as superoxide and H_2O_2. Moreover, the enzyme DHA reductase, that could reduce DHA using reduced glutathione (GSH), had been described by Szent-Gyorgyi (1931) and in 1936, Hopkins and Morgan demonstrated that ASC in carrot juice was protected from oxidation by GSH. With the discovery of GSH and glutathione reductase (GR) in chloroplasts (Foyer and Halliwell, 1976), and a highly active chloroplastic ASC peroxidase (APX; Groden and Beck, 1979; Kelly and Latzko, 1979), the experimental basis for the ASC–glutathione cycle in chloroplasts, and later in other compartments of the plant cell, was established. Soon after, it was shown that isolated chloroplasts metabolize hydrogen peroxide at high rates (Nakano and Asada, 1980). This process was shown to be tightly coupled to the turnover of the endogenous chloroplastic ASC and glutathione pools and driven by photosynthetic electron flow (Anderson *et al.*, 1983). In addition, monodehydroascorbate (MDHA) was found to be an extremely powerful electron acceptor that was re-reduced to ASC *via* reduced ferredoxin (Miyake and Asada, 1994). It was soon clear that chloroplasts could not only detoxify ROS produced via photosynthesis but they could also participate in the removal of H_2O_2 that escaped destruction in the peroxisome (Foyer *et al.*, 1994). During the last 10 years the roles of ASC as the major redox buffer of plant cells with functions not only in plant defense responses but also possibly in control of growth, and most importantly, as a signal transducing metabolite have been elucidated and characterized. These will be considered in the following discussion.

4.2 Ascorbate and chloroplasts

Chloroplasts contain a hierarchy of defence mechanisms in which ASC plays a central role (Noctor and Foyer 1998; Foyer and Harbinson, 1999; Foyer and

Noctor, 2000). These include avoidance defences such as the xanthophyll cycle and antioxidant defences such as the Mehler-peroxidase (water–water) cycle and the ASC–glutathione cycle (Noctor and Foyer, 1998). In the xanthophyll cycle ASC takes part in the de-epoxidation step, providing the reducing power necessary to convert violaxanthin to zeaxanthin. Zeaxanthin is an integral part of the photosystem II excess energy dissipation system that prevents over-reduction and hence singlet oxygen generation (Foyer and Harbinson, 1999). Chloroplasts contain high concentrations of ascorbate and glutathione as well as stromal and thylakoid-bound superoxide dismutases, stromal, thylakoid and lumen APX, stromal MDHA reductases, DHA reductases and glutathione reductases (Foyer and Noctor, 2000). APX was first reported in chloroplasts by Groden and Beck (1979) and Kelly and Latzko (1979), but cytosolic, mitochondrial, glyoxysomal and leaf peroxisomal isoforms have now been characterized (Chen and Asada, 1989, Amako *et al.*, 1994; Yamaguchi *et al.*, 1995; Bunkelmann and Trelease, 1996; Mittler and Zillinskas, 1991). The chloroplast APX, which is found in both stroma-soluble, thylakoid-bound and lumen forms, is rapidly inactivated at ASC concentrations below 20 μM (Hossain and Asada, 1984a). In contrast, the nonchloroplastic isoforms are more stable in the absence of ASC (Chen and Asada 1989; Yoshimura *et al.*, 1998). As the ASC concentrations in the chloroplast and cytosol are likely to be 100–1000 times higher than that necessary to prevent APX inactivation (Foyer *et al.*, 1983; Foyer and Lelandais 1996), the physiological significance of this process is not clear.

The final step of ASC biosynthesis is mitochondrial (Mapson *et al.*, 1954; Bartoli *et al.*, 2000, Siendones *et al.*, 1999; see also Chapter 1 by Smirnoff *et al.*) and is transported into the chloroplast stroma through the envelope membrane *via* carrier-mediated facilitated diffusion (Horemans *et al.*, 1999). The low affinity of the envelope carrier for ASC indicates that ASC regeneration is vital in maintaining high concentrations of this antioxidant in the chloroplasts. In contrast, ASC transport through the thylakoid membrane in to the lumen occurs by simple diffusion (Foyer and Lelandais, 1996).

4.2.1 The extent of coupling between the ascorbate and glutathione pools

The relative redox potentials of the ASC and glutathione couples favour net electron flow from GSH to dehydroascorbate (DHA; Foyer and Noctor, 2000). The capacity for re-cycling of the oxidized forms, DHA and glutathione disulphide (GSSG) clearly affects the sizes of the ASC and glutathione pools since over-expression of either GR or DHA reductase (DHAR; Chen *et al.*, 2003) leads to increased ASC accumulation. The nonenzymatic reduction of DHA by GSH can occur at significant rates even in the absence of DHAR, particularly at alkaline pH values (Winkler, 1992). In conditions that require high fluxes through APX, however, DHAR activity could be necessary to ensure effective maintenance of the reduced ASC pool. In animals, this activity can be catalyzed by proteins such as glutaredoxins and protein disulphide isomerases (Wells *et al.*, 1990, see chapter by May and Asard). These proteins, along with certain types of trypsin inhibitor (Trumper *et al.*, 1994), might also catalyze the reaction in plants. DHA reductase activity has been purified from several plant tissues (Foyer and Halliwell, 1977;

Hossain and Asada, 1984b; Dipierro and Borranccino, 1991; Kato *et al.*, 1997). The amino acid sequence of the purified rice enzyme suggests that it is a specific DHAR (Kato *et al.*, 1997) containing domains that match those encoded by several ESTs from *Arabidopsis* (Foyer and Mullineaux, 1998). Variations in DHAR activity, however, between the organelles within the plant cell could mean that the tightness of coupling between the glutathione and ASC pools might differ considerably between compartments (Baier *et al.*, 2000; Noctor *et al.*, 2000). For example, the apoplast contains abundant amounts of ASC but little or no GSH (Vanacker *et al.*, 1998). Thus, although the ASC–glutathione cycle is a major route for peroxide detoxification in chloroplasts (Foyer and Halliwell, 1976), it may be that during oxidative stress the two antioxidants are coupled less tightly in this organelle than in the cytosol (Baier *et al.*, 2000; Noctor *et al.*, 2000). It is important to note that chloroplasts have a high capacity of direct ASC regeneration from MDHA through reduced ferredoxin. The Mehler-peroxidase pathway functions alongside the enzymic route involving GSH and GR. Glutathione has many functions in plants and is important in other detoxification processes as well as the regeneration of ASC (Foyer and Noctor, 2000). We can conclude that while the ASC and glutathione pools operate together in many instances as the ASC–glutathione cycle, these antioxidants are not always tightly coupled and can function independently in different cellular compartments giving rise to independent paths of signal transduction (Foyer and Noctor, 2000; Pastori *et al.*, 2003).

4.2.2 Interactions between ascorbate and peroxiredoxins

H_2O_2 and organic peroxides formed during chloroplast metabolism are involved in signal transduction (Neill *et al.*, 2002a; Baier and Dietz, 1999a). The lifetime of these species must be rigorously controlled, as chain-type reactions, particularly involving lipid hydroperoxides, can lead to severe disruption of membranes. In addition to the enzymes of the ASC–glutathione cycle, several other types of plant proteins have been identified that might carry out hydrogen peroxide detoxification and also the important task of reducing organic peroxides to the corresponding harmless alcohols. These include glutathione peroxidases (GPX: Eshdat *et al.*, 1997), glutathione *S*-transferases (Bartling *et al.*, 1993), and 2-cys peroxiredoxins (PRX; Baier and Dietz, 1997). GPXs and PRXs differ from other known higher-plant peroxidases in that they do not contain heme (Baier and Dietz, 1999b). The plant GPXs also differ from animal GPXs as the selenocysteine residue at the active site of the latter is replaced by cysteine, which diminishes the nucleophilic properties of the enzyme and probably accounts for its much lower activity with H_2O_2. Although both plant GPX and PRXs can reduce H_2O_2 to water, their low activities mean that APX is by far the most important H_2O_2-removing peroxidase in plant tissue. Nevertheless, PRXs are an important part of the intricate web of chloroplast antioxidant defences (Baier *et al.*, 1999c; 2000). Four PRXs in *Arabidopsis* are predicted to be targeted to the chloroplasts (Horling *et al.*, 2003). To date, only the physiological role of the 2-Cys PRX has been studied in detail in photosynthesizing cells (Baier and Dietz, 1999c, Baier *et al.*, 2000). 2-Cys PRX protects photosynthesis from photo-damage and loss of this antioxidant could not be fully compensated by other components of the antioxidant network (Baier *et al.*, 1999c; 2000). ASC-regenerating enzymes were induced in antisense *A. thaliana* plants with decreased 2-Cys PRX contents and the ASC pool became more oxidized indicating that loss of 2-Cys PRX placed a severe oxidative load on the ASC system. The antisense plants had greater amounts of mRNAs encoding thylakoid APX, stroma-soluble APX, and stromal MDHA reductase, and this was reflected in markedly enhanced activity of the latter enzyme in leaf extracts (Baier *et al.*, 2000). These changes in expression were correlated with increased oxidation of leaf ASC, but not glutathione (Baier *et al.*, 2000). A chloroplast 2-Cys PRX/thioredoxin system might operate in a similar manner to the Mehler-peroxidase cycle in the removal of H_2O_2 but it would have a lower turnover. The expression of the gene encoding 2-Cys PRX (*prx*) is markedly inhibited by ASC (Horling *et al.*, 2003). This would suggest that in many growth conditions where leaf ASC is very high, 2-Cys-PRx expression would be largely suppressed.

4.2.3 Regulation of transcription of chloroplast proteins

The concept that, like hydrogen peroxide, antioxidants could be signal-transducing molecules leading to specific changes in gene expression has only been established relatively recently (Hérouart *et al.*, 1993; Karpinski *et al.*, 1997, 1999; Foyer and Noctor, 2000). However, GSH had been shown to regulate defense gene expression as long ago as 1988 (Wingate *et al.*, 1988; Dron *et al.*, 1988), and it is now considered to be a physiological regulator of processes such as chloroplast transcription (Link *et al.*, 1997). The study of catalase deficient mutants revealed not only that the ASC–glutathione cycle was tightly coupled in the cytosol and peroxisomes (Kendall *et al.*, 1983; Smith *et al.*, 1984, 1985; Noctor *et al.*, 2000) but also that loss of catalase and concomitant oxidation of the glutathione pool leads to activation of genes involved in pathogen resistance (Takahashi *et al.*, 1997; Willekens *et al.*, 1997) and induction of cell death responses (Dat *et al.*, 2003). Low ASC has relatively little effect on the level of expression of antioxidant defense genes but causes strong activation of pathogen defense genes (Kiddle *et al.*, 2003; Pastori *et al.*, 2003). High ASC favors the repression of pathogenesis-related protein 1 (PR1) and other pathogenesis-related proteins (Kiddle *et al.*, 2003; Pastori *et al.*, 2003). These observations demonstrate the existence of extensive metabolic crosstalk between the different defense processes in plants (*Figure 4.1*). ASC also modulates the abundance of transcripts encoding chloroplast proteins. These include electron transport chain components such as those of the light harvesting system, photosystem (PS)I, PSII and the oxygen-evolving complex (Kiddle *et al.*, 2003). Similarly, the abundance of transcripts encoding enzymes found in the stroma is also modulated by leaf ASC content. Glucose 6-phosphate dehydrogenase (G6PDH) transcripts, for example, are increased by high ASC. The activity of G6PDH is decreased by reduced thioredoxin and increased by oxidized thioredoxin. Chloroplast protein (CP) 12 is an intriguing regulatory protein that is involved in the formation of a high molecular weight complex with certain enzymes of the CO_2 assimilation cycle in the chloroplast including G6PDH. CP12 transcripts are also increased in tissues supplied with ASC. Conversely, high ASC decreases the level of mRNAs encoding fructose-1,6-bisphosphatase (FBPase) and sedoheptulose-1,7-bisphosphatase (SBPase; Kiddle *et al.*, 2003). Both FBPase and SBPase undergo rapid reductive activation via reduced thioredoxin and conversely they are inactivated by oxidized thioredoxin. These observations suggest that ASC could be involved in the control of the transcription of at least some photosynthetic enzymes. FBPase and SBPase cannot undergo reductive activation by ASC; this process requires reduced thioredoxin. Overall, the redox state of a compartment such as the chloroplast is governed by the sum of the ASC, glutathione and pyridine nucleotide pools. The thioredoxin pool is very small in comparison and it makes only a minor contribution to overall redox state. Unlike the ASC pool, which is largely reduced in both light and dark acting as a redox buffer, the thioredoxin pool goes from being completely oxidized in the dark to a variable, but relatively highly reduced state in the light. The absence of evidence for crosstalk between ASC and thioredoxin-mediated pathways perhaps reflects this rather different contribution to chloroplast redox reactions. These results suggest a two-tier level of redox control of redox-modulated chloroplast proteins. The first involves ASC-mediated control of transcription, which senses the overall background or 'dark-level' of reductant in the chloroplast. Through the abundance of ASC, this controls the transcription of the genes encoding Benson–Calvin cycle proteins. The second tier of control involves short-term reductive activation of the enzyme proteins by reduced thioredoxin, which allows activation of the Benson–Calvin cycle irrespective of the background redox buffering capacity. The physiological significance of such results remains to be demonstrated but these data might infer that a high antioxidant defense capacity tends to favor a lower photosynthetic capacity in terms of the mRNA pool available for photosynthetic protein formation.

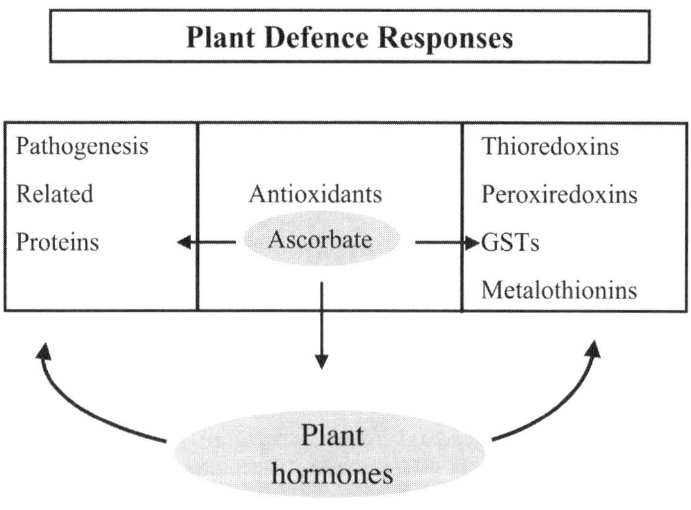

Figure 4.1: Metabolic crosstalk between the different elements of the plant defense network mediated by ASC.

4.3 Regulation of ascorbate synthesis in mitochondria

The enzymatic oxidation of L-galactono-1,4-lactone to ASC was demonstrated in plants half a century ago (Mapson *et al.*, 1954). Despite much research on the pathway of ASC production in plants in the intervening years, it is only relatively recently that a degree of harmonization has been achieved (Wheeler *et al.*, 1998). There is now a general consensus concerning overall preservation of the carbon skeleton during ASC synthesis and that the final step in ASC biosynthesis involves the cytochrome *c*-dependent oxidation of L-galactono-1,4-lactone (L-GalL) by L-galactono-1,4-lactone dehydrogenase (GalLDH; EC 1.3.2.3). This enzyme is localized in the inner mitochondrial membrane (Bartoli *et al.*, 2000; Siendones *et al.*, 1999). Except for this last step, the pathway is believed to be cytosolic (Wheeler *et al.*, 1998, see Chapter 1 by Smirnoff, Running and Gatzek). These authors proposed that the major route of L-galactono-1,4-lactone synthesis is from L-galactose *via* a novel higher-plant enzyme: L-galactose dehydrogenase. In some species or tissues an additional pathway of L-GalL synthesis *via* an NADPH-dependent galacturonate reductase and galacturonic acid could also contribute to ASC synthesis (Agius *et al.*, 2003).

In animals, L-gulono-1,4-lactone, a stereo-isomer of L-GalL, is the preferential precursor of ASC. Moreover, the enzyme producing ASC, L-gulono-1,4-lactone oxidase, is localized in microsomes and produces H_2O_2 as a by-product. In contrast, the plant L-GalLDH occurs only in mitochondria (Mapson *et al.*, 1954; Bartoli *et al.*, 2000) and catalysis does not lead to H_2O_2 production. Mitochondrial ASC synthesis is inhibited by KCN and is stimulated by antimycin A, suggesting that L-GalLDH activity was restricted by the availability of oxidized cytochrome *c* (Bartoli *et al.*, 2000).

ASC synthesis is not only dependent on the availability of cytochrome $c_{(ox)}$ but it also requires electron flow through complex I. Studies using rotenone, together with the tricarboxylic acid (TCA) cycle substrates, pyruvate and malate, have shown that this specific inhibitor of Complex I, inhibits ASC synthesis rate to the same degree as KCN (Millar *et al.*, 2003). As rotenone oxidizes the cytochrome *c* pool, this inhibition of ASC synthesis cannot be explained by the simple substrate kinetics invoked to interpret the effects of KCN and antimycin A. When electrons enter the mitochondrial electron transport chain through either succinate dehydrogenase (by the addition of succinate) or the external NADH dehydrogenases (by addition of exogenous NADH), no effect of rotenone on ASC synthesis was observed (Millar *et al.*, 2003). This means that electron flow through complex I is required for optimal rates of ASC synthesis. Proteomic techniques have recently revealed the structural basis for this functional relationship. A combination of Blue Native PAGE coupled to a second dimension of denaturing SDS/PAGE and mass spectrometry has been used to identify many of the components of complexes I, III and V (Heazlewood *et al.*, 2003). There are two types of complex I in *Arabidopsis thaliana* mitochondria: a minor high mobility form and a major lower mobility form. The higher mobility form contains an additional subunit of approximately 62 kDa that is absent from the major lower mobility form (Millar *et al.*, 2003). Sequencing of peptides derived from this protein revealed seven separate peptides that matched to the sequence of the

Arabidopsis thaliana L-GalLDH (GenBank Accession BAA95212). Hence L-GalLDH is physically associated with complex I in *Arabidopsis*. Rotenone is believed to act by binding to the 20 kDa, 4Fe–4S binding subunit of complex I, preventing electron transport to the membrane arm of complex I leading ultimately to ubiquinone reduction. It might be that redox poise regulates a component of complex I that modifies ASC synthesis. This could be a Fe–S center or the flavin center upstream or downstream of the 4Fe–4S binding subunit. Alternatively, conformational changes induced in complex I by rotenone binding alter the redox poise in such as way as to prevent GLDH catalysis (Millar *et al.*, 2003). Respiration and ASC synthesis appear to be intricately co-ordinated and mitochondria might therefore have a profound influence over the ability of the plant to produce ASC during stress.

Like the enzymes of the Benson–Calvin cycle in the chloroplasts, the enzymes of the TCA cycle in the mitochondrial matrix are particularly sensitive to oxidative inhibition. Superoxide, formed during electron transport through mitochondrial complexes I and III is therefore potentially harmful and must be rapidly dismuted to H_2O_2 by matrix Mn-superoxide dismutase. However, H_2O_2 is a potent inhibitor of aconitase and has also been shown to damage other TCA cycle components. The matrix therefore requires robust detoxification systems for H_2O_2. The mitochondrial electron transport system contains an alternative oxidase, internal and external NADH dehydrogenases and uncoupling proteins that together serve to limit superoxide generation. A number of recent reports have indicated that the enzymes of the ASC–glutathione cycle are present in the mitochondrial matrix. A complete ASC–glutathione enzyme cycle and membrane-bound APX isoenzymes have been detected in pea leaf mitochondria (Jiménez *et al.*, 1997) and tomato leaf mitochondria (Mittova *et al.*, 2000). Moreover, matrix-associated APX isoforms and other mitochondrial antioxidants act in concert to protect the TCA cycle enzymes against oxidative stress.

Until recently, the role of mitochondria as a ROS source in plant cells was largely unexplored. However, superoxide radicals produced during respiration make an important contribution to the oxidative load experienced by the whole plant cell. As discussed above, respiratory complexes I and III are considered to be major sites of mitochondrial superoxide radical production. Impaired complex I function in the cytoplasmic male sterile (CMS)II mutant, for example, results in specific changes in antioxidant gene expression and alters antioxidant enzyme activities in the cytosol, peroxisomes and chloroplasts as well as in mitochondria (Dutilleul *et al.*, 2003). Mitochondrial redox signals therefore play a crucial role in the regulation of plant defense processes.

4.4 Ascorbate and the apoplast

ASC is the only antioxidant in the apoplast, where it can be present at values of 1–2 mM (Sanmartin *et al.*, 2003; Pignocchi and Foyer, 2003). ASC oxidase (AO) is also localized in the apoplast (Diallinas *et al.*, 1997; Lin and Varner, 1991). Although there is some evidence to suggest that the enzyme is also present in organelles (Yamaguchi *et al.*, 1995), a recent study in tobacco leaves found AO activity only in the apoplast (Sanmartin *et al.*, 2003). AO catalyzes the oxidation of ASC to DHA with the concomitant four-electron reduction of oxygen to water. AO is a homodimeric glycosylated, copper-containing protein. Together with laccase and ceruplasmin it belongs to the 'blue oxidase' multi-copper protein family (Messerschmidt and Huber, 1990). Each AO monomer has three disulphide bridges, due to three pairs of Cys residues: one between domain 1 and domain 2, one between domain 1 and 3 and one within domain 2. These disulphide bridges are believed to contribute to the high stability of the tertiary structure of the protein (Messerschmidt and Huber, 1990). The domain that binds ASC and the copper binding centers are highly conserved between AOs isolated from different species. Melon AO exhibits optimal activity at pH 5.5 and 37°C, with a K_m for ASC of 50 μM (Moser and Kanellis, 1994). Two studies have recently explored the function of AO in tobacco leaves using sense and antisense technologies (Sanmartin *et al.*, 2003; Pignocchi *et al.*, 2003; Pignocchi and Foyer, 2003). Overexpression of AO was found to greatly decrease the ratio of reduced to oxidized ASC in the apoplast while having only a minimal effect on the amount and redox state of the whole leaf ASC pool (Pignocchi *et al.*, 2003). The high AO-mediated oxidation of the apoplastic ASC pool greatly increased the sensitivity of the transgenic leaves to ozone (Sanmartin *et al.*, 2003). Moreover, high AO activity favored enhanced growth (Pignocchi *et al.*, 2003) and modified the responses of the transformed plants to plant growth regulators such as auxin. Such observations led to the hypothesis that control of the redox state of the apoplastic ASC pool by AO might provide a mechanism for control of signal transduction (Pignocchi and Foyer, 2003). High AO activity favors oxidation of the ASC pool and modifies the apoplastic redox state. This might affect defense responses and possibly hormone responses (Pignocchi and Foyer, 2003).

4.5 Ascorbate signaling

ASC acts as a cofactor and modulator of the activities of many enzymes (Noctor and Foyer, 1998; Arrigoni and De Tullio, 2000). It is involved as an enzyme cofactor in the synthesis of plant hormones such as abscisic acid (ABA), gibberellic acid (GA) and ethylene (Arrigoni and De Tullio, 2000). As discussed above, it also has roles in the control of growth and the mitotic cell cycle (Noctor and Foyer, 1998; Smirnoff and Wheeler, 2000). As shown in *Figure 4.1* plants sense tissue ASC content and use this information in the metabolite crosstalk that modulates different pathways involved in the defense network. Leaf ASC contents are markedly dependent on the position on the stem (*Figure 4.2*) and environmental conditions such as growth light intensity (Gillham and Dodge 1987; Grace and Logan 1995; Conklin *et al.*, 1996). In many plants, such as tobacco (*Figure 4.2*) there is a marked loss of leaf ASC at night with net synthesis occurring in the light. Light appears to be essential for the expression of some of the genes encoding enzymes of ASC metabolism (Tabata *et al.*, 2002) but there is little evidence to suggest a diurnal rhythm of transcript abundance that underpins the diurnal rhythm of ASC abundance as shown in *Figure 4.2* (Pignocchi *et al.*, 2003). In both light and dark the leaf ASC is largely in the reduced form (*Figure 4.2*) in most leaves. A marked exception is the old senescent leaves where the total ASC pool is low and is largely in the form of DHA (*Figure 4.2*). Relatively little ASC can be detected in older senescent leaves and in these leaves L-GalLDH activity is also low (Bartoli *et al.*, 2002).

The ASC-deficient mutant of *A. thaliana*, vtc1 (Conklin *et al.*, 1996; 1997) has proved to be a useful tool in the analysis of how varying levels of ASC modulate gene expression to control plant growth and metabolism (Pastori *et al.*, 2003). Initially identified via its sensitivity to elevated ozone concentrations, this mutant was instrumental in the elucidation of the pathway of ASC biosynthesis (Conklin *et al.*, 1999). The mutant possesses decreased activity of GDP-mannose pyrophosphorylase, and as a result accumulates ASC to only about 30% of wild-type levels (Conklin *et al.*, 1996; 1999). Vtc1 leaves have similar rates of photosynthesis and energy dissipation to the wild-type (Col0) but the mutant is substantially smaller and grows more slowly than the Col0 (Veljovic-Jovanovic *et al.*, 2001). Moreover, there is no evidence of significant upregulation of other elements of the antioxidant system to compensate for decreased ASC concentrations (Conklin *et al.*, 1997; Veljovic-Jovanovic *et al.*, 2001; Pastori *et al.*, 2003). The ratios of leaf ASC/DHA or GSH/GSSG and the total leaf H_2O_2 contents are similar in the mutant and wild-type suggesting that the mutant does not suffer a higher degree of oxidative stress than the wild-type (Veljovic-Jovanovic *et al.*, 2001). Evidence was found, however, for an effect on the distribution of APX activity between chloroplastic and nonchloroplastic APX activities (Veljovic-Jovanovic *et al.*, 2001). Using H_2O_2 sensitivity as a means of distinguishing between the APX activities in the chloroplastic and cytosolic compartments (Amako *et al.*, 1994), nonchloroplastic APX activity was found to make only a minor contribution to the total APX activity in the wild-type but to be enhanced more than sixfold in the mutant (Veljovic-Jovanovic *et al.*, 2001). Similarly, guaiacol-type peroxidase activity was significantly increased in the mutant and also in plants with low ASC content as a result of antisense L-GalDH expression

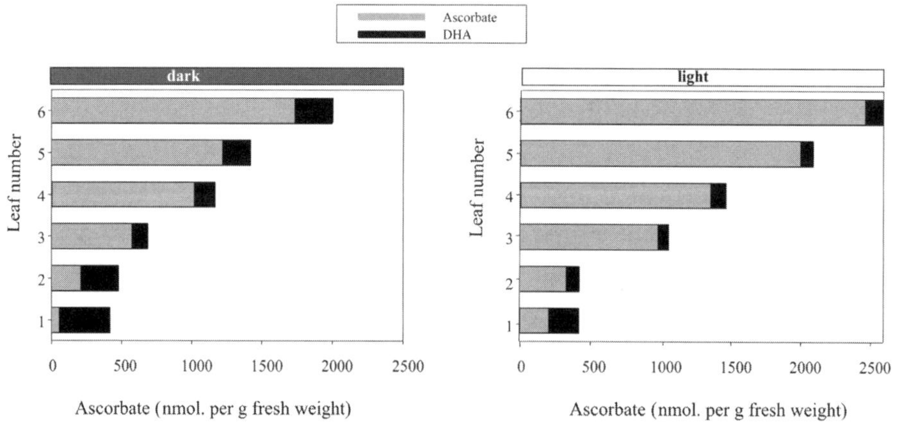

Figure 4.2: The amount of total ASC and the relative content of reduced ASC (gray areas) and DHA (dark areas) in the different leaves on a tobacco plant in the light and in the dark.

(Gatzek *et al.*, 2002). These and other effects observed in the mutant are linked to ASC deficiency and not to the ASC redox state, since this is similar in the wild-type and mutant (Conklin *et al.*, 1996; Veljovic-Jovanovic *et al.*, 2001; Pastori *et al.*, 2003; Kiddle et al 2003). Vtc1 leaves were used to investigate the effect of ASC deficiency on the transcriptome of *Arabidopsis thaliana* (Pastori *et al.*, 2003), where evidence for regulation of growth via changes in plant hormone abundance and related signaling was obtained (Pastori *et al.*, 2003), as well as upregulation of pathogenesis-related proteins (Kiddle *et al.*, 2003). Taken together, the evidence obtained from studies on (i) ASC-mediated changes in PRX expression (Horling *et al.*, 2003), (ii) the analysis of mutants in antioxidant metabolism (Baier *et al.*, 2000; Noctor *et al.*, 2000) and (iii) from transcriptome analysis (Pastori *et al.*, 2003; Kiddle et al, 2003) demonstrates that ASC is potentially a signal transducing molecule in plants.

4.6 Interaction between ascorbate and plant hormones

Plant cell growth is the result of cell elongation and cell division. ASC, DHA, MDHA and AO could be key factors regulating cell elongation (Pignocchi *et al.*, 2003). The first indication that ASC could also be involved in the regulation of the cell cycle was obtained in onion (*Allium cepa* L.) and broad bean (*Vicia faba* L.) root meristems (Liso *et al.*, 1984). Moreover, ASC was below the level of detection in the quiescent center of maize roots (Kerk and Feldman, 1995). The quiescent center cells do not divide and the mitotic cycle is interrupted in the G1 phase (Kerk and Feldman, 1995). The quiescent center has high levels of AO mRNA, protein and activity, consistent with the hypothesis that a high ASC/DHA ratio is necessary to drive the cell cycle from G1 to S phase (Kerk and Feldman, 1995).

While the mechanisms by which ASC regulates the cell cycle remain to be elucidated, it is clear that the amount of this antioxidant is low in dormant tissues and is increased markedly in conditions that favor rapid metabolism and growth. It is probable that the basis for these observations is that the ASC pool is a crucial buffer against the high oxidative load that accompanies rapid metabolism. The ASC pool plays a role in the control of plant growth and development by modulation of hormone signaling Pastori *et al.*, 2003; *Figure 4.3*). Low ASC contents promote slow growth via a stimulation of ABA synthesis and signaling and a concomitant inhibition of GA acid synthesis and signaling (*Figure 4.3*). The constitutive upregulation of ABA-associated pathways and simultaneous inhibition of GA-related pathways explains why the *vtc1* mutant is slower growing than the wild-type. ABA is a regulator of metabolic arrest and is a key inducer of cell survival pathways particularly in stress conditions. It functions in an inverse fashion to GA that stimulates growth and can leave plants more open to oxidative damage as observed for example in the regulated death of barley aleurone cells (Bethke and Jones, 2001).

In addition to modulating the sensitivity of plant cells to growth hormones, as described above (Pignocchi *et al.*, 2003), ASC has other roles in the regulation of cell expansion (Smirnoff, 2000). The cell wall is considered to grow by a process of

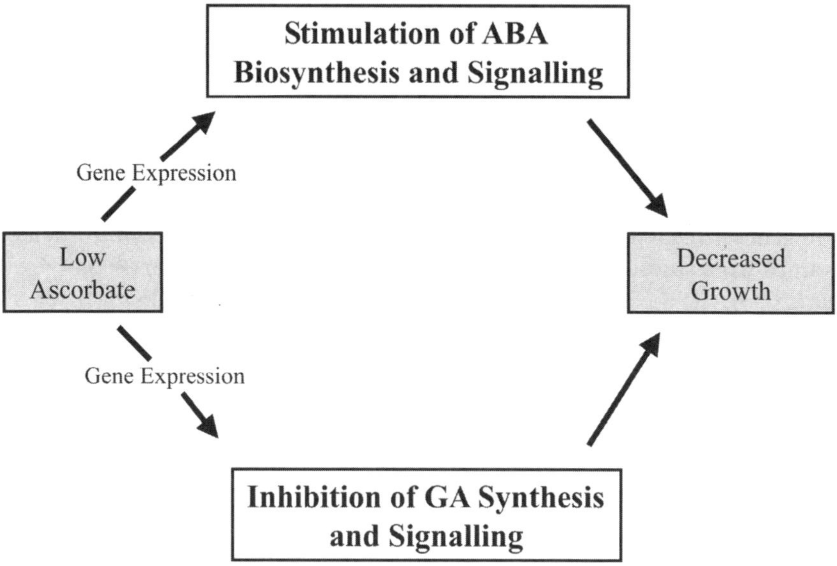

Figure 4.3: The co-ordinate regulation of growth by ASC involving modulation of abscisic acid (ABA) and gibberellic acid (GA) signaling.

continuous oxidative cleavage and re-sealing with new material (Schopfer *et al.*, 2002). The growing meristem thus has a sheath of ROS surrounding the expanding cells. As an apoplastic antioxidant, ASC can detoxify hydroxyl radicals and other ROS, and may therefore tend to reduce the rate of cell wall cleavage during extension growth. In addition, ASC is a cofactor for prolyl hydroxylase (De Tullio *et al.*, 1999), and is therefore involved in extensin synthesis. Extensins are involved in cell wall synthesis and assembly during mitosis. The oxidation of these proteins results in enhancement of wall strength, a key defense mechanism against pathogen attack. ASC also potentially inhibits the formation of lignin polymers by apoplastic peroxidases using extracellular H_2O_2 (Takahama, 1998). The inhibition is due to the ASC-dependent reduction of hydroxyl radicals and quinones (Takahama and Oniki, 1994), but there is no evidence to date this process regulates lignification *in vivo*.

4.7 Conclusions and perspectives

ASC has crucial roles in the plant defense network. ASC deficiency in *vtc1* leads to constitutive expression of pathogen resistance genes. It is interesting to note that the mutant is highly sensitive to ozone despite the expression of many proteins involved in responses to biotic stress. Ozone-induced changes in gene expression and those associated with biotic stress responses share many common features. However, the observation that the constitutive expression of the latter in *vtc1* cannot compensate for the loss of ASC, indicate the crucial role of this antioxidant in determining ozone tolerance.

Plants are sessile organisms that have evolved a highly flexible and adaptable metabolic network to achieve an appropriate balance between growth, development and defense against environmental threats. Plants induce defense pathways in response to environmental stress but they also simultaneously downregulate growth and development. The inverse relationship between defense gene activation is evident in *vtc1*, the coordinate regulation of gene expression by ASC providing clues as to the mechanisms that underpin this response (Pastori *et al.*, 2003). Activation of plant pathogenesis related defenses could have a high cost in terms of decreased growth and competitiveness with neighbors in the plant community. The signal transduction processes that underpin the switch from primary to secondary metabolism are poorly understood but it is clear that ROS and the major redox buffer of plant cells, ASC, are part of the metabolite signaling network that orchestrates this response. ASC signaling could also have profound effects on gene expression in each of the individual leaves of a plant since the amount of ASC decreases significantly from growing shoot meristem to the oldest senescing leaves to the growing shoot meristem (*Figure 4.4*).

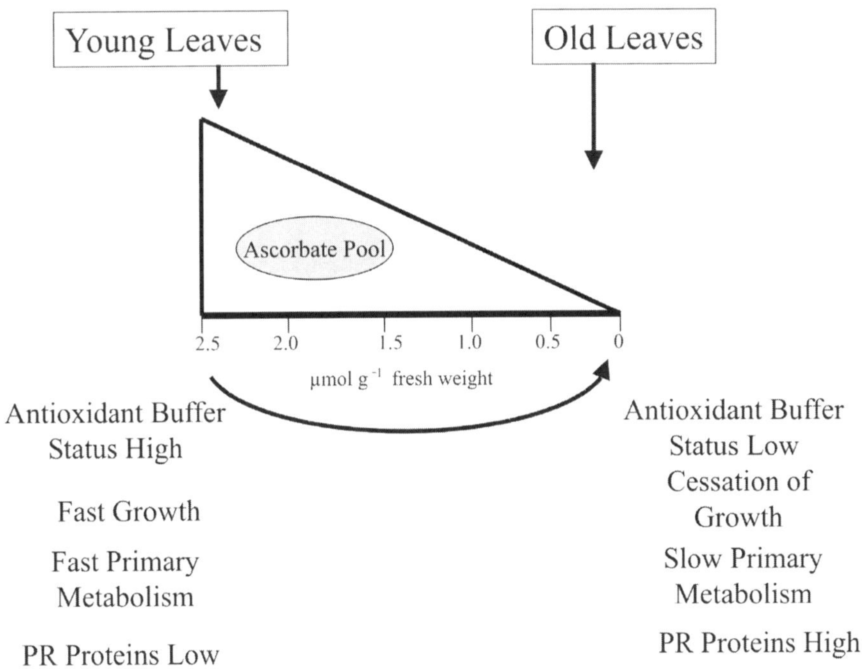

Figure 4.4: Possible relationships between the amount of ASC in leaves and their metabolism and physiology.

References

Agius F, Gonzalez-Lamothe R, Caballero JL, Munoz-Blanco J, Botella MA, Valpuesta V (2003) Engineering increased vitamin C levels in plants by overexpression of a D-galacturonic acid reductase. *Nature Biotechnol.* **21**, 177–181.

Amako K, Chen G-X, Asada K (1994) Separate assays specific for ascorbate peroxidase and guaiacol peroxidase and for the chloroplastic and cytosolic isozymes of ascorbate peroxidase in plants. *Plant Cell Physiol.* **35**, 497–504.

Anderson JW, Foyer CH, Walker DA (1983) Light-dependent reduction of hydrogen peroxide by intact spinach chloroplasts. *Biochim. Biophys. Acta* **724**, 69–74.

Arrigoni O, Arrigoni-Liso R, Calabrese G (1975) Lycorine as an inhibitor of ascorbic acid biosynthesis. *Nature* **256**, 513–514.

Arrigoni O, De Tullio MC (2000) The role of ascorbic acid in cell metabolism: between gene-directed functions and unpredictable chemical reactions. *J. Plant Physiol.* **157**, 481–488.

Azevedo RA, Alas RM, Smith RJ, Lea PJ (1998) Response of antioxidant enzymes to transfer from elevated carbon dioxide to air and ozone fumigation, in the leaves and roots of wild-type and a catalase-deficient mutant of barley. *Physiol. Plant* **104**, 280–292.

Baier M, Dietz KJ (1997) The plant 2-cys peroxiredoxin BAS1 is a nuclear-encoded chloroplast protein: its expressional regulation, phylogenetic origin, and implications for its specific physiological function in plants. *Plant J.* **12**, 179–190.

Baier M, Dietz KJ (1999a) The costs and benefits of oxygen for photosynthesizing plant cells. *Progr. Bot.* **60**, 282–314.

Baier M, Dietz KJ (1999b) Alkyl hydroperoxide reductases: the way out of the oxidative breakdown of lipids in chloroplasts. *Trends Plant Sci.* **4**, 166–168.

Baier M, Dietz KJ (1999c) Protective function of chloroplast 2-cysteine peroxiredoxin in photosynthesis. Evidence from transgenic *Arabidopsis*. *Plant Physiol.* **119**, 1407–1414.

Baier, M, Noctor G, Foyer CH, Dietz KJ (2000) Antisense suppression of 2-cysteine peroxiredoxin in *Arabidopsis* specifically enhances the activities and expression of enzymes associated with ascorbate metabolism but not glutathione metabolism. *Plant Physiol.* **124**, 823–832.

Bartling D, Radzio R, Steiner U, Weiler EW (1993) A glutathione-S-transferase with glutathione peroxidase activity from *Arabidopsis thaliana* – molecular cloning and functional characterization. *Eur. J. Biochem.* **216**, 579–586.

Bartoli C, Pastori GM, Foyer CH (2000) Ascorbate biosynthesis in mitochondria is linked to the electron transport chain between complexes III and IV. *Plant Physiol.* **123**, 335–343.

Bethke PC, Jones RL (2001) Cell death of barley aleurone protoplasts is mediated by reactive oxygen species. *Plant J.* **25**, 19–29.

Brisson LF, Zelitch I, Havir EA (1998) Manipulation of catalase levels produces altered photosynthesis in transgenic tobacco plants. *Plant Physiol.* **116**, 259–269.

Bunkelmann JR, Trelease RN (1996) Ascorbate peroxidase. A prominent membrane protein in oilseed glyoxysomes. *Plant Physiol.* **110**, 589–598.

Chen G, Asada K (1989) Ascorbate peroxidase in tea leaves: occurrence of two isozymes and the differences in their enzymatic and molecular properties. *Plant Cell Physiol.* **30**, 987–998.

Chen Z, Young TE, Ling J, Chang S-C, Gallie DR (2003) Increasing vitamin C content of plants through enhanced ascorbate cycling. *Proc. Natl Acad. Sci. USA* **100**, 3525–3530.

Conklin PL, Williams EH, Last RL (1996) Environmental stress sensitivity of an ascorbic acid-deficient *Arabidopsis* mutant. *Proc. Natl Acad. Sci. USA* **93**, 9970–9974.

Conklin PL, Pallanca J, Last RL, Smirnoff N (1997) L-Ascorbic acid metabolism in the ascorbate-deficient *Arabidopsis* mutant *vtc1*. *Plant Physiol.* **115**, 1277–1285.

Conklin PL, Norris SR, Wheeler GL, Williams EH, Smirnoff N, Last RL (1999) Genetic evidence for the role of GDP-mannose in plant ascorbic acid (vitamin C) biosynthesis. *Proc. Natl Acad. Sci. USA* **96**, 4198–4203.

Dat JF, Pellinen R, Beeckman T, Van De Cotte B, Langebartels C, Kangasjarvi J, Inzé D, Van Breusegem F (2003) Changes in hydrogen peroxide homeostasis trigger an active cell death process in tobacco. *Plant J.* **33**, 1–12.

De Tullio MC, Paciolla C, Dalla Veechia F, Rascio N, D'Emerico S, De Gara L, Liso R, Arrigoni O (1999) Changes in onion root development induced by the inhibition of peptidyL-prolyl hydroxylase and influence of the ascorbate system on cell division and elongation. *Planta* **209**, 424–434.

Diallinas G, Pateraki I, Sanmartin M, Scossa A, Stilianou E, Panopoulos NJ, Kanellis AK (1997) Melon ascorbate oxidase: cloning of a multigene family, induction during fruit development and repression by wounding. *Plant Mol. Biol.* **34**, 759–770.

Dipierro S, Borranccino G (1991) Dehydroascorbate reductase from potato tubers. *Phytochemistry* **30**, 427–429.

Dron M, Clouse SD, Dixon RA, Lawton MA, Lamb CJ (1988) Glutathione and fungal elicitor regulation of a plant defense gene promoter in electroporated protoplasts. *Proc. Natl Acad. Sci. USA* **85**, 6738–6742.

Dutilleul C, Garmier M, Mathieu C, Chetrit P, Noctor G, Foyer CH, De Paepe R (2003) Mitochondria participate in whole plant cell antioxidant cross-talk and maintenance of redox homeostasis. *Plant Cell* **15**, 1212–1226.

Egneus H, Heber U, Matthiesen U, Kirk M (1975) Reduction of oxygen by the electron transport chain of chloroplasts during assimilation of carbon dioxide. *Biochim. Biophys. Acta* **408**, 252–268.

Eshdat Y, Holland D, Faltin Z, Ben-Hayyim G (1997) Plant glutathione peroxidases. *Physiol. Plant* **100**, 234–240.

Foyer CH (1977) The function and relationship of ascorbic acid and glutathione in spinach leaves. PhD thesis, Kings College, London, UK.

Foyer CH, Descourvières P, Kunert KJ (1994). Protection against oxygen radicals: an important defense mechanism studied in transgenic plants. *Plant, Cell and Environ.* **17**, 507–524.

Foyer CH, Halliwell B (1976) The presence of glutathione and glutathione reductase in chloroplasts: a proposed role in ascorbic acid metabolism. *Planta* **133**, 21–25.

Foyer CH, Halliwell B (1977) Purification and properties of dehydroascorbate reductase from spinach leaves. *Phytochemistry* **16**, 1347–1350.

Foyer CH, Lelandais M (1996) A comparison of the relative rates of transport of ascorbate and glucose across the thylakoid, chloroplast and plasma membranes of pea leaf mesophyll cells. *J. Plant Physiol.* **148**, 391–398.

Foyer CH, Lopez-Delgado H, Dat JF, Scott IM (1997) Hydrogen peroxide- and glutathione-associated mechanisms of acclimatory stress tolerance and signalling. *Physiol. Plant* **100**, 241–254.

Foyer CH, Mullineaux PM (1998) The presence of dehydroascorbate and dehydroascorbate reductase in plant tissues. *FEBS Lett.* **425**, 528–529.

Foyer CH, Harbinson J (1999) Relationships between antioxidant metabolism and carotenoids in the regulation of photosynthesis. In: *The Photochemistry of Carotenoids* (eds HA Frank, AJ Young, G Britton, RJ Cogdell). Kluwer Academic, Dortrecht/Boston/London, pp. 305–325.

Foyer CH, Noctor G (2000) Oxygen processing in photosynthesis: a molecular approach. *New Phytol.* **146**, 359–388.

Foyer CH, Rowell J, Walker D (1983) Measurements of the ascorbate content of spinach leaf protoplasts and chloroplasts during illumination. *Planta* **157**, 239–244.

Gatzek S, Wheeler GL, Smirnoff N (2002) Antisense suppression of L-galactose dehydrogenase in *Arabidopsis thaliana* provides evidence for its role in ascorbate synthesis and reveals light modulated L-galactose synthesis. *Plant J.* **30**, 541–553.

Gerhardt B (1964) Untersuchungen uber beziehungen zwishen ascorbinsaure und photosynthese. *Planta* **61**, 101–129.

Gillham DJ, Dodge AD (1987) Chloroplast superoxide and hydrogen peroxide scavenging systems from pea leaves: seasonal variations. *Plant Sci.* **50**, 105–109.

Grace SC, Logan BA (1996) Acclimation of foliar antioxidant systems to growth irradiance in three broad-leaved evergreen species. *Plant Physiol.* **112**, 1631–1640.

Groden D, Beck E (1979) H_2O_2 destruction by ascorbate-dependent systems from chloroplasts. *Biochim. Biophys. Acta* **546**, 426–435.

Halliwell B (1981) Toxic effects of oxygen on plant tissues. In: *Chloroplast Metabolism*. Clarendon Press, Oxford, pp. 179–205.

Halliwell B, Gutteridge JMC (1989) *Free Radicals in Biology and Medicine*, 2nd Edn. Oxford University Press, Oxford.

Heazlewood JA, Howell KA, Millar AH (2003) Mitochondrial complex I from *Arabidopsis* and rice: orthologs of mammalian and yeast components coupled to plant specific subunits. *BBA Bioenergetics* (in press).

Hérouart D, Van Montagu M, Inzé D (1993) Redox-activated expression of the cytosolic copper–zinc superoxide dismutase gene in *Nicotiana*. *Proc. Natl Acad. Sci. USA* **90**, 3108–3112.

Hopkins FG, Mogan EJ (1936) Some relations between ascorbic acid and glutathione. *Biochem. J.* **30**, 1446–1462.

Horemans N, Foyer CH, Asard H (1999) The functions of ascorbate and ascorbate transport systems in plant membranes. In: *Different Pathways Through Life. Biochemical Aspects of Plant Biology and Medicine* (eds A Denke, K Dornisch, F Fleischmann, J Graßmann, I Heiser, S Hippeli, W Oßwald, H Schempp). Lincom Europa, Munich, pp. 217–238.

Horling F, Lamkemeyer P, Konig J, Finkemeier I, Kandlbinder A, Baier M, Dietz K-J (2003) Divergent light-, ascorbate- and oxidative stress-dependent regulation of expression of the peroxiredoxin gene family in *Arabidopsis*. *Plant Physiol.* **131**, 317–325.

Hossain MA, Asada K, (1984a) Inactivation of ascorbate peroxidase in spinach chloroplasts on dark addition of hydrogen peroxide: its protection by ascorbate. *Plant Cell Physiol.* **25**, 1285–1295.

Hossain MA, Asada K (1984b) Purification of dehydroascorbate reductase from spinach and its characterisation as a thiol enzyme. *Plant Cell Physiol.* **25**, 85–92.

Jiménez A, Hernández JA, del Río L, Sevilla F (1997) Evidence for the presence of the ascorbate–glutathione cycle in mitochondria and peroxisomes of pea leaves. *Plant Physiol.* **114**, 275–284.

Kaiser WM (1979) Reversible inhibition of the Calvin cycle and activation of oxidative pentose phosphate cycle in isolated intact chloroplasts by hydrogen peroxide. *Planta* **145**, 377–382.

Karpinski S, Escobar C, Karprinska B, Creissen G, Mullineaux PM (1997) Photosynthetic electron transport regulates the expression of cytosolic ascorbate peroxidase genes in *Arabidopsis* during excess light stress. *Plant Cell* **9**, 627–640.

Karpinski S, Reynolds H, Karpinska B, Wingsle G, Creissen G, Mullineaux P (1999) Systemic signaling and acclimation in response to excess excitation energy in *Arabidopsis*. *Science* **284**, 654–657.

Kato Y, Urano J, Maki Y, Ushimaru T (1997) Purification and characterisation of dehydroascorbate reductase from rice. *Plant Cell Physiol.* **38**, 173–178.

Kelly GJ, Latzko E (1979) Soluble ascorbate peroxidase. *Naturwissenschaften* **66**, 377–382.

Kendall AC, Keys AJ, Turner JC, Lea PJ, Miflin BJ (1983) The isolation and characterisation of a catalase-deficient mutant of barley (*Hordeum vulgare* L). *Planta* **159**, 505–511.

Kerk NM, Feldman LJ (1995) A biochemical-model for the initiation and maintenance of the quiescent center – implications for organization of root meristems. *Development* **121**, 2825–2833.

Kiddle G, Pastori GM, Bernard S, Pignocchi C, Antoniw J, Verrier PJ, Foyer CH (2003) Effects of leaf ascorbate content on defense and photosynthesis gene expression in *Arabidopsis thaliana*. *Antioxid. Redox Signal* 5, 23–32.

Levine, A, Tenhaken, R, Dixon, R, Lamb, C (1994) H_2O_2 from the oxidative burst orchestrates the plant hypersensitive disease resistance response. *Cell* **79**, 583–593.

Lin LS, Varner JE (1991) Expression of ascorbic–acid oxidase in zucchini squash (*Cucurbita pepo* L). *Plant Physiol.* **96**, 159–165.

Link G, Tiller K, Baginsky S (1997) Glutathione: a regulator of chloroplast transcription. In: *Regulation of Enzymatic Systems Detoxifying Xenobiotics in Plants*. NATO ASI Series (ed. KK Hatzios). Kluwer Academic, Dordrecht/Boston/London, pp. 125–137.

Liso R, Calabrese G, Bitonti MB, Arrigoni O (1984) Relationship between ascorbic acid and cell division. *Exp. Cell Res.* **150**, 314–320.

Mapson LW (1958) Metabolism of ascorbic acid in plants: Part I. Function. *Annu. Rev. Plant Physiol.* **9**, 119–150.

Mapson LW, Isherwood FA, Chen YT (1954) Biological synthesis of L-ascorbic acid: the conversion of L-galactono-γ-lactone into L-ascorbic acid by plant mitochondria. *Biochem. J.* **56**, 21–28.

Meister A (1994) Glutathione–ascorbic acid antioxidant system in animals. *J. Biol. Chem.* **269**, 9397–9400.

Messerschmidt A, Huber R (1990) The blue oxidases, ascorbate oxidase, laccase and ceruloplasmin – modeling and structural relationships. *Eur. J. Biochem.* **187**, 341–352.

Millar AH, Mittova V, Kiddle G, Heazlewood JL, Bartoli CG, Guiamet JJ, Theodoulou FL, Foyer CH (2003) Control of ascorbate synthesis by respiration and its implications for stress responses. *Plant Physiol.* (in press).

Mittova V, Volokita M, Guy M, Tal M (2000) Activities of SOD and the ascorbate–glutathione cycle enzymes in subcellular compartments in leaves and roots of the cultivated tomato and its wild salt-tolerant relative *Lycopersicon pennellii*. *Physiol. Plant* **110**, 42–51.

Mittler R, Zilinskas BA (1991) Purification and characterization of pea cytosolic ascorbate peroxidase. *Plant Physiol.* **97**, 962–968.

Miyake C, Asada K (1994) Ferredoxin-dependent photoreduction of the monodehydroascorbate radical in spinach thylakoids. *Plant Cell Physiol.* 35, 539–549.

Moser O, Kanellis AK (1994) Ascorbate oxidase of *Cucumis melo* L var reticulatus – purification, characterization and antibody-production. *J. Exp. Bot.* 45, 717–724.

Nakano Y, Asada K (1980) Spinach chloroplasts scavenge H_2O_2 on illumination. *Plant Cell Physiol.* **21**, 1295–1307.

Neill SJ, Desikan R, Clarke A, Hancock JT (2002a) Nitric oxide is a novel component of abscisic acid signaling in stomatal guard cells. *Plant Physiol.* **128**, 13–16.

Neill SJ, Desikan R, Clarke A, Hurst RD, Hancock JT (2002b) Hydrogen peroxide and nitric oxide as signalling molecules in plants. *J. Exp. Bot.* **53**, 1237–1247.

Noctor G, Foyer CH (1998) Ascorbate and glutathione: keeping active oxygen under control. *Annu. Rev. Plant Physiol. Plant Mol. Biol.* **49**, 249–279.

Novitskaya L, Trevanion SJ, Driscoll S, Foyer CH, Noctor G (2002) How does photorespiration modulate leaf amino acid contents? A dual approach through modelling and metabolite analysis. *Plant Cell Environ.* 25, 821–835.

Pauling L (1971) *Vitamin C and the Common Cold.* W.H. Freeman and Co., San Franscisco, CA.

Pastori GM, Kiddle G, Antoniw J, Bernard S, Veljovic-Jovanovic S, Verrier PJ, Noctor G, Foyer CH (2003) Leaf vitamin C contents modulate plant defense transcripts and regulate genes that control development through hormone signaling. *Plant Cell* **15**, 939–951.

Pignocchi C, Fletcher JE, Barnes J, Foyer CH (2003) The function of ascorbate oxidase (AO) in tobacco (*Nicotiana tabacum* L.) *Plant Physiol.* **132**, 1631–1641.

Pignocchi, C, Foyer CH (2003) Apoplastic ascorbate metabolism and its role in the regulation of cell signaling. *Curr. Opin. Plant Biol.* **6**, 379–389.

Sanmartin M, Drogoudi PA, Lyons T, Pateraki I, Barnes J, Kanellis AA (2003) Over-expression of ascorbate oxidase in the apoplast of transgenic tobacco results in altered ascorbate and glutathione redox states and increased sensitivity to ozone. *Planta* **216**, 918–928.

Schopfer P, Liszkay A, Bechtold M, Frahry G, Wagner A (2002) Evidence that hydroxyl radicals mediate auxin-induced extension growth. *Planta* **214**, 821–828.

Siendones E, Gonzalez-Reyes JA, Santos-Ocana C, Navas PFCR (1999) Biosynthesis of ascorbic acid in kidney bean. L-galactono–γ–lactone dehydrogenase is an intrinsic protein located at the mitochondrial inner membrane. *Plant Physiol.* **120**, 907–912.

Smith IK, Kendall AC, Keys AJ, Turner JC, Lea PJ (1984) Increased levels of glutathione in a catalase-deficient mutant of barley (*Hordeum vulgare* L.). *Plant Sci. Lett.* **37**, 29–33.

Smith IK, Kendall AC, Keys AJ, Turner JC, Lea PJ (1985) The regulation of the biosynthesis of glutathione in leaves of barley (*Hordeum vulgare* L.). *Plant Sci.* **41**, 11–17.

Smirnoff N (2000) Ascorbic acid: metabolism and functions of a multi-facetted molecule. *Curr. Opin. Plant Biol.* **3**, 229–235.

Smirnoff N, Wheeler GL (2000) Ascorbic acid in plants: biosynthesis and function. *Crit. Rev. Biochem. Mol. Biol.* **35**, 291–314.

Szent-Gyorgyi A (1931) On the function of hexuronic acid in the respiration of the cabbage leaf. *J. Biol. Chem.* **90**, 385–393.

Tabata K, Takaoka T, Esaka M (2002) Gene expression of ascorbic acid-related enzymes in tobacco, *Phyochemistry* **61**, 631–635.

Takahama U (1998) Ascorbic acid-dependent regulation of redox levels of chlorogenic acid and its isomers in the apoplast of leaves of *Nicotiana tabacum* L. *Plant Cell Physiol.* **39**, 681–689.

Takahama U, Oniki T (1994) The association of ascorbate and ascorbate oxidase in the apoplast with IAA-enhanced elongation of epicotyls from *Vigna angularis*. *Plant Cell Physiol.* **35**, 257–266.

Takahashi H, Chen Z, Du H, Liu Y, Klessig DF (1997) Development of necrosis and activation of disease resistance in transgenic tobacco plants with severely reduced catalase levels. *Plant J.* **11**, 993–1005.

Trumper S, Follmann H, Haberlein I (1994) A novel dehydroascorbate reductase from spinach chloroplasts homologous to plant trypsin inhibitor. *FEBS Lett.* **352**, 159–162.

Vanacker H, Carver TLW, Foyer CH (1998) Pathogen-induced changes in the antioxidant status of the apoplast in barley leaves. *Plant Physiol.* **117**, 1103–1114.

Veljovic-Jovanovic SD, Pignocchi C, Noctor G, Foyer CH (2001) Low ascorbic acid in the vtc1 mutant of *Arabidopsis* is associated with decreased growth and intracellular redistribution of the antioxidant system. *Plant Physiol.* **127**, 426–435.

Wells WW, Xu DP, Yang Y, Rocque PA (1990) Mammalian thioltransferase (glutaredoxin) and protein disulfide isomerase have dehydroascorbate reductase activity. *J. Biol. Chem.* **265**, 15361–15364.

Wheeler GL, Jones MA, Smirnoff N (1998) The biosynthetic pathway of vitamin C in higher plants. *Nature* **393**, 365–369.

Willekens H, Chamnongpol S, Davey M, Schraudner M, Langebartels C, Van Montagu M, Inzé D, Van Camp, W (1997) Catalase is a sink for H_2O_2 and is indispensable for stress defense in C3 plants. *EMBO J.* **16**, 4806–4816.

Wingate VPM, Lawton MA, Lamb CJ (1988) Glutathione causes a massive and selective induction of plant defense genes. *Plant Physiol.* **87**, 206–210.

Winkler BS (1992) Unequivocal evidence in support of the non-enzymatic redox coupling between glutathione/glutathione disulfide and ascorbic acid/dehydroascorbic acid. *Biochim. Biophys. Acta* **1117**, 287–290.

Yamaguchi K, Mori H, Nishimura M (1995) A novel isozyme of ascorbate peroxidase localized on glyoxysomal and leaf peroxisomal membranes in pumpkin. *Plant Cell Physiol.* **36**, 1157–1162.

Yoshimura K, Ishikawa T, Nakamura Y, Tamoi M, Takeda T, Tada T, Nishimura K, Shigeoka S (1998) Comparative study on recombinant chloroplastic and cytosolic ascorbate peroxidase isozymes of spinach. *Arch. Biochem. Biophys.* **353**, 55–63.

5

Ascorbate and plant growth: from germination to cell death

Laura De Gara

5.1 The ascorbate system: related molecules with different properties, linked by a network of redox reactions

Three different interconvertible metabolites form the ascorbate system: ascorbic acid (ASC), ascorbate free radical (AFR) and dehydroascorbate (DHA). ASC acts as a reductant in both animal and plant cells. Its best-known role is the detoxification of reactive oxygen species (ROS); in particular, ASC peroxidase (APX) is a key ROS scavenger in plant cells. Several APX isoenzymes, differing in their cellular localization and kinetic properties, have been identified in plant cells (for a review see Shigeoka *et al.*, 2002).

ASC also regenerates the lipophilic antioxidant α-tocopherol from its α-tocopheroxyl radical (Asada, 1999), and intervenes in the nonphotochemical dissipation of energy, the mechanism that allows chloroplasts to minimize the formation of ROS in the presence of excess light energy. However, the reducing capability of ASC is not limited to maintaining cellular redox balance. ASC is also required as co-substrate for the catalytic activity of several enzymes. This is the case for the Fe(II)ASC oxidase superfamily, a widely divergent family of structurally related proteins, mainly mono- and di-oxygenase, that comprises enzymes involved in the biosynthesis of plant hormones, such as ethylene and gibberellin (1-aminocyclopropane-1-carboxylic acid (ACC) oxidase and gibberellin 3-β-dioxygenase, respectively) and secondary metabolites (e.g., flavone synthase, naringenin 3-dioxygenase), as well as in the post-transcriptional hydroxylation of amino acid residues (e.g., peptidylprolyl hydroxylase; see De Tullio, Chapter 9).

Multicopper ascorbate oxidases (CuAOX) are another different class of ASC oxidizing enzymes. They catalyze a monovalent ASC oxidation with the concomitant four-electron reduction of di-oxygen to water.

The first product of ASC oxidation is AFR, also known as monodehydroascorbate, semidehydroascorbate, or ascorbyl. AFR has a very high dismutation rate

Vitamin C, edited by Han Asard, James M May and Nicholas Smirnoff

(Asada, 1999). Its spontaneous dismutation is the only characterized reaction generating DHA within cells. In spite of AFR dismutation not being enzymatically regulated, several factors can affect DHA generation, for example the level of ASC oxidation or the pH of subcellular environments where AFR is formed. AFR dismutation is in fact promoted in acidic environments, such as the cell wall, vacuole or thylakoid lumen during photosynthetic electron flow, whereas it is slowed in alkaline compartments, such as the stroma of photosynthetically active chloroplasts. DHA generation is also decreased by AFR reduction to ASC, a reaction catalyzed by NAD(P)H-dependent oxido-reductases (AFRR). Moreover, NADPH or ferredoxin transfers electrons directly to AFR, at least within chloroplasts (Asada, 1999).

DHA production not only decreases the antioxidant power of cells or cellular compartments, but also affects specific metabolic pathways, probably also acting as a redox sensor of environmental conditions (De Pinto *et al.*, 1999, see Section 5.3.1). The DHA levels within cells are under enzymatic control, as DHA is reduced to ASC by a GSH-dependent DHA reductase (DHAR). Indeed, under normal conditions, the DHA level is much lower than that of ASC. Several proteins with redox properties, such as thioredoxin, glutaredoxin and protein disulfide isomerase, are also able to reduce DHA and/or AFR to ASC, and the presence of two opportunely-spaced cysteine residues seems to be the only factor required for catalyzing DHA reduction (Potters *et al.*, 2002). Moreover, it has been proposed that the same multimeric prolyl hydroxylase uses ASC as a reductant in a catalytic domain and performs AFR/DHA reduction in another catalytic domain (De Gara and Tommasi, 1999 and references therein), thus increasing ASC reducing efficiency. Finally, DHA is reduced by ASC, in a reaction forming two molecules of AFR, with a constant of equilibrium of 10^{-8}, and a second order rate of $10^5 \cdot M^{-1} \cdot s^{-1}$. This means that the increase of one of the three chemical species alters the balance between the three molecules, in a nonenzymatically controlled way.

5.2 Ascorbate metabolism and cellular compartmentalization

The final step of ASC biosynthesis occurs in mitochondria (see Smirnoff *et al.*, Chapter 1). However, ASC is widely utilized in all cell organelles, a complete set of its redox enzymes being present in mitochondria (De Leonardis *et al.*, 1995), microbodies (Del Rio *et al.*, 2002), chloroplasts and cytosol (De Gara and Tommasi, 1999) of almost all analyzed tissues. The cell wall, endoplasmic reticulum, plasma membrane and nucleus also contain ASC, but the distribution of its redox enzyme in these cellular compartments is not so well defined. The redox state of ASC (i.e., the ratio between ASC and ASC + DHA) and the levels of ASC redox enzymes are often indicative of the physiological phase or health of a tissue, since high levels characterize young cells, whereas they are lower in quiescent or senescing tissues (De Gara, *et al.* 1996; De Pinto, *et al.* 2000).

5.2.1 Ascorbate in nonphotosynthetic cells

In photosynthetic cells, the presence of high concentrations of ASC and high activity of stromal and thylakoid-bound forms of APX may be related to the production of superoxide, hydrogen peroxide and singlet oxygen by photosynthesis and photorespiration. Its function in relation to photosynthesis and protection against photo-oxidative stress is discussed in the chapter by Foyer (Chapter 4) and is reviewed by Noctor and Foyer (1998).

ASC metabolism is also active in nongreen cells. ASC-redox enzymes have been purified from both cytosol and mitochondria of potato tubers (Borraccino *et al.*, 1986; De Leonardis *et al.*, 1995, 2000); and their activities have been detected in the roots of different species (Arrigoni *et al.*, 1997; De Tullio *et al.*, 1999; Córdoba-Pedregosa *et al.*, 2003), in nonphotosynthetic tobacco cultured cells (De Pinto *et al.*, 2000) as well as in mature sieve tube (Walz *et al.*, 2002). Interestingly, the steady-state transcription levels of total plastidic APX mRNA has been reported to be almost equal in leaves and roots. However, the ratio between stromal and thylakoidal forms of APX changes in the two organs. It is around 1 : 1 in leaves, whereas in roots only the stromal APX is expressed and active. This could be an obvious consequence of the lack of organized thylakoids in the nonphotosynthetic plastid (Yoshimura *et al.*, 2002). However, the consistent APX expression within root plastids could be a sign of H_2O_2 production within nongreen plastids, through metabolic pathways which are still not well-characterized.

Studies performed with mutants or transgenic plants, in which the enzymes that regulate redox balance in specific organelles are selectively affected, have given further information on the roles that various antioxidants play in different cellular compartments or physiological processes. In plants lacking catalase, the main H_2O_2 detoxification system in microbodies, GSH synthesis, but not ASC biosynthesis, is increased in order to balance the decreased ROS scavenging capability (Smith *et al.*, 1984; Willekens *et al.*, 1997). In contrast, the antisense suppression of 2-cysteine peroxiredoxin, that is part of the chloroplastic antioxidant network, enhances the expression and activity of ASC redox enzymes but does not affect GSH redox enzymes (Baier *et al.*, 2000). These different homeostatic responses could indicate that these two redox systems are differently regulated in different organelles. They also underline that GSH and ASC redox pairs, in spite of cooperating in ROS scavenging, are not perfectly interchangeable.

5.3 Ascorbate involvement in the cellular mechanisms of plant growth

The high level of ASC content and utilization that characterizes meristematic tissues (De Gara *et al.*, 1996) has often been correlated with its involvement in cell division and elongation, the two processes on which plant growth is based.

5.3.1 Ascorbate and the cell cycle

ASC involvement in cell division has been recently reviewed (Potters *et al.*, 2002). The first evidence for ASC requirement in the cell cycle was elegantly reported nearly 20 years ago in a series of studies in which the alteration in ASC levels was correlated with the cell cycle progression of root meristem (Liso *et al.*, 1984, 1998). These studies demonstrated that ASC is required for G1 to S transition, and that depletion of ASC contributes to reduce the mitotic activity of the quiescent center, the cells of which have a prolonged G1 phase. In the maize root quiescent center, ASC is maintained at low levels by a localized expression of CuAOX (Kerk and Feldman, 1995), an observation that suggests that this enzyme regulates ASC levels in specific cells. The increase in ASC content that follows the exogenous application of ASC or L-galactono-1,4-lactone, the immediate precursor of ASC, also stimulates mitotic activity in secondary meristems as well as the production of lateral roots in lupine plantlets (Arrigoni *et al.*, 1997).

The ASC requirement for cell cycle progression has been further confirmed by using tobacco cell cultures. ASC content transiently increases during the exponential phase, mirroring the behavior of mitotic activity. Moreover, stimulation of ASC biosynthesis increases the mitotic index (De Pinto *et al.*, 1999, 2000). In contrast, tobacco cells, in which ASC content has been decreased by antisense suppression of L-galactono-1,4-lactone dehydrogenase (GalLDH), the last enzyme in the D-mannose/L-galactose ASC biosynthetic pathway, have a reduced mitotic index and a prolonged cell cycle with respect to control cells (Tabata *et al.*, 2001). Analysis of synchronized tobacco cells also indicates that cellular changes in the ASC redox balance mark different phases in the cell cycle: a transient increase in DHA occurs during M phase; whereas, the G1 phase has minimal DHA levels. CuAOX regulates ASC/DHA levels during the cell cycle even in cultured cells (Kato and Esaka, 1999). Exogenous addition of DHA to the culture medium increases DHA uptake and slows down cell proliferation by reducing the mitotic index and extending the cell cycle (De Pinto *et al.*, 1999; Potters *et al.*, 2000). In synchronized cells, DHA treatment prolongs S phase (M.C. De Pinto and L. De Gara, unpublished results). Intriguingly, the DHA that enters cells is promptly reduced to ASC, since the net increase in cellular DHA is very low, compared with the increase in ASC. DHA reduction raises the cellular levels of glutathione disulphide and NAD^+, thus suggesting that both these redox pairs are involved in DHA reduction (De Pinto *et al.*, 1999; Paciolla *et al.*, 2001). However, their oxidation is not sufficient to explain the massive DHA reduction observed. Another interesting candidate that is able to shift part of the cellular reducing power towards DHA reduction is the thioredoxin–thiredoxin reductase system. This system maintains proteins involved in the progression of the cell cycle (enzymes, such as ribonucleotide reductase, or redox regulating transcription factors) in the active state (De Pinto *et al.*, 1999; Potters *et al.*, 2002). Indeed, cellular increases in DHA content could slow down cell cycle progression by competing with other pathways for reducing power. It is worth noting that an increase in the ASC oxidation, with a consequent DHA production often occurs under stress conditions (Kampfenkel *et al.*, 1995; Di Cagno *et al.*, 2001; Hernández *et al.*, 2001), during which cell division is blocked (Francis, 1998).

ASC could also affect cell division as it is the co-substrate of peptidyl-prolyl hydroxylase. It has been reported that the activity of this 2-oxoglutarate-dependent di-oxygenase is influenced by ASC availability in the cells, and also that ASC levels are significantly influenced by the activity of this enzyme (Arrigoni *et al.*, 1977; De Gara *et al.*, 1991). When this di-oxygenase is inhibited, cell division is also blocked (Cooper *et al.*, 1994, De Tullio *et al.*, 1999). The other symptoms induced by under-hydroxylation of proteins (alteration in cell wall texture, vesicle deposition, cell plate formation, and cell wall adhesion to plasma membrane as well as an increase in the number of cells halted in metaphase) suggest that, in this case, the block in the cell cycle is correlated to cytoskeleton alteration.

5.3.2 Cell elongation and differentiation

In the last decade particular attention has been paid to the role of the apoplastic ASC system in cell elongation. Different mechanisms through which ASC regulates this process have been proposed. It is generally assumed that wall extensibility is reduced by the secretory peroxidases (POD) involved in the cross-linking of polysaccharides and structural proteins. An intermediate step in the POD-catalyzed crosslinks is the formation of phenoxy-radicals. Since ASC reduces these radicals, its presence in the apoplast could nullify POD activity. Indeed, a negative correlation between ASC and POD has been reported: when an increase in ASC is experimentally induced in plant tissues, a decrease in POD activity also occurs; moreover, in developing organs, cells or apoplastic fluids having high ASC content have low POD activity (De Gara *et al.*, 1996; Sánchez *et al.* 1997; Gonzáles-Reyes *et al.*, 1998, De Tullio *et al.*, 1999; Córdoba-Pedregosa *et al.*, 2003). ASC itself induces formation of hydroxyl radicals promoting the cleavage of cell wall polysaccharides and favoring cell extension (Fry, 1998). An alternative model of ASC-dependent promotion of cell elongation is based on the evidence that AFR and, to a lesser extent, ASC induce energization of the plasma membrane as well as cell wall acidification by H^+-ATPase stimulation. These events increase nutrient uptake with a consequent decrease in cellular osmotic potential that drives cell elongation. Vacuolar enlargement within cells treated with ASC or AFR-generating medium has also been reported (Gonzales-Reyes *et al.*, 1998).

The activity of the apoplastic CuAOX also seems to be positively correlated with cell elongation. Its activity rises in cowpea epicotyls under auxin-induced elongation (Takahama and Oniki, 1994), or under treatments that increase elongation in cultured cells (Kato and Esaka, 1999). More direct evidence for its role in cell elongation is seen in tobacco cells overexpressing pumpkin CuAOX. These cells show a more rapid expansion than wild-type (Kato and Esaka, 2000). However, transgenic tobacco plants overexpressing the cucumber gene have the same size as the wild-type (A. Kanellis, personal communication). Therefore, at this point, further work is needed to clarify the role of CuAOX in cell expansion. Assuming it might have a role, the molecular mechanisms by which CuAOX affects cell elongation are not clear. CuAOX could act as an AFR producer. The production of DHA, occurring by AFR dismutation, could also stimulate cell elongation. DHA could prevent the formation of crosslinks between structural proteins and hemicelluloses by reacting with lysine and arginine residues. Moreover, oxalic acid, produced by DHA degradation reduces the number of calcium bridges between pectin chains, thus increasing wall plasticity, by promoting the formation of calcium oxalate crystals (Lin and Varner, 1991). Since CuAOX reduces molecular oxygen to water, its activity could also hinder the oxidation of phenolic compounds needed for wall stiffening (Takahama and Oniki, 1994).

ASC could also be involved in cell elongation by intervening in peptidyl proline hydroxylation. When structural proteins of cell wall are underhydroxylated, as in the case of ASC depletion or under direct inhibition of peptidyl-proline hydroxylase, they cannot be glycosylated correctly. This causes abnormal cellular or tissue growth because it decreases the interaction between proteins and other structural components of the cell wall (Cooper *et al.*, 1994; De Tullio *et al.*, 1999).

5.3.3 Cell death

As in all multicellular organisms, programmed cell death (PCD) is a key event for normal reproductive and vegetative development of plants. Gamete development, embryogenesis, and formation of tracheary elements are examples of developmental processes that require the activation of this genetically orchestrated program of cellular suicide. PCD is also part of plant defense strategies against external negative factors: it is activated in the hypersensitive response against pathogens or under different abiotic stresses. Disparate signals trigger PCD and PCD shows different phenotypes (cell walls are completely degraded during the PCD forming aerenchyma, but not during vessel formulation; and lytic processes are only very modest in hypersensitive PCD). In spite of this, common events characterize the transduction pathway leading to PCD (Dangl *et al.*, 1996; Fukuda, 2000).

A growing body of data suggests the involvement of ASC metabolisms in this process. In the hypersensitive response, the decrease in APX is one of the metabolic changes associated with PCD activation (Mittler, 2002; De Gara *et al.*, 2003b). Decrease in the ASC and glutathione pools and their shift toward the oxidized forms also characterize the PCD triggered in tobacco cells by simultaneous generation of nitric oxide and hydrogen peroxide (De Pinto *et al.*, 2002). Heat shock induces PCD in the same cells. Also in this case, APX downregulation seems to be one of the earliest steps, as 10 min after PCD induction, APX activity is almost halved (De Pinto and De Gara unpublished results). PCD occurring under hormonal control in aleurone cells after the onset of germination is mediated by ROS and is delayed by treatments with ASC (Bethke and Jones, 2001). During the dehydration phase of kernel maturation, PCD occurs in endosperm cells (Young and Gallie, 2000). During this phase remarkable decreases in the ASC pool and redox state also occur (see Section 5.4.3). Interestingly, when ASC depletion is experimentally delayed in drying kernels, death of endosperm cells is also delayed (A. Paradiso and L. De Gara, unpublished results).

The alterations in the ASC system occurring during PCD could be aimed to allow ROS to increase within cells condemned to death. However, changes in the ASC levels and redox state could also alter the expression of genes involved in PCD. The expression of hundreds of genes is altered by changes in ASC level in the leaves of *vtc1 A. thaliana* mutants (Pastori *et al.*, 2003). Moreover, in animal cells changes in ASC levels seem to alter apoptosis-associated gene expression (Catani *et al.*, 2001).

5.4 Ascorbate and plant growth

5.4.1 Seed germination

The most dramatic changes in ASC metabolism characterizing plant development probably occur during germination of the so-called orthodox seeds. During germination of these seeds, which reach maturity in a highly dehydrated state, cells switch from a quiescent to a very active metabolism. In dry seeds the ASC pool is generally very low and its oxidized form, DHA, is predominant. APX, and other ROS scavenging enzymes, such as superoxide dismutase and catalase, are also absent or have a very low activity. On the other hand, dry seeds are able to reduce DHA and AFR, since they are provided with AFRR and DHAR (De Gara *et al.*, 1997, 2003a; Tommasi *et al.*, 2001). ROS production starts at the very beginning of germination, for both recovery of respiratory activity and mobilization of fatty acid storage. Indeed, efficiency and correct timing in restoring ROS scavenging systems are of great importance in ensuring seedling development. ASC biosynthesis does not start for some hours after imbibition of dry seeds, in spite of GalLDH being active in dry seeds (De Gara *et al.*, 1997; Pallanca and Smirnoff, 1999; Tommasi *et al.*, 2001). This lag phase is probably required for supplying ASC precursors (De Gara *et al.*, 1997). In this period, the required ASC is supplied by an initial reduction of the DHA stored in the seed and by a further efficient recycling of ASC oxidized forms. As soon as ASC biosynthesis is restored, DHA reduction loses relevance because the DHAR activity and the number of proteins performing DHA reduction decrease during germination of both grass kernels and pine seeds (Cakmak *et al.*, 1993; De Gara *et al.*, 1997; Tommasi *et al.* 2001). However, GSH-dependent DHAR activity increases in germinating pea seedlings (Pallanca and Smirnoff, 1999), thus suggesting that the relevance of DHA recycling might be different during the germination of different species. When ASC biosynthesis activation is anticipated by treating wheat kernels with ASC precursors, the decrease in DHAR activity is speeded up (De Gara *et al.*, 1997). APX activity promptly increases during germination in order to cope with ROS production. In wheat, the three different cytosolic APX isoforms that are present in the seedlings progressively appear during germination. Interestingly these APXs have different kinetic characteristics allowing them to better utilize the amount of ASC present when they are activated. The first APX form, appearing when the ASC content is still very low, has the highest affinity for ASC, whereas the other two, activated when the ASC availability is increased, have a lower affinity for the substrate.

5.4.2 Vegetative development

Plant growth is under hormonal control. However, several data collected during the last decade indicate that ASC availability also affects vegetative development. The seedlings of maize single-cross hybrid B73 × Mo17 have a higher rate of growth and productivity than the parental lines. They also have higher activity of all the enzymes of the ASC oxido-reduction, as well as that of GalLDH (De Gara *et al.*, 2000). An increase in growth rate occurs when lupine seedlings are grown in a hydroponic medium containing ASC or galactono-1,4-lactone; under this condition APX activity is also increased. On the other hand, the seedling size is remarkably reduced when the endogenous ASC level is decreased by treating seedlings with lycorine, an inhibitor of ASC biosynthesis (Arrigoni *et al.*, 1997; Imai *et al.*, 1998). A relation between ASC availability and plant growth has been demonstrated using the *vtc1 A. thaliana* mutant. This mutant, having only 30% of the wild-type ASC content, grows more slowly and flowers later than the wild-type (Veljovic-Jovanovic *et al.*, 2001). Besides affecting cell division and elongation, ASC could influence plant growth by altering hormone homeostasis, as it is required as a co-substrate of the Fe(II)ASC oxidases involved in hormone biosynthesis and metabolism (see Section 5.1). The utilization of ASC by these enzymes does not mean in itself that the ASC level regulates the hormone levels, since it is not known whether ASC is the limiting factor in the reactions catalyzed by the Fe(II)ASC oxidase. However, recent data do support ASC involvement in such regulation. During ripening of climacteric fruit, ASC availability seems to be a regulatory factor of ethylene production (Liu *et al.*, 1999). Leaves of the *vtc1* mutant have more abscissic acid (ABA), a growth-inhibiting hormone, than the wild-type. The expression of several genes regulated by ABA-dependent signaling pathways is also altered in *vtc1* mutant (Pastori *et al.*, 2003). Moreover, comparison of transcriptome analyses of *vtc1* and wild-type indicates that ASC levels also affect the metabolism of other hormones by altering the expression of genes involved in their biosynthesis (an ethylene-responsive transcription factor is up-regulated in *vtc1* mutants and the expression of Fe(II)ASC oxidases is sensitive to changes in ASC levels). Alteration in the biosynthesis of a single hormone might be sufficient to modify the ratio between hormones antagonistically acting on plant development (Pastori *et al.*, 2003), as discussed in more detail in the chapter by Foyer (Chapter 4).

5.4.3 Fruit and seed development

Biosynthesis and storage of ASC in fruits is of great relevance to the nutritional properties of this metabolite. Several papers describe changes in ASC levels during maturation of fruits from different species. However, discordance among available data make it difficult to draw a clear picture of ASC behavior during this process. Differences in ASC metabolism during fruit ripening might also be due to the fact that different species follow different maturation models that are genetically determined, but also strongly influenced by the climatic and edaphic conditions in which plants grow.

Oxidative processes accompany fruit ripening as they are required to confer organoleptic characteristics to mature fruits. An increase in ROS, particularly hydrogen peroxide, has often been correlated with the onset of ripening. The ASC system seems to oversee ROS turnover during fruit ripening (Jimenez *et al.*, 2002). There are indications that immature green fruits contain the highest amount of ASC in their tissue. For example, a decrease in ASC content has been reported to coincide with changes in color both in tomato and bell pepper (Yahia *et al.*, 2001; Jimenez *et al.*, 2002). In climacteric fruits, which are characterized by a peak in ethylene production at the onset of climacteric respiration, the decrease in ASC downregulates ACC oxidase, an Fe(II)ASC oxidase responsible for ethylene biosynthesis. In banana fruit a tight correlation between ACC oxidase activity and ASC levels has been reported, both during physiological ripening and in experimental conditions where ASC levels have been altered within ripening tissues (Liu *et al.*, 1999). CuAOX activity also decrease ASC levels. In melon fruit CuAOX is very active in the ovaries, but falls during the first days after anthesis, the period of the most intense cellular proliferation and expansion. In the last ripening phase of these fruits a transient increase in CuAOX expression and activity again occurs (Diallinas *et al.*, 1997). In contrast, a progressive decrease in CuAOX has been described during fruit ripening of zucchini, another Cucurbitaceae (Lin and Varner, 1991). Changes in ASC redox enzymes have been described during ripening of tomato and bell pepper fruits (Schantz *et al.*, 1995; Jimenez *et al.*, 2002). However, understanding the roles played by the ASC system in fruit ripening requires more systematic studies on the changes in redox enzymes occurring in the different cellular compartments that play key roles during the process.

ASC metabolism has also been studied during seed maturation. In this process, ASC utilization varies according to seed development patterns followed by different species. In recalcitrant seeds, which shed from the mother plant with a high water content and are produced mainly by tropical species, the ASC/DHA redox balance and the activities of ASC redox enzymes are maintained at high levels, similar to those characterizing vegetative tissues, throughout the maturation period (Tommasi *et al.*, 1999). On the other hand, in orthodox seeds, the ASC content and redox state are high only until the start of the dehydration process, after which they progressively decrease. During the drying phase of seed maturation the ASC pool shifts towards the oxidized form, DHA becoming predominant and, often, at the end of the process, it is the only form present in the seeds (Arrigoni *et al.*, 1992; De Gara *et al.*, 2003a). During orthodox seed development, the ASC redox enzymes behave differently. In wheat kernels APX seems to be required to the same extent throughout the period of increase in fresh weight (during which a phase of intense mitotic activity precedes cell enlargement and storage of reserve polymers). When kernels activate the dehydration process, APX starts to decrease, until it is not detectable in dry kernels. AFR reductase has the highest activity during the middle period of kernel development. After drying, its activity is still comparable with that found at the beginning of kernel development. DHA reducing activity remarkably increases during kernel development, reaching its maximum level when the storage activity is more intense. During kernel development new DHA reducing proteins appear, as detected by native polyacrylamide gel electrophoresis (De Gara *et al.*, 2003a). It is known that DHA is reduced by protein disulphide isomerase (PDI), this reaction allows PDI to restore two cysteine residues of the catalytic domain in the active oxidized form. Since the activity of PDI is particularly high during the phase of seed storage protein deposition, at least part of the DHA reducing activity detected in this period could be correlated with PDI activity. The maintenance of ASC recycling enzymes during drying maturation is also of physiological relevance. As already mentioned, the fact that dry seeds contain DHA and the ASC recycling enzymes guarantees the ASC supply during germination, before the activation of ASC biosynthesis. ASC recycling enzymes also seem to be involved in the acquisition of the drying tolerance characterizing orthodox seeds. Indeed, when *Vicia faba* orthodox seeds are collected from the mother plants immediately before the start of drying maturation and are experimentally dehydrated, the activity of the ASC recycling enzymes increases. In contrast, in *Ginkgo biloba* seeds (that, like all recalcitrant seeds, are quite sensitive to dehydration) forced drying causes a strong decrease in the activities of these enzymes and a loss in viability (Tommasi *et al.*, 2003). A loss of ROS scavenging capability also seems to be a key point for desiccation sensitivity, at least in some species of recalcitrant seeds (Li and Sun, 1999).

An interesting point concerning the relationship between ASC and seed or fruit maturation is the transport at long distance of ASC. Both seeds and fruits act as sink organs for their ASC supply (Franceschi and Tarlyn, 2002). Mature seeds are endowed with GalLDH (De Gara *et al.*, 1997; Tommasi *et al.*, 2001) but the ASC biosynthetic capability, at least through this enzyme, seems to be acquired by seeds only before their dehydration (Arrigoni *et al.*, 1992). Long distance transport of ASC through phloem from source to sink organs has been recently reported (Franceschi and Tarlyn, 2002); the regulation of this transport towards a specific organ could be another factor that strongly affects plant development.

5.5 Future perspectives and conclusions

Increasing ASC biosynthetic capability in plants has potential benefits for human nutrition and possibly also for plant growth. The evidence presented in this chapter suggests multiple roles for ASC in plant growth and development. However, at this stage, it is not clear if increased ASC synthesis would have an impact on any of these processes and by which molecular mechanisms it acts. Plant genetic manipulation, by which specific enzymes in ASC biosynthesis or utilization pathways have been switched off or increased, as well as the use of *A. thaliana* mutants will provide stronger evidence for the relevance of this molecule in controlling plant growth and will probably be the source of a new vision of ASC involvement in plant cell metabolism.

References

Arrigoni O, Arrigoni-Liso R, Calabrese G (1977) Ascorbic acid requirement for biosynthesis of hydroxyproline proteins in plants. *FEBS Lett.* **82**, 135–138.

Arrigoni O, Calabrese G, De Gara L, Bitonti MB, Liso R (1997) Correlation between changes in cell ascorbate and growth of *Lupinus albus* seedlings. *J. Plant Physiol.* **150**, 302–308.

Arrigoni O, De Gara L, Tommasi F, Liso R (1992) Changes in the ascorbate system during seed development of *Vicia faba* L. *Plant Physiol.* **99**, 235–238.

Asada K (1999) The water–water cycle in chloroplasts: scavenging of active oxygens and dissipation of excess photons. *Annu. Rev Plant Physiol. Plant Mol. Biol.* **50**, 601–639.

Baier M, Noctor G, Foyer CH, Dietz KJ (2000) Antisense suppression of 2-cysteine peroxiredoxin in *Arabidopsis* specifically enhances the activities and expression of enzymes associated with ascorbate metabolisms but not glutathione metabolism. *Plant Physiol.* **124**, 823–832.

Bethke PC, Jones R (2001) Cell death of barley aleurone protoplasts is mediated by reactive oxygen species. *Plant J.* **25**, 19–29.

Borraccino G, Dipierro S, Arrigoni O (1986) Purification and properties of ascorbate free-radical reductase from potato tubers. *Planta* **167**, 521–526.

Cakmak I, Strbac D, Marschner H (1993) Activities of hydrogen peroxide-scavenging enzymes in germinating wheat seeds. *J. Exp. Bot.* **44**, 127–132.

Catani MV, Rossi A, Sabatini S, Levriero M, Melino G, Avigliano L (2001) Induction of gene expression via activator protein-1 in the ascorbate protection against UV-induced damage. *Biochem. J.* **356**, 77–85.

Cooper JB, Heuse JE, Varner JE (1994) 3–4dehydroproline inhibits cell wall assembly and cell division in tobacco protoplasts. *Plant Physiol.* **104**, 747–752.

Córdoba-Pedregosa MC, Córdoba F, Villalba JM, Gonzáles-Reyes JA (2003) Zonal changes in ascorbate and hydrogen peroxide contents, peroxidase, and ascorbate-related enzymes activities in onion roots. *Plant Physiol.* **131**, 1–10.

Dangl JL, Dietrich RA, Richberg MH (1996) Death don't have no mercy: cell death programs in plant–microbe interaction. *Plant Cell* **8**, 1793–1807.

De Gara L, De Pinto MC, Arrigoni O (1997) Ascorbate biosynthesis and ascorbate peroxidase activity during the early stage of wheat germination. *Physiol. Plant.* **100**, 894–900.

De Gara L, De Pinto MC, Paciolla C, Cappetti V, Arrigoni O (1996) Is ascorbate peroxidase only a scavenger of hydrogen peroxide? In *Plant Peroxidases: Biochemistry and Physiology* (eds C Obinger, O Burner, R Ederman, C Penel, H Greppen). University of Geneve, Geneve, pp. 157–162.

De Gara L, De Pinto MC, Moliterni VM, D'Egidio MG (2003a) Redox regulation and storage processes during maturation in kernels of *Triticum durum*. *J. Exp. Bot.* **54**, 249–258.

De Gara L, De Pinto MC, Tommasi F (2003b) The antioxidant systems vis à vis reactive oxygen species during plant–pathogen interaction. *Plant Physiol. Biochem.* (in press).

De Gara L, Paciolla C, De Tullio MC, Motto M, Arrigoni O (2000) Ascorbate-dependent hydrogen peroxide detoxification and ascorbate regeneration during germination of a highly productive maize hybrid: evidence of an improved detoxification mechanism against reactive oxygen species. *Physiol. Plant.* **109**, 7–13.

De Gara L, Tommasi F (1999) Ascorbate redox enzymes: a network of reactions involved in plant development. *Recent Res. Devel. Phytochem.* **3**, 1–15.

De Gara L, Tommasi F, Liso R, Arrigoni O (1991) Ascorbic acid utilization by prolyl hydroxylase *in vivo*. *Phytochem.* **30**, 1397–1399.

De Leonardis S, De Lorenzo G, Borraccino G, Dipierro S (1995) A specific ascorbate free radical in regeneration of ascorbate for scavenging toxic oxygen species in potato tuber mitochondria. *Plant Physiol.* **109**, 847–851.

De Leonardis S, Dipierro N, Dipierro S (2000) Purification and characterization of an ascorbate peroxidase from potato tuber mitochondria. *Plant Physiol. Biochem.* **38**, 773–779.

Del Rio LA, Corpas FJ, Sandalio LM, Palma JM, Gomez M, Barroso JB (2002) Reactive oxygen species, antioxidant systems and nitric oxide in peroxisome. *J. Exp. Bot.* **53**, 1255–1272.

De Pinto MC, Francis D, De Gara L (1999) The redox state of the ascorbate–glutathione pairs as a specific sensor of cell division in tobacco BY-2 cells. *Protoplasma* **209**, 90–97.

De Pinto MC, Tommasi F, De Gara L (2000) Enzymes of ascorbate biosynthesis and ascorbate–glutathione cycle in cultured cells of tobacco Bright Yellow 2. *Plant Physiol. Biochem.* **38**, 541–550.

De Pinto MC, Tommasi F, De Gara L (2002) Changes in the antioxidant systems as part of the signaling pathway responsible for programmed cell death activated by nitric oxide and reactive oxygen species in tobacco Bright-Yellow 2 cells. *Plant Physiol.* **130**, 698–708.

De Tullio MC, Paciolla C, Della Vecchia F, Rascio N, D'Emerico S, De Gara L, Liso R, Arrigoni O (1999) Changes in onion root development induced by the inhibition of peptidyl-prolyl hydroxylase and influence of the ascorbate system on cell division and elongation. *Planta* **209**, 424–434.

Diallinas G, Pateraki I, Sanmartin M, Scossa A, Stilianou E, Panopoulos NJ, Kanellis AK (1997) Melon ascorbate oxidase: cloning of a multigene family, induction during fruit development and repression by wounding. *Plant Mol. Biol.* **34**, 759–770.

Di Cagno R, Guidi L, De Gara L, Soldatini F (2001) Combined cadmium and ozone treatments affect photosynthesis and ascorbate-dependent defences in sunflower. *New Phytol.* **151**, 627–636.

Franceschi VR, Tarlyn NM (2002) L-ascorbic acid is accumulated in source leaf phloem and tranported to sink tissues in plants. *Plant Physiol.* **130**, 649–656.

Francis D (1998) Environmental control of the cell cycle in higher plants. In: *Plant Cell Proliferation and its Regulation in Growth and Development* (eds JA Bryant, D Chiatante). Wiley, New York, pp. 79–98.

Fry SC (1998) Oxidative scission of plant cell wall polysaccharides by ascorbate-induced hydroxyl radicals. *Biochem. J.* **50**, 931–937.

Fukuda H. (2000) Programmed cell death of tracheary elements as a paradigm in plants. *Plant Mol. Biol.* **44**, 245–253.

González-Reyes JA, Córdoba F, Navas P (1998) Involvement of plasma membrane redox systems in growth control of animal and plant cells. In: *Plasma Membrane Redox Systems and their Role in*

Biological Stress and Disease (ed. H Asard, A Bérczi, RJ Caubergs). Kluwer Academic Publishers, Dordrecht, pp. 193–213.

Hernández JA, Ferrer MA, Jiménez A, Ros Barcelò A, Sevilla F (2001) Antioxidant systems and $O_2^{\cdot-}/H_2O_2$ production in the apoplast of pea leaves. Its relation with salt-induced necrotic lesions in minor veins. *Plant Physiol.* **127**, 817–831.

Imai T, Karita S, Shiratori G, Hattori M, Nunome T, Oba K, Hirai M (1998) L-galactono-gamma-lactone dehydrogenase from sweet potato: purification and cDNA sequence analysis. *Plant Cell Physiol.* **39**, 1350–1358.

Jimenez A, Creissen G, Firmin J, Robinson S, Verhoeyen M, Mullineaux P (2002) Changes in oxidative processes and components of the antioxidant system during tomato fruit ripening. *Planta* **214**, 751–758.

Kampfenkel K, Van Montagu M, Inzé D (1995) Effects of iron excess on *Nicotiana plumbaginifolia* plants. *Plant Physiol.* **107**, 725–735.

Kato N, Esaka M (1999) Changes in ascorbate oxidase gene expression and ascorbate levels in cell division and cell elongation in tobacco cells. *Physiol. Plant* **105**, 321–329.

Kato N, Esaka M (2000) Expansion of transgenic tobacco protoplasts expressing pumpkin ascorbate oxidase is more rapid than that of wild-type protoplasts. *Planta* **210**, 1018–1022.

Kerk NM, Feldman LJ (1995) A biochemical model for initiation and maintenance of quiescent centre: implication for organization of root meristems. *Development* **121**, 2825–2833

Li C, Sun WQ (1999) Desiccation sensitivity and activities of free radical-scavenging enzymes in recalcitrant *Theobroma cacao* seeds. *Seed Sci. Res.***9**, 209–217.

Lin L, Varner JE (1991) Expression of ascorbic acid oxidase in zucchini squash (*Cucurbita pepo* L.). *Plant Physiol.* **101**, 969–976.

Liso R, Calabrese G, Bitonti MB, Arrigoni O (1984) Relationship between ascorbic acid and cell division. *Exp. Cell Res.* **150**, 314–320.

Liso R, Innocenti AM, Bitonti MB, Arrigoni O (1988) Ascorbic acid-induced progression of quiescent centre cells from G_1 to S phase. *New Phytol.* **110**, 469–471.

Liu X, Shiomi S, Nakatsuka A, Kubo Y, Nakamura R, Inaba A (1999) Characterization of ethylene biosynthesis associated with ripening in banana fruit. *Plant Physiol.* **121**, 1257–1265.

Mittler R (2002) Oxidative stress, antioxidants and stress tolerance. *Trends Plant Sci.* **7**, 405–410.

Noctor G, Foyer CH (1998) Ascorbate and glutathione: keeping active oxygen under control. *Annu. Rev Plant Physiol. Plant Mol. Biol.* **49**, 249–279.

Pallanca JE, Smirnoff N (1999) Ascorbic acid metabolism in pea seedlings. A comparison of D-glucosone, L-sorbosone, and L-galactono-1,4-lactone as ascorbate precursors. *Plant Physiol.* **120**, 453–461.

Pastori GM, Kiddle G, Antoniw J, Bernard S, Veljovic-Jovanovic S, Verrier PJ, Foyer CH (2003) Leaf vitamin C contents modulate plant defense transcripts and regulate genes controlling development through hormone signaling. *Plant Cell* **15**, 939–951.

Paciolla C, De Tullio MC, Chiappetta A, Innocenti AM, Bitonti B, Liso R, Arrigoni O (2001) Short- and long-term effects of dehydroascorbate in *Lupinus albus* and *Allium cepa* roots. *Plant Cell Physiol.* **42**, 857–863.

Potters G, De Gara L, Asard H, Horemans N (2002) Ascorbate and glutathione: guardians of the cell cycle, partners in crime? *Plant Physiol. Biochem.* **40**, 537–548.

Potters G, Horemans N, Caubergs RJ, Asard H (2000) Ascorbate and dehydroascorbate influence cell cycle progression in *Nicotiana tabaccum* cell suspension. *Plant Physiol.* **124**, 17–20.

Sánchez M, Queijeiro E, Revilla G, Zarra I (1997) Changes in ascorbic acid levels in apoplastic fluid during growth of pine hypocotyls. Effect on peroxidase activities associated with cell walls. *Physiol. Plant* **101**, 815–820.

Schantz ML, Schreiber H, Guillemaut P, Schantz R (1995) Changes in ascorbate peroxidase activities during fruit ripening in *Capsicum annuum*. *FEBS Lett.* **358**, 149–152.

Shigeoka S, Ishikawa T, Tamoi M, Miyagawa Y, Takeda T, Yabuta Y, Yoshimura K (2002) Regulation and function of ascorbate peroxidase isoenzymes. *J. Exp. Bot.* **53**, 1305–1319.

Smith IM, Kendall AC, Keys AJ, Turner CJ, Lea, PJ (1984) Increased levels of glutathione in a catalase-deficient mutant of barley (*Hordeum vulgare* L.). *Plant Sci. Lett.* **37**, 29–33.

Tabata K, Ôba K, Suzuki K, Esaka M (2001) Generation and properties of ascorbic acid-deficient transgenic tobacco cells expressing antisense RNA for L-galactono-1,4-lactone dehydrogenase. *Plant J.* **27**, 139–148.

Takahama U, Oniki T (1994) The association of ascorbate and ascorbate oxidase in the apoplast with IAA-enhanced elongation of epicotyl from *Vigna angularis*. *Plant Cell Physiol.* **35**, 257–266.

Tommasi F, Paciolla C, Arrigoni O (1999) The ascorbate system in recalcitrant and orthodox seeds. *Physiol. Plant* **105**, 193–198.

Tommasi F, Paciolla C, De Pinto MC, De Gara L (2001) A comparative study of glutathione and ascorbate metabolism during germination of *Pinus pinea* L. seeds. *J. Exp. Bot.* **52**, 1647–1654.

Tommasi F, Paciolla C, De Pinto MC, De Gara L (2003) Relationships between water levels and ascorbate peroxidase–ascorbate recycling enzymes in recalcitrant and orthodox seeds. In *Plant Peroxidase Biochemistry and Physiology* (eds M Acosta, JN Rodríguez-López, MA Pedreño). Press Service, University of Murcia, Mursia, pp. 209–213.

Veljovic-Jovanovic SD, Pignocchi C, Noctor G, Foyer C (2001) Low ascorbic acid in the *vtc-1* mutants of *Arabidopsis* is associated with decreased growth and intracellular redistribution of the antioxidant system. *Plant Physiol.* **127**, 426–435.

Walz C, Juenger M, Schad M, Kehr J (2002) Evidence for the presence and activity of a complete antioxidant defence system in mature sieve tubes. *Plant J.* **31**: 189–197.

Willekens H, Chammongpol S, Davey M, Schraudner M, Langebartels C, Van Montagu M, Inzè D, Van Camp W (1997) Catalase is sink for H_2O_2 and it is indispensable for stress defence in C_3 plants. *EMBO J.* **16**, 4806–4816.

Yahia EM, Contreras-Padilla M, González-Aquilar G (2001) Ascorbic acid content in relation to ascorbic acid oxidase activity and polyamine content in tomato and bell pepper fruits during development, maturation and senescence. *Food Sci. Technol.* **34**, 452–457.

Yoshimura K, Yabuta Y, Ishikawa T, Shigeoka S (2002) Identification of a cis element for tissue-specific alternative splicing of chloroplast ascorbate peroxidase pre-mRNA in higher plants. *J. Biol. Chem.* **277**, 40623–40632.

Young TE, Gallie DR (2000) Programmed cell death during endosperm development. *Plant Mol. Biol.* **44**, 283–301.

6

Vitamin C transport in animals and plants

John X. Wilson

6.1 Introduction

The aim of this chapter is to describe and compare the vitamin C transport mechanisms of higher plants and vertebrate animals. Some species, such as humans, must acquire vitamin C from food and all may absorb the vitamin when it is available in the environment. Transport systems associated with the plasma membrane of cells mediate absorption. Additional transport systems distribute ascorbic acid (ASC, the reduced form of vitamin C) and its reversibly oxidized metabolite, dehydroascorbic acid (DHA), among a large number of cells and subcellular compartments, thereby regulating the concentration and redox state of vitamin C in each one. Membranes act as selective permeability barriers around cells and organelles. The selectivity of the membranes depends on the solubility properties of their lipid bilayers and activities of protein-based transport systems. Simple diffusion, facilitated diffusion and active mechanisms all contribute to the membrane transport of vitamin C.

6.2 Simple diffusion

6.2.1 Animals

ASC is a lactone ($C_6H_8O_6$) with a molecular weight of 176. In solution the hydroxyl groups at positions 2 and 3 ionize with pK values of 4.17 and 11.57. Therefore reduced vitamin C exists predominantly as the ascorbate (ASC) anion in most plant and animal fluids. A consequence of ionization is high solubility in water. Molecules that are comparably water-soluble but smaller and nonionized, such as ethanol and glycerol, diffuse quickly through nonspecific pathways in cell membranes, especially through the lipid bilayer. In contrast ASC, because of its size and charge, does not readily permeate the lipid bilayer.

Weak organic acids such as propionic acid enter cells rapidly by simple diffusion of their undissociated forms. Once in the cytoplasm, these acids dissociate into organic ions (e.g., propionate) and proton. This mechanism is often referred to as 'ion

Vitamin C, edited by Han Asard, James M May and Nicholas Smirnoff

trapping'. To examine whether appreciable uptake of AA might occur this way, rat bone-derived osteoblasts were incubated with 5 mM sodium ASC or sodium propionate at pH 7.3 (Wilson and Dixon, 1989). Cytoplasmic pH was monitored to detect uptake of the acids. Sodium propionate was observed to cause rapid acidification of the cytoplasm whereas ASC had little effect. In contrast to the failure of osteoblasts to take up appreciable amounts of undissociated AA, these cells rapidly increase their intracellular concentration of ASC through plasma membrane transport systems that translocate either ASC anion (Wilson and Dixon, 1989) or DHA (Qutob *et al.*, 1998). Therefore simple diffusion across animal cells' plasma membranes at physiological pH comprises only a slow component of vitamin C uptake, a component which may be negligibly small in those cells that contain ASC or DHA transport systems of high affinity and/or capacity. Exceptions may occur where membranes are exposed to substantial amounts of nonionized (protonated) AA in highly acidic fluids such as gastric juice.

DHA differs from AA in being unable to ionize. Since the oil : water distribution coefficient of DHA is similar to that of mannitol, which is excluded from most animal cells so effectively that it can be used as an osmotic diuretic, DHA is evidently a hydrophilic molecule that cannot readily permeate lipid bilayers (Rose, 1987).

6.2.2 Plants

ASC is present in all green plants except dry seeds and dormant buds (Davey *et al.*, 2000; Potters *et al.*, 2002). Plant intracellular oxygen concentrations are high, because of the activity of chloroplasts, and ASC functions as an essential antioxidant. It is found in many intracellular compartments: the mitochondrion, cytoplasm, chloroplast, peroxisome, vacuole and extracellular matrix (apoplast). The ASC concentration in the cytoplasm and chloroplast may reach as high as 10–50 mM, while the apoplast contains 0.2–2 mM (Foyer and Lelandais, 1996; Luwe *et al.*, 1993; Rautenkranz *et al.* 1994; Takahama, 1993).

ASC is synthesized in mitochondria and then distributed to other subcellular compartments. The mechanism of vitamin C transport out of mitochondria has not been characterized. The mechanisms of vitamin C flux through vacuolar, chloroplast and plasma membranes have been partially elucidated. Regulation of the contents and volume of vacuoles in plant cells depends on the coordinated activities of transporters and channels located in the vacuolar membrane (tonoplast) (Maeshima, 2001). Accumulation of ASC in the vacuole lumen is achieved by simple diffusion through the surrounding tonoplast membrane (Rautenkranz *et al.*, 1994). Chloroplasts contain multiple grana and each granum consists of 5–30 membrane discs or sacks called thylakoids. Accumulation of ASC in the lumen of these sacks is achieved by simple diffusion through the thylakoid membranes (Foyer and Lelandais, 1996). It is likely that vitamin C is in the form of AA when it permeates through the tonoplast and thylakoid membranes.

Exogenous vitamin C has been shown to enter plants and increase intracellular ASC content (Maggio *et al.*, 2002; Paciolla *et al.*, 2001). Additionally, the vitamin C that is synthesized by source cells may eventually be absorbed into other cells within the plant. Membrane transport and bulk flow are required for this redistribution. The cytoplasmic compartments of many cells in each individual plant are connected by plasmodesmata to form a continuous pathway called the symplast. Thus vitamin C may diffuse from one connected cell to another. Additionally, the nutrient may travel to distant parts of the plant if it is carried there by the phloem. This vascular tissue distributes dissolved molecules from the places where the molecules are produced (sources) to the places where they are utilized or stored (sinks). Radioactivity quickly becomes distributed to many parts of a plant after radiolabeled vitamin C is applied to roots or leaves (Mozafar and Oertli, 1993). Although facilitative transporters for vitamin C may exist in phloem, none have yet been identified there. It can be inferred that vitamin C moves by diffusion and bulk flow from plant's source tissues – where it is absorbed from the environment or synthesized – to the growth and storage tissues that function as sinks.

Sinks receive dissolved assimilates by bulk flow through the phloem that is driven by hydrostatic pressure gradients (Patrick *et al.*, 2001). These gradients along the phloem pathway are created by osmotic water movement, which is driven by membrane transport of solutes. It is not known to what extent vitamin C transport by cells may contribute to the osmotic gradient.

The long distance transport of vitamin C is directed to specific targets because the ASC synthesized in source leaves is transported through the phloem to sink tissues initially (Franceschi and Tarlyn, 2002). Radiolabeled ASC that was applied initially to the mature (source) leaves of plants (*A. thaliana*, *Impatiens walleriana*, *Medicago sativa*) was found to accumulate in the phloem. Whole-plant autoradiography demonstrated that over 12–24 h, radioactivity from source leaves was translocated to young, rapidly growing sink tissues but not to adjacent source leaves. The same pattern of targeted delivery was seen for [1-^{14}C]- and [6-^{14}C]-labeled ASC, indicating that the transported molecule was not a degradation product of vitamin C. Similarly, application of L-galactono-1,4-lactone (the immediate precursor of AA) to source leaves led to an elevation of ASC content in the adjacent young – but not mature – tissues. L-Galactono-1,4-lactone apparently was not transported, since it was not detectable in stems or sink tissue. Delivery of ASC by phloem may regulate sink tissues' development since this antioxidant is required for cell division and growth (Franceschi and Tarlyn, 2002).

Efflux of reduced vitamin C from cells must occur as part of the redistribution of the antioxidant from source to sink tissues. Aquatic plants also release vitamin C to their aqueous environment, where it is found in the form of ASC (Running *et al.*, 2002). ASC may move from the cytoplasm to apoplast by simple diffusion through the plasma membrane and then dissociate to ASC – indeed this pH-sensitive mechanism of ion trapping appears to be stimulated by alkalinization of the apoplast when bean plants are exposed to ozone (Bichele *et al.*, 2000). However, in addition to simple diffusion, plant membranes exhibit carrier-mediated transport of ASC and DHA that is described in Section 6.3.2.

6.3 Facilitated diffusion

6.3.1 Animals

Vitamin C enters and exits cells and subcellular compartments by means of interactions with specific proteins as well as by simple diffusion. These proteins are, or are component parts of, transport systems that work by facilitated diffusion or active transport. Facilitated diffusion achieves net movement only in the direction of a chemical or electrochemical gradient of the transported solute. Active transport can move solute against this gradient by using energy derived from cellular metabolism. The following paragraphs review facilitated diffusion of DHA and ASC.

Utilization of ASC as an antioxidant and enzyme cofactor causes its oxidation to DHA in extracellular fluid and cells. DHA has been demonstrated to play important roles in animal and plant cells because it can be used to regenerate ASC and, directly or indirectly, change the redox state of many other molecules (Wilson, 2002). For instance in humans, who cannot synthesize vitamin C and depend on dietary sources, ingesting either ASC or DHA raises serum ASC concentration to similar extents. Thus DHA can serve as a dietary source of vitamin C. Indeed, there is a widely held understanding that dietary ASC and DHA possess roughly equivalent bioavailability (i.e., the fraction of the ingested amount that has the potential to meet the functional requirements of tissues for vitamin C). Evidently cellular mechanisms of transport and metabolism convert DHA to ASC at appropriate locations and rates to fulfill the human requirement for vitamin C. It is because both molecules are commonly thought to be bioavailable that the vitamin C content of foods is commonly reported as the sum of ASC and DHA contents. The bioavailability of nutrients depends on their absorption, distribution and metabolism within the body, and excretion. These processes are discussed next.

The absorptive epithelium of the intestine is comprised of polarized cells. These cells remove vitamin C from the intestinal lumen and release it toward the blood. The human small intestine absorbs both reduced and oxidized vitamin C. DHA can be degraded by bicarbonate at alkaline pH but is unlikely to meet these destabilizing conditions in the lumen of the small intestine, where the alkaline secretions from the pancreas and duodenal glands mix with the acidic gastric juice to form a slightly acidic fluid.

Absorption sites for ASC and DHA are found along the entire length of the small intestine, as has been shown by measuring transport activities in luminal (brush border) membrane vesicles (Malo and Wilson, 2000). The mechanism of DHA uptake by luminal membranes of human jejunum has pharmacological characteristics that differ distinctly from those of ASC uptake. However, for both the DHA and ASC transport systems, initial rates of uptake saturate with increasing external substrate concentration, reflecting high affinity interactions that can be described by Michaelis–Menten kinetics. Sodium-independent carriers take up DHA by facilitated diffusion and these are distinct from the sodium-dependent transporters of ASC. The transport systems' affinities are reflected in the substrate concentrations required for half-maximal flux, which are the apparent K_m (i.e., Michaelis constant). Although DHA is taken up with lower affinity than ASC (the apparent K_m values are 0.8 and 0.2 mM, respectively), the maximal rates of uptake are similar for both when glucose is absent. Furthermore glucose inhibits ASC but not DHA uptake, so that glucose derived from food may increase the relative bioavailability of the oxidized form of vitamin C (Malo and Wilson, 2000). Human enterocytes contain enzymes that convert DHA to ASC (Buffinton and Doe, 1995). These enzymes keep intracellular DHA at low levels and thereby maintain a gradient favoring its continued uptake across the plasma membrane.

The role of intestinal DHA absorption may be especially important during gastritis. Chronic gastritis decreases the ASC concentration in gastric juice but does not cause scurvy (Schorah, 1992). It seems likely that inflammation in the gastric mucosa accelerates oxidation of ASC to DHA but intestinal absorption of the latter prevents the patients from becoming scorbutic. Furthermore, any DHA that escapes reduction in enterocytes and enters the blood may be taken up and reduced to ASC by other cell types.

Numerous cell types can clear DHA from the extracellular fluid. Among those that take up DHA and reduce it to ASC are adipocytes, neutrophils, erythrocytes, smooth muscle cells, granulosa cells, hepatocytes, astrocytes and osteoblasts (Daskalopoulos *et al.*, 2002; Himmelreich *et al.*, 1998; Holmes *et al.*, 2002; Kodaman and Behrman, 1999; Mendiratta *et al.*, 1998; Qutob *et al.*, 1998; Upston *et al.*, 1999; Vera *et al.*, 1998). DHA uptake is not mediated by sodium–ASC cotransporters (described below) because it is neither dependent on sodium nor blocked by an antagonist of sodium–ASC cotransport (4,4′-diisothiocyanatostilbene-2,2′-disulfonic acid) (Daskalopoulos *et al.*, 2002). Many cells possess both sodium–ASC cotransporters and the ability to take up and reduce DHA. In contrast, human neutrophils lack sodium–ASC cotransporters but incubation of these myeloid cells with DHA *in vitro* increases their intracellular ASC concentration to levels equal to those found in mature neutrophils *in vivo* (Vera *et al.*, 1998). Similarly, in rat astrocyte cultures exogenous DHA and ASC are capable of raising intracellular ASC concentration to the same level (Daskalopoulos *et al.*, 2002; Siushansian *et al.*, 1997).

In bone the oxidation and regeneration of ASC may be an important way for the osteoid-resorbing activity of osteoclasts to stimulate the osteoid-forming activity of osteoblasts (Wilson, 2002). Osteoclasts are sources of reactive oxygen species that oxidize extracellular vitamin C. As DHA is produced in the extracellular fluid, it may be taken up by a sodium-independent, facilitated transport system in neighboring osteoblasts and reduced to ASC just in time and place to support the synthesis of extracellular matrix by those cells (Qutob *et al.*, 1998). Intracellular ASC stimulates the osteoblasts to produce collagenous osteoid that subsequently becomes calcified (Franceschi *et al.*, 1995). Thus DHA uptake may couple bone resorption and formation and thereby contribute to the remodeling and repair of this tissue.

In animals, the influence of glucose on DHA uptake varies between cell types. Some are acutely inhibited by physiological concentrations of glucose (adipocytes, erythrocytes, granulosa cells, neutrophils, osteoblasts, smooth muscle cells), while others are less sensitive (erythrocytes, astrocytes), and some cells' DHA uptake is not changed detectably by glucose (luminal membranes of intestinal enterocytes and renal tubular cells) (Daskalopoulos *et al.*, 2002; Himmelreich *et al.*, 1998; Holmes *et al.*, 2002; Malo and Wilson, 2000, 2001; Qutob *et al.*, 1998; Rumsey *et al.*, 2000; Kodaman and Behrman 1999; Vera *et al.*, 1998).

DHA competes with glucose for uptake through the mammalian facilitative glucose transporters GLUT1, GLUT3 and GLUT4 (Rumsey *et al.*, 1997, 2000; Vera *et al.*, 1998; Wilson *et al.*, 1999). GLUT proteins do not transport ASC or AA. Generally the results obtained with mammalian transporters expressed experimentally in *Xenopus* oocytes are consistent with the hypothesis that GLUT1 and GLUT3 transport DHA with apparent affinities and maximal capacities similar to their transport of glucose. Agents that enhance the activity of endogenous facilitative glucose transporters in mammalian cells also increase the initial rate of DHA uptake. Examples are cyclic AMP in astrocytes (Siushansian *et al.*, 1997), insulin and insulin-like growth factor I in osteoblasts (Qutob *et al.*, 1998), insulin-like growth factor I and follicle stimulating hormone in granulosa cells (Kodaman and Behrman, 1999), and colony-stimulating factors in neutrophils (Vera *et al.*, 1998).

It is clear DHA and ASC transporters can be regulated independently of each other. For instance, colony-stimulating factors enhance DHA uptake in human neutrophils that lack sodium–ASC cotransporters (Vera *et al.*, 1998). Bone-derived osteoblasts provide another intriguing example: transforming growth factor-β increases the maximal rate of ASC uptake through sodium–ASC cotransporters, but does not stimulate DHA uptake (Wilson and Dixon, 1995) while insulin increases the maximal rate of DHA uptake through glucose transporters without changing sodium–ASC cotransport activity (Qutob *et al.*, 1998). Thus transforming growth factor-β and insulin both increase intracellular ASC concentration in osteoblasts but act through different vitamin C transport systems.

An excess of glucose, such as occurs during the hyperglycemia of uncontrolled diabetes or sepsis, may competitively block most DHA uptake through facilitative glucose transporters and thus impair the clearance of DHA by some cell types. However the glucose analogs and glucose transporter antagonists (blockers) used experimentally to characterize DHA transport also have confounding effects. Glucose analogs and antagonists may interfere with the reduction of DHA to ASC that normally keeps the cytoplasmic concentration of DHA low and thus maintains a gradient favoring DHA uptake. For instance a nonmetabolizable glucose analog, 2-deoxyglucose, inhibits DHA uptake by THP-1 monocytic cells in a noncompetitive manner (Laggner *et al.*, 1999). 2-Deoxyglucose blocks glucose uptake and metabolism by the cells and may thereby diminish the production of NADPH that is required for DHA reduction. Consequently DHA might be taken up through a glucose-insensitive transporter, fail to be converted to ASC, and then exit the cells during the wash (with DHA-free solution) at the end of the uptake period. It is consistent with this hypothesis that DHA has been shown to exit erythrocytes much faster than ASC (May *et al.*, 1995), indicating that oxidized vitamin C can be lost relatively easily from the cells.

In addition to direct competition between glucose and DHA for binding sites on transporters, chronic changes in glucose supply can have noncompetitive effects on the amount of DHA uptake mediated by glucose transporters. For example, glucose pretreatment slows DHA uptake (measured during 30–90 s transport assays in glucose-free medium) by the L6 skeletal muscle cell line by downregulating facilitative glucose transporters in the plasma membrane, principally GLUT1 (Wilson *et al.*, 1999). Conversely, overnight incubation in low glucose medium upregulates the facilitative glucose transporters and increases the cells' DHA transport capacity (Wilson *et al.*, 1999).

Hormonal dysregulation may also contribute to local deficiencies in DHA recycling within tissues. Normally insulin increases the maximal rate of DHA transport by facilitative glucose transporters in target cells such as osteoblasts, thereby raising intracellular ASC concentration (Qutob *et al.*, 1998). Insulin insufficiency may impair DHA uptake through facilitative glucose transporters during type I diabetes. Insulin-like growth factor I can activate insulin receptors to stimulate cellular uptake of DHA but has only one-tenth the potency of insulin (Qutob *et al.*, 1998). Indeed the maximal rate of DHA uptake is decreased in lymphoblasts from patients with type I diabetes and nephropathy (Ng *et al.*, 1998). Slowing of DHA uptake impairs regeneration of ASC. For instance ASC concentration is decreased and DHA is increased in the sciatic nerve of rats made diabetic by streptozotocin (Obrosova *et al.*, 2001). Moreover, because intracellular ASC is required for collagen synthesis by osteoblasts (Franceschi *et al.*, 1995), deficient recycling of DHA to ASC in this bone cell type may contribute to the development of osteopenia. Support for this hypothesis comes from the observation that feeding ASC to diabetic pregnant rats improves skeletal development in their offspring (Braddock *et al.*, 2002).

DHA uptake occurs through a glucose-insensitive mechanism in rat astrocytes (Daskalopoulos *et al.*, 2002) and human erythrocytes, enterocytes and renal epithelial cells (Himmelreich *et al.*, 1998; Malo and Wilson, 2000, 2001). This pathway is discussed next.

DHA is filtered from the plasma at the renal glomerular capillaries and then reabsorbed across the luminal membranes of the epithelial cells that comprise the renal proximal tubules. These cells are the principal sites at which vitamin C is removed from the lumen of the renal tubule and translocated toward the blood. Studies of brush border membrane vesicles prepared from rat renal cortex, which are representative of the luminal membranes of proximal tubule cells, have shown that ASC uptake occurs through sodium–ASC cotransport while DHA uptake occurs by a sodium-independent process (*Table 6.1*) (Malo and Wilson, 2001). Neither ASC nor DHA uptake is inhibited by glucose; hence the facilitative and sodium-dependent glucose transporters present in the luminal membranes cannot be the routes of vitamin C uptake. Perhaps a previously unidentified, specific DHA transporter exists in epithelial cells.

Table 6.1: Pharmacological properties of ASC and DHA transport by brush border membrance vesicles isolated from renal cortex.

Treatment	[^{14}C]ASC uptake	[^{14}C]DHA uptake
Vehicle control	100	100
ASC (1 mM)	30 ± 5*	92 ± 14
DHA (5 mM)	Not determined	35 ± 3*
Sodium-free	19 ± 5*	93 ± 27
Sulfinpyrazone (3 mM)	22 ± 13*	85 ± 14
Phloretin (1 mM)	15 ± 6*	87 ± 18
Phlorizin (1 mM)	93 ± 7	72 ± 17
Glucose (5 mM)	90 ± 23	105 ± 26
Fructose (50 mM)	107 ± 3	95 ± 27

Brush border membrane vesicles were prepared from the renal cortex of adult male rats. DHA and [^{14}C]DHA were synthesized by incubating ASC and [^{14}C]ASC, respectively, with ASC oxidase. Initial rates of uptake of [^{14}C]ASC (200 μM) and [^{14}C]DHA (200 μM) were determined by measuring uptake at nine time points during 30-s incubations (37°C). The uptake buffer was glucose-free except where otherwise indicated. Shown are mean ± standard deviation values of initial rates expressed as percentages of control rates [100% = 29 ± 6 pmol ASC·mg^{-1} protein·s^{-1} (n = 5) and 17 ± 5 pmol DHA·mg^{-1} protein·s^{-1} (n = 7)]. *$P < 0.05$ compared to control (data from Malo and Wilson, 2001).

Uptake of DHA (5–200 μM) by brain-derived astrocytes is inhibited only partially by a high concentration of glucose (10 mM) (Daskalopoulos *et al.*, 2002). The remaining, glucose-insensitive accumulation of intracellular ASC from DHA is blocked by phloretin and cytochalasin B, which are antagonists of facilitative glucose transporters, but it is also inhibited reversibly by sulfinpyrazone. Astrocytes are the most numerous cells in animal brain. These non-neuronal, glial cells regulate the composition of extracellular fluid and thus influence the environment and activity of neurons. DHA is lethal to neurons in the absence of astrocytes (Song *et al.*, 2001), whereas the latter cells are capable of regenerating ASC from DHA without ill effect (Daskalopoulos *et al.*, 2002; Siushansian *et al.*, 1997). Brain cells produce cyclic AMP in response to neurotransmitters and ischemia and this intracellular messenger stimulates astrocytic uptake of DHA and accumulation of ASC (Siushansian *et al.*, 1997). Thus astrocytes may clear neurotoxic DHA from extracellular fluid using a transport system that is insensitive to inhibition by glucose and upregulated by cyclic AMP.

Little is known about how vitamin C is transported out of cells. However, efflux from enterocytes and renal tubular cells, directed to the blood, is essential for intestinal absorption and renal conservation of vitamin C. One hypothesis is that ASC diffuses from the cytoplasm to the extracellular fluid through volume-sensitive anion channels in the epithelial cells' basolateral membrane and then enters the blood plasma through discontinuities in the capillary wall. There are two lines of evidence in support of this idea. First, epithelial cells swell markedly during transepithelial transport of nutrients; for example, sodium-dependent absorption of glucose or alanine causes a sustained increase in cell volume in enterocytes and renal tubular cells (Beck and Potts, 1990; Breton *et al.*, 1996; Macleod *et al.*, 1992). Second, volume-sensitive anion channels are permeant to ASC (Furst *et al.*, 2002). However, the molecular identities of the proteins mediating ASC transport across the basolateral membranes of intestinal and renal epithelia have yet to be determined.

ASC efflux also occurs from nonepithelial animal cells. Brain astrocytes rapidly release ASC when stimulated appropriately *in vitro*. For instance swelling astrocytes, with either hypotonic medium or glutamate, transiently and reversibly permits large quantities of cytoplasmic ASC to diffuse to the extracellular fluid (Siushansian *et al.*, 1996; Wilson *et al.*, 2000). It has been inferred from the osmotic and pharmacological characteristics that volume-sensitive anion channels in the plasma membrane mediate this efflux. Thus astrocytes condition the extracellular fluid for neurons by clearing DHA and replacing ASC. Similarly, adrenal chromaffin cells release vitamin C from multiple intracellular compartments to the extracellular fluid when incubated with high-potassium solutions that depolarize and likely also swell the cells (Diliberto *et al.*, 1987).

Glutamate uptake into brain cells triggers ASC efflux (Rebec and Pierce, 1994; Yusa, 2001). *In vitro* experiments have shown that astrocytes release ASC when exposed to glutamate (Wilson *et al.*, 2000). Astrocytes are in close contact with neuronal synapses and respond to the glutamate released there. Thus a demand-driven component of ASC efflux from astrocytes may appear just in time and place to reach active neurons. A putative glutamate–ASC heteroexchanger has been proposed to mediate the counter movements of glutamate and ASC (Rebec and

Pierce, 1994). However it is more likely that glutamate uptake through sodium–glutamate cotransporters, located in astrocyte processes (endfeet), causes swelling of the endfeet and thereby activates a volume-sensitive efflux pathway for ASC (Wilson *et al.*, 2000).

Incubation of cells with extracellular vitamin C stimulates efflux of preloaded [^{14}C]vitamin C. This effect has been interpreted as transtimulation of ASC efflux by extracellular ASC, which is supposed to enter cells through a homeoexchange transport system. Examples have been reported for various animal cells of both epithelial and nonepithelial types (Finn and Johns, 1980; Khatami *et al.*, 1986; Socci and Delamere, 1988). However the existence of ASC homeoexchange systems has not been demonstrated conclusively and the molecular identities of the membrane proteins responsible for homeoexchange have not been determined.

Finally, cellular uptake and reduction of DHA leads to ASC efflux. Human erythrocytes (Mendiratta *et al.*, 1998), HepG2 liver cells (Upston *et al.*, 1999) and astrocytes (J. X. Wilson, unpublished results) take up DHA, reduce it intracellularly and subsequently release ASC to the extracellular fluid. It is not known if a single transporter exchanges extracellular DHA for intracellular ASC or if different transporters mediate uptake than efflux. Nevertheless this recycling mechanism allows the reducing equivalents derived from cell metabolism to be transferred to DHA and carried into the extracellular fluid as ASC, thus becoming available to neighboring cells and blood.

6.3.2 Plants

Although the molecular identities of plant vitamin C transporters have not been determined, substantial information exists about the mechanisms of ASC and DHA transport. In addition to engineering increased ASC synthesis by overexpression of enzymes (Agius *et al.*, 2003; Jain and Nessler, 2000), interventions that modulate ASC transport might be used to raise ASC content in the sink tissues, such as potato tubers and fruits, which are ingested as food by humans (Franceschi and Tarlyn, 2002). Furthermore, treatments that accelerate ASC transport into the extracellular matrix (apoplast) might increase plant resistance to pathogens and pollutants.

The apoplast's ASC participates in reactions that detoxify ozone and other reactive oxygen species (Burkey and Eason, 2002; Castillo and Greppin 1988; Plöchl *et al.*, 2000; Vanacker *et al.*, 1998). ASC acts as a substrate for ASC peroxidase and is converted by this enzyme to DHA. Transport of ASC and DHA across the plasma membrane affects the redox state of vitamin C in the apoplast. Therefore transport may influence the apoplast's ability to prevent damage to the plasma membrane by the prooxidant components of environmental insults.

Incubation of protoplasts and purified plasma membrane vesicles with [^{14}C]ASC leads to uptake of radioactivity that depends on time, temperature and substrate concentration (Foyer and Lelandais, 1996; Horemans *et al.*, 1997, 1998a, 1998b, 1998c; Rautenkranz *et al.*, 1994). DHA uptake across the plasma membrane results in an increased intracellular ASC content in cultured cells (de Pinto *et al.*, 1999; Potters *et al.*, 2000). Administration of DHA also results in a rapid and large increase in cellular ASC content in isolated root tips (Paciolla *et al.*, 2001). DHA does not become highly concentrated in roots because it is transported inside the cells and reduced to ASC rapidly (Paciolla *et al.*, 2001). DHA generated in the apoplast during oxidative stress may create a concentration gradient, across the plasma membrane, that favors net DHA uptake even if the transporter in the membrane is capable of bi-directional operation mediating both uptake and efflux. Plants possess monosaccharide transporters that translocate many simple sugars (Williams *et al.*, 2000). However DHA uptake across the plasma membrane of plants is not slowed by glucose and therefore is not mediated by monosaccharide transporters (Foyer and Lelandais, 1996; Horemans *et al.*, 1997; 1998a, 1998b).

There have been attempts to determine if the plant plasma membrane contains separate DHA and ASC transporters. Some plant plasma membrane vesicles rapidly oxidize ASC (Horemans *et al.*, 1997). This confounds attempts to distinguish between DHA and ASC carriers on the basis of kinetic experiments alone. For example, external ASC may appear to compete with [^{14}C]DHA for a putative common transporter (Rautenkranz *et al.*, 1994), but in fact the ASC may be oxidized to DHA that competes with [^{14}C]DHA for a specific transporter.

On the one hand, the observation that DHA fails to inhibit ASC uptake by pea leaf protoplasts argues for the existence of an ASC transport system (Foyer and Lelandais, 1996). Attempts to inhibit ASC oxidation with the strong reductant dithiothreitol failed to completely block uptake and thus support the view that ASC transport is significant (Foyer and Lelandais, 1996). On the other hand, the low affinity of ASC uptake in these cells (apparent K_m of 6.7 mM) must limit this system's effectiveness (Foyer and Lelandais, 1996). Barley protoplasts have higher affinity for DHA (20 μM) than ASC (90 μM) (Rautenkranz *et al.*, 1994). Similarly DHA appears to be the preferred form of vitamin C taken up by tobacco BY-2 protoplasts (apparent K_m of 139 μM) and bean plasma membrane vesicles (apparent K_m of 14–79 μM) (Horemans *et al.*, 1997; 1998a, 1998b, 1998c).

Horemans *et al.* (1998b) suggested that the uptake of DHA across the plasma membrane involves exchange for ASC. But how tightly the efflux of ASC from cytoplasm to apoplast is related to influx of DHA from apoplast to cytoplasm remains to be determined and the actual existence of an ASC–DHA exchanger protein has not been established conclusively. An alternative possibility is that DHA uptake into the cytoplasm is followed by reduction and the product is then translocated to the apoplast by a DHA-independent ASC transporter and/or simple diffusion of AA across the plasma membrane (Bichele *et al.*, 2000). Evidence for secondary active transport of ASC and DHA into plant cells is discussed below (Section 6.4.2).

The mechanism by which chloroplasts accumulate ASC has also been examined. Chloroplasts, isolated from pea and spinach plants, take up ASC by a glucose-insensitive system of facilitated diffusion (Anderson *et al.*, 1983; Beck *et al.*, 1983; Foyer and Lelandais, 1996). Although the affinity of this chloroplast system is very low (apparent K_m ~ 20 mM ASC) compared to the vitamin C transporters in the plasma membrane, it appears physiologically relevant considering the high concentration of ASC often reported for plant cytoplasm. Chloroplasts are capable of saturable uptake and efflux of glucose (Servaites and Geiger, 2002). However ASC must not be taken into chloroplasts through glucose carriers because glucose does not alter the rate of ASC uptake (Foyer and Lelandais, 1996).

6.4. Active transport

6.4.1 Animals

The large electrochemical gradient of sodium across the plasma membrane of animal cells is coupled to cotransport systems for numerous organic substrates. ASC is absorbed from the lumen of the human intestine by sodium–ASC cotransport into enterocytes, as has been shown by measuring transport activities in luminal (brush border) membrane vesicles (Malo and Wilson, 2000). This is an example of secondary active transport because it couples ASC uptake to the concentration gradient of sodium ion across the plasma membrane that is maintained by sodium/potassium-ATPase. It is likely, because of the limited capacity of enterocytes for sodium–ASC cotransport, that large oral doses of ASC are absorbed less completely than small doses *in vivo*. The absorption sites are found along the entire length of the small intestine. In vesicles prepared from the jejunum, intravesicular (cytoplasmic) glucose inhibits ASC uptake (Malo and Wilson, 2000). This observation does not implicate sodium–glucose cotransporters (i.e., SGLT proteins) as carriers of ASC because only glucose on the cytoplasmic side of the plasma membrane slows sodium-dependent ASC uptake. If trans-inhibition by glucose occurs *in vivo* then it may permit ingested glucose and hyperglycemia to decrease ASC bioavailability by slowing intestinal absorption of the vitamin. Other ingredients in foods and drugs may have comparable effects. For instance the inhibition by salicylate of sodium–ASC cotransport that has been demonstrated *in vitro* (Dixon *et al.*, 1991) may explain the decrease in vitamin C bioavailability caused by aspirin (Loh *et al.*, 1973).

Secondary active transport in the kidney is another important determinant of bioavailability. Most vitamin C circulates in the blood in the form of the ASC anion. The ASC in the blood plasma is freely filtered at the renal glomerulus, but much of it is reabsorbed in the proximal tubule. ASC uptake across the luminal membranes of renal proximal tubule cells occurs through sodium–ASC cotransport. The amount of ASC lost in the urine rises when the plasma ASC concentration exceeds the renal threshold. Above this threshold the tubular reabsorptive capacity is overwhelmed. A small study found the renal threshold for vitamin C to be slightly higher in men than in women (plasma ASC concentrations of 86 and 71 μM, respectively), but the physiological importance of this difference is unknown (Oreopoulos *et al.*, 1993).

The renal threshold may be lowered by an increased glomerular filtration rate or by the presence in the filtrate of molecules that interfere with the tubule's cotransporters (Davey *et al.*, 2000). For instance hyperglycemia may stimulate glomerular filtration, thus increasing the total filtered load of ASC and saturating the sodium–ASC cotransporters in the proximal tubules, leading to more ASC excretion. However, unlike the sodium-dependent ASC transport system in the luminal membranes of human jejunum that are inhibited by cytoplasmic glucose (Malo and Wilson, 2000), the initial rate of ASC uptake by the luminal membranes of rat proximal tubule is not retarded by glucose (*Table 6.1*) (Malo and Wilson, 2001). An early report found a small, low-potency inhibitory action of glucose (30% inhibition by 100 mM glucose), during relatively long uptake periods, which was attributable to sodium loading of the renal tubule membrane vesicles by prolonged, maximal operation of the sodium–glucose cotransporter (Toggenburger *et al.*, 1981). The briefer uptake period used by Malo and Wilson (2001) did not allow glucose-coupled sodium entry to increase intracellular (intravasicular) sodium concentration to levels that inhibit ASC uptake during the transport assay. Since ASC and DHA uptakes are not inhibited by glucose directly, it is evident that both occur independently of glucose transporters in these rat kidney membranes. Moreover the SGLT blocker phlorizin (1 mM) abolished glucose uptake but failed to slow either ASC or DHA transport (*Table 6.1*).

Excessive excretion of ASC in the urine may occur in patients with diabetes mellitus even if the cause is not a direct interaction of sodium–ASC cotransporters with glucose. For instance, patients with diabetic nephropathy have lower plasma ASC concentration and higher renal clearance of ASC than do control subjects (Hirsch *et al.*, 1998). This observation is consistent with the hypothesis that patients with diabetic nephropathy have diminished plasma ASC levels due to increased ASC loss in urine. Since tubular atrophy and interstitial fibrosis are common in the diabetic kidney, it is likely that structural damage impairs tubular reabsorption of ASC in these patients (Hirsch *et al.*, 1998).

Many nonepithelial cell types are capable of high affinity vitamin C uptake through sodium–ASC cotransporters. Mammalian examples include bone osteoblasts, cerebral astrocytes, ovarian granulosa cells, and vascular smooth muscle and endothelial cells (Holmes *et al.*, 2000; Korcok *et al.*, 2000; Wilson and Dixon, 1995; Wilson *et al.*, 1996; Zreik *et al.*, 1999). Sodium–ASC cotransport has also been found in nonmammalian animals' organs including chicken brain (Wilson and Jaworski, 1992; Wilson, 1990) and teleost fish intestine (Maffia *et al.*, 1993).

Sodium–ASC cotransporters are highly specific for L-ASC (Liang *et al.*, 2001; Dixon *et al.*, 1991; Franceschi *et al.*, 1995; Malo and Wilson, 2000; Wilson, 1989). Among the molecules that have been tested and found not to be substrates for these cotransporters are: ASC-2-O-phosphate, DHA, glucose, 2-deoxyglucose, xanthine, hypoxanthine, L-gulono-lactone, formate, lactate, pyruvate, gluconate, oxalate, malonate, succinate and an assortment of nucleosides and nucleotides. The cotransporters' stereoselectivity has been demonstrated as a greater affinity for L-ASC over the epimer D-isoascorbate, which leads to higher intracellular ASC concentrations of L-ASC than D-isoascorbate at steady state (Franceschi *et al.*, 1995).

The cotransporters are absolutely dependent on sodium (Dixon *et al.*, 1991; Liang *et al.*, 2001; Malo and Wilson, 2000; Wilson, 1989). They translocate at least two sodium cations with each ASC anion and thus use the electrochemical gradient of sodium ion across the plasma membrane to provide the energy required for concentrative uptake of ASC.

The apparent K_m values for sodium–ASC cotransport systems range from 200 μM L-ASC in absorptive epithelia [e.g., luminal membranes of human jejunum (Malo and Wilson, 2000) and rodent renal cortex (Toggenburger *et al.*, 1981)] to 20 μL-ASC in nonepithelial cells (e.g., osteoblasts derived from bone; Dixon and Wilson, 1992a). Thus, in animal cells, the secondary active transport of ASC occurs with higher affinity than does facilitated transport of DHA.

Expression cloning has led to the molecular identification of several sodium–ASC cotransporters. Two isoforms (i.e., sodium-dependent vitamin C transporters SVCT1 and SVCT2) have been cloned from rats and humans and an ortholog of SVCT2 has been cloned from mice (Daruwala *et al.*, 1999; Liang *et al.*, 2001,. 2002; Rajan *et al.*, 1999; Sotiriou *et al.*, 2002; Tsukaguchi *et al.*, 1999; Wang *et al.*, 1999, 2000). The HUGO gene names for SVCT1 and SVCT2 were revised in 2003. SVCT1 is now assigned to SLC23A1 and SVCT2 is now assigned to SLC23A2 (see http://www.gene.ucl.ac.uk/nomenclature). These isoforms have extensive sequence identity with each other but do not share structural homology with other families of sodium cotransporters. Concerning the rat isoforms, SVCT1 is a 604-amino acid protein and SVCT2 is a 592-amino acid protein. Hydropathy plots of the amino acid sequences are consistent with 12 transmembrane domains. Both SVCT isoforms are glycoproteins containing *N*-linked oligosaccharides when expressed in heterologous systems such as *Xenopus* oocytes, HRPE cells and COS-1 cells (Liang *et al.*, 2002).

Expression of SVCT and SVCT2 confers concentrative ASC uptake in heterologous cell systems. Both SVCT isoforms are specific for L-ASC. There is no evidence that they transport glucose. Moreover, their stereoselectivity for L-ASC over the epimer D-isoascorbate (erythorbate) likely accounts for the low antiscorbutic activity of the latter. It is not clear if the SVCT isoforms differ markedly in affinity for ASC because hSVCT1 has a threefold lower affinity than hSVCT2 in COS-1 but not HRPE cells (Liang *et al.*, 2001; Rajan *et al.*, 1999; Wang *et al.*, 1999). SVCT1 has consistently been found to have a higher capacity for ASC than does SVCT2. It has not been determined if this greater capacity is an inherent property of the SVCT1 protein (i.e., a larger turnover number) or simply reflects a larger abundance of SVCT1 than SVCT2 in the plasma membrane of the transfected cells (Liang *et al.*, 2002).

Endogenous mRNA for one or both of the SVCT isoforms has been found in most organs (Berger and Hediger, 2000; Daruwala *et al.*, 1999; Rajan *et al.*, 1999; Tsukaguchi *et al.*, 1999; Wang *et al.*, 1999, 2000). SVCT1 is localized mostly in epithelial tissues. SVCT2 has a wider distribution and has been detected in most organs, with the exception of lung and skeletal muscle. On the one hand, the high capacity of SVCT1 is appropriate for epithelial cells that transport much more ASC than required for their own internal use. On the other hand, the affinities of both SVCT1 and SVCT2 are sufficiently high to enable cells to absorb ASC effectively from extracellular fluid where – in most tissues except the stomach, eye and central nervous system – the concentration of ASC may approximate the 25–100 μM typically found in plasma.

In human populations, elderly subjects require more vitamin C in their diet than do young subjects to reach a desired plasma ASC concentration (Brubacher *et al.*, 2000). Animal studies have revealed an age-related decline in the capacity of cells to absorb vitamin C (Michels *et al.*, 2003). The ASC concentration in the liver of male rats decreases with age even though the rate of *de novo* ASC synthesis does not (Michels *et al.*, 2003). When incubated with ASC, isolated hepatocytes from old as compared to young rats show decreased maximal rate of ASC uptake and lower steady-state intracellular ASC concentration. Sodium-free media significantly reduces ASC uptake, implicating sodium–ASC cotransporters. Hepatic SVCT1 mRNA levels decline 45% with age, with no significant changes in SVCT2 mRNA abundance. It thus appears likely that a fall in SVCT1 expression changes hepatic

ASC concentration. If confirmed, this finding may indicate a dominant role for plasma membrane transporters in the regulation of ASC levels, even in cells capable of synthesizing the vitamin. Moreover the deficit in intracellular ASC concentration can be overcome by increasing the external supply of the vitamin, which suggests that intervention to increase plasma ASC concentrations might be beneficial for hepatic function in elderly subjects (Michels *et al.*, 2003). Another possible therapeutic approach is to elevate the activity of sodium–ASC cotransporters. The age-related decline in SVCT1 mRNA levels would be consistent with either a decline in RNA stability or alterations in transcriptional regulation of the gene. No specific transcription factors have been linked to changes in either SVCT1 or SVCT2 mRNA levels, a lack that is due primarily to the absence of sequence data on the 5′-untranslated regions of SVCT1 and SVCT2 (Michels *et al.*, 2003).

Much research has been directed towards identifying mechanisms that control the ASC transport systems in various animal cell types. Transport activity may be altered by changes in the affinity of substrate binding, the translocation capacity (i.e., turnover number) of each transporter protein or the number of transporter proteins present in the plasma membrane.

Sodium–ASC cotransport may be regulated kinetically by changes in either the concentrations of the transported solutes or membrane potential. ASC downregulates the maximal rate of sodium–ASC cotransport in animals. This has been shown for absorptive epithelia by feeding excess ASC to guinea pigs and afterwards measuring the rate of sodium-dependent uptake of vitamin C into the intestinal mucosa (Karasov *et al.*, 1991). Downregulation of SVCT1 by ASC may limit its usefulness for raising intracellular ASC concentration. Using the Caco-2 TC7 cell model of small intestinal enterocytes, MacDonald *et al.* (2002) measured the effects of 24-h pretreatment with ASC (4.5 $mg \cdot ml^{-1}$) on subsequent [^{14}C]ASC uptake and SVCT1 mRNA expression (determined by reverse transcription-polymerase chain reaction). [^{14}C]ASC uptake was decreased by 50% and expression of SVCT1 was decreased by 77% following ASC pretreatment. Thus it appears that the activity of SVCT1 in enterocytes is regulated to adjust for the recent history of ASC absorption.

Substrate downregulation of SVCT2 has been elucidated by preincubating cultured astrocytes and osteoblasts with varying concentrations of ASC for 1–24 h and subsequently determining the kinetic properties of the initial rate of [^{14}C]ASC uptake (Dixon and Wilson, 1992b; Wilson *et al.*, 1990). Changes in SVCT2 activity are rapid and large. The observation that SVCT2 activity varies inversely with intracellular ASC concentration is consistent with the hypothesis that this transporter regulates the intracellular concentration of its organic substrate. Upregulation of SVCT2 in cells depleted of ASC leads to more efficient absorption of extracellular ASC and tends to restore the intracellular concentration of the vitamin (Dixon and Wilson, 1992b; Wilson *et al.*, 1990). Thus, in the absence of neurohormonal or paracrine signals, SVCT2 acts to maintain intracellular ASC concentration constant. Experiments with transgenic mice lacking SVCT2 have confirmed that this transporter's activity contributes to the high ascorbate concentration found in brain normally (Sotiriou *et al.*, 2002). However, because sodium–ASC cotransporters become downregulated when intracellular ASC concentration is high (Wilson *et al.*, 1990), they may not be suitable targets for therapeutic strategies that attempt to raise intracellular ASC to supraphysiological levels. Instead it has been suggested that DHA be injected as a pro-drug to increase tissue – particularly brain – ASC concentration (Agius *et al.*, 1997; Huang *et al.*, 2001).

It is interesting to consider how the limited capacities of SVCT isoforms and their susceptibility to downregulation by ASC may have influenced the many human clinical trials with oral vitamin C supplements that failed to confer antioxidant protection or clinical benefit. Optimization of the dosing regime may be critical to the success of future intervention studies using vitamin C (Padayatty *et al.*, 2003). An important consequence of substrate regulation of SVCT1 and SVCT2 activities may be more efficient absorption by the intestine, conservation by the kidney and uptake into target cells of intermittent doses than of continuously ingested doses of ASC. This phenomenon may account for the finding in rainbow trout (which require vitamin C from food) that hepatic ASC concentration at the end of a 3-month experiment was higher if ASC was fed at 10-day intervals (i.e., pulse and withdrawal) instead of continuously (Blom and Dabrowski, 1998). Although ASC transport activity was not measured, the intermittent withdrawal of dietary ASC may have caused a compensatory upregulation of sodium–ASC cotransporters in these fish.

Because SVCT transporters are electrogenic (at least two sodium cations are translocated with each ASC anion), changes in membrane potential affect ASC flux (Malo and Wilson, 2000; Wilson *et al.*, 1991). This can be understood from a thermodynamic viewpoint to result from changes in membrane potential altering the sodium electrochemical gradient and thus the free energy of the transport system. In particular, membrane depolarization slows ASC uptake into cells reliant on SVCT.

ASC transport by membranes expressing SVCT1 or SVCT2 occurs with a pH optimum of approximately 7.5 (Liang *et al.*, 2001; Malo and Wilson, 2001). Acidification of the aqueous compartments on either the extracellular or cytoplasmic side of the plasma membrane inhibits ASC transport. The potency of protons for this effect greatly exceeds that which can be explained by protonation of ASC to AA. Thus protons inhibit the ASC transport systems and may act directly on the SVCT proteins. It has been suggested that protonation of histidine residues in the transporters decreases ASC binding affinity (Liang *et al.*, 2001).

Hormones, paracrine factors and intracellular signaling molecules regulate the expression of sodium–ASC cotransporters. For example, agents that elevate cyclic AMP levels in cultures of rat brain astrocytes increase SVCT2 mRNA levels and subsequently elevate the maximal rate of high affinity, sodium-dependent ASC uptake (Korcok *et al.*, 2000; Siushansian *et al.*, 1997). The stimulation of ASC transport capacity appears to require *de novo* protein synthesis because it is inhibited by cycloheximide. Another example comes from studies of cultured bone cells. Transforming growth factor-β (Wilson and Dixon, 1995) and glucocorticoids (Fujita *et al.*, 2001; Pandipati *et al.*, 1998) similarly increase the maximal rate of ASC uptake through sodium–ASC cotransporters in osteoblasts. Fujita *et al.* (2002) showed that the glucocorticoid agonist dexamethasone increases expression of mRNA for a mouse ortholog of SVCT2 in an osteoblastic cell line derived from mouse bone.

Redistribution or post-translational modification of the SVCT proteins may also regulate sodium–ASC cotransport. This may explain how drugs that activate protein kinase C (PKC) inhibit sodium-dependent ASC uptake by rabbit nonpigmented ciliary epithelial cells (Liang *et al.*, 2001), monkey kidney COS-1 cells (Liang *et al.*, 2002) and *Xenopus* oocytes heterologously expressing either hSVCT1 or hSVCT2 (Daruwala *et al.*, 1999). The mechanisms of PKC regulation of the human isoforms hSVCT1 and hSVCT2 have been studied in monkey kidney COS-1 cells (Liang *et al.*, 2002). Transient transfection of recombinant carboxyl-terminal V5 epitope-tagged forms of the transporters was used to permit their immunodetection. The PKC activator phorbol 12-myristate 13-acetate (PMA) induced rapidly (5–80 min) a fall in the maximal velocity of ASC uptake through hSVCT1 and hSVCT2. Western blot and confocal microscopy analyses indicated that the total amounts of hSVCT1 or hSVCT2 proteins in the transfected COS-1 cells were not altered by PMA treatment. However PMA caused a net redistribution of hSVCT1 protein from the cell surface to intracellular membranes. PMA did not induce a detectable change in the amount of hSVCT2 protein in the plasma membrane, so it appears that activation of PKC decreases the catalytic transport efficiency (i.e., translocation capacity, turnover number) of this isoform (Liang *et al.*, 2002). A possible confound is that transfection in COS-1 cells achieves higher protein levels and transport activities for hSVCT1 than hSVCT2 prior to PMA exposure (Liang *et al.*, 2002).

6.4.2 Plants

There is evidence that vitamin C uptake across plant plasma membranes is coupled to the electrochemical gradient of protons by a process of secondary active transport. The roots of soybean plants take up radiolabeled vitamin C by a system that is sensitive to temperature, oxygen supply and metabolic inhibitors (Mozafar and Oertli, 1993). Similarly vitamin C uptake by barley leaf protoplasts depends on cell metabolism, since it is inhibited by the mitochondrial uncoupler carbonyl-cyanide-3-chlorophenylhydrazone (Rautenkranz *et al.*, 1994). This system operates with high affinity because its K_m values are 90 μM for ASC and 20 μM for DHA (Rautenkranz *et al.*, 1994).

Active transport may also characterize a low affinity DHA uptake system. In a recent study by Kollist *et al.* (2001), ASC was applied to the leaves of *Betula pendula* and the rate of disappearance of total vitamin C (ASC and/or DHA) was recorded as an indirect measure of cellular uptake. Observing that extracellular ASC was rapidly converted to DHA and inhibition of this oxidation (with dithiothreitol) retarded the disappearance of total vitamin C, the authors inferred that only DHA was taken up into cells. The H^+ ATPase inhibitor erythrosin B also inhibited uptake, consistent with dependence of DHA uptake on the plasma membrane proton gradient (Kollist *et al.*, 2001). However the transport system's very low affinity for DHA (apparent K_m of 12.8 mM) relative to the concentration of total vitamin C measured in the apoplast (0.73 mM), indicates it may not be capable of regulating DHA levels in the apoplast during stress.

6.5 Conclusion

Animals and plants have evolved independently and their vitamin C transport systems are quite different. Concerning flux across the plasma membrane of animal cells, simple diffusion of ASC plays only a small or negligible role. More significant are the facilitated diffusion of ASC through channels, facilitated diffusion of DHA through glucose-sensitive and -insensitive transporters, and secondary active transport of ASC through SVCT proteins. For plants, simple diffusion of ASC across the plasma membrane may be an important way of supplying the antioxidant to the apoplast during oxidative stress. Plant cells absorb ASC and DHA by glucose-insensitive, facilitated diffusion and may also take up these molecules by secondary active transport coupled to the plasma membrane's proton gradient.

Future research may lead to gene engineering of vitamin C transport systems that will increase the nutritional value and resistance to stress of agricultural plants and animals. Although acceleration of ASC synthesis has already been achieved by overexpression of enzymes in some plants, manipulation of transport systems may be necessary to maximize vitamin C levels in the sink tissues of plants that animals and humans ingest for food. Enhancing the release of ASC from aquatic plants may facilitate the harvesting of vitamin C from these sources for industrial use. Finally, discovering ways to alter vitamin C transport systems therapeutically in humans may create new treatments for symptoms of aging and disease.

Acknowledgments

I thank Drs. S. Jeffrey Dixon, Christiane Malo and Richard Rose for past collaborations and discussions of vitamin C transport. Original research in my laboratory was supported by grants from the Canadian Institutes of Health Research, the Heart and Stroke Foundation of Ontario, and the Natural Sciences and Engineering Research Council of Canada.

References

Agius F, Gonzalez-Lamothe R, Caballero JL, Munoz-Blanco J, Botella MA, Valpuesta V (2003) Engineering increased vitamin C levels in plants by overexpression of a D-galacturonic acid reductase. *Nature Biotechnol.* **21**, 177–181.

Agus DB, Gambhir SS, Pardridge WM, Spielholz C, Baselga J, Vera JC, Golde DW (1997) Vitamin C crosses the blood–brain barrier in the oxidized form through the glucose transporters. *J. Clin. Invest.* **100**, 2842–2848.

Anderson JW, Ushimaru T, Asada K (1983) Light-dependent reduction of dehydroascorbic acid and uptake of exogenous ascorbate by spinach chloroplasts. *Planta* **158**, 442–450.

Beck E, Bukert A, Hofmann M (1983) Uptake of L-ascorbate by intact spinach chloroplasts. *Plant Physiol.* **73**, 41–45.

Beck JS, Potts DJ (1990) Cell swelling, co-transport activation and potassium conductance in isolated perfused rabbit kidney proximal tubules. *J. Physiol.* **425**, 369–378.

Berger UV, Hediger MA (2000) The vitamin C transporter SVCT2 is expressed by astrocytes in culture but not *in situ*. *Neuroreport* **11**, 1395–1399.

Bichele I, Moldau H, Padu E (2000) Estimation of plasmalemma conductivity to ascorbic acid in intact leaves exposed to ozone. *Physiol. Plantarum* **108**, 405–412.

Blom JH, Dabrowski K. (1998) Continuous or 'pulse-and-withdraw' supply of ascorbic acid in the diet: a new approach to altering the bioavailability of ascorbic acid, using teleost fish as a scurvy-prone model. *Int. J. Vitam. Nutr. Res.* **68**, 88–93.

Braddock R, Siman CM, Hamilton K, Garland HO, Sibley CP (2002) Gamma-linoleic acid and ascorbate improves skeletal ossification in offspring of diabetic rats. *Pediatr. Res.* **51**, 647–652.

Breton S, Marsolais M, Lapointe JY, Laprade R (1996) Cell volume increases of physiologic amplitude activate basolateral K and CI conductances in the rabbit proximal convoluted tubule. *J. Am. Soc. Nephrol.* **7**, 2072–2087.

Brubacher D, Moser U, Jordan P (2000) Vitamin C concentrations in plasma as a function of intake: a meta-analysis. *Int. J. Vitam. Nutr. Res.* **70**, 226–237.

Buffinton GD, Doe WF (1995) Altered ascorbic acid status in the mucosa from inflammatory bowel disease patients. *Free Radic. Res.* **22**, 131–143.

Burkey KO, Eason G (2002) Ozone tolerance in snap bean is associated with elevated ascorbic acid in the leaf apoplast. *Physiol. Plantarum* **114**, 387–394.

Castillo FJ, Greppin H (1988) Extracellular ascorbic acid and enzyme activities related to ascorbic acid metabolism in *Sedum album L.* leaves after ozone exposure. *Environ. Exp. Bot.* **28**, 231–238.

Daruwala R, Song J, Koh WS, Rumsey SC, Levine M (1999) Cloning and functional characterization of the human sodium-dependent vitamin C transporters hSVCT1 and hSVCT2. *FEBS Lett.* **460**, 480–484.

Daskalopoulos R, Korcok J, Tao L, Wilson JX (2002) Accumulation of intracellular ascorbate from dehydroascorbic acid by astrocytes is decreased after oxidative stress and restored by propofol. *Glia* **39**, 124–132.

Davey MW, Van Montagu M, Inze D, Sanmartin M, Kanellis A, Smirnoff N, Benzie IJJ, Strain JJ, Favell D, Fletcher J (2000) Plant L-ascorbic acid: chemistry, function, metabolism, bioavailability and effects of processing. *J. Sci. Food Agric.* **80**, 825–860.

De Pinto MC, Dennis F, De Gara L (1999) The redox state of the ascorbate–dehydroascorbate pair as a specific sensor of cell division in tobacco BY-2 cells. *Protoplasma* **209**, 90–97.

Diliberto EJ Jr, Menniti FS, Knoth J, Daniels AJ, Kizer JS, Viveros OH (1987) Adrenomedullary chromaffin cells as a model to study the neurobiology of ascorbic acid: from monooxygenation to neuromodulation. *Ann. N. Y. Acad. Sci.* **498**, 28–53.

Dixon SJ, Kulaga A, Jaworski EM, Wilson JX (1991) Ascorbate uptake by ROS 17/2.8 osteoblast-like cells: substrate specificity and sensitivity to transport inhibitors. *J. Bone Miner. Res.* **6**, 623–629.

Dixon SJ, Wilson JX (1992a) Transforming growth factor-β stimulates ascorbate transport activity in osteoblastic cells. *Endocrinology* **130**, 484–489.

Dixon SJ, Wilson JX (1992b) Adaptive regulation of ascorbate transport in osteoblastic cells. *J. Bone Miner. Res.* **7**, 675–681.

Finn FM, Johns PA (1980) Ascorbic acid transport by isolated bovine adrenal cortical cells. *Endocrinology* **106**, 811–817.

Foyer CH, Lelandais M (1996) A comparison of the relative rates of transport of ascorbate and glucose across the thylakoid, chloroplast and plasmalemma membranes of pea leaf mesophyll cells. *J. Plant Physiol.* **148**, 391–398.

Franceschi VR, Tarlyn NM (2002) L-ascorbic acid is accumulated in source leaf phloem and transported to sink tissues in plants. *Plant Physiol.* **130**, 649–656.

Franceschi RT, Wilson JX, Dixon SJ (1995) Requirement for Na^+-dependent ascorbic acid transport in osteoblast function. *Am. J. Physiol.* **268**, C1430–C1439.

Fujita I, Hirano J, Itoh N, Nakanishi T, Tanaka K. (2001) Dexamethasone induces sodium-dependent vitamin C transporter in a mouse osteoblastic cell line MC3T3-E1. *Br. J. Nutr.* **86**, 145–149.

Furst J, Gschwentner M, Ritter M, Botta G, Jakab M, Mayer M, *et al.* (2002) Molecular and functional aspects of anionic channels activated during regulatory volume decrease in mammalian cells. *Pflug. Arch. Eur. J. Physiol.* **444**, 1–25.

Himmelreich U, Drew K.N, Serianni AS, Kuchel PW (1998) ^{13}C NMR studies of vitamin C transport and its redox cycling in human erythrocytes. *Biochemistry* **37**, 7578–7588.

Hirsch IB, Atchley DH, Tsai E, Labbe RF, Chait A (1998) Ascorbic acid clearance in diabetic nephropathy. *J. Diabetes Complications* **12**, 259–263.

Holmes ME, Samson SE, Wilson JX, Dixon SJ, Grover AK (2000) Ascorbate transport in pig coronary artery smooth muscle: Na^+-removal and oxidative stress increase loss of accumulated cellular ascorbate. *J. Vasc. Res.* **37**, 390–398.

Holmes ME, Mwanjewe J, Samson SE, Haist JV, Wilson JX, Dixon SJ, Karmazyn M, Grover AK (2002) Dehydroascorbic acid uptake by coronary artery smooth muscle: effect of intracellular acidification. *Biochem. J.* **362**, 507–512.

Horemans N, Asard H, Caubergs RJ (1997) The ascorbate carrier of higher plant plasma membranes preferentially translocates the fully oxidized (dehydroascorbate) molecule. *Plant Physiol.* **114**, 1247–1253.

Horemans N, Asard H, Van Gestelen P, Caubergs RJ (1998a) Facilitated diffusion drives transport of oxidised ascorbate molecules into purified plasma membrane vesicles of *Phaseolus vulgaris*. *Physiol. Plantarum* **104**, 783–789.

Horemans N, Asard H, Caubergs RJ (1998b) Carrier mediated uptake of dehydroascorbate into higher plant plasma membrane vesicles shows trans-stimulation. *FEBS Lett.* **421**, 41–44.

Horemans N, Potters G, Caubergs RJ, Asard H (1998c) Transport of ascorbate into protoplasts of *Nicotiana tabacum* Bright Yellow-2 cell line. *Protoplasma* **205**, 114–121.

Huang J, Agus DB, Winfree CJ, Kiss S, Mack WJ, McTaggart RA, *et al.* (2001) Dehydroascorbic acid, a blood–brain barrier transportable form of vitamin C, mediates potent cerebroprotection in experimental stroke. *Proc. Natl Acad. Sci. USA* **98**, 11720–11724.

Jain AK, Nessler CL (2000) Metabolic engineering of an alternative pathway for ascorbic acid biosynthesis in plants. *Mol. Breed.* **6**, 73–78.

Karasov WH, Darken BW, Bottum MC (1991) Dietary regulation of intestinal ascorbate uptake in guinea pigs. *Am. J. Physiol.* **260**, G108–G118.

Khatami M, Stramm LE, Rockey JH (1986) Ascorbate transport in cultured cat retinal pigment epithelial cells. *Exp. Eye Res.* **43**, 607–615.

Kodaman PH, Behrman HR (1999) Hormone-regulated and glucose-sensitive transport of dehydroascorbic acid in immature rat granulosa cells. *Endocrinology* **140**, 3659–3665.

Kollist H, Moldau H, Oksanen E, Vapaavuori E (2001) Ascorbate transport from the apoplast to the symplast in intact leaves. *Physiol. Plant* **113**, 377–383.

Korcok J, Yan R, Siushansian R, Dixon SJ, Wilson JX (2000) Sodium-ascorbate cotransport controls intracellular ascorbate concentration in primary astrocyte cultures expressing the SVCT2 transporter. *Brain Res.* **881**, 144–151

Laggner H, Besau V, Goldenberg H (1999) Preferential uptake and accumulation of oxidized vitamin C by THP-1 monocytic cells. *Eur. J. Biochem.* **262**, 659–665.

Liang WJ, Johnson D, Jarvis SM (2001) Vitamin C transport systems of mammalian cells. *Mol. Membr. Biol.* **18**, 87–95.

Liang WJ, Johnson D, Ma LS, Jarvis SM, Wei-Jun L (2002) Regulation of the human vitamin C transporters expressed in COS-1 cells by protein kinase C. *Am. J. Physiol.* **283**, C1696–C1704.

Loh HS, Watters K, Wilson CW (1973) The effects of aspirin on the metabolic availability of ascorbic acid in human beings. *J. Clin. Pharmacol.* **13**, 480–486.

Luwe MWF, Takahama U, Heber U (1993) Role of ascorbate in detoxifying ozone in the apoplast of spinach (*Spinacia oleracea* L) leaves. *Plant Physiol.* **101**, 969–976.

Maffia M, Ahearn GA, Vilella S, Zonno V, Storelli C (1993) Ascorbic acid transport by intestinal brush–border membrane vesicles of the teleost *Anguilla anguilla*. *Am. J. Physiol.* **264**, R1248–R1253.

Malo C, Wilson JX (2000) Glucose modulates vitamin C transport in adult human small intestinal brush border membrane vesicles. *J. Nutr.* **130**, 63–69.

Malo C, Wilson JX (2001) Transport of ascorbic acid and dehydroascorbic acid in rat kidney cortex. *FASEB J.* **15**, A838.

MacDonald L, Thumser AE, Sharp P (2002) Decreased expression of the vitamin C transporter SVCT1 by ascorbic acid in a human intestinal epithelial cell line. *Br. J. Nutr.* **87**, 97–100.

MacLeod RJ, Lembessis P, Hamilton JR (1992) Differences in Ca^{2+}-mediation of hypotonic and Na(+)-nutrient regulatory volume decrease in suspensions of jejunal enterocytes. *J. Membr. Biol.* **130**, 23–31.

Maeshima M (2001) Tonoplast transporters: organization and function. *Ann. Rev. Plant Physiol.* **52**, 469–497.

Maggio A, McCully MG, Kerdnaimongkol K, Bressan RA, Hasegawa PM, Joly RJ (2002) The ascorbic acid cycle mediates signal transduction leading to stress-induced stomatal closure. *Func. Plant Biol.* **29**, 845–852.

May JM, Qu ZC, Whitesell RR (1995) Ascorbate is the major electron donor for a transmembrane oxidoreductase of human erythrocytes. *Biochim. Biophys. Acta* **1238**, 127–136.

Mendiratta S, Qu ZC, May JM (1998) Erythrocyte ascorbate recycling: antioxidant effects in blood. *Free Rad. Biol. Med.* **24**, 789–797.

Michels AJ, Joisher N, Hagen TM (2003) Age-related decline of sodium-dependent ascorbic acid transport in isolated rat hepatocytes. *Arch. Biochem. Biophys.* **410**, 112–120.

Mozafar A, Oertli JJ (1993) Vitamin C (ascorbic acid): uptake and metabolism by soybean. *J. Plant Physiol.* **141**, 316–321.

Ng LL, Ngkeekwong FC, Quinn PA, Davies JE (1998) Uptake mechanisms for ascorbate and dehydroascorbate in lymphoblasts from diabetic nephropathy and hypertensive patients. *Diabetologia* **41**, 435–442.

Obrosova IG, Fathallah L, Stevens MJ (2001) Taurine counteracts oxidative stress and nerve growth factor deficit in early experimental diabetic neuropathy. *Exp. Neurol.* **172**, 211–219.

Oreopoulos DG, Lindeman RD, VanderJagt DJ, Tzamaloukas AH, Bhagavan HN, Garry PJ (1993) Renal excretion of ascorbic acid: effect of age and sex. *J. Am. Coll. Nutr.* **12**, 537–542.

Paciolla C, De Tullio MC, Chiappetta A, Innocenti AM, Bitonti MB, Liso R, Arrigoni O. (2001) Short- and long-term effects of dehydroascorbate in *Lupinus albus* and *Allium cepa* roots. *Plant Cell Physiol.* **42**, 857–863.

Padayatty SJ, Katz A, Wang Y, Eck P, Kwon O, Lee JH, Chen S, Corpe C, Dutta A, Dutta SK, Levine M (2003) Vitamin C as an antioxidant: evaluation of its role in disease prevention. *J. Am. Coll. Nutr.* **22**, 18–35.

Pandipati S, Driscoll JE, Franceschi RT (1998) Glucocorticoid stimulation of Na^+-dependent ascorbic acid transport in osteoblast-like cells. *J. Cell. Physiol.* **176**, 85–91.

Patrick JW, Zhang WH, Tyerman SD, Offler CE, Walker NA (2001) Role of membrane transport in phloem translocation of assimilates and water. *Aust. J. Plant Physiol.* **28**, 695–707.

Plöchl M, Lyons T, Ollerenshaw J, Barnes J (2000) Simulating ozone detoxification in the leaf apoplast through the direct reaction with ascorbate. *Planta* **210**, 454–467.

Potters G, Horemans N, Caubergs RJ, Asard H (2000) Ascorbate and dehydroascorbate influence cell cycle progression in a tobacco cell suspension. *Plant Physiol.* **124**, 17–20.

Potters G, De Gara L, Asard H, Horemans N (2002) Ascorbate and glutathione: guardians of the cell cycle, partners in crime? *Plant Physiol. Biochem.* **40**, 537–548.

Qutob S, Dixon SJ, Wilson JX (1998) Insulin stimulates vitamin C recycling and ascorbate accumulation in osteoblastic cells. *Endocrinology* **139**, 51–56.

Rajan DP, Huang W, Dutta B, Devoe LD, Leibach FH, Ganapathy V, Prasad PD (1999) Human placental sodium-dependent vitamin C transporter (SVCT2): molecular cloning and transport function. *Biochem. Biophys. Res. Commun.* **262**, 762–768.

Rautenkranz AAF, Li LJ, Machler F, Martinoia E, Oertli JJ (1994) Transport of ascorbic and dehydroascorbic acids across protoplast and vaculoe membranes isolated from barley (*Hordeum–vulgare L CV Gerbel*) leaves. *Plant Physiol.* **106**, 187–193.

Rebec GV, Pierce RC (1994) A vitamin as neuromodulator: ascorbate release into the extracellular fluid of the brain regulates dopaminergic and glutamatergic transmission. *Prog. Neurobiol.* **43**, 537–565.

Rose RC (1987) Solubility properties of reduced and oxidized ascorbate as determinants of membrane permeation. *Biochim. Biophys. Acta* **924**, 254–256.

Rumsey SC, Kwon O, Xu GW, Burant CF, Simpson I, Levine M (1997) Glucose transporter isoforms GLUT1 and GLUT3 transport dehydroascorbic acid. *J. Biol. Chem.* **272**, 18982–18989.

Rumsey SC, Daruwala R, Al-Hasani H, Zarnowski MJ, Simpson IA, Levine M (2000) Dehydroascorbic acid transport by GLUT4 in *Xenopus* oocytes and isolated rat adipocytes. *J. Biol. Chem.* **275**, 28246–28253.

Running JA, Severson DK, Schneider KJ (2002) Extracellular production of L-ascorbic acid by *Chlorella protothecoides*, *Prototheca* species, and mutants of *P. moriformis* during aerobic culturing at low pH. *J. Ind. Microbiol. Biot.* **29**, 93–98.

Schorah CJ (1992) The transport of vitamin C and effects of disease. *Proc. Nutr. Soc.* **51**, 189–198.

Servaites JC, Geiger DR (2002) Kinetic characteristics of chloroplast glucose transport. *J. Exp. Bot.* **53**, 1581–1591.

Siushansian R, Dixon SJ, Wilson JX (1996) Osmotic swelling stimulates ascorbate efflux from cerebral astrocytes. *J. Neurochem.* **66**, 1227–1233.

Siushansian R, Tao L, Dixon SJ, Wilson JX (1997) Cerebral astrocytes transport ascorbic acid and dehydroascorbic acid through distinct mechanisms regulated by cyclic AMP. *J. Neurochem.* **68**, 2378–2385.

Socci RR, Delamere NA (1988) Characteristics of ascorbate transport in the rabbit iris-ciliary body. *Exp. Eye Res.* **46**, 853–861.

Song JH, Shin SH, Ross GM (2001) Oxidative stress induced by ascorbate causes neuronal damage in an *in vitro* system. *Brain Res.* **895**, 66–72.

Sotiriou S, Gispert S, Cheng J, Wang Y, Chen A, Hoogstraten-Miller S, Miller GF, Kwon O, Levine M, Guttentag SH, Nussbaum RL (2002) Ascorbic-acid transporter Slc23a1 is essential for vitamin C transport into the brain and for perinatal survival. *Nature Med.* **8**, 514–517.

Takahama U (1993) Regulation of peroxidase-dependent oxidation of phenolics by ascorbic acid – different effects of ascorbic acid on the oxidation of coniferyl alcohol by the apoplastic soluble and cell wall-bound peroxidases from epicotyls of *Vigna angularis*. *Plant Cell Physiol.* **34**, 809–817.

Toggenburger G, Hausermann M, Mutsch B, Genoni G, Kessler M, Weber F, Hornig D, O'Neill B, Semenza G (1981) Na^+-dependent, potential-sensitive L-ascorbate transport across brush border membrane vesicles from kidney cortex. *Biochim. Biophys. Acta* **646**, 433–443.

Tsukaguchi H, Tokui T, Mackenzie B, Berger UV, Chen XZ, Wang Y, Brubaker RF, Hediger MA (1999) A family of mammalian Na^+-dependent L-ascorbic acid transporters. *Nature* **399**, 70–75.

Upston JM, Karjalainen A, Bygrave FL, Stocker R (1999) Efflux of hepatic ascorbate: a potential contributor to the maintenance of plasma vitamin C. *Biochem. J.* **342**, 49–56.

Vanacker HT, Carver L, Foyer CH (1998) Pathogen-induced changes in the antioxidant status of the apoplast in barley leaves. *Plant Physiol.* **117**, 1103–1114.

Vera JC, Rivas CI, Zhang RH, Golde DW (1998) Colony-stimulating factors signal for increased transport of vitamin C in human host defense cells. *Blood* **91**, 2536–2546.

Wang H, Dutta B, Huang W, Devoe LD, Leibach FH, Ganapathy V, Prasad PD (1999) Human Na^+-dependent vitamin C transporter 1 (hSVCT1): primary structure, functional characteristics and evidence for a non-functional splice variant. *Biochim. Biophys. Acta* **1461**, 1–9.

Wang Y, Mackenzie B, Tsukaguchi H, Weremowicz S, Morton CC, Hediger MA (2000) Human vitamin C (L-ascorbic acid) transporter SVCT1. *Biochem. Biophys. Res. Commun.* **267**, 488–494.

Williams LE, Lemoine R, Sauer N (2000) Sugar transporters in higher plants – A diversity of roles and complex regulation. *Trends Plant Sci.* 5, 283–290.

Wilson JX (1989) Ascorbic acid uptake by a high-affinity sodium-dependent mechanism in cultured rat astrocytes. *J. Neurochem.* **53**, 1064–1071.

Wilson JX (1990) Regulation of ascorbic acid concentration in embryonic chick brain. *Dev. Biol.* **139**, 292–298.

Wilson JX (2002) The physiological role of dehydroascorbic acid. *FEBS Lett.* **527**, 5–9.

Wilson JX, Dixon SJ (1989) High-affinity sodium-dependent uptake of ascorbic acid by rat osteoblasts. *J. Membrane Biol.* **111**, 83–91.

Wilson JX, Dixon SJ (1995) Ascorbate concentration in osteoblastic cells is elevated by transforming growth factor-β. *Am. J. Physiol.* **268**, E565–E571.

Wilson JX, Dixon SJ, Yu J, Nees S, Tyml K (1996) Ascorbate uptake by microvascular endothelial cells of rat skeletal muscle. *Microcirculation* **3**, 211–221.

Wilson JX, Jaworski EM (1992) Effect of oxygen on ascorbic acid uptake and concentration in embryonic chick brain. *Neurochem. Res.* **17**, 571–576.

Wilson JX, Jaworski EM, Dixon SJ (1991) Evidence for electrogenic sodium-dependent ascorbate transport in rat astroglia. *Neurochem. Res.* **16**, 73–78.

Wilson JX, Jaworski EM, Kulaga A, Dixon SJ (1990) Substrate regulation of ascorbate transport activity in astrocytes. *Neurochem. Res.* **15**, 1037–1043.

Wilson JX, Lo TCY, Dixon SJ (1999) Glucose regulates dehydroascorbic acid uptake by skeletal muscle cells. *FASEB J.* **13**, A701.

Wilson JX, Peters C, Sitar SM, Daoust P, Gelb AW (2000) Glutamate stimulates ascorbate transport by astrocytes. *Brain Res.* **858**, 61–66.

Yusa T. (2001) Increased extracellular ascorbate release reflects glutamate re-uptake during the early stage of reperfusion after forebrain ischemia in rats. *Brain Res.* **897**, 104–113.

Zreik TG, Kodaman PH, Jones EE, Olive DL, Behrman H. (1999) Identification and characterization of an ascorbic acid transporter in human granulosa-lutein cells. *Mol. Human Reprod.* 5, 299–302.

7

Membrane redox proteins involved in ascorbate-mediated reactions

José Manuel Villalba, María del Carmen Córdoba-Pedregosa and José Antonio González-Reyes

7.1 Introduction

Redox enzymes are widely distributed among cellular membranes. Some catalyze *cis*-electron transport, when both donor and acceptor sites are located on the same side of the membrane, whereas others catalyze the reduction of acceptors inside the membrane and/or to a transmembrane flow of electrons. Ascorbate (ASC) is an efficient electron donor and it is used by some of these membrane enzymes. On the other hand, different redox forms of ASC exist, namely monodehydroascorbate (MDHA) and the fully oxidized dehydroascorbate (DHA), and both can act as electron acceptors to regenerate and maintain ASC. Enzymes that reduce DHA are mainly cytosolic (see Chapter 8 by May and Asard), whereas membrane proteins related to the regeneration of ASC use MDHA as acceptor. Regeneration of ASC may be important to maintain cellular levels of this compound, especially in those animals unable to synthesize it *de novo*. Furthermore, ASC regeneration at or near the membranes may be useful for cells because membranes are major sites of oxygen radical production and these reactions may favor the interaction of ASC/MDHA with lipophilic antioxidants such as α-tocopherol and coenzyme Q. In this chapter, we will review the current knowledge of redox proteins present in the different cellular membranes (from both animals and plants) using ASC or MDHA as substrates. We will also include new data obtained by the use of the yeast model, which have given new insights about the interactions between ASC/MDHA and some of these membrane proteins. On the other hand, it is important to consider that the enzymes catalyzing the last step of ASC biosynthesis (microsomal L-gulono-1,4-lactone oxidase of mammals, and L-galactono-1,4-γ-lactone dehydrogenase

Vitamin C, edited by Han Asard, James M May and Nicholas Smirnoff

located at the inner mitochondrial membrane in plants) are also membrane redox proteins that yield ASC as their catalysis product (see Smirnoff *et al.*, Chapter 1).

7.2 ASC and membrane oxidoreductases in animal cells

7.2.1 Cytochrome b$_{561}$ *of secretory vesicles*

Cytochrome b_{561} is a transmembrane electron carrier located in secretory vesicles of many endocrine tissues, including the chromaffin vesicles of adrenal medulla cells, neurohypophyseal secretory vesicles, and adrenergic synaptic vesicles in the central and peripheral nervous systems. Extravesicular ASC is the electron donor and MDHA is the intravesicular electron acceptor, which replenishes the high concentrations of intravesicular ASC needed as a cofactor for the activity of different monooxygenases (Njus *et al.*, 1990; *Figure 7.1*). However, cytochrome b_{561} is not restricted to neuroendocrine cells, since significant levels of cytochrome b_{561} mRNA expression have been also detected in colon cancer cell lines, T-cell lymphomas, and undifferentiated cell lines such as K-562 cells (Perin *et al.*, 1988; Srivastava, 1995).

Cytochrome b_{561} is characterized by a rather high redox midpoint potential and an asymmetric absorption peak in the α-band with a maximum at 561 nm and a shoulder at about 558 nm. This protein is the best characterized membrane redox enzyme that is involved in ASC-related reactions in animal cells. It is a highly hydrophobic hemoprotein of about 28 kDa. Bovine cytochrome b_{561} has six transmembrane helices with both N- and C-terminal ends facing the cytosol (Perin *et al.*, 1988). However, the human protein lacks the first putative 22 amino acids at the N terminus predicted from the bovine sequence to be in the cytoplasmic side, and a revised hypothesis has been proposed in which cytochrome b_{561} is inserted in the membrane with five transmembrane helices (Srivastava *et al.*, 1994). The genomic structure of the human gene encoding cytochrome b_{561}, with five consecutive exons encoding the five transmembrane segments, also supports this interpretation (Srivastava, 1995).

Cytochrome b_{561} has two hemes *b* per molecule (Tsubaki *et al.*, 1997), which are expected to be located on both sides of vesicular membranes (Okuyama *et al.*, 1998). Presence of two independent heme *b* centers is consistent with the observation of two different forms of cytochrome b_{561} with midpoint potentials of + 170 and + 70 mV respectively (Apps *et al.*, 1984). The two hemes can be also distinguished by different electron paramagnetic resonance spectroscopy (EPR) signals (Tsubaki *et al.*, 1997). The primary sequence of bovine cytochrome b_{561} contains six totally conserved histidine residues (His^{54}, His^{88}, His^{92}, His^{110}, His^{122}, and His^{161}). It is accepted that the heme iron of the extravesicular side is coordinated with the pair, either His^{88} (helix 3)–His^{161} (helix 5) or His^{92} (helix 3)–His^{161} (helix 5), whereas the heme iron oriented to the intravesicular side is coordinated with the pair His^{54} (helix 2)–His^{122} (helix 4) (Tsubaki *et al.*, 1997). The two hemes *b* of cytochrome b_{561} are not functionally equivalent. It is very likely that the heme with lower redox potential (EPR signal at g_z = 3.70) participates in accepting an electron from extravesicular ASC, whereas the heme with higher redox potential (EPR signal at g_z = 3.14) donates an electron to intravesicular MDHA (Tsubaki *et al.*, 1997; Kobayashi *et al.*, 1998).

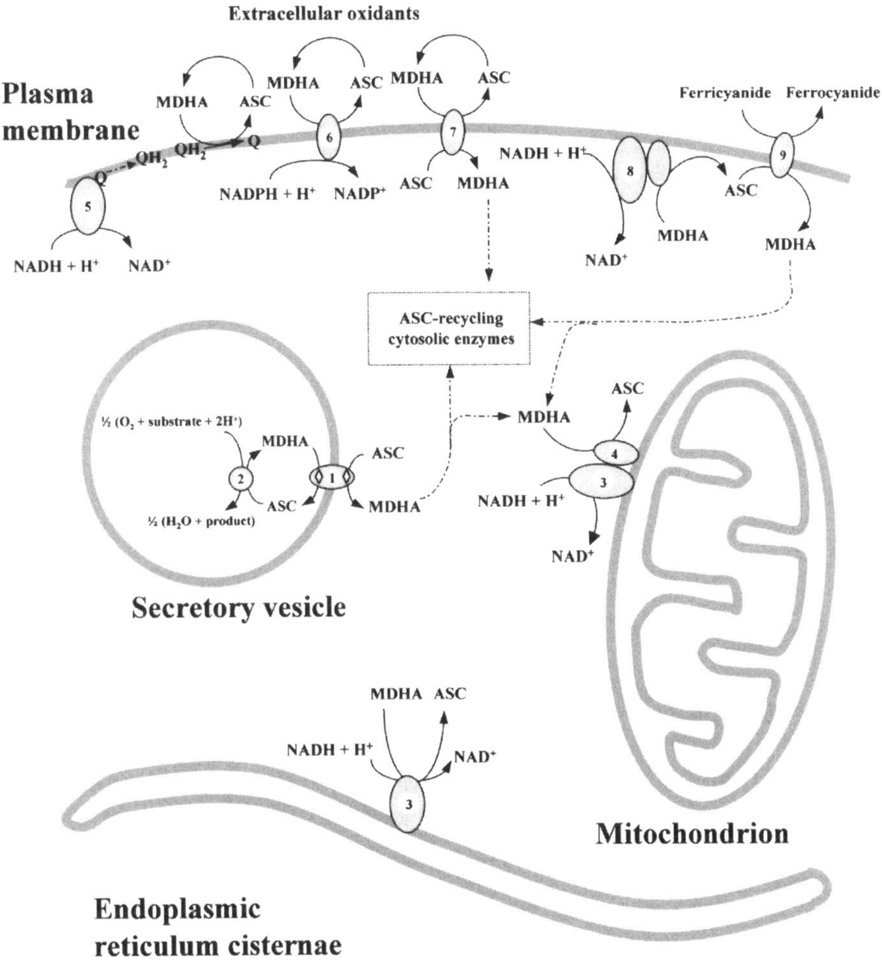

Figure 7.1: Membrane redox proteins involved in ascorbate-mediated reactions in mammals. (1) Cytochrome b_{561} showing its intra and extravesicular hemes. (2) Monooxygenases of secretory vesicles. (3) NADH-cytochrome b_5 reductase in outer mitochondrial membranes and ER. (4) Cytochrome b_5 of outer mitochondrial membranes. (5) PM quinone reductase and transmembrane quinone-dependent MDHA reductase. (6) Ferrireductase of yeast PM. (7) Transmembrane ASC-MDHA reductase. (8) Endofacial NADH-MDHA reductase in PM. (9) Transmembrane ASC-ferricyanide reductase. See text for further details.

On the other hand, based on deduced amino acid sequences of seven species, two fully conserved regions have been recognized as putative binding sites for ASC and MDHA respectively. The first conserved sequence (^{69}ALLVYRVFR^{77}) is located on the extravesicular side of an α-helical segment and likely constitutes the binding site for ASC. The second one (^{120}SLHSW^{124}) is located in an intravesicular loop and probably constitutes the binding site for MDHA (Okuyama *et al.*, 1998). The two extravesicular heme-coordinating histidine residues may be also a part of the ASC binding site and additional residues (Tyr^{218} and Lys^{85}) have been proposed to play a role in the recognition/binding process for ASC and the fast electron transfer reaction (Takeuchi *et al.*, 2001).

A homologue of cytochrome b_{561} (Dcytb) is also present in brush border plasma membranes (PM) from duodenal enterocytes. Dcytb is also a di-heme cytochrome but is involved in iron reduction. This protein also contains the conserved His residues involved in heme binding, and shares a similar membrane topology with six predicted transmembrane spanning regions. Dcytb lacks sequence homology with ferric reductases from plants and yeasts, and also lacks any conventional NADH, NADPH or flavin binding motifs. Interestingly, putative binding sites for ASC and MDHA were found to be partially conserved in Dcytb, suggesting that ASC might be an electron donor for this cytochrome (McKie *et al.*, 2001).

7.2.2 The NADH-cytochrome b_5 *reductase/cytochrome* b_5 *system of outer mitochondrial membranes*

NADH-cytochrome b_5 reductase (NADH : ferricytochrome b_5 oxidoreductase, Cytb_5R) is an integral membrane enzyme of about 35 kDa located mainly in the endoplasmic reticulum (ER) and outer mitochondrial (OM) membranes (Borgese and Pietrini, 1986). The enzyme is also located in PM from erythrocytes (Kitajima *et al.*, 1981) and nucleated cells such as hepatocytes (Navarro *et al.*, 1995). In addition to these membrane-bound forms, soluble forms have been also described in erythrocytes and nonerythroid cells (Leroux *et al.*, 2001). Membrane-bound Cytb_5R contains a large catalytic domain exposed to the cytosol and a short N-terminal membrane anchor which is usually myristylated (Pietrini *et al.*, 1992), and is absent in soluble forms. Biochemical and molecular studies have shown that Cytb_5R in its various subcellular and tissue locations derives from the translation of various transcripts generated by alternative promoters and splicing pattern (Pietrini *et al.*, 1988, 1992; Leroux *et al.*, 2001).

The reductase was known primarily by its function in the reduction of microsomal cytochrome b_5 (Cytb_5), a small heme protein that consists of a cytoplasmic domain of 90 amino acids and a C-terminal membrane anchor. Its molecular mass is about 23 kDa and its oxidation-reduction potential is –44.0 mV (Nishino and Ito, 1986; D'Arrigo *et al.*, 1993). Most cellular Cytb_5 localizes to the ER and, to a minor extent, to OM membranes. These two membranes contain distinct Cytb_5 isoforms that are specifically targeted through post-translational pathways (D'Arrigo *et al.*, 1993). The Cytb_5R/Cytb_5 system participates in various aspects of lipid metabolism, whereas erythrocyte Cytb_5R is involved in the reduction of methemoglobin (Kitajima *et al.*, 1981; Shirabe *et al.*, 1995).

Cytb_5R in OM membranes plays an important role in ASC metabolism (*Figure 7.1*). OM Cytb_5R is involved in the reduction of MDHA (OM NADH-MDHA reductase) for regeneration of cytosolic ASC (Ito *et al.*, 1981; Diliberto *et al.*, 1982). Besides Cytb_5R, NADH-MDHA reductase involves an OM-specific Cytb_5 (Nishino and Ito, 1986). Consistent with the involvement of more than one component in the electron transfer between NADH and MDHA, the activity does not survive detergent solubilization of membranes (Lee *et al.*, 2001). Similar to the ER Cytb_5R/Cytb_5 system, the OM components also constitute the rotenone-insensitive NADH-cytochrome *c* reductase (Ito, 1980). It is noteworthy that the ratio of OM cytochrome *b*-mediated

NADH-MDHA reductase activity to the corresponding NADH-cytochrome *c* reductase activity varied among various tissues and subcellular fractions. Moreover, 2-thenoyl trifluoroacetone (TTFA) inhibited NADH-MDHA reductase, but not NADH-cytochrome *c* reductase activity. Together, these results suggest that a third component may be involved between OM Cytb_5 and the MDHA (Nishino and Ito, 1986). However, such a component has not been identified to date.

The specific activities of OM MDHA reductase in various rat and guinea pig tissues appear to correlate positively with their ASC contents. Thus, the OM MDHA reductase is probably one of the major factors that regulate ASC levels in mammalian tissues (Diliberto *et al.*, 1982). The role of Cytb_5R in the metabolism of cellular ASC in humans has been clearly demonstrated from a genetic point of view. A genetically transmitted Cytb_5R deficit is a cause of hereditary methemoglobinemia. Fibroblasts without immunologically undetectable Cytb_5R derived from a patient with generalized methemoglobinemia had severely reduced MDHA reductase activity. Whereas altered lipid metabolism has been suggested to be the cause of the highly pleiotropic effects of generalized methemoglobinemia, a reduced capacity to regenerate ASC could be the cause of some of the clinical manifestations of this syndrome (Shirabe *et al.*, 1995).

7.2.3 Plasma membrane MDHA reductases

The PM also contains NADH-MDHA reductase activity that is mostly oriented to the cytosol, although a minor portion is transmembranous (Goldenberg *et al.*, 1983; May *et al.*, 2001). Similar to the activity present in OM membranes, the PM NADH-MDHA reductase is inhibited by solubilization, indicating that more than one component may be involved (Villalba *et al.*, 1993; May *et al.*, 2001). PM NADH-MDHA reductase has a high affinity for both NADH and MDHA, with a K_m value of about 1 μM for NADH, and 0.75 μM for MDHA in erythrocyte membranes (Goldenberg *et al.*, 1983; May *et al.*, 2001).

The Cytb_5R is a candidate to mediate *cis*-oriented PM MDHA reductase, given the role played by this enzyme in the reduction of MDHA in intracellular membranes. Addition of purified Cytb_5R to isolated liver PM stimulated NADH-MDHA reductase activity, and the effect was not attributable to the direct reduction of MDHA by the added Cytb_5R (Villalba *et al.*, 1993, 1995; Gómez-Díaz *et al.*, 1997a). However, removal of Cytb_5R from erythrocyte PMs by proteolytic digestion with cathepsin D did not affect NADH-MDHA reductase, whereas the ferricyanide reductase activity was severely reduced (May *et al.*, 2001). Failure of cathepsin D to produce a decrease in NADH-MDHA reductase argues against the participation of Cytb_5R in the reduction of MDHA in erythrocyte membranes (May *et al.*, 2001). Additional research is required to elucidate the putative role of Cytb_5R in *cis*-oriented NADH-MDHA reductase at the PM.

The systems we have described for OM membranes and the cytosolic face of the PM, together with additional cytosolic systems (see chapter by May and Asard), maintain ASC levels inside the cells. However, animal cells also stabilize extracellular ASC (Alcaín *et al.*, 1991; Van Duijn *et al.*, 2000), which has been recognized as an important cellular function for the maintenance of an antioxidant system that protects the PM from oxidative damage (Villalba *et al.*, 1998; May *et al.*, 2001). The activity has been explained by transmembrane ASC regeneration via the reduction of extracellular MDHA (Rodríguez-Aguilera and Navas, 1994; May, 1999; Van Duijn *et al.*, 2000). This is further evidenced by the fact that HL-60 cells (Van Duijn *et al.*, 1998), K-562 cells (Arroyo *et al.*, 2000) and human erythrocytes (May *et al.*, 2000) decrease steady-state levels of MDHA generated from ASC with ASC oxidase.

Reduction of oxidized species of ASC outside the cell is a more complicated process than regeneration of intracellular ASC (Van Duijn *et al.*, 2000), and the mechanisms involved are not completely understood. In general, two different enzymatic systems have been recognized that can account for MDHA reduction outside the cell, one relying on NADH and another one based on intracellular ASC as electron donor.

Transplasma membrane NADH-MDHA reductase

The existence of a transmembrane MDHA reductase that reduces extracellular MDHA using intracellular NADH was early recognized by Goldenberg *et al.* (1983) in erythrocyte membranes, and later confirmed by Villalba *et al.* (1993, 1995) in liver cell PM. This activity is considered to account for the ASC stabilization displayed by both nucleated human cells in culture (Alcaín *et al.*, 1991; Gómez-Díaz *et al.*, 1997a) and whole erythrocytes (May *et al.*, 2000, 2001). Transmembrane NADH-MDHA reductase in erythrocyte membranes constitutes about 12% of the total NADH-MDHA reductase. Also, this activity can be separated from NADH-ferricyanide oxidoreductase (Villalba *et al.* 1993; May *et al.*, 2001; *Figure 7.1*).

Transmembrane NADH-MDHA reductase activity does not involve a single transmembrane polypeptide, but more likely requires different redox carriers. Data support a role for coenzyme Q_{10} in mediating electron transport for transmembrane reduction of MDHA in PM (*Figure 7.1*). In this way, NADH-MDHA reductase from PM of rat liver and K562 cells, as well as ASC stabilization and MDHA scavenging by K562 cells, are sensitive to the coenzyme Q_{10} status and can be inhibited by quinone antagonists (Gómez-Díaz *et al.*, 1997a; Arroyo *et al.*, 2000; Villalba *et al.*, 1995). In addition, a higher amount of coenzyme Q_{10} in the PM determines a higher activity of ASC stabilization in whole HL-60 cells (Gómez-Díaz *et al.*, 1997b).

Participation of the quinone implies that reductases acting on PM coenzyme Q are important determinants in the reduction of extracellular MDHA from intracellular NADH. PM Cytb_5R can function as a quinone reductase reducing ubiquinone *in situ* or in reconstituted liposomes (Navarro *et al.*, 1995; Villalba *et al.*, 1997). However, it is not clear if ubiquinol can reduce MDHA directly, or if another carrier is required. Lectin inhibition of NADH-MDHA reductase in isolated PMs from rat liver (Villalba *et al.*, 1993) and scavenging of MDHA by intact cells (Arroyo *et al.*, 2000) suggests that some component of the glycocalix may favor the access of MDHA to the bilayer surface, where it could be reduced by ubiquinol.

Transplasma membrane ASC-MDHA reductase

In addition to NADH, ASC is also an electron donor for a transplasma membrane reductase that reduces MDHA outside the cell (May *et al.*, 1996; Van Duijn *et al.*, 2000; *Figure 7.1*). By loading human erythrocytes with different amounts of ASC, it was found that the intracellular ASC concentration correlated with the ability of erythrocytes to decrease the oxidation rate of external ASC. Exposure of ASC-loaded erythrocytes to higher levels of extracellular MDHA in the presence of Tris-(ethylenediamine)-nickel (II) chloride 2-hydrate ($Ni(en)_3 2+$) allowed the detection of intracellular MDHA. $Ni(en)_3 2+$ is a compound that cannot cross the PM, and thus broadens the EPR signal of MDHA only in the extracellular space. Taken together, these results are indicative of extracellular MDHA reduction using intracellular ASC. This system coexists with the transmembrane NADH-dependent MDHA reductase (Van Duijn *et al.*, 2000). Reduction of extracellular MDHA by intracellular ASC is an electrogenic process, indicating vectorial electron transport. This observation argues against participation of redox carriers such a coenzyme Q or α-tocopherol, because they also transport a proton together with the electron (Van Duijn *et al.*, 2001a). Moreover, capsaicin and dicumarol, inhibitors of coenzyme Q-mediated electron transport, did not inhibit the decrease in the level of MDHA in the presence of ASC-loaded erythrocytes (see Van Duijn *et al.*, 2000).

It has been proposed that a single protein or a protein complex is involved in the electron transport from intracellular ASC to extracellular MDHA (Van Duijn *et al.*, 2000). This system appears to be mechanistically equivalent to the trans-membrane ASC-MDHA electron transport of secretory vesicles, that involves cytochrome b_{561} (see above and *Figure 7.1*). Erythrocyte membranes contain a cytochrome *b* with spectral properties similar to cytochrome b_{561}. However, the existence of cytochrome b_{561} in erythrocyte membranes was not confirmed by immunoblotting techniques nor by reverse transcriptase polymerase chain reaction (RT-PCR) analysis in peripheral reticulocytes, the last erythrocyte progenitor cells still containing mRNA.

It remains for further investigation to elucidate whether or not the unknown *b*-cytochrome present in erythrocyte membranes plays a role in transmembrane ASC-dependent reduction of MDHA (Van Duijn *et al.*, 2001b). In *Figure 7.1* we represent different MDHA reductases acting at the PM, and their possible links with other redox constituents.

7.2.4 Transplasma membrane ASC-ferricyanide reductase

ASC is also a primary electron donor for the reduction of other impermeable electron acceptors such as ferricyanide (*Figure 7.1*). This transplasma membrane activity, recognized in erythrocytes (May *et al.*, 1996; May and Qu, 1999), HL-60 cells (Van Duijn, 1998) and pulmonary arterial endothelial cells (Merker *et al.*, 1998), is limited by the ability of the cells to recycle intracellular ASC (May and Qu, 1999) and can explain the long known effect of DHA-stimulated ferricyanide reduction by cells (Orringer and Roer, 1979). This stimulatory effect relies on the cellular uptake of DHA by the glucose transporter, its intracellular reduction to ASC, and a transmembrane electron transport to reduce extracellular ferricyanide (May *et al.*, 1996).

α-Tocopherol in PM appears to play a role shuttling electrons from intracellular ASC to extracellular ferricyanide. In this mechanism, increasing the amount of α-tocopherol in erythrocyte PMs also increased the rate of ferricyanide reduction, whereas depleting erythrocyte PMs from α-tocopherol inhibited the reaction, and this activity was restored by α-tocopherol reconstitution (May *et al.*, 1996). However, changing the α-tocopherol content of HL-60 cells had no effect on the ASC-stimulated ferricyanide reduction, showing that a nonenzymatic redox system using α-tocopherol is not involved (Van Duijn *et al.*, 1998). Enzyme(s) involved in transmembrane ASC-ferricyanide reductase activity have not yet been characterized, although participation of endofacial NADH-cytochrome b_5 reductase has been excluded. Studies with the impermeable thiol reagent *para*-(chloromercuri)benzenesulfonic acid (*p*CMBS) supported the idea that the enzyme involved in this reaction is a transmembrane protein that contains sensitive thiol groups exposed to both the extracellular and cytoplasmic faces (May and Qu, 1999).

Inhibitor studies have also shown that the enzyme is different from the transmembrane ASC-MDHA reductase. Thus, *p*CMBS strongly affected the reduction of ferricyanide in the absence of ASC in HL-60 cells, and partially inhibited the ASC-stimulated reaction (Van Duijn *et al.*, 1998). However, *p*CMBS did not inhibit ASC-MDHA reductase (Van Duijn *et al.*, 2000). Since *p*CMBS did not affect reduction of ferric iron in erythrocytes (May and Qu, 1999), these results could indicate that the erythrocyte transmembrane MDHA reductase also catalyzes ferric iron reduction (Van Duijn *et al.*, 2000). Existence of at least two separate transplasma membrane redox systems in mammalian cells is also supported by the fact that DHA stimulated ferricyanide reduction had little effect on methylene blue reduction by pulmonary arterial endothelial cells (Merker *et al.*, 1998). Different mechanisms involving ASC and redox membrane proteins in mammals are depicted in *Figure 7.1*.

7.3 ASC and membrane oxidoreductases in plant cells

In higher plants, the best characterized membrane proteins using ASC as an electron donor or MDHA as an electron acceptor are mainly involved in reactive oxygen species (ROS) scavenging. Different protein isoforms have been found attached to membranes in which redox reactions lead to relatively high amounts of ROS. These include peroxisomal and glyoxysomal membranes, the thylakoid membrane, inner mitochondrial membrane and PM (Del Rio *et al.*, 2002; Bérczi and Møller, 2000; Asada, 1999; Møller, 2001).

7.3.1 Membrane-bound ASC peroxidase

With the exception of the bovine eye (Wada *et al.*, 1998), ASC peroxidase (APX) constitutes a family of proteins exclusively found in higher plants, eukaryotic algae and cyanobacteria (Shigeoka *et al.*, 2002). Besides the soluble cytosolic, chloroplastic stroma and mitochondrial matrix isoforms, three different membrane-bound APXs have been well characterized in plants. These are: peroxisome-, glyoxisome- and thylakoid-bound APX (pAPX, gAPX and tAPX respectively). In addition, two more membrane-attached isoforms of unknown identity (Jespersen *et al.*, 1997; Ishikawa *et al.*, 1995) and a mitochondrial membrane-bound (mAPX) isoform have been reported (Jiménez *et al.*, 1997; De Leonardis *et al.*, 2000; Iturbe-Ormaetxe *et al.*, 2001).

Both soluble and membrane-bound, APXs catalyze the reaction:

$$2ASC + 2H_2O_2 \rightarrow 2MDHA + 2H_2O + O_2$$

It is interesting to note that some of the compartments showing APX activity also contain catalase, making the presence of APX apparently redundant. However, the higher affinity of APX for H_2O_2 compared to catalase suggests different levels of detoxification in these organelles.

Peroxisome and glyoxisome APXs

Peroxisomes are single membrane-bound organelles of about 1 μm diameter found in almost all eukaryotic cells. Typical peroxisomes contain H_2O_2-producing oxidases, catalase, enzymes for the glyoxylate cycle, as well as enzymes involved in lipid metabolism (e.g., biosynthesis, cholesterol and dolichol metabolism, β-oxidation of fatty acids) (Bunkelmann and Trelease, 1996; Nito *et al.*, 2001; Del Rio *et al.*, 2002). The peroxisomes involved in the glycoxylate cycle are specifically called glyoxysomes, and are present in germinating seedlings (see Del Rio *et al.*, 2002). Another specialized peroxisome is found in root nodules from certain tropical legumes (Iturbe-Ormaetxe *et al.*, 2001). All these different forms of peroxisomes are generally called microbodies, so that the term microbody-bound APX is widely used.

Peroxisome and glyoxysome membrane-bound APXs are synthesized at the rough ER, accumulated at specific subdomains of the organelle and sorted to the peroxisome membrane (Mullen *et al.*, 2001a). The protein spans the membrane by a C-terminal amino acid residue sequence, and as such is an integral 'tail-anchored' type protein. The catalytic domain of the 31–35 kDa enzyme faces the cytosolic surface (Mullen and Trelease, 2000; Mullen *et al.*, 2001a, 2001b; Nito *et al.*, 2001).

Although peroxisomes and glyoxysomes contain catalase, some authors have suggested that H_2O_2 escaping from the catalase action could permeate the microbody membrane reaching the cytosol where it could be scavenged by the membrane-attached APX or by the cytosolic isoform. Interestingly, the microbody membrane contains an MDHA reductase as well (Bunkelmann and Trelease, 1996). Thus, MDHA formed by the action of APX could be reduced back to ASC by the microbody-bound MDHA reductase in the cytosol next to the membrane. DHA originating from MDHA disproportionation is used as substrate for the ASC–glutathione cycle in the cytosol, thus contributing to ASC regeneration (Noctor and Foyer, 1998; see also *Figure 7.2*).

Thylakoid membrane-bound APX

A thylakoid membrane-specific isoform has been demonstrated in several species. In spinach the protein has a molecular mass of about 37.7 kDa (Ishikawa *et al.*, 1996). Similar to the microbody APX, the thylakoid isoform contains a membrane spanning C-terminal hydrophobic domain (Jespersen *et al.*, 1997; Ishikawa *et al.*, 1996). The catalytic site faces the chloroplast stroma (Miyake and Asada, 1992; Shigeoka *et al.*, 2002). In many species, such as pumpkin, spinach, tobacco or *Mesembryanthemum crystalinum*, the tAPX originates from alternative splicing of a common pre-mRNA that encodes for both the soluble stromal APX and tAPX

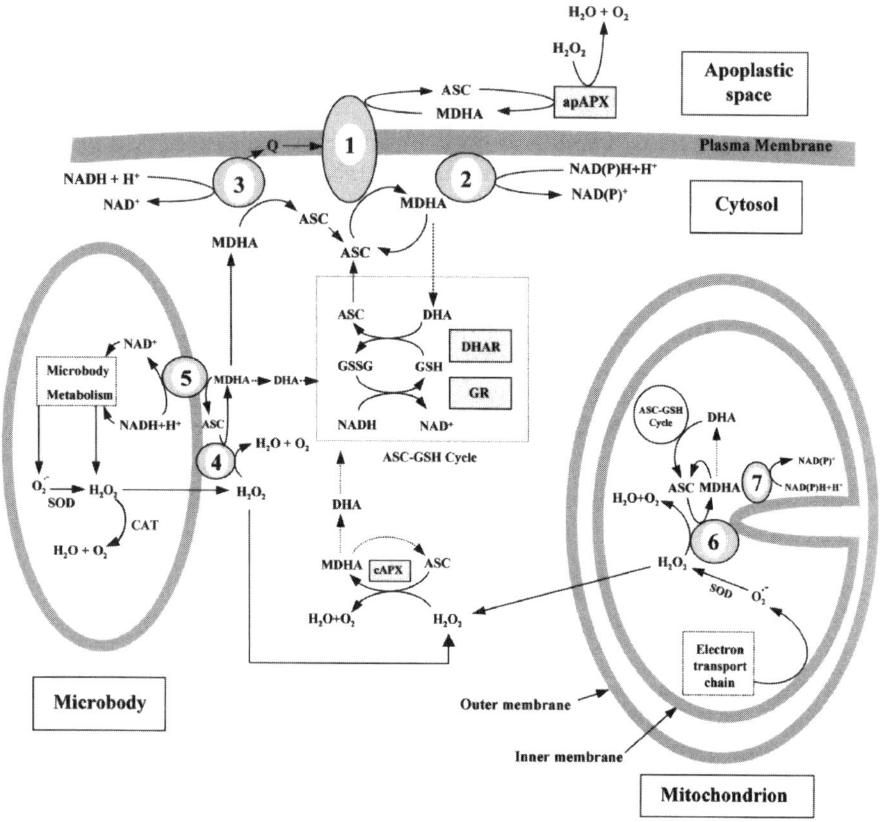

Figure 7.2: Integration of cytosolic and plasma membrane, mitochondrial and microbody membrane events in the regulation of ascorbate-mediated processess in higher plant cells. (1) Cytochrome b_{561}. (2) NAD(P)H-MDHA oxidoreductase. (3) NADH-quinone/MDHA reductase. (4) and (5), microbody (peroxisome and glyoxysome) APX and MDHA reductase, respectively. (6) and (7), inner mitochondrial membrane bound APX and MDHA reductase, respectively. Possible links between organella scavenging systems and cytosolic detoxifying pathways are included. ApAPX, Apoplastic ascorbate peroxidase; cAPX, cytosolic ascorbate peroxidase; DHAR, DHA reductase; GR, glutathione reductase.

(Ishikawa *et al.*, 1996; Yoshimura *et al.*, 2002). However, in *Arabidopsis* the chloroplast isoforms are encoded by different genes (Jespersen *et al.*, 1997).

Photosynthetic electron transport in chloroplasts is accompanied by the reduction of O_2 to $O_2^{\cdot-}$ and susbsequent dismutation of $O_2^{\cdot-}$ to yield H_2O_2 and other ROS in the thylakoid environment. Thus, specific mechanisms for controling ROS levels in the chloroplast have been reported. Higher plant chloroplasts lack catalase, but use the so called 'water–water cycle' as the main mechanism to scavenge ROS (see Asada, 1999). This cycle involves a number of soluble stromal and thylakoidal membrane-bound enzymes, such as the tAPX, and is connected to the ASC–glutathione cycle, which also operates in the chloroplast stroma (Asada, 1999; Shigeoka *et al.*, 2002). These processes are summarized in *Figure 7.3*.

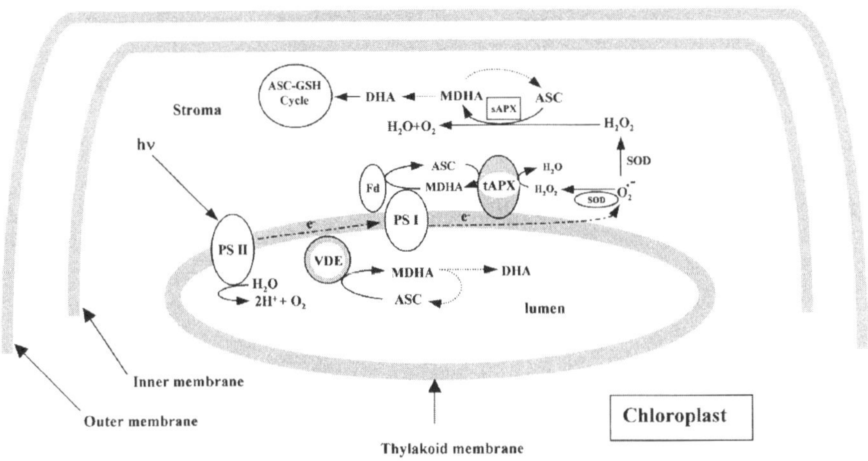

Figure 7.3: Redox membrane-bound proteins involved in ascorbate metabolism and ROS detoxification in higher plant chloroplasts.
Dotted line is photosynthetic electron transport. tAPX, Thylakoidal membrane bound ascorbate peroxidase; Fd, ferredoxin; sAPX, stromal soluble ascorbate peroxidase; SOD, superoxide dismutase; VDE, violaxanthin de-epoxidase. Soluble stromal MDHA reductase and the reaction violaxanthin–zeaxanthin inside the thylakoidal membrane, are not shown.

The importance of tAPX in the control of chloroplast ROS has been demonstrated using transgenic tobacco plants overexpressing the enzyme and showed an increased tolerance to oxidative stress (Yabuta *et al.*, 2002). On the other hand, it was not possible to obtain transgenic tobacco and spinach plants with lowered tAPX (Yabuta *et al.*, 2002). Thus, tAPX is apparently a limiting factor of the antioxidative system under photo-oxidative stress. Further, enhanced tAPX activity in transgenic plants functions to maintain an appropriate content and redox status of ASC under stress conditions.

Mitochondrial membrane-bound APX

As a compartment in which ROS are produced at high rates (Møller, 2001), mitochondria contain highly effective scavenging systems. These include Mn-superoxide dismutase (SOD), a matrix soluble APX, the enzymes involved in the ASC–glutathione cycle, as well as relatively high amounts of ASC and glutathione (Dalton *et al.*, 1993, De Leonardis *et al.*, 2000; Jiménez *et al.*, 1997, 1998). APX activity attached to the inner membrane has been reported as well (Jiménez *et al.*, 1997; Iturbe-Ormaetxe *et al.*, 2001). However, data concerning its molecular structure and/or topology are not available.

7.3.2 Membrane-bound MDHA reductase

Soluble MDHA reductase is a well-known enzyme, and has been detected in several subcellular compartments in plant cells such as cytosol, mitochondria, peroxisome/glyoxysomes, and chloroplasts (Hossain and Asada, 1985; Jiménez *et al.*, 1997). This FAD-containing enzyme catalyzes the following reaction:

$$2MDHA + NAD(P)H + H^+ \rightarrow 2ASC + NAD(P)^+$$

The enzyme prefers NADH as the electron donor but may work with ferricyanide as the acceptor. In membrane-bound form, this enzyme has been detected in PMs (Bérczi and Møller, 1998, 2000), peroxisomes and glyoxysomes (Bunkelmann and Trelease, 1996). A mitochondrial membrane-bound enzyme activity has been reported as well (Jiménez *et al.*, 1997; see *Figure 7.2*).

The PM enzyme has a molecular mass of 45 kDa and is attached to the membrane at the cytosolic surface. In this location the MDHA reductase could transfer reducing equivalents from ASC to the cell wall via the cytochrome b_{561} (see below, and chapter by May and Asard) for apoplastic ASC regeneration (*Figure 7.2*). The cytochrome b_{561}-mediated ASC–ASC cycle at the PM can thus play a role in the maintenance of the ASC redox status at the extracellular matrix.

Peroxisomal and glyoxysomal membranes contain MDHA reductase forms of about 32 kDa molecular mass (*Figure 7.2*). This protein is suggested to be transmembrane or facing the organelle matrix (Bunkelmann and Trelease, 1996; López-Huertas *et al.*, 1999). A dual, complementary function for the microbody membrane-bound APX and MDHA reductase has been proposed. The first function could be the reoxidation of NADH to NAD^+ at the luminal side of the organelle, which is required for peroxisomal metabolism. The second function could be related to the protection against H_2O_2 and other ROS leaking out of peroxisomes (Jiménez *et al.*, 1997; Del Rio *et al.*, 2002).

7.3.3 Cytochrome b_{561} *of plant plasma membranes*

Originally described in mammalian secretory chromaffin granules (see above), cytochrome b_{561} has been also found in plant cell membranes, particularly in PM (Asard *et al.*, 2001). Despite some biochemical similarities with animal cells, cytochrome b_{561} isolated from plants shows specific differential characteristics: it is probably a glycosylated protein of about 50–55 kDa (Trost *et al.*, 2000). This contrasts with the mean molecular mass reported for animal cells, which is 25–28 kDa. However, the proposed molecular architecture is very similar to that reported for animals (see Asard *et al.*, 2001). In plant PM, the protein is supposed to transfer electrons from cytosolic ASC to apoplastic MDHA (*Figure 7.2*). MDHA generated at the cytosolic face can be reduced back to ASC by the MDHA reductase (see above) at the expense of NADH (Bérczi and Møller, 2000). The protein can also accept electrons from hydroquinones as juglonol (Preger *et al.*, 2001). More details on its structure and functions are reviewed in the chapter by May and Asard.

7.3.4 Other ASC-mediated redox processes in plant membranes

In addition to the membrane proteins discussed above, it is known that ASC may interact with other membrane-bound enzymes. However, in most of the cases neither the protein itself nor the biological significance of such interactions are adequately understood. For example, several interactions of ASC and thylakoidal membranes have been reported. It has been shown that, under certain circumstances, photosystem I (PSI) can accept electron from ASC – via plastocyanin or

P700 – in C4 plants (Ivanov *et al.*, 2001). This reaction is involved either in the energization of the thylakoidal membrane or in priming of electron transport (Ivanov *et al.*, 2001). Also, ferredoxin, a soluble protein, can reduce MDHA with higher efficiency than MDHA reductase, and its presence in the immediate vicinity of PSI seems to be necessary as part of the thylakoidal scavenging system. The soluble chloroplastic MDHA reductase isoform is apparently most involved in the stromal ROS scavenging system (Asada, 1999; see *Figure 7.3*).

The thylakoid lumen also contains violaxanthin de-epoxidase (VDE), an enzyme involved in the dissipation of excess energy (*Figure 7.3*). VDE catalyzes the de-epoxidation of violaxanthin to zeaxanthin inside the membrane using ASC as a reductant (Asada, 1999; Noctor and Foyer, 1998). At the acid pH in the thylakoid lumen (about pH 5), the enzyme binds to the inner surface of the membrane by a 12-amino acid sequence at the C-terminal end (Hieber *et al.*, 2002). In plant PMs the presence of a 31-kDa protein with NADH-MDHA oxido-reductase activity facing the cytosol has also been reported (Serrano *et al.*, 1994; Córdoba *et al.*, 1995). This protein can also use quinones as electron acceptors, and could therefore be involved in the reduction cytochrome b_{561} as described above. Different mechanisms involving ASC and redox membrane proteins in plants are depicted in *Figures 7.2* and *7.3*.

7.4 The yeast model as a tool for the study of ASC-related membrane oxidoreductases

7.4.1 Mitochondrial NADH-cytochrome b_5 *reductase and the reduction of D-erythroascorbate free radical*

Saccharomyces cerevisiae lacks ASC but instead contains D-erythroascorbate (EASC), a C5 analog of ASC. EASC is very similar to ASC in its structure and physicochemical properties (see Chapter 1 by Smirnoff *et al.*). EASC is also readily oxidized in aqueous systems and has a redox system similar to that of ASC, consisting of EASC, D-erythromonodehydroascorbate (EMDHA), and dehydro-D-erythroascorbate. Regeneration and maintenance of EASC in yeasts is largely dependent on a EMDHA reductase (Lee *et al.*, 2001).

Genetic evidence has clearly proven the involvement of OM Cytb_5R in EMDHA reductase. Disruption of the *MCR1* gene, which encodes mitochondrial Cytb_5R in yeasts, resulted in a dramatic decrease of both NADH-Cytb_5 reductase and EMDHA reductase activities (Lee *et al.*, 2001). Under these conditions, intracellular levels of EASC in cells also decreased significantly. On the contrary, over-expression of the *MCR1* gene resulted in significant increases of NADH-Cytb_5 reductase, EMDHA reductase, and higher levels of intracellular EASC in cells. Maintenance of EASC by OM Cytb_5R is important for antioxidant protection in yeast cells because the *mcr1* disruptant was highly sensitive to oxidative stress and conversely, *MCR1*-overexpressing cells were more resistant than parental cells (Lee *et al.*, 2001). Importantly *mcr1* disruptants are even more sensitive to oxidative stress than yeast without a functional *ALO1* gene, which encodes D-arabinono-1,4-lactone oxidase and completely lack EASC (Kim *et al.*, 1998; Huh *et al.*, 1998). This indicates that hypersensitivity of *mcr1* disruptants to oxidative stress is not caused merely by their low EASC levels. It also implies a more general function for Cytb_5R in antioxidant protection, for instance through the regeneration of lipophilic antioxidants (Lee *et al.*, 2001).

Whereas such genetic approaches have clearly established the participation of Cytb_5R in the regeneration of EASC in yeasts, these investigations have not supported a role for Cytb_5 in this function. This is because neither overexpression of the *CYB5* gene, the only reported gene for Cytb_5 in *Saccharomyces cerevisiae* (Truan *et al.*, 1994), nor disruption of this gene, had an effect on NADH-EMDHA reductase activity. Since NADH-EMDHA reductase of OM membranes was also lost after membrane solubilization, another component must be required to reduce EMDHA (Lee *et al.*, 2001).

7.4.2 Coenzyme Q, the ferrireductase system, and the stabilization of extracellular ASC

Yeast cells decrease the oxidation rate of extracellular ASC and this activity is probably related to a transmembrane NADH-MDHA reductase. Sensitivity to quinone antagonists suggested the participation of a quinoid compound. The genetic approach using *S. cerevisiae* has provided strong evidence for the participation of coenzyme Q in MDHA reduction. Several mutant strains (*coq2*Δ, *coq3*Δ and *coq7*Δ) that are defective in coenzyme Q biosynthesis and completely lack coenzyme Q_6, the natural coenzyme Q homologue in *S. cerevisiae*, are also defective in ASC stabilization. PMs isolated from strain *coq3*Δ are also defective in NADH-MDHA reductase activity. Both transformation of mutant strains with plasmids bearing the corresponding wild-type genes, and culturing cells in the presence of coenzyme Q_6 restored these ASC-dependent activities (Santos-Ocaña *et al.*, 1995, 1998a, 1998b).

Participation of coenzyme Q_6 in electron transport from intracellular NADH to extracellular MDHA supports an important role for quinone reductases. A Cytb_5R could be involved in the reduction of PM coenzyme Q_6 in yeasts, as deduced from inhibition of the NADH-MDHA reductase by the thiol reagent *p*-hydroxymercuribenzoate, as well as the one-electron reaction mechanism of the quinone reductase present in yeast PMs. Since dicumarol was without a significant effect on either ASC stabilization by whole yeast cells, or on NADH-MDHA reductase in isolated PMs, a role for NAD(P)H:quinone oxidoreductase 1 (NQO1) in these functions is unlikely. Whatever the nature of the quinone reductase that is involved in the coenzyme Q_6-mediated reduction of extracellular MDHA by yeast cells, indirect evidence suggests that the enzyme system uses cytosolic NADH as electron donor *in vivo*. The activity is increased by treating cells with ethanol, which increases NADH in yeasts, and can be suppressed by inhibition of alcohol dehydrogenase by pyrazole (Santos-Ocaña *et al.*, 1995, 1998a).

Whether or not reduced coenzyme Q_6 interacts directly with MDHA, or additional electron carriers are required, remains to be established. Interestingly, little recovery of MDHA reductase was achieved by direct supplementation of PMs derived from a *coq3*Δ mutant strain with coenzyme Q_6. However, full restoration of the activity was obtained by either transformation with a single-copy plasmid containing the wild-type *COQ3* gene, or by culturing yeasts in the presence of coenzyme Q_6 (Santos-Ocaña *et al.*, 1998a). An unidentified additional component might be required for proper integration of coenzyme Q_6 into the PM in order to function as a transmembrane electron carrier to reduce MDHA. In addition, the presence of coenzyme Q_6 might be necessary for the correct assembly of a functional MDHA reductase enzyme system, as demonstrated recently for activity and stability of the bc_1 complex at the inner mitochondrial membrane (Santos-Ocaña *et al.*, 2002).

Whereas NADH-MDHA reductase in yeast PMs requires coenzyme Q_6, some quinone-independent activity is also present (Santos-Ocaña *et al.*, 1998a). Furthermore, a significant portion (~ 65%) of MDHA reduction by whole yeast cells does not require coenzyme Q_6. Studies on the sensitivity to various effectors and inhibitors also supports that a separate enzyme system is involved in quinone-independent NADH-MDHA reductase. Interestingly, the stabilization of extracellular ASC was increased when cells were cultured in iron-free medium. Moreover, the ASC stabilization activity of *coq3*Δ mutant cells was completely abolished when cells were cultured in medium supplemented with 2 mM FeEDTA (Santos-Ocaña *et al.*, 1998a, 1998b). These findings suggest that the system responsible for the reduction of ferric iron is a possible candidate to catalyze coenzyme Q_6-independent reduction of extracellular MDHA (*Figure 7.1*). This possibility appears likely, given the structural similarities (two *b*-type hemes bound to His residues on their transmembrane helices) between reductases of the *FRE* family, that reduce extracellular iron, and cytochrome b_{561} that reduces MDHA (Finegold *et al.*, 1996; Tsubaki *et al.*, 1997). In addition, di-heme Dcytb, which is responsible for iron reduction in duodenal enterocytes, also displays a high similarity to the cytochrome b_{561} family of PM reductases (McKie *et al.*, 2001, see above and chapter by May and Asard).

7.5 Concluding remarks

From the data presented above, it is possible to deduce that those membrane-bound enzymes using ASC as an electron donor, or involved in its regeneration, or simply interacting with ascorbic acid, function in a coordinated way to maintain the redox status of different compartments. It is well known that different stress conditions may affect some organelles without appreciable changes in others. Thus, each compartment seems to have specific defense mechanisms and probably different regulatory systems. However, the multiple possible links between the scavenging system, pathways operating on the cytosol, and membrane-bound systems in subcellular compartments, lead us to hypothesize that there is complex crosstalk between the different organelles required to control ASC homeostasis and, therefore, all the biological functions in which ASC is involved.

Acknowledgments

The author's work is supported by the Spanish MEC and MCyT (grants. PB98–0329-CO2–02 and BMC2002–01078) and Junta de Andalucía (CVI-276). M.C.C.-P. is supported by CVI 276.

References

Alcain FJ, Burón MI, Villalba JM, Navas P (1991) Ascorbate is regenerated by HL-60 cells through the transplasmalemma redox system. *Biochim. Biophys. Acta* **1073**, 380–385.

Apps DK, Boisclair MD, Gavine FS, Pettigrew GW (1984) Unusual redox behaviour of cytochrome b_{561} from bovine chromaffin granule membranes. *Biochim. Biophys. Acta* **764**, 8–16.

Arroyo A, Navarro F, Gómez-Díaz C, Crane FL, Alcaín FJ, Navas P, Villalba JM (2000) Interactions between ascorbyl free radical and coenzyme Q at the plasma membrane. *J. Bioenerg. Biomembr.* **32**, 199–210.

Asada K (1999) The water–water cycle in chloroplasts: scavenging of active oxygens and dissipation of excess photons. *Annu. Rev. Plant Physiol. Plant Mol. Biol.* **50**, 601–639.

Asard H, Kapila J, Verelst W, Bérczi A (2001) Higher-plant plasma membrane cytocrome b561: a protein in search of a function. *Protoplasma* **217**, 77–93.

Bérczi A, Møller IM (1998) NADH-monodehydroascorbate oxido-reductase is one of the redox enzymes in spinach leaf plasma membrane. *Plant Physiol.* **116**, 1029–1036.

Bérczi A, Møller IM (2000) Redox enzymes in the plant plasma membrane and their possible roles. *Plant Cell Environ.* **23**, 1287–1302.

Borgese N, Pietrini G (1986) Distribution of the integral membrane protein NADH-cytochrome b_5 reductase in rat liver cells, studied with a quantitative radioimmunoblotting assay. *Biochem. J.* **239**, 393–403.

Bunkelmann JR, Trelease RN (1996) Ascorbate peroxidase: a prominent membrane protein in oilseed glyoxysomes. *Plant Physiol.* **110**, 589–598.

Córdoba MC, Serrano A, Córdoba F, González-Reyes JA, Navas P, Villalba JM (1995) Topography of the 27- and 31-kDa electron transport proteins in the onion root plasma membrane. *Biochem. Biophys. Res. Commun.* **216**, 1054–1059.

Dalton DA, Baird LM, Langeberg L, Taugher CY, Anyan WR, Vance CP, Sarath G (1993) Subcellular localization of oxygen defense enzymes in soybean (*Glycine max* [L.] Merr.) root nodules. *Plant Physiol.* **102**, 481–489.

D'Arrigo A, Manera E, Longhi R, Borgese N (1993) The specific subcellular localization of two isoforms of cytochrome b_5 suggests novel targeting pathways. *J. Biol. Chem.* **268**, 2802–2808.

De Leonardis S, Dipierro N, Dipierro S (2000) Purification and characterization of an ascorbate peroxidase from potato tuber mitochondria. *Plant Physiol. Biochem.* **38**, 773–779.

Del Rio LA, Corpas FJ, Sandalio LM, Palma JM, Gómez M, Barroso JB (2002) Reactive oxygen species, antioxidant systems and nitric oxide in peroxisomes. *J. Exp. Bot.* **53**, 1255–1272.

Diliberto EJ Jr, Dean G, Cartyer C, Allen PL (1982) Tissue, subcellular, and submitochondrial distributions of: Possible role of semidehydroascorbate reductase in cofactor regeneration. *J. Neurochem.* **39**, 563–568.

Finegold AA, Shatwell KP, Segal AW, Klausner RD, Dancis A (1996) Intramembrane bis-heme motif for transmembrane electron transport conserved in a yeast iron reductase and the human NADPH oxidase. *J. Biol. Chem.* **271**, 31021–31024.

Goldenberg H, Grebing C, Löw H (1983) NADH-monodehydroascorbate reductase in human erythrocyte membranes. *Biochem. Int.* **6**, 1–9.

Gómez-Díaz C, Rodríguez-Aguilera JC, Barroso MP, Villalba JM, Navarro F, Crane FL, Navas P (1997a) Antioxidant ascorbate is stabilized by NADH-coenzyme Q_{10} reductase in the plasma membrane. *J. Bioenerg. Biomembr.* **29**, 251–257.

Gómez-Díaz C, Villalba JM, Pérez-Vicente R, Crane FL, Navas P (1997b) Ascorbate stabilization is stimulated in ρ^0HL-60 cells by CoQ_{10} increase at the plasma membrane. *Biochem. Biophys. Res. Commun.* **234**, 79–81.

Hieber ADVerhoeven AS, Yamamoto HY (2002) Overexpression of violaxanthin de-epoxidase: properties of C-terminal deletions an activity and pH-dependent lipid binding. *Planta* **214**, 476–483.

Hossain MA, Asada K (1985) Monodehydroascorbate reductase from cucumber is a flavin adenine dinucleotide enzyme. *J. Biol. Chem.* **260**, 12920–12926.

Huh W-K, Lee B-H, Kim S-T, Kim Y-R, Rhie G-E, Baek Y-W, Hwang C-S, Lee J-S, Kang S-O (1998) D-Erythroascorbic acid is an important antioxidant molecule in *Saccharomyces cerevisiae*. *Mol. Microbiol.* **30**, 895–903.

Ishikawa T, Sakai K, Takeda T, Shigeoka S (1995) Cloning and expression of cDNA encoding a new type of ascorbate peroxidase from spinach. *FEBS Lett.* **367**, 28–32.

Ishikawa T, Sakai K, Yoshimura K, Takeda T, Shigeoka S (1996) cDNAs encoding spinach stromal and thylakoid-bound ascorbate peroxidase, differing in the presence or absence of their 3′-coding regions. *FEBS Lett.* **384**, 289–293.

Ito A (1980) Cytochrome b_5-like hemoprotein of outer mitochondrial membrane: OM cytochrome b. II. Contribution of OM cytochrome b to rotenone-insensitive NADH-cytochrome c reductase activity. *J. Biochem.* **87**, 73–80.

Ito A, Hayashi S-I, Yoshida T (1981) Participation of a cytochrome b_5-like hemoprotein of outer mitochondrial membrane (OM cytochrome b) in NADH-semidehydroascorbic acid reductase activity of rat liver. *Biochem. Biophys. Res. Commun.* **101**, 591–598.

Iturbe-Ormaetxe I, Matamoros MA, Rubio MC, Dalton DA, Becana M (2001) The antioxidants of legume nodule mitochondria. *Mol. Plant-Microb. Interact.* **14**, 1189–1196.

Ivanov BN, Sacksteder CA, Kramer DM, Edwards GE (2001) Light-induced ascorbate-dependent electron transport and membrane energization in chloroplasts of bundle sheath cells of the C4 plant maize. *Arch. Biochem. Biophys.* **385**, 145–153.

Jespersen HM, Kjaersgård IVH, Østergaard L, Welinder KG (1997) From sequence analysis of three novel ascorbate peroxidases from *Arabidopsis thaliana* to structure, function and evolution of seven types of ascorbate peroxidase. *Biochem. J.* **326**, 305–310.

Jiménez A, Hernández JA, del Río LA, Sevilla F (1997) Evidence for the presence of the ascorbate-glutathione cycle in mitochondria and peroxisomes of pea leaves. *Plant Physiol.* **114**, 275–284.

Jiménez A, Hernández JA, Pastori G, del Rio LA, Sevilla F (1998) Role of the ascorbate-glutathione cycle of mitochondria and peroxisomes in the senescence of pea leaves. *Plant Physiol.* **118**, 1327–1335.

Kim S-T, Huh W-K, Lee B-H, Kang S-O (1998) D-arabinose dehydrogenase and its gene from *Saccharomyces cerevisiae*. *Biochim. Biophys. Acta* **1429**, 29–39.

Kitajima S, Yasukochi Y, Minakami S (1981) Purification and properties of human erythrocyte membrane NADH-cytochrome b_5 reductase. *Arch. Biochem. Biophys.* **210**, 330–339.

Kobayashi K, Tsubaki M, Tagawa S (1998) Distinct roles of two heme centers for transmembrane electron transfer in cytochrome b_{561} from bovine adrenal chromaffin vesicles as revealed by pulse radiolysis. *J. Biol. Chem.* **273**, 16038–16042.

Lee JS, Huh WK, Lee BH, Baek YU, Hwang CS, Kim ST, Kim YR, Kang SO (2001) Mitochondrial NADH-cytochrome b_5 reductase plays a crucial role in the reduction of D-erythroascorbyl free radical in *Saccharomyces cerevisiae*. *Biochim. Biophys. Acta* **1527**, 31–38.

Leroux A, Mota-Vieira LM, Kahn A (2001) Transcriptional and translational mechanisms of cytochrome b_5 reductase isoenzyme generation in humans. *Biochem. J.* **355**, 529–535.

López-Huertas E, Corpas FJ, Sandalio LM, Del Rio LA (1999) Characterization of membrane polypeptides from pea peroxisomes involved in superoxide radical generation. *Biochem. J.* **337**, 531–536.

May JM (1999) Is ascorbic acid an antioxidnt for the plasma membrane? *FASEB J.* **13**, 995–1006.

May JM, Qu Z-c (1999) Ascorbate-dependent electron transfer across the human erythrocyte membrane. *Biochim. Biophys. Acta* **1421**, 19–31.

May JM, Qu Z-c, Morrow JD (1996) Interaction of ascorbate and α-tocopherol in resealed human erythrocyte ghosts. Transmembrane electron transfer and protection from lipid peroxidation. *J. Biol. Chem.* **271**, 10577–10582.

May JM, Qu Z-c, Cobb CE (2000) Extracellular reduction of ascorbate free radical by human erythrocytes. *Biochem. Biophys. Res. Commun.* **267**, 118–123.

May JM, Qu Z-c, Cobb CE (2001) Recycling of the ascorbate free radical by human erythrocyte membranes. *Free Radic. Biol. Med.* **31**, 117–124.

McKie AT, Barrow D, Latunde-Dada GO, Rolfs A, Sager G, Mudaly E, *et al.* (2001) An iron-regulated ferric reductase associated with the absorption of dietary iron. *Science* **291**, 1755–1759.

Merker MP, Olson LE, Bongard RD, Patel MK, Linehan JH, Dawson CA (1998) Ascorbate-mediated transplasma membrane electron transport in pulmonary arterial endothelial cells. *Am. J. Physiol.* **274** (*Lung Cell. Mol. Physiol.* 18), L685–L693.

Miyake C, Asada K (1992) Thylakoid-bound ascorbate peroxidase in spinach chloroplasts and photoreduction of its primary product monodehydroascorbate radicals in thylakoids. *Plant Cell Physiol.* **33**, 541–553.

Møller IM (2001) Plant mitochondria and oxidative stress: Electron transport, NADPH turnover, and metabolism of reactive oxygen species. *Annu. Rev. Plant Physiol. Plant Mol. Biol.* **52**, 561–591.

Mullen RT, Trelease RN (2000) The sorting signal for peroxisomal membrane-bound ascorbate peroxidase are within its C-terminal tail. *J. Biol. Chem.* **275**, 16337–16344.

Mullen RT, Flynn CR, Trelease RN (2001a) How are peroxisomes formed? The role of the endoplasmic reticulum and peroxins. *Trends Plant Sci.* **6**, 256–261.

Mullen RT, Lisenbee CS, Flynn CR, Trelease RN (2001b) Stable and transient expression of chimeric peroxisomal membrane proteins induced an independent "zippering" of peroxisomes and an endoplasmic reticulum subdomain. *Planta* **213**, 849–863.

Navarro F, Villalba JM, Crane FL, Mackellar WC, Navas P (1995) A phospholipid-dependent NADH-coenzyme Q reductase from liver plasma membrane. *Biochem. Biophys. Res. Commun.* **212**, 138–143.

Nishino H, Ito A (1986) Subcellular distribution of OM cytochrome b-*mediated* NADH-semidehydroascorbate reductase activity in rat liver. *J. Biochem.* **100**, 1523–1531.

Nito K, Yamaguchi K, Kondo M, Hayashi M, Nishimura M (2001) Pumpkin peroxisomal ascorbate peroxidase is localized on peroxisomal membranes and unknown structures. *Plant Cell Physiol.* **42**, 20–27.

Njus D, Kelley PM, Harnadek GJ, Jalukar V (1990) Oxidation–reduction reactions in secretory vesicles. In: *Oxidoreduction at the Plasma Membrane: Relation to Growth and Transport*, Vol. I. *Animals*. (eds FL Crane, DJ Morré, HE Löw). CRC Press, Boca Raton, pp. 85–99.

Noctor G, Foyer CH (1998) Ascorbate and glutathione: keeping active oxygen under control. *Annu. Rev. Plant Physiol. Plant Mol. Biol.* **49**, 249–279.

Okuyama E, Yamamoto R, Ichikawa Y, Tsubaki M (1998) Structural basis for the electron transfer across the chromaffin vesicle membranes catalyzed by cytochrome b_{561}: analyses of cDNA nucleotide sequences and visible absorption spectra. *Biochim. Biophys. Acta* **1383**, 269–278.

Orringer EP, Roer MES (1979) An ascorbate-mediated transmembrane reducing system of the human erythrocyte. *J. Clin. Invest.* **63**, 53–58.

Perin MS, Fried VA, Slaughter CA, Sudhof TC (1988) The structure of cytochrome b_{561}, a secretory vesicle-specific electron transport protein. *EMBO J.* **7**, 2697–2703.

Pietrini G, Carrera P, Borgese N (1988) Two transcripts encode rat cytochrome b_5 reductase. *Proc. Natl Acad. Sci. USA* **85**, 7246–7250.

Pietrini G, Aggujaro D, Carrera P, Malyszko J, Vitale A, Borgese N (1992) A single mRNA, transcribed from an alternative, erythroid-specific, promoter, codes for two non-myristylated forms of NADH-cytochrome b_5 reductase. *J. Cell Biol.* **117**, 975–986.

Preger V, Pesaresi A, Pupillo P, Trost P (2001) Ascorbate-independent electron transfer between cytochrome b561 and a 27 kDa ascorbate peroxidase of bean hypocotyls. *Protoplasma* **217**, 137–145.

Rodríguez-Aguilera JC, Navas P (1994) Extracellular ascorbate stabilization: enzymatic or chemical process? *J. Bioenerg. Biomembr.* **26**, 379–384.

Santos-Ocaña C, Navas P, Crane FL, Navas P (1995) Extracellular ascorbate stabilization as a result of transplasma electron transfer in *Saccharomyces cerevisiae*. *J. Bioenerg. Biomembr.* **27**, 597–603.

Santos-Ocaña C, Villalba JM, Córdoba F, Padilla S, Crane FL, Clarke CF, Navas P (1998a) Genetic evidence for coenzyme Q requirement in plasma membrane electron transport. *J. Bioenerg. Biomembr.* **30**, 465–475.

Santos-Ocaña C, Córdoba F, Crane FL, Clarke CF, Navas P (1998b) Coenzyme Q_6 and iron reduction are responsible for the extracellular ascorbate stabilization at the plasma membrane of *Saccharomyces cerevisiae*. *J. Biol. Chem.* **273**, 8099–8105.

Santos-Ocaña C, Do TQ, Padilla S, Navas P, Clarke CF (2002) Uptake of exogenous coenzyme Q and transport to mitochondria is required for bc_1 complex stability in yeast coq mutants. *J. Biol. Chem.* **277**, 10973–10981.

Serrano A, González-Reyes JA, Navas P, Villalba JM (1994) Purification and characterization of two distinct NAD(P)H dehydrogenases from onion (*Allium cepa* L.) root plasma membrane. *Plant Physiol.* **106**, 87–96.

Shigeoka S, Ishikawa T, Tamoi M, Miyagawa Y, Takeda T, Yabuta Y, Yoshimura K (2002) Regulation and function of ascorbate peroxidase isoenzymes. *J. Exp. Bot.* **53**, 1305–1319.

Shirabe K, Landi MT, Takeshita M, Uziel G, Fedrizzi E, Borgese N (1995) A novel point mutation in a 3′ splice site of the NADH-cytochrome b_5 reductase gene results in immunologically undetectable enzyme and impaired NADH-dependent ascorbate regeneration in cultured fibroblasts of a patient with type II hereditary methemoglobinemia. *Am. J. Hum. Genet.* 57, 302–310.

Srivastava M (1995) Genomic structure and expression of the human gene encoding cytochrome b_{561}, an integral protein of the chromaffin granule membrane. *J. Biol. Chem.* **270**, 22714–22720.

Srivastava M, Gibson KR, Pollard HB, Fleming PJ (1994) Human cytochrome b_{561}: a revised hypothesis for conformation in membranes which reconciles sequence and functional information. *Biochem. J.* **303**, 915–921.

Takeuchi F, Kobayashi K, Tagawa S, Tsubaki M (2001) Ascorbate inhibits the carbethoxylation of two histidyl and one tyrosyl residues indispensable for the transmembrane electron transfer reaction of cytochrome b_{561}. *Biochemistry* **40**, 4067–4076.

Trost P, Bérczi A, Sparla F, Sponza G, Marzadori B, Asard H, Pupillo P (2000) Purification of cytochrome b_{561} from bean hypocotyls plasma membrane. Evidence for the presence of two heme centers. *Biochim. Biophys. Acta* **1468**, 1–5.

Truan G, Epinat J-C, Rougeulle C, Cullin C, Pompon D (1994) Cloning and characterization of a yeast cytochrome b_5-encoding gene which suppresses ketoconazole hypersensitivity in a NADPH-P-450 reductase-deficient strain. *Gene* **149**, 123–127.

Tsubaki M, Nakayama M, Okuyama E, Ichikawa Y, Hori H (1997) Existence of two heme B centers in cytochrome b_{561} from bovine adrenal chromaffin vesicles as revealed by a new purification procedure and EPR spectroscopy. *J. Biol. Chem.* **272**, 23206–23210.

Van Duijn MM, Van der Zee J, Van Steveninck J, Van den Broek PJA (1998) Ascorbate stimulates ferricyanide reduction in HL-60 cells through a mechanism distinct from the NADH-dependent plasma membrane reductase. *J. Biol. Chem.* **273**, 13415–13420.

Van Duijn MM, Tijssen K, Van Steveninck J, Van den Broek PJA, Van der Zee J (2000) Erythrocytes reduce extracellular ascorbate free radicals using intracellular ascorbate as an electron donor. *J. Biol. Chem.* **275**, 27720–27725.

Van Duijn MM, Van der Zee J, Van den Broek PJA (2001a) The ascorbate-driven reduction of ascorbate free radicals by the erythrocyte is an electrogenic process. *FEBS Lett.* **491**, 67–70.

Van Duijn MM, Buijs JT, Van der Zee J, Van den Broek, PJA (2001b) The ascorbate:ascorbate free radical oxidoreductase from the erythrocyte membrane is not cytochrome b_{561}. *Protoplasma* **217**, 94–100.

Villalba JM, Canalejo A, Rodríguez-Aguilera JC, Burón MI, Morré DJ, Navas P (1993) NADH-ascorbate free radical and -ferricyanide reductase activities represent different levels of plasma membrane electron transport. *J. Bioenerg. Biomembr.* **25**, 411–417.

Villalba JM, Navarro F, Cordoba F, Serrano A, Arroyo A, Crane FL, Navas P (1995) Coenzyme Q reductase from liver plasma membrane: purification and role in trans-plasma-membrane electron transport. *Proc. Natl Acad. Sci. USA* **92**, 4887–4891.

Villalba JM, Navarro F, Gómez-Diaz C, Arroyo A, Bello RI, Navas P (1997) Role of cytochrome b5 reductase on the antioxidant function of coenzyme Q in the plasma membrane. *Mol. Aspects Med.* **18**, s7–s13.

Villalba JM, Crane FL, Navas P (1998) Antioxidative role of ubiquinone in animal plasma membrane. In: *Plasma Membrane Redox Systems and Their Role in Biological Stress and Disease* (eds H Asard, A Bérczi, RJ Caubergs). Kluwer Academic Publishers, Dordrecht, pp. 247–265.

Wada N, Kinoshita S, Matsuo M, Amako K, Miyake C, Asada K (1998) Purification and molecular properties of ascorbate peroxidase from bovine eye. *Biochem. Biophys. Res. Commun.* **242**, 256–261.

Yabuta Y, Motoki T, Yoshimura K, Takeda T, Ishikawa T, ShigeOka S (2002) Thylakoid membrane-bound ascorbate peroxidase is a limiting factor of antioxidative systems under photo-oxidative stress. *Plant J.* **32**, 915–925.

Yoshimura K, Yabuta Y, Ishikawa T, Shigeoka S (2002) Identification of a cis element for tissue-specific alternative splicing of chloroplast ascorbate peroxidase pre-mRNA in higher plants. *J. Biol. Chem.* **277**, 40623–40632.

8

Ascorbate recycling

James May and Han Asard

8.1 Introduction

Ascorbate (ASC) redox chemistry is complex (see for example the chapter by Buettner). It becomes even more so in cells, where ASC in different forms may be compartmentalized, and where it will have variable access to oxidants, recycling enzymes, and reducing co-factors. Nonetheless, there has been progress in unraveling many features of cellular ASC redox metabolism. In this chapter we will review what is known about the mechanisms and significance of ASC recycling in plant and animal cells.

8.2 Chemistry of cellular ascorbate recycling in cells

Most oxidation of ASC in and around cells likely involves loss of a single electron to a radical species, such as hydroperoxyl radicals, superoxide, carbon-centered radicals on lipids or proteins, or the α-tocopheroxyl radical (Buettner, 1993). This manifests as appearance of the ascorbate free radical (MDHA or monodehydroascorbate), which is readily measured by electron paramagnetic resonance spectroscopy (EPR). In fact, it is likely that all ASC oxidation proceeds through MDHA as an intermediate step. In cells and tissues, which generally contain low levels of catalytic transition metals, dehydroascorbate (DHA) is generated by a disproportionation reaction of MDHA rather than by oxidation. The favored reversible steps of ASC recycling are a single electron reduction of MDHA and a two-electron reduction of DHA, but not reduction of DHA to MDHA (*Figure 8.1*).

8.2.1 Ascorbate oxidation

As noted above, ASC appears capable of losing only one electron at a time (Bielski, 1982). That is, a one-electron oxidation of the ASC monoanion to MDHA occurs even with a two-electron oxidant, such as 2,6-dichlorophenolindolphenol (Iyanagi *et al.*, 1985). MDHA then either undergoes a further one-electron oxidation to DHA, or two molecules of MDHA dismutate to form one molecule of ASC and one of DHA (*Figure 8.1*). MDHA is a stronger reducing agent than ASC, since the

Vitamin C, edited by Han Asard, James M May and Nicholas Smirnoff

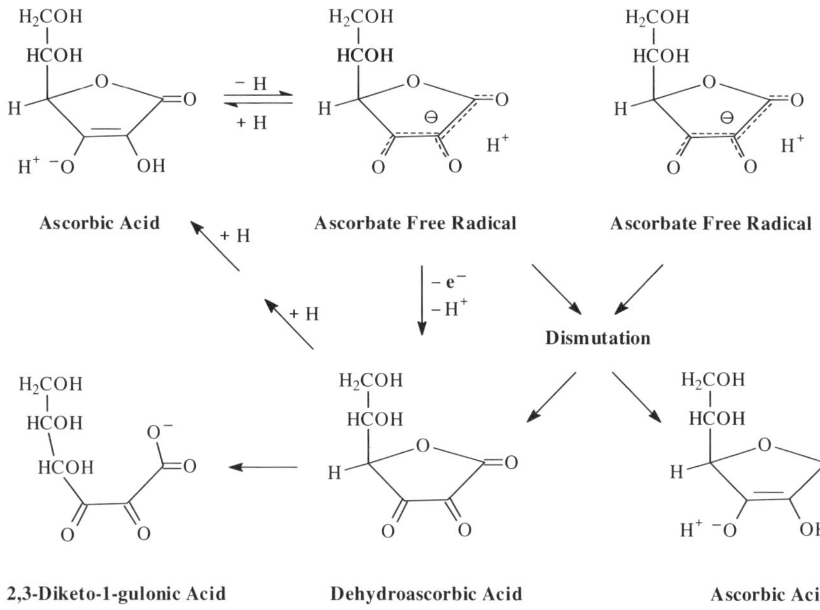

Figure 8.1: Redox chemistry of ASC.

mid-point reduction potential of MDHA/DHA couple (–174 to –210 mV) is lower than that of the ASC/MDHA (+282 to +330 mV) (Buettner, 1993; Iyanagi *et al.*, 1985). Therefore, MDHA is oxidized more rapidly than is the ASC monoanion by a large molar excess of oxidant such as ferricyanide or 2,6-dichlorophenol-indolphenol (Iyanagi *et al.*, 1985). The second-order rate constant for oxidation of MDHA by Fe^{3+}/EDTA is $4 \times 10^{6} \cdot M^{-1} \cdot s^{-1}$ (Kobayashi *et al.*, 1991). This is more than an order of magnitude greater than that of the dismutation reaction (rate constant $= 2\text{–}3 \times 10^{5} \cdot M^{-1} \cdot s^{-1}$ (Bielski *et al.*, 1975), and two orders of magnitude greater than reduction of cytochrome *c* by MDHA (rate constant $= 4 \times 10^{4} \cdot M^{-1} \cdot s^{-1}$) (Ohnishi *et al.*, 1969). However, under physiologic conditions, in which reactive Fe^{3+} or other transition metals are absent, the dismutation reaction is highly favored (Bielski *et al.*, 1975; Roginsky and Stegmann, 1994). As the pH decreases, so does the rate of dismutation, so that at pH 4.8 it is $> 10^{8} \cdot M^{-1} \cdot s^{-1}$ (Bielski *et al.*, 1975). Still, MDHA is surprisingly stable, in part because the unpaired electron distributes over three vicinal carbonyl groups (Bielski *et al.*, 1975) (*Figure 8.1*). MDHA, missing a single electron, still retains a negative charge (Laroff *et al.*, 1972) and thus like ASC is trapped in the cell compartment of its formation. It does not reduce ferric hemoproteins (Kobayashi *et al.*, 1991), and can co-exist with phenoxyl (Roginsky and Stegmann, 1994), nitroxyl (Minetti *et al.*, 1992), and other radicals (Roginsky and Stegmann, 1994). MDHA is so stable that it has been proposed as a terminal marker for oxidant stress (Buettner and Jurkiewicz, 1993; Koyama *et al.*, 1994; Pietri *et al.*, 1990, 1994).

With loss of the second electron, the uncharged tricarbonyl DHA is produced. DHA is more lipophilic than ASC, primarily because ASC carries a negative charge at physiologic pH (the pK_a of the carbon-3 hydroxyl is 4.2). DHA is quite unstable, with a half-life of about 6 min at physiologic pH and temperature (Drake *et al.*, 1942; Koshiishi *et al.*, 1998; Winkler, 1987). Hydrolysis of the lactone ring converts DHA to 2,3-diketo-1-gulonic acid (*Figure 8.1*) (Bode *et al.*, 1990; Chatterjee, 1970). Although this conversion can be partially reversed by high concentrations of mercaptoethanol (Deutsch and Santhosh-Kumar, 1996), it is probably irreversible in cells (see Chapter 2 by Banhegyi and Loewus). Loss of ASC through decomposition of DHA is obviously wasteful, and most cells utilize redundant mechanisms to recycle DHA back to ASC.

8.3 Mechanisms of ascorbate recycling

8.3.1 Reduction of MDHA in animal cells

Recycling of ASC from MDHA makes sense from the viewpoint of the ASC economy of the cell, since it avoids generation of DHA, which might ring-open or leave the cell on glucose transporters. There is no evidence that cellular thiols reduce MDHA to ASC. For example, the steady-state concentration of MDHA generated from 1 mM ASC during oxidation by either 20 μM methylene blue or 2 μM Fe^{3+}/EDTA is unaffected by concentrations of reduced glutathione (GSH) of 2–4 mM (Roginsky and Stegmann, 1994). Nor is MDHA easily reduced by ferrous cytochrome b_5 (Kobayashi *et al.*, 1991). On the other hand, ASC is rapidly regenerated from MDHA in the presence of either the fully reduced or semiquinone forms of cytochrome b_5 reductase, with rate constants of 4.3×10^6 and 3.7×10^5 $M^{-1} \cdot s^{-1}$, respectively (Kobayashi *et al.*, 1991). Note that MDHA reductase (MDHAR) of cucumber may be 10 × faster in reducing MDHA (Kobayashi *et al.*, 1991).

Mitochondrial and microsomal MDHA reduction

In mammalian cells, NADH-dependent cytochrome *b*-mediated MDHA reductase activity (MDHAR, EC 1.6.5.4) has been localized to the outer membranes of mitochondria (Diliberto *et al.*, 1982; Ito *et al.*, 1981) and to microsomes (Hara and Minakami, 1971; Schulze and Staudinger, 1971). Most of the activity is considered to reside in the outer membrane mitochondrial enzyme in liver, kidney, adrenal gland, heart, and brain (Nishino and Ito, 1986). The kinetics of MDHA reduction by the liver microsomal NADH-dependent cytochrome b_5 reductase show a high affinity for MDHA (apparent K_m = 4 μM), as expected if there is physiologic relevance for the enzyme. With regard to the latter, there is a report of a patient with Type II hereditary methemoglobinemia with a mutation in the NADH-cytochrome b_5 reductase gene. This patient's erythrocytes and fibroblasts had only 15% of normal MDHAR activity (Shirabe *et al.*, 1995). It was hypothesized that lack of MDHA reduction might contribute to some of the developmental abnormalities of this syndrome (Shirabe *et al.*, 1995).

Cytochrome b_{561}*-dependent MDHA reduction*

Perhaps the best characterized role for MDHAR is to maintain cytoplasmic ASC concentrations for use in neurosecretory cells, such as those in the adrenal medulla (Njus *et al.*, 1983, 2001; Wakefield *et al.*, 1986). In the chromaffin granules of adrenal medullary cells, ASC functions as a co-factor for dopamine β-hydroxylase in the hydroxylation of dopamine to norepinephrine. This generates MDHA in the granules, which is reduced back to ASC by a unique trans-membrane redox system that uses cytochrome b_{561} to transfer electrons across the granule membrane. The energy for this system is provided by proton pumping into the granule by a membrane-bound H^+-ATPase. The electrons for the trans-membrane transfer are derived from cytoplasmic ASC, which in turn is recycled by the cytochrome b_5 reductase of the outer mitochondrial membrane (Wakefield *et al.*, 1986). This function of MDHA may extend to other neurosecretory cells, since islet secretory granules have also been shown to have cytochrome b_{561} (Mackin *et al.*, 1986).

Cytochrome b_{561} has been purified from bovine adrenal gland chromaffin granules (Silsand and Flatmark, 1974; Tsubaki *et al.*, 1997), and its primary sequence has been determined. Based on this sequence, cytochrome b_{561} homologues have been identified in a large number of species including mammals, insects, worms and plants (Asada *et al.*, 2002; Asard *et al.*, 2001; Bashtovyy *et al.*, 2003). Interestingly, no cytochrome b_{561} proteins have so far been identified in yeast or prokaryotes (W. Verelst and H. Asard, unpublished results). Since yeast cells contain erythroascorbate instead of ASC (Spickett *et al.*, 2000), the absence of cytochrome b_{561} apparently correlates with the absence of ASC as a primary antioxidant.

The sequence homologies between cytochromes b_{561} from different species are generally not very high. However, several structural features are highly conserved among all species, suggesting similar catalytic functions for each of these isoforms (Asard *et al.*, 2001). Secondary structure prediction for all cytochromes b_{561} invariably suggests the presence of six transmembrane helices. The location of these helices in multiple alignments is very well conserved. All cytochromes b_{561} contain four strictly conserved histidine residues. It is generally believed that these provide the coordination sites for the two protoporphyrine IX (heme) molecules in the protein. Cytochromes b_{561} from all species also contain two well conserved motifs that possibly constitute the binding sites for ASC and MDHA respectively (Asard *et al.*, 2001; Okuyama *et al.*, 1998, see also Chapter 7 by Villalba *et al.*).

These conserved histidine residues and putative substrate binding sites all occur within the four central helices of the predicted six trans-membrane helices of the protein. The presence of this core structure has recently also been identified as part of a number of other proteins in plants and animals (Ponting, 2001; W. Verelst and H. Asard, unpublished results). In these proteins the so called cytochrome b_{561} domain is often found in combination with a dopamine β-hydroxylase-like domain. The physiological function of these proteins is not yet known. However, given the conservation of potential ASC and MDHA binding sites, they may constitute an interesting new family of proteins involved in ASC metabolism.

At present no information is available on the tertiary structure of the cytochromes b_{561}. Based on the prediction of the location and properties of the trans-membrane helices and the presence of heme-coordinating histidine residues, a three-dimensional model of a human and *A. thaliana* cytochrome b_{561} has recently been presented (Bashtovyy *et al.*, 2003). This model provides an interesting tool for the prediction and subsequent testing of the involvement of amino acid residues involved in the binding of the substrates and transport of electrons across the protein.

All species for which considerable DNA sequence information is available contain several cytochrome b_{561} homologues. For example, in humans and mice, three cytochrome b_{561} isoforms have been identified and *A. thaliana* contains at least four different genes putatively encoding cytochrome b_{561} (H. Asard, unpublished results). Since each of these sequences contains the conserved putative ASC and MDHA binding sites, they are likely to participate in ASC recycling through trans-membrane electron transport, similar to the chromaffin granule cytochrome b_{561}. Interestingly however, one of the mammalian isoforms, located in the duodenal brush border cell membrane, is suggested to be involved in reduction of iron chelates prior to uptake of the Fe^{2+} (McKie *et al.*, 2001). The physiological function and subcellular localization of the third mammalian isoform remains to be determined.

In plants, at least one of the cytochromes b_{561} is located at the plasma membrane (Asard *et al.*, 2001; Bérczi *et al.*, 2001; Trost *et al.*, 2000) and has been suggested to be involved in the regeneration of extracellular ASC (Asard *et al.*, 1992, 2001). Recent evidence suggests that another plant cytochrome b_{561} homologue is likely located at the vacuolar membrane (H. Asard, D. Griesen and D. Su, unpublished results). However, the catalytic role and physiological function of each of these isoforms remains to be determined.

Cytosolic MDHAR activities

Whereas membrane bound MDHARs are considered the major routes for recycling of ASC from MDHA, cytosolic enzymes have also been described that can act as MDHARs. For example, MDHA is reduced by the thioredoxin reductase system, which uses NADPH as a reducing cofactor (May *et al.*, 1998). This was demonstrated with purified rat liver thioredoxin reductase, which had an apparent K_m for MDHA of about 3 μM (May *et al.*, 1998), which is in line with that observed for the NADH-dependent cytochrome b_{561} reductase. NADPH-dependent reduction of MDHA by thioredoxin reductase was also observed in dialysates of cytosolic fractions from rat liver. Dietary selenium deficiency in the rats decreased NADPH-dependent MDHA reduction to about 10% of control. However, comparisons of microsomal NADH-dependent MDHA reduction with those of cytosolic thioredoxin reductase suggested that the former was more than 20-fold more active at the same MDHA concentration. A soluble NADH-dependent MDHAR has also been demonstrated in extracts of human lenses (Bando and Obazawa, 1988, 1994) that may differ from the membrane-bound forms.

Plasma membrane MDHA activities

Reduction of MDHA to ASC has also been described to occur on both the inner erythrocyte membrane (Goldenberg *et al.*, 1983; May *et al.*, 2001a) and on the outer membrane surface of K562 erythroleukemic cells (Schweinzer and Goldenberg, 1993) and human erythrocytes (May *et al.*, 2000; VanDuijn *et al.*, 2000, 2001). An outer membrane MDHAR might help preserve ASC in blood plasma, where ASC concentrations are much lower than in nucleated cells. MDHA reduction by intact erythrocytes was half-maximal at about 6 μM MDHA, and was 30-fold greater than measured release of ASC itself from the cells (May *et al.*, 2000). The mechanism of the cell-surface MDHAR has not been resolved. Rather than NADH, the intracellular electron donor for this activity may be ASC itself (VanDuijn *et al.*, 2000). Transfer of ASC-derived electrons depends on membrane polarization, suggesting an energy-dependent, electrogenic process (VanDuijn *et al.*, 2001). The mechanism of electron transfer across the membrane remains to be determined. It could involve a plasma membrane cytochrome b_5 reductase coupled to ubiquinone (see Chapter 7 by Villaba *et al.*).

8.3.2 Reduction of MDHA in plant cells

Recycling of MDHA to ASC in plants is mediated by a variety of enzymatic systems in different cell compartments. ASC regeneration has been most thoroughly studied in the chloroplast, a major source of reactive oxygen species (ROS). MDHA in chloroplasts is mainly generated by the scavenging of H_2O_2 originating from the enzymatic or nonenzymatic dismutation of superoxide. H_2O_2 scavenging is mediated by ASC peroxidases (APX), heme proteins with a high specificity towards ASC as an electron donor (for review see Asada, 1999). Chloroplasts contain membrane-bound (thylakoid APX) and soluble (stroma APX) isoforms.

Two major pathways for ASC regeneration in chloroplasts have been identified; the so called water–water cycle (Asada, 1999) and the ASC–glutathione or Halliwell–Foyer–Asada cycle (Halliwell and Foyer, 1976). In the water–water cycle, electrons are derived from the photooxidation of water at photosystem II and transferred to MDHA by reduced ferredoxin (Fd) at the site of photosystem I (Asada, 1999; Miyake and Asada, 1994). The rate constant for this reaction is $10^7 \cdot M^{-1} \cdot s^{-1}$, significantly higher than the reduction rate of the Fd-mediated reduction of $NADP^+$. The rate limiting step for the water–water cycle is the photoreduction of O_2 (Asada, 1999). EPR analysis of healthy leaves indicates very low levels of MDHA, supporting the efficiency of this pathway (Heber *et al.*, 1996; Hideg *et al.*, 1997).

In the ASC–glutathione cycle, MDHA is reduced to ASC by an NAD(P)H-dependent MDHAR. Other components of the ASC–glutathione cycle include a stromal glutathione-dependent DHAR and an NAD(P)H-dependent glutathione reductase (Noctor and Foyer, 1998). The DHAR functions to reduce DHA generated through disporportionation of MDHA that has escaped the MDHAR activity (see *Section 8.3.4*). MDHAR was first isolated from cucumber roots (Hossain and Asada, 1985). It is a flavin (FAD)-containing enzyme with high specificity towards MDHA. The affinity of the enzyme towards NADH is higher than towards NADPH (5 and 20–200 μM, respectively) (Sano *et al.*, 1995). The plant enzymes shows homology to bacterial flavoenzymes but not to other eukaryotic reductases (Sano and Asada, 1994). Despite the high activity of MDHAR, the high reaction rate of Fd with MDHA in the water–water cycle suggests that Fd-mediated MDHA reduction may be physiologically more important than reduction through the ASC–G5H cycle. However, reduction of MDHA by MDHAR may be effective in specific locations in the chloroplast stroma, with low Fd concentrations (Asada, 1999).

MDHA reductases have also been demonstrated in other plant cell compartments such as the cytosol, mitochondria and peroxisomes (Boraccino *et al.*, 1986; Dalton *et al.*, 1993; Jimenez *et al.*, 1997; Koshiba, 1993; Yamaguchi *et al.*, 1995). Peroxisomes are involved in various oxidative, ROS-producing metabolic reactions such as photorespiration, fatty acid β-oxidation and the glyoxylate cycle (Corpas *et al.*, 2001). The recent demonstration of nitric oxide production in peroxisomes underlines their importance in the overall regulation of the cellular redox status (Barroso *et al.*, 1999; del Rio *et al.*, 2002). The high latency of the MDHAR activity in mitochondria and peroxisomes suggests that these enzymes are located on the inside of the organelles, possibly associated with the membrane (Jimenez *et al.*, 1997). However, little is known about the nature and detailed enzymatic properties of the mitochondrial and peroxisomal MDHARs. Interestingly, in each of these organelles, the presence of other components of the ASC–glutathione cycle, such as a DHAR and glutathione reductase has also been demonstrated, suggesting a similar functional coupling of these enzymes as in the chloroplast.

In addition to the MDHARs from chloroplasts, mitochondria and peroxisomes, the plant plasma membrane has been demonstrated to contain a membrane-associated MDHAR (Bérczi and Møller, 1998, see also chapter by Villaba *et al.*). This 45-kDa enzyme shows a slight preference for NADH over NADPH, is specific for the β-hydrogen on NADH (Møller *et al.*, 1995), and appears tightly bound to the cytosolic face of the membrane.

In plant cells, millimolar concentrations of ASC are also present in vacuoles and in the extracellular matrix (apoplast) (Foyer and Lelandais, 1996; Rautenkranz *et al.*, 1994). In the apoplast, ASC plays an essential role in regulation of cell growth and cell wall formation through its fine regulation of ROS levels. However, the mechanism(s) of ASC regeneration in these compartments remains to be elucidated, and may involve cytochromes b_{561} (see above), specific ASC transporters (Horemans *et al.*, 2000, see also chapter by Wilson), or yet unidentified MDHAR activities.

8.3.3 Cellular uptake of DHA

Whereas the reduction of DHA has been described on the cell surface of K562 erythroleukemic cells (Schweinzer *et al.*, 1996), most DHA reduction occurs within cells. This is because of efficient reducing mechanisms present in cells, and because DHA readily traverses the cell membrane. As detailed in the chapter by Wilson, most DHA enters and leaves cells by facilitative diffusion on one or more of the GLUT-type glucose transporters. ASC, probably because of its negative charge at physiologic pH, is not a substrate for these transporters. For the same reason, and because of its short half-life, MDHA is also not likely a substrate for glucose transporters, although this has not been tested. The interaction of glucose and DHA during transport of the latter has been known for some time (Bianchi and Rose, 1986b; Bigley and Stankova, 1974). However, it was not proven that DHA uses glucose transporters to enter cells until Vera *et al.* (1993) expressed mRNA for the various GLUT isoforms in *Xenopus* oocytes and showed that DHA was readily taken up by the cells. *Xenopus* oocytes normally lack glucose transporters, and thus provided a clear answer to the question of whether DHA simply inhibited glucose transporters, or was actually a substrate. Facilitated diffusion is also the primary source of DHA uptake in plant cells, although the transporter in this case appears different from a GLUT-type carrier (Horemans *et al.*, 2000; see also Chapter 6 by Wilson).

Once DHA enters cells it is rapidly reduced to ASC. This coupled transport and reduction complicates kinetic analysis of DHA transport. For example, the apparent K_m for DHA transport and reduction derived from initial rates of transport is 100–500 μM (May *et al.*, 1999, 2001b; Mendiratta *et al.*, 1998a), which is considerably less than expected for the transport step. This indicates that the reduction step occurs with a higher affinity for DHA, such that the net or overall apparent affinity is in between those of the two steps. The second consequence of rapid DHA reduction in cells is that the ASC generated is then trapped in the cell. This is especially apparent in the human erythrocyte. In contrast to other cells in which intracellular ASC concentrations are many times greater than those in plasma due to transport of ASC on the high affinity SVCT-type transporter (see Chapter 6 by Wilson), the intracellular ASC concentration in erythrocytes is the same as that in the surrounding plasma from which it was taken (Evans *et al.*, 1982; Mendiratta *et al.*, 1998b). However, when allowed to take up DHA, erythrocytes at least temporarily concentrate ASC to levels of several millimolar due to the fact that ASC escapes only slowly (Christine *et al.*, 1956; Hughes and Maton, 1968; Mendiratta *et al.*, 1998b).

8.3.4 Reduction of DHA

A role for GSH in recycling of ASC from DHA was first demonstrated *in vivo* by Meister and colleagues. They showed that GSH deficiency induced by inhibition of γ-glutamylcysteine synthetase with buthionine sulfoximine increased mortality in guinea pigs with dietary deficiency of ASC (Mårtensson and Meister, 1992). ASC at 176 $mg \cdot kg^{-1}$ twice a day reversed this mortality, but the same dose of DHA failed to protect. This suggests that GSH deficiency impaired the ability of the animals to reduce DHA to ASC. They next showed that providing ASC-deficient guinea pigs with GSH as its monoethyl ester delayed onset of scurvy (Mårtensson *et al.*, 1993). These results led to the conclusion that GSH helped to recycle and thus maintain ASC in the animals.

Subsequent studies have clearly documented the ability of mammalian cells to carry out GSH-dependent reduction of DHA. The direct reduction of DHA by GSH likely involves a nucleophilic addition of the thiolate anion of GSH to carbon-3 of DHA, followed by a reducing step mediated by a second molecule of GSH, as shown in *Figure 8.2*. Direct DHA reduction by GSH has been conclusively demonstrated in cell-free systems (Winkler, 1987; Winkler, 1992). It is likely that a similar mechanism pertains to enzyme-dependent DHA reduction, except that the first attack is from an active thiolate anion in the enzyme (Washburn and Wells, 1999). Reduction of DHA to ASC does not involve MDHA as an intermediate in cells. Although circumstantial evidence has been presented that cellular reducing agents and protein thiols in particular reduce the DHA to MDHA (Harwood *et al.*, 1986), this has not been proven. On the contrary, during the first few minutes of uptake and reduction of 1 mM DHA by intact erythrocytes, there was no increase in MDHA

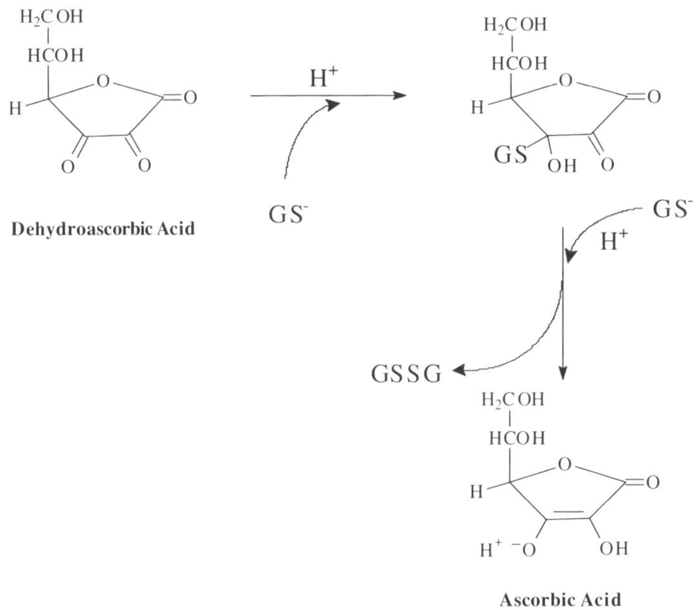

Figure 8.2: Reduction of DHA to ascorbic acid by GSH.

measured by EPR (May *et al.*, 1996). In this cell type, GSH is the primary reducing agent for DHA (May *et al.*, 1996; Mendiratta *et al.*, 1998a). If MDHA is an intermediate in the recycling of DHA to ASC, it should have been apparent in this experiment. Rather, it is likely that DHA is reduced without MDHA as an intermediate, either directly by cellular thiols, or by any of several enzymes.

Demonstrating that GSH is necessary for DHA reduction in cells requires showing that loading cells with DHA decreases cellular GSH content (especially in the absence of an energy source such as glucose), and that lowering GSH, either by direct reaction or by inhibition of glutathione synthetase, impairs the ability of the cells to reduce DHA to ASC. Such dependence of cellular DHA reduction on GSH, whether direct or enzyme-dependent, has been documented in a variety of cell types. These include lens epithelial cells (Giblin *et al.*, 1993; Winkler *et al.*, 1994), erythrocytes (May *et al.*, 1996; Mendiratta *et al.*, 1998a), cultured liver cell lines (Li *et al.*, 2001b), and endothelial cells (May *et al.*, 2001b). However, other cell types do not require GSH for DHA reduction, such as cultured tumor cell lines (HL-60; Guaiquil *et al.*, 1997) and U-937 cells (May *et al.*, 1999), as well as skin keratinocytes (Savini *et al.*, 2000). These results suggest that cells have a variety of options for enzyme-dependent DHA reduction.

Several mammalian GSH-dependent DHA reductases (EC 1.8.5.1) have been described (*Table 8.1*). These include the thioltransferase glutaredoxin (Wells *et al.*, 1990), protein disulfide isomerase (PDI, Wells *et al.*, 1990), and a 30–32-kDa enzyme purified initially from rat liver (Maellaro *et al.*, 1994) and subsequently from a variety of tissues including brain (Fornai *et al.*, 1999, 2001), most solid organs (Paolicchi *et al.*, 1996), and human erythrocytes (Mieyal *et al.*, 1991; Xu *et al.*, 1996).

It was recently shown (M. F. Lou, personal communication) that DHA reductase activity catalyzed by a thioltransferase isolated from human lens epithelial cells (HLE-B3) was stronger than that of protein disulfide isomerases, albumin and thioredoxin, suggesting an important physiological role of these enzymes in the regulation of the cellular ASC redox status.

The kinetic features of these enzymes, where available, are summarized in *Table 8.1*. Except for protein disulfide isomerase, which is located in the endoplasmic reticulum, the DHA reductases purified thus far appear to be located in the cytosol. However, rat liver mitochondria, which contain ASC in the matrix, also reduce DHA to ASC, and this reduction is largely GSH dependent (Li *et al.*, 2001a). Since the GSH concentrations of the mitochondrial matrix are 5–10 mM (Jocelyn, 1975; Li *et al.*, 2001a; Wahlländer *et al.*, 1979), DHA reduction might be direct. However, a novel mammalian glutaredoxin, Grx-2, has been described that is localized to mitochondria (Gladyshev *et al.*, 2001). By analogy with its cytosolic isoform, with which it shows 36% identity (Gladyshev *et al.*, 2001), this enzyme might also participate in ASC recycling from DHA in mitochondria.

NADPH-dependent reduction of DHA has also been reported for a 3̣-hydroxysteroid dehydrogenase purified from rat liver (Del Bello *et al.*, 1994), and for thioredoxin reductase (May *et al.*, 1997). The kinetics for these two enzymes are also listed in *Table 8.1*. Based on the dogma that the physiologic substrate range of an enzyme should bracket its K_m, the enzymes listed in *Table 8.1* qualify as a DHA reductase only if intracellular DHA concentrations rise above 0.1 mM. DHA concentrations above 100 μM may be toxic to cells (Bianchi and Rose, 1986a; Fiorani *et al.*, 2000). Thus, in addition to regenerating ASC, these enzymes might also serve to prevent even transient accumulation of DHA to toxic concentrations.

Table 8.1: Kinetic properties of enzymes with DHA reductase activities.

Enzyme	Co-factor	K_m (mM)[a]	k_{cat} (s^{-1})	Reference
Glutaredoxin (pig liver)	GSH	0.2	374	Wells *et al.*, 1990
Glutaredoxin (neutrophils)	GSH	0.25	68	Park and Levine, 1996
Protein disulfide isomerase	GSH	2.8	16	Wells *et al.*, 1995
DHA reductase (erythrocytes, 32-kDa protein)	GSH	0.21	316	Wells *et al.*, 1995
DHA reductase (liver, 31-kDa protein)	GSH	0.25	1.0 [b]	Maellaro *et al.*, 1997
3̣-Hydroxysteroid dehydrogenase (liver)	NADPH	4	59	Del Bello *et al.*, 1994
Thioredoxin reductase (rat liver)	NADPH	0.7	71	May *et al.*, 1997
DHA reductase (spinach leaves)	GSH	0.34	ND	Foyer and Halliwell, 1977
DHA reductase (spinach leaves)	GSH	0.07	ND	Hossain and Asada, 1984
DHA reductase (potato tubers)	GSH	0.39	ND	Dipierro and Borraccino, 1991
DHA reductase (spinach chloroplasts)	GSH	0.053	490	Shimaoka *et al.*, 2003
DHA reductase (rice bran)	GSH	0.35	ND	Urano *et al.*, 2000

[a]For DHA reduction. [b]Calculated from an M_r of 31 300, enzyme may not be pure.

Plants also contain DHAR activity. In the chloroplast ASC–glutathione pathway for ASC regeneration, DHA is reduced by a GSH-dependent reductase, and glutathione disulphide (GSSG) is subsequently re-reduced by an NAD(P)H-dependent glutathione reductase. A chloroplast DHAR has been purified from spinach leaves. The enzyme is a 23-kDa monomer with affinities of 50–80 μM for DHA and 1.1 mM for GSH (Shimaoka *et al.*, 2000). The much lower affinities towards DHA previously reported (*Table 8.1*) possibly reflect lower levels of purity or nonchloroplast origins of these enzymes. The affinity for DHA of the chloroplast DHARs is also significantly higher than those of mammalian proteins catalyzing a similar activity. A recent analysis of the activity of the spinach chloroplast DHAR suggests a ping-pong mechanism in which DHA binds to the free, reduced form of the enzyme followed by the binding of GSH (Shimaoka *et al.*, 2003).

DHAR activity is not limited to the chloroplast. A second DHAR activity was isolated from spinach leaves, with similar kinetic parameters as the chloroplast enzyme, for which the subcellular localization remained unclear (Shimaoka *et al.*, 2000). DHARs have also been identified in mitochondria and peroxisomes (del Rio *et al.*, 2002; Jimenez *et al.*, 1997, 2002; Mittova *et al.*, 2000).

Another protein reported to possess DHAR activity is the Kunitz-trypsin inhibitor (Trumper *et al.*, 1994). This protein is very heat stable, whereas the GSH-dependent DHARs are readily inactivated by heat treatment (Shimaoka *et al.*, 2000). DHAR and the trypsin inhibitor have been suggested to occur as a multi-enzyme complex in mung bean mitochondria (Hou *et al.*, 2000), however, the physiological relevance, if any, of this association is not yet clear.

The importance of DHARs in ASC recycling is still a matter of debate. It has been argued that DHA levels in chloroplasts, even under oxidative stress conditions, are very low and that proteins with *in vitro* DHAR activity are not effective enough to regenerate these low levels (Morell *et al.*, 1997). In a recent search in *A. thaliana* for human omega class-like glutathione transferases, four proteins were identified with glutathione-dependent DHAR activity (Dixon *et al.*, 2002). Two of these were predicted to be cytosolic, whereas another was likely targeted to the chloroplast. However, these proteins also showed thiol transferase activity with the model substrate 2-hydroxyethyl disulfide, but showed no glutathione conjugation activity. These findings add to the uncertainty of whether proteins with DHAR activity also perform this function in the cell. Overexpression of DHAR in *A. thaliana* and maize results in two- to fourfold higher levels of reduced ASC, and, interestingly, also in improved levels and redox state of glutathione (Chen *et al.*, 2003). However, whether the increase of ASC levels is due to increased DHAR-mediated recycling, or to other changes in thiol transferase-dependent pathways remains to be demonstrated.

Evidence for the presence of other DHA recycling pathways in plant cells was recently obtained using tobacco cell cultures. DHA administered to cultured plant cells is rapidly taken up and reduced to ASC (Potters *et al.*, 2000; see also Chapter 5 by De Gara). Changes in internal ASC concentrations obtained in this way affect cell cycle progression (Potters *et al.*, 2000). However, the mechanism of the rapid reduction of DHA remains unclear. Application of diethylmaleate to deplete the cells of GSH did not inhibit uptake and intracellular DHA reduction. Moreover, addition of DHA did not affect the endogenous GSH levels of the tobacco cells, suggesting that other electron donors may play an essential role in DHAR activity in these cells (G. Potters, personal communication).

8.4 Relative importance of recycling of ASC from MDHA or DHA

From the foregoing it is clear that animal and plant cells have multiple redundant mechanisms for preserving intracellular ASC. Whether reduction from MDHA or DHA predominates likely depends on the cell type and the region of the cell involved. On the whole, the ASC economy of the cell would benefit more from recycling at the MDHA stage, since this would minimize the risk of loss of DHA by ring-opening or exit from the cell on glucose transporters. There is also the evidence discussed earlier from *in vitro* studies that MDHA is relatively unreactive, which might favor its recycling. Further, enzymes described thus far with MDHAR capacity show high affinity for MDHA (typically 10 μM or less). This MDHA concentration is in the range expected for cells undergoing oxidant stress (Buettner and Jurkiewicz, 1993; Pietri *et al.*, 1990, 1994). Coassin *et al.* (1991) compared MDHA and DHA reductase activities in homogenates of several pig tissues. They found that MDHAR activity was highest when normalized to protein in liver, kidney, and heart, with lesser activities in spleen, brain, and lung. Not surprisingly, most of the activity localized to mitochondria and microsomes in heart tissue. In contrast, enzyme-dependent reduction of 1 mM DHA by 2 mM GSH occurred at rates about 10–30% that of MDHAR. The DHA reductase activity had to be corrected for a rate of nonenzymatic DHA reduction that was typically greater than the enzymatic component. These results led the authors to conclude that in most tissues, reduction of MDHA predominates over enzyme-dependent DHA reduction.

MDHA reduction is a high affinity system that is well suited to recycling ASC in certain organelles, especially to carry out hydroxylation reactions. However, under severe oxidant stress, such as ischemia-reperfusion injury (Pietri *et al.*, 1990) or cardiopulmonary bypass (Clermont *et al.*, 2002; Pietri *et al.*, 1994), MDHA is generated so rapidly that it overwhelms the capacity of the reductases and accumulates to levels easily measured by EPR. The latter is capable of detecting MDHA at concentrations of 10 nM in biological fluids (Buettner and Jurkiewicz, 1993; Mehlhorn, 1991). Under these conditions, either dismutation or direct oxidation of MDHA may lead to substantial concentrations of DHA that will require reduction by one or more of the relatively low affinity DHA reductases.

Whether DHA accumulates in significant amounts in nonstressed tissues or cells is uncertain. The major problem is that DHA is difficult to measure in tissues, given the tendency of ASC to oxidize during sample work-up. In plasma, DHA concentrations are 0–2 μM, or at most 1–2% of ASC concentrations (Banerjee, 1982; Dhariwal *et al.*, 1991; Lykkesfeldt *et al.*, 1995; Schorah *et al.*, 1996). Several-fold increases in plasma DHA concentrations have been found in conditions manifesting oxidant stress, such as uncontrolled diabetes mellitus (Banerjee, 1982) and smoking (Lykkesfeldt *et al.*, 1997). In human lymphocytes isolated in oxygenated medium, intracellular DHA concentrations may be as much as 20% those of ASC concentrations (Farber *et al.*, 1983). DHA concentrations in homogenates of rat liver, heart, kidney, brain, and other organs were quite variable with regard to ASC, ranging from 10 to 100% of total ASC (Behrens and Madère, 1989). This variability, as well as the high levels of DHA relative to ASC, likely reflect oxidation of ASC during sample work-up. Nonetheless, the ability to readily detect DHA even when care is taken with sample preparation, suggests that DHA may be generated in substantial amounts, especially when tissues or cells are exposed to ambient oxygen concentrations. ASC recycling from MDHA may suffice under conditions of low (i.e., normal) oxidant stress, but when excess reactive oxygen or nitrogen species are present, such as in inflammation and recovery from ischemia, direct and enzyme-dependent reduction of DHA may play a significant role. Further, under such conditions ASC recycling may be very rapid. The question arises as to what might be the maximal capacity for ASC recycling in mammalian tissues, which is discussed in the next section.

8.5 Role of ascorbate recycling in cellular homeostasis of ascorbate

The capacity for cells to recycle ASC may not only reflect their ability to combat oxidant stress, but also provide an estimate of the extent to which ASC recycling can maintain intracellular ASC homeostasis. The capacity for ASC recycling can be measured as the ability to take up and reduce DHA to ASC, although this necessarily measures only DHA reductase activity. By taking advantage of a ubiquitous trans-membrane ferricyanide oxidoreductase that prefers intracellular ASC as an electron donor, it is possible to derive an integrated measure of the capacity of a cell to recycle ASC. Orringer and Roer (1979) first showed that the ability of intact human erythrocytes to reduce extracellular ferricyanide was proportional to the intracellular ASC concentration. Because of its size and charge, ferricyanide does not penetrate the cell membrane, and thus can be used to nondestructively oxidize ASC in the cell. Further, the amount of extracellular ferrocyanide (i.e., the reduced form) can be quantified over time in a manner that reflects the ability of the cell to recycle intracellular ASC. The basal rate of ferricyanide reduction in the absence of ASC must be subtracted from the total rate to obtain the ASC-dependent rate. This is typically less than 30% of the ASC-dependent rate in erythrocytes, and decreases if the cells are loaded with ASC (May *et al.*, 1995a, 1995b). In human erythrocytes containing about 40 μM ASC, the initial rate of ferricyanide reduction is about 40 nmol·(ml erythrocytes)$^{-1}$·min^{-1}. From this rate, it can be calculated that erythrocytes at a 45% hematocrit can regenerate a 40 μM concentration of ASC in the corresponding plasma every 3 min (May *et al.*, 1995b). The question arises as to whether this recycling contributes to ASC homeostasis in cells. This can be addressed in erythrocytes, which have the same intracellular concentration of ASC as exists in the surrounding plasma (Evans *et al.*, 1982; Mendiratta *et al.*, 1998b). When incubated briefly with 100 μM DHA, the cells accumulate ASC to about 300 μM, and this level is sustained for several hours, even in ASC-free buffer (Mendiratta *et al.*, 1998b). This suggests that a cell lacking an SVCT-type transporter can accumulate and retain ASC against a gradient. Thus, ASC recycling (in this instance from DHA) can help to maintain intracellular ASC. On the other hand, the fact that erythrocytes do not concentrate ASC in blood shows the critical function of the SVCT in nucleated cells to generate intracellular ASC concentrations 20–40 times those outside of cells.

8.6 Conclusion

Most cells tested thus far have robust and redundant mechanisms for recycling ASC from its oxidized forms. This may indicate how important ASC recycling is in the ASC economy of the cell, and the need to maintain ASC concentrations for crucial enzyme and antioxidant functions. Many unanswered question remain, such as whether recycling from MDHA or DHA is most important, whether DHA accumulates in cells and tissues to a significant extent, and whether differences in ASC recycling capacity correlate with the ability of cells to withstand an oxidant stress.

References

Asada A, Kusakawa T, Orii H, Agata K, Watanabe K, Tsubaki M (2002) Planarian cytochrome b(561): conservation of a six transmembrane structure and localization along the central and peripheral nervous system. *J. Biochem.* (Tokyo) **131**, 175–182.

Asada K (1999) The water–water cycle in chloroplasts: Scavenging of active oxygens and dissipation of excess photons. *Ann. Rev. Plant Physiol. Plant Mol. Biol.* **50**, 601–639.

Asard H, Horemans N, Caubergs RJ (1992) Transmembrane electron transport in ascorbate-loaded plasma membrane vesicles from higher plants involves a b-type cytochrome. *FEBS Lett.* **306**, 143–146.

Asard H, Kapila J, Verelst W, Bérczi A (2001) Higher-plant plasma membrane cytochrome b561: a protein in search of a function. *Protoplasma* **217**, 77–93.

Bando M, Obazawa H (1988) Ascorbate free radical reductase and ascorbate redox cycle in the human lens. *Jpn. J. Ophthalmol.* **32**, 176–186.

Bando M, Obazawa H (1994) Soluble ascorbate free radical reductase in the human lens. *Jpn. J. Ophthalmol.* **38**, 1–9.

Banerjee A (1982) Blood dehydroascorbic acid and diabetes mellitus in human beings. *Ann. Clin. Biochem.* **19**, 65–70.

Barroso JB, Corpas FJ, Carreras A, Sandalio LM, Valderrama R, Palma JM, Lupianez JA, del Rio LA (1999) Localization of nitric-oxide synthase in plant peroxisomes. *J. Biol. Chem.* **274**, 36729–36733.

Bashtovyy D, Bérczi A, Asard H, Páli T (2003) Structure prediction for the di-heme cytochrome b-561 protein family. *Protoplasma* **221**, 31–40.

Behrens WA, Madère R (1989) Ascorbic and dehydroascorbic acids status in rats fed diets varying in vitamin E levels. *Int. J. Vitam. Nutr. Res.* **59**, 360–364.

Bérczi A, Luthje S, Asard H (2001) b-type cytochromes in plasma membranes of *Phaseolus vulgaris* hypocotyls, *Arabidopsis thaliana* leaves, and *Zea mays* roots. *Protoplasma* **217**, 50–55.

Bérczi A, Møller IM (1998) Characterization and solubilization of residual redox activity in salt-washed and detergent-treated plasma membrane vesicles from spinach leaves. *Protoplasma* **205**, 59–65.

Bianchi J, R (1986a) Dehydroascorbic acid and cell membranes: possible disruptive effects. *Toxicology* **40**, 75–82.

Bianchi J, Rose RC (1986b) Glucose-independent transport of dehydroascorbic acid in human erythrocytes. *Proc. Soc. Exp. Biol. Med.* **181**, 333–337.

Bielski BH, Richter HW, Chan PC (1975) Some properties of the ascorbate free radical. *Ann. NY Acad. Sci.* **258**, 231–237.

Bielski BHJ (1982) Chemistry of ascorbic acid radicals. In: *Ascorbic Acid: Chemistry, Metabolism, and Uses* (eds PA Seib, BM Tolbert). American Chemical Society, Washington, D.C., pp. 81–100.

Bigley RH, Stankova L (1974) Uptake and reduction of oxidized and reduced ascorbate by human leukocytes. *J. Exp. Med.* **139**, 1084–1092.

Bode AM, Cunningham L, R (1990) Spontaneous decay of oxidized ascorbic acid (dehydro-L-ascorbic acid) evaluated by high-pressure liquid chromatography. *Clin. Chem.* **36**, 1807–1809.

Boraccino G, Dipierro S, Arrigoni O (1986) Purification and properties of ascorbate free-radical reductase from potato tubers. *Planta* **167**, 521–526.

Buettner GR (1993) The pecking order of free radicals and antioxidants: lipid peroxidation, alpha-tocopherol, and ascorbate. *Arch. Biochem. Biophys.* **300**, 535–543.

Buettner GR, Jurkiewicz BA (1993) Ascorbate free radical as a marker of oxidative stress: an EPR study. *Free Radic. Biol. Med.* **14**, 49–55.

Chatterjee IB (1970) Biosynthesis of L-ascorbate in animals. *Methods Enzymol.* **18**, 28–34.

Chen Z, Young TE, Ling J, Chang SC, Gallie DR (2003) Increasing vitamin C content of plants through enhanced ascorbate recycling. *Proc. Natl Acad. Sci. USA* **100**, 3525–3530.

Christine L, Thomson G, Iggo B, Brownie AC, Stewart CP (1956) The reduction of dehydroascorbic acid by human erythrocytes. *Clin. Chim. Acta* **1**, 557–569.

Clermont G, Vergely C, Jazayeri S, Lahet JJ, Goudeau JJ, Lecour S, David M, Rochette L, Girard C (2002) Systemic free radical activation is a major event involved in myocardial oxidative stress related to cardiopulmonary bypass. *Anesthesiology* **96**, 80–87.

Coassin M, Tomasi A, Vannini V, Ursini F (1991) Enzymatic recycling of oxidized ascorbate in pig heart: one- electron vs two-electron pathway. *Arch. Biochem. Biophys.* **290**, 458–462.

Corpas FJ, Barroso JB, del Rio LA (2001) Peroxisomes as a source of reactive oxygen species and nitric oxide signal molecules in plant cells. *Trends Plant. Sci.* **6**, 145–150.

Dalton DL, Baird LM, Langeber L, Taugher CY, Anyan WR, Vance CP, Sarath G (1993) Subcellular localisation of oxygen defense enzymes in soybean (*Glycine max* [L] Merr.) root nodules. *Plant Physiol.* **102**, 481–489.

Del Bello B, Maellaro E, Sugherini L, Santucci A, Comporti M, Casini AF (1994) Purification of NADPH-dependent dehydroascorbate reductase from rat liver and its identification with 3a-hydroxysteroid dehydrogenase. *Biochem. J.* **304**, 385–390.

del Rio LA, Corpas FJ, Sandalio LM, Palma JM, Gomez M, Barroso JB (2002) Reactive oxygen species, antioxidant systems and nitric oxide in peroxisomes. *J. Exp. Bot.* **53**, 1255–1272.

Deutsch JC, Santhosh-Kumar CR (1996) Dehydroascorbic acid undergoes hydrolysis on solubilization which can be reversed with mercaptoethanol. *J. Chromatogr. A* **724**, 271–278.

Dhariwal KR, Hartzell WO, Levine M (1991) Ascorbic acid and dehydroascorbic acid measurements in human plasma and serum. *Am. J. Clin. Nutr.* **54**, 712–716.

Diliberto EJ Jr, Dean G, Carter C, Allen PL (1982) Tissue, subcellular, and submitochondrial distributions of semidehydroascorbate reductase: possible role of semidehydroascorbate reductase in cofactor regeneration. *J. Neurochem.* **39**, 563–568.

Dipierro S, Borraccino G (1991) Dehydroascorbate reductase from potato-tubers. *Phytochemistry* **30**, 427–429.

Dixon DP, Davis BG, Edwards R (2002) Functional divergence in the glutathione transferase superfamily in plants. Identification of two classes with putative functions in redox homeostasis in *Arabidopsis thaliana*. *J. Biol. Chem.* **277**, 30859–30869.

Drake BB, Smythe CV, King CG (1942) Complexes of dehydroascorbic acid with three sulfhydryl compounds. *J. Biol. Chem.* **143**, 89–98.

Evans RM, Currie L, Campbell A (1982) The distribution of ascorbic acid between various cellular components of blood, in normal individuals, and its relation to the plasma concentration. *Br. J. Nutr.* **47**, 473–482.

Farber CM, Kanengiser S, Stahl R, Liebes L, Silber R (1983) A specific high-performance liquid chromatography assay for dehydroascorbic acid show an increased content in CLL lymphocytes. *Anal. Biochem.* **134**, 355–360.

Fiorani M, De Sanctis R, Scarlatti F, Vallorani L, De Bellis R, Serafini G, Bianchi M, Stocchi V (2000) Dehydroascorbic acid irreversibly inhibits hexokinase activity. *Mol. Cell. Biochem.* **209**, 145–153.

Fornai F, Piaggi S, Gesi M, Saviozzi M, Lenzi P, Paparelli A, Casini AF (2001) Subcellular localization of a glutathione-dependent dehydroascorbate reductase within specific rat brain regions. *Neuroscience* **104**, 15–31.

Fornai F, Saviozzi M, Piaggi S, Gesi M, Corsini GU, Malvaldi G, Casini AF (1999) Localization of a glutathione-dependent dehydroascorbate reductase within the central nervous system of the rat. *Neuroscience* **94**, 937–948.

Foyer C, Halliwell B (1977) Purification and properties of dehydroascorbate reductase from spinach leaves. *Phytochemistry* **16**, 1347–1350.

Foyer CH, Lelandais M (1996) A comparison of the relative rates of transport of ascorbate and glucose across the thylakoid, chloroplast and plasmalemma membranes of pea leaf mesophyll cells. *J. Plant Physiol.* **148**, 391–398.

Gladyshev VN, Liu AM, Novoselov SV, Krysan K, Sun QA, Kryukov VM, Kryukov GV, Lou MF (2001) Identification and characterization of a new mammalian glutaredoxin (thioltransferase), Grx2. *J. Biol. Chem.* **276**, 30374–30380.

Goldenberg H, Grebing C, Löw H (1983) NADH-monodehydroascorbate reductase in human erythrocyte membranes. *Biochem. Int.* **6**, 1–9.

Guaiquil VH, Farber CM, Golde DW, Vera JC (1997) Efficient transport and accumulation of vitamin C in HL-60 cells depleted of glutathione. *J. Biol. Chem.* **272**, 9915–9921.

Halliwell B, Foyer CH (1976) Ascorbic acid, metal ions and the superoxide radical. *Biochem. J.* **155**, 697–700.

Hara T, Minakami S (1971) On functional role of cytochrome b5. II. NADH-linked ascorbate radical reductase activity in microsomes. *J. Biochem. (Tokyo)* **69**, 325–330.

Harwood HJ Jr, Greene YJ, Stacpoole PW (1986) Inhibition of human leukocyte 3-hydroxy-3-methylglutaryl coenzyme A reductase activity by ascorbic acid. An effect mediated by the free radical monodehydroascorbate. *J. Biol. Chem.* **261**, 7127–7135.

Heber U, Miyake C, Mano J, Ohno C, Asada K (1996) Monodehydroascorbate radical detected by electron paramagnetic resonance spectrometry is a sensitive probe of oxidative stress in intact leaves. *Plant Cell. Physiol.* **37**, 1066–1072.

Hideg E, Mano J, Ohno C, Asada K (1997) Increased levels of monodehydroascorbate radical in UV-B-irradiated broad bean leaves. *Plant Cell. Physiol.* **38**, 684–690.

Horemans N, Foyer CH, Asard H (2000) Transport and action of ascorbate at the plant plasma membrane. *Trends Plant. Sci.* 5, 263–267.

Hossain M, Asada A (1984) Purification of dehydroascorbate reductase from spinach and its characterization as a thiol enzyme. *Plant Cell. Physiol.* **25**, 85–92.

Hossain MA, Asada K (1985) Monodehydroascorbate reductase from cucumber is a flavin adenine dinucleotide enzyme. *J. Biol. Chem.* **260**, 12920–12926.

Hou WC, Wang YT, Lin YH, Hsiao LJ, Chen TE, Wang CW, Dai H (2000) A complex containing both trypsin inhibitor and dehydroascorbate reductase activities isolated from mitochondria of etiolated mung bean (*Vigna radiata* L. (Wilczek) cv. Tainan No. 5) seedlings. *J. Exp. Bot.* **51**, 713–719.

Hughes RE, Maton SC (1968) The passage of vitamin C across the erythrocyte membrane. *Br. J. Haematol.* **14**, 247–253.

Ito A, Hayashi S, Yoshida T (1981) Participation of a cytochrome b5-like hemoprotein of outer mitochondrial membrane (OM cytochrome b) in NADH-semidehydroascorbic acid reductase activity of rat liver. *Biochem. Biophys. Res. Commun.* **101**, 591–598.

Iyanagi T, Yamazaki I, Anan KF (1985) One-electron oxidation-reduction properties of ascorbic acid. *Biochim. Biophys. Acta* **806**, 255–261.

Jimenez A, Gomez JM, Navarro E, Sevilla F (2002) Changes in the antioxidative systems in mitochondria during ripening of pepper fruits. *Plant Physiol. Biochem.* **40**, 515–520.

Jimenez A, Hernandez JA, del Rio LA, Sevilla F (1997) Evidence for the presence of the ascorbate-glutathione cycle in mitochondria and peroxisomes of pea leaves. *Plant Physiol.* **114**, 275–284.

Jocelyn PC (1975) Some properties of mitochondrial glutathione. *Biochim. Biophys. Acta* **396**, 427–436.

Kobayashi K, Harada Y, Hayashi K (1991) Kinetic behavior of the monodehydroascorbate radical studied by pulse radiolysis. *Biochemistry* **30**, 8310–8315.

Koshiba T (1993) Cytosolic ascorbate peroxidase in seedlings and leaves of maize (*Zea mays*). *Plant Cell. Physiol.* **34**, 713–721.

Koshiishi I, Mamura Y, Liu J, Imanari T (1998) Degradation of dehydroascorbate to 2,3-diketogulonate in blood circulation. *Biochim. Biophys. Acta* **1425**, 209–214.

Koyama K, Takatsuki K, Inoue M (1994) Determination of superoxide and ascorbyl radicals in the circulation of animals under oxidative stress. *Arch. Biochem. Biophys.* **309**, 323–328.

Laroff GP, Fessenden RW, Schuler RH (1972) The electron spin resonance spectra of radical intermediates in the oxidation of ascorbic acid and related substances. *J. Am. Chem. Soc.* **94**, 9062–9073.

Li X, Cobb CE, Hill KE, Burk RF, May JM (2001a) Mitochondrial uptake and recycling of ascorbic acid. *Arch. Biochem. Biophys.* **387**, 143–153.

Li X, Hill KE, Burk RF, May JM (2001b) GSH is required to recycle ascorbic acid in cultured liver cell lines. *Antioxid. Red. Signal.* **3**, 1089–1097.

Lykkesfeldt J, Loft S, Nielsen JB, Poulsen HE (1997) Ascorbic acid and dehydroascorbic acid as biomarkers of oxidative stress caused by smoking. *Am. J. Clin. Nutr.* **65**, 959–963.

Lykkesfeldt J, Loft S, Poulsen HE (1995) Determination of ascorbic acid and dehydroascorbic acid in plasma by high-performance liquid chromatography with coulometric detection – Are they reliable biomarkers of oxidative stress. *Anal. Biochem.* **229**, 329–335.

Mackin RB, Jones DP, Noe BD (1986) Islet secretory granules contain cytochrome b561. Diabetes **35**, 881–885.

Maellaro E, Del Bello B, Sugherini L, Comporti M, Casini AF (1997) Purification and characterization of glutathione-dependent dehydroascorbate reductase from rat liver. *Methods Enzymol.* **279**, 30–35.

Maellaro E, Del Bello B, Sugherini L, Santucci A, Comporti M, Casini AF (1994) Purification and characterization of glutathione-dependent dehydroascorbate reductase from rat liver. *Biochem. J.* **301**: 471–476.

Mårtensson J, Han J, Griffith OW, Meister A (1993) Glutathione ester delays the onset of scurvy in ascorbate-deficient guinea pigs. *Proc. Natl Acad. Sci. USA* **90**, 317–321.

Mårtensson J, Meister A (1992) Glutathione deficiency increases hepatic ascorbic acid synthesis in adult mice. *Proc. Natl Acad. Sci. USA* **89**, 11566–11568.

May JM, Cobb CE, Mendiratta S, Hill KE, Burk RF (1998) Reduction of the ascorbyl free radical to ascorbate by thioredoxin reductase. *J. Biol. Chem.* **273**, 23039–23045.

May JM, Mendiratta S, Hill KE, Burk RF (1997) Reduction of dehydroascorbate to ascorbate by the selenoenzyme thioredoxin reductase. *J. Biol. Chem.* **272**, 22607–22610.

May JM, Mendiratta S, Qu ZC, Loggins E (1999) Vitamin C recycling and function in human monocytic U-937 cells. *Free Radic. Biol. Med.* **26**, 1513–1523.

May JM, Qu ZC, Cobb CE (2000) Extracellular reduction of the ascorbate free radical by human erythrocytes. *Biochem. Biophys. Res. Commun.* **267**, 118–123.

May JM, Qu ZC, Cobb CE (2001a) Recycling of the ascorbate free radical by human erythrocyte membranes. *Free Radic. Biol. Med.* **31**, 117–124.

May JM, Qu ZC, Li X (2001b) Requirement for GSH in recycling of ascorbic acid in endothelial cells. *Biochem. Pharmacol.* **62**, 873–881.

May JM, Qu ZC, Whitesell RR, Cobb CE (1996) Ascorbate recycling in human erythrocytes: Role of GSH in reducing dehydroascorbate. *Free Radic. Biol. Med.* **20**, 543–551.

May JM, Qu Z-C, Whitesell RR (1995a) Ascorbate is the major electron donor for a transmembrane oxidoreductase of human erythrocytes. *Biochim. Biophys. Acta* **1238**, 127–136.

May JM, Qu Z-C, Whitesell RR (1995b) Ascorbic acid recycling enhances the antioxidant reserve of human erythrocytes. *Biochemistry* **34**, 12721–12728.

McKie AT, Barrow D, Latunde-Dada GO, Rolfs A, Sager G, Mudaly E, *et al* (2001) An iron-regulated ferric reductase associated with the absorption of dietary iron. *Science* **291**, 1755–1759.

Mehlhorn RJ (1991) Ascorbate- and dehydroascorbic acid-mediated reduction of free radicals in the human erythrocyte. *J. Biol. Chem.* **266**, 2724–2731.

Mendiratta S, Qu Z-C, May JM (1998a) Enzyme-dependent ascorbate recycling in human erythrocytes: role of thioredoxin reductase. *Free Radic. Biol. Med.* **25**, 221–228.

Mendiratta S, Qu Z-C, May JM (1998b) Erythrocyte ascorbate recycling: Antioxidant effects in blood. *Free Radic. Biol. Med.* **24**, 789–797.

Mieyal JJ, Starke DW, Gravina SA, Dothey C, Chung JS (1991) Thioltransferase in human red blood cells: purification and properties. *Biochemistry* **30**, 6088–6097.

Minetti M, Forte T, Soriani M, Quaresima V, Menditto A, Ferrari M (1992) Iron-induced ascorbate oxidation in plasma as monitored by ascorbate free radical formation. No spin-trapping evidence for the hydroxyl radical in iron-overloaded plasma. *Biochem. J.* **282**, 459–465.

Mittova V, Volokita M, Guy M, Tal M (2000) Activities of SOD and the ascorbate-glutathione cycle enzymes in subcellular compartments in leaves and roots of the cultivated tomato and its wild salt-tolerant relative *Lycopersicon pennellii*. *Physiol. Plant* **110**, 42–51.

Miyake C, Asada K (1994) Ferredoxin-dependent photoreduction of the monodehydroascorbate radical in spinach thylakoids. *Plant Cell Physiol.* **35**, 539–549.

Møller IM, Fredlund KM, Bérczi A (1995) The stereospecificity, purification, and characterization of an NADH-ferricyanide reductase from spinach leaf plasma membrane. *Protoplasma* **184**, 124–132.

Morell S, Follmann H, De Tullio M, Haberlein I (1997) Dehydroascorbate and dehydroascorbate reductase are phantom indicators of oxidative stress in plants. *FEBS Lett.* **414**, 567–70.

Nishino H, Ito A (1986) Subcellular distribution of OM cytochrome b-mediated NADH-semidehydroascorbate reductase activity in rat liver. *J. Biochem. (Tokyo)* **100**, 1523–1531.

Njus D, Knoth J, Cook C, Kelley PM (1983) Electron transfer across the chromaffin granule membrane. *J. Biol. Chem.* **258**, 27–30.

Njus D, Wigle M, Kelley PM, Kipp BH, Schlegel HB (2001) Mechanism of ascorbic acid oxidation by cytochrome b561. *Biochemistry* **40**, 11905–11911.

Noctor G, Foyer CH (1998) Ascorbate and glutathione: Keeping active oxygen under control. *Ann. Rev. Plant Physiol. Plant Mol. Biol.* **49**, 249–279.

Ohnishi T, Yamazaki H, Iyanagi T, Nakamura T, Yamazaki I (1969) One-electron-transfer reactions in biochemical systems. II. The reaction of free radicals formed in the enzymic oxidation. *Biochim. Biophys. Acta* **172**, 357–369.

Okuyama E, Yamamoto R, Ichikawa Y, Tsubaki M (1998) Structural basis for the electron transfer across the chromaffin vesicle membranes catalyzed by cytochrome b(561): analyses of cDNA nucleotide sequences and visible absorption spectra. *Biochim. Biophys. Acta* **1383**, 269–278.

Orringer EP, Roer ME (1979) An ascorbate-*mediated* transmembrane-reducing system of the human erythrocyte. *J. Clin. Invest.* **63**, 53–58.

Paolicchi A, Pezzini A, Saviozzi M, Piaggi S, Andreuccetti M, Chieli E, Malvaldi G, Casini AF (1996) Localization of a GSH-dependent dehydroascorbate reductase in rat tissues and subcellular fractions. *Arch. Biochem. Biophys.* **333**, 489–495.

Park JB, Levine M (1996) Purification, cloning and expression of dehydroascorbic acid-reducing activity from human neutrophils: Identification as glutaredoxin. *Biochem. J.* **315**, 931–938.

Pietri S, Culcasi M, Stella L, Cozzone PJ (1990) Ascorbyl free radical as a reliable indicator of free-radical-mediated myocardial ischemic and post-ischemic injury. A real-time continuous-flow ESR study. *Eur. J. Biochem.* **193**, 845–854.

Pietri S, Seguin JR, d'Arbigny PD, Culcasi M (1994) Ascorbyl free radical: a noninvasive marker of oxidative stress in human open-heart surgery. *Free Radic. Biol. Med.* **16**, 523–528.

Ponting CP (2001) Domain homologues of dopamine beta-hydroxylase and ferric reductase: roles for iron metabolism in neurodegenerative disorders? *Hum. Mol. Genet.* **10**, 1853–1858.

Potters G, Horemans N, Caubergs RJ, Asard H (2000) Ascorbate and dehydroascorbate influence cell cycle progression in a tobacco cell suspension. *Plant Physiol.* **124**, 17–20.

Rautenkranz AAF, Li LJ, Machler F, Martinoia E, Oertli JJ (1994) Transport of ascorbic and dehydroascorbic acids across protoplast and vacuole membranes isolated from barley (*Hordeum vulgare* L Cv Gerbel) leaves. *Plant Physiol.* **106**, 187–193.

Roginsky VA, Stegmann HB (1994) Ascorbyl radical as natural indicator of oxidative stress: Quantitative regularities. *Free Radic. Biol. Med.* **17**, 93–103.

Sano S, Asada K (1994) cDNA cloning of monodehydroascorbate radical reductase from cucumber – a high-degree of homology in terms of amino-acid sequence between this enzyme and bacterial flavoenzymes. *Plant Cell. Physiol.* **35**, 425–437.

Sano S, Miyake C, Mikami B, Asada K (1995) Molecular characterization of monodehydroascorbate radical reductase from cucumber highly expressed in *Escherichia coli*. *J. Biol. Chem.* **270**, 21354–21361.

Sasaki H, Giblin FJ, Winkler BS, Chakrapani B, Leverenz V, Shu CC (1995) A protective role for glutathione-dependent reduction of dehydroascorbic acid in lens epithelium. *Invest. Ophthalmol. Vis. Sci.* **36**, 1804–1817.

Savini I, Duflot S, Avigliano L (2000) Dehydroascorbic acid uptake in a human keratinocyte cell line (HaCaT) is glutathione-independent. *Biochem. J.* **345**, 665–672.

Schorah CJ, Downing C, Piripitsi A, Gallivan L, Al-Hazaa AH, Sanderson MJ, Bodenham A (1996) Total vitamin C, ascorbic acid, and dehydroascorbic acid concentrations in plasma of critically ill patients. *Am. J. Clin. Nutr.* **63**, 760–765.

Schulze HU, Staudinger H (1971) Untersuchungen über die Verteilung von Enzymproteinen in den endoplasmatischen Membranen der Leberzelle, I. Trennung von Rattenlebermikrosomen durch DifferentiaL- und Zonenzentrifugation. *Hoppe Seylers. Z. Physiol. Chem.* **352**, 1659–1674.

Schweinzer E, Goldenberg H (1993) Monodehydroascorbate reductase activity in the surface membrane of leukemic cells. Characterization by a ferricyanide-driven redox cycle. *Eur. J. Biochem.* **218**, 1057–1062.

Schweinzer E, Mao Y, Krajnik P, Getoff N, Goldenberg H (1996) Reduction of extracellular dehydroascorbic acid by K562 cells. *Cell Biochem. Funct.* **14**, 27–31.

Shimaoka T, Miyake C, Yokota A (2003) Mechanism of the reaction catalyzed by dehydroascorbate reductase from spinach chloroplasts. *Eur. J. Biochem.* **270**, 921–8.

Shimaoka T, Yokota A, Miyake C (2000) Purification and characterization of chloroplast dehydroascorbate reductase from spinach leaves. *Plant Cell. Physiol.* **41**, 1110–1118.

Shirabe K, Landi MT, Takeshita M, Uziel G, Fedrizzi E, Borgese N (1995) A novel point mutation in a 3′ splice site of thc NADII-cytochrome b5 reductase gene results in immunologically undetectable enzyme and impaired NADH-dependent ascorbate regeneration in cultured fibroblasts of a patient with type II hereditary methemoglobinemia. *Am. J. Hum. Genet.* 57, 302–310.

Silsand T, Flatmark T (1974) Purification of cytochrome b-561. An integral heme protein of the adrenal chromaffin granule membrane. *Biochim. Biophys. Acta* **359**, 257–266.

Spickett CM, Smirnoff N, Pitt AR (2000) The biosynthesis of erythroascorbate in *Saccharomyces cerevisiae* and its role as an antioxidant. *Free Radic. Biol. Med.* **28**, 183–192.

Trost P, Bérczi A, Sparla F, Sponza G, Marzadori B, Asard H, Pupillo P (2000) Purification of cytochrome b-561 from bean hypocotyls plasma membrane. Evidence for the presence of two heme centers. *Biochim. Biophys. Acta* **1468**, 1–5.

Trumper S, Follmann H, Haberlein I (1994) A novel-dehydroascorbate reductase from spinach chloroplasts homologous to plant trypsin inhibitor. *FEBS Lett.* **352**, 159–162.

Tsubaki M, Nakayama M, Okuyama E, Ichikawa Y, Hori H (1997) Existence of two heme B centers in cytochrome b(561) from bovine adrenal chromaffin vesicles as revealed by a new purification procedure and EPR spectroscopy. *J. Biol. Chem.* **272**, 23206–23210.

Urano J, Nakagawa T, Maki Y, Masumura T, Tanaka K, Murata N, Ushimaru T (2000) Molecular cloning and characterization of a rice dehydroascorbate reductase. *FEBS Lett.* **466**, 107–111.

VanDuijn MM, Tijssen K, Van Steveninck J, van den Broek PJA, Van der Zee J (2000) Erythrocytes reduce extracellular ascorbate free radicals using intracellular ascorbate as an electron donor. *J. Biol. Chem.* **275**, 27720–27725.

VanDuijn MM, Van der Zee J, van den Broek PJA (2001) The ascorbate-driven reduction of extracellular ascorbate free radical by the erythrocyte is an electrogenic process. *FEBS Lett.* **491**, 67–70.

Vera JC, Rivas CI, Fischbarg J, Golde DW (1993) Mammalian facilitative hexose transporters mediate the transport of dehydroascorbic acid. *Nature* **364**, 79–82.

Wahlländer A, Soboll S, Sies H, Linke I, Muller M (1979) Hepatic mitochondrial and cytosolic glutathione content and the subcellular distribution of GSH-S-transferases. *FEBS Lett.* **97**, 138–140.

Wakefield LM, Cass AEG, Radda GK (1986) Functional coupling between enzymes of the chromaffin granule membrane. *J. Biol. Chem.* **261**, 9739–9745.

Washburn MP, Wells WW (1999) The catalytic mechanism of the glutathione-dependent dehydroascorbate reductase activity of thioltransferase (glutaredoxin). *Biochemistry* **38**, 268–274.

Wells WW, Xu DP, Washburn MP (1995) Glutathione: Dehydroascorbate oxidoreductases. *Methods Enzymol.* **252**, 30–38.

Wells WW, Xu DP, Yang YF, Rocque PA (1990) Mammalian thioltransferase (glutaredoxin) and protein disulfide isomerase have dehydroascorbate reductase activity. *J. Biol. Chem.* **265**, 15361–15364.

Winkler BS (1987) In vitro oxidation of ascorbic acid and its prevention by GSH. *Biochim. Biophys. Acta* **925**, 258–264.

Winkler BS (1992) Unequivocal evidence in support of the nonenzymatic redox coupling between glutathione/glutathione disulfide and ascorbic acid/dehydroascorbic acid. *Biochim. Biophys. Acta* **1117**, 287–290.

Winkler BS, Orselli SM, Rex TS (1994) The redox couple between glutathione and ascorbic acid: A chemical and physiological perspective. *Free Radic. Biol. Med.* **17**, 333–349.

Xu DP, Washburn MP, Sun GP, Wells WW (1996) Purification and characterization of a glutathione dependent dehydroascorbate reductase from human erythrocytes. *Biochem. Biophys. Res. Commun.* **221**, 117–121.

Yamaguchi K, Mori H, Nishimura M (1995) A novel isoenzyme of ascorbate peroxidase localized on glyoxysomal and leaf peroxisomal membranes in pumpkin. *Plant Cell. Physiol.* **36**, 1157–1162.

9

How does ascorbic acid prevent scurvy? A survey of the nonantioxidant functions of vitamin C

Mario C. De Tullio

9.1 Introduction

If humans had not lost the ability to synthesize ascorbate (ASC), probably there would be less need now to write books on vitamin C. Research on ASC has its roots in the history of medicine, namely in the inborn error in the pathway of ASC biosynthesis leading to the disease known as scurvy (Nishikimi and Udenfriend, 1976, 1977). About 70 years after the identification of ASC as the antiscorbutic factor vitamin C, there is general agreement on ASC importance for human health, yet its role(s) in cell metabolism is not fully understood. Both ASC-synthesizing and nonsynthesizing organisms use ASC in different ways. There is an increasing propensity to divide the biological role(s) of ASC into the two distinct categories of 'antioxidant' and 'nonantioxidant' functions. The former group includes ASC reactions with reactive oxygen species (ROS) such as singlet oxygen, superoxide anion, hydroxyl radical and hydrogen peroxide. Nonantioxidant functions include almost any other ASC-related phenomenon. Such a distinction could generate the misconception that ASC has different forms of reactivity. It should be borne in mind that the molecular mechanism underlying a large part of the many different functions of ASC known so far, is 'simply' its capability to transfer reducing equivalents to an electron acceptor. As it will be described below, ASC specific reactivity, depending on the context, can have many different biological consequences.

Nowadays, a standard procedure that is used widely in biology to understand the function of a gene is the generation of knockout mutants, followed by the observation of the consequences of the inactivation of the target gene. Whether we like it or not, mutation of the gene encoding gulono-lactone oxidase during

Vitamin C, edited by Han Asard, James M May and Nicholas Smirnoff

evolution made humans, some other primates, guinea pigs and some birds the perfect experimental subjects with which to study the function of ASC. Scurvy is the consequence of dietary deprivation of ASC in humans and other nonsynthesizing organisms. Analysis of scurvy symptoms and the biochemical basis of this disease can lead to surprising insights.

9.2 Scurvy

It is not clear if scurvy was already known by the ancients. Indeed, some of the symptoms reported in the Ebers papyrus (an Egyptian document of ca. 1500 B.C. reporting medical descriptions), Hippocrates writings (ca. 400 B.C.), the Old Testament, and Pliny's *Historia naturalis* could be related to scurvy, but detailed description of the disease occurred only later. Scurvy became endemic in northern Europe during the Middle Ages. In Old Norse, a medieval Scandinavian language, the disease was called *skyrbjugr*, literally 'a swelling (*bjugr*) from drinking sour milk (*skyr*) on long sea voyages'. Hence probably the Dutch word *scheurbuik*, German *Scharbock* (then *Skorbut*), English scurvy, and medieval Latin *scorbutus*.

Between the fifteenth and eighteenth centuries, scurvy became a major threat for sailors during long voyages of discovery. The disease also caused many losses among troops (e.g., the Hungarian and Russian armies) during war campaigns.

James Lind's *A Treatise of the Scurvy*, published in 1753, is an invaluable source of information to understand the extent and impact of scurvy in those days (full text of Lind's book is available on the website of the Cornell University Digital Library Collections at http://cdl.library.cornell.edu/cgi-bin/chla/chla-cgi?notisid=AFM4117). In his book, Lind reports several medical descriptions of scorbutic symptoms, the earliest being made in 1541 by the Dutch physician Johannes Echtius:

> *He [Echtius] distinguishes the symptoms into two classes. The first contains such as appear at the beginning and are common to it with other diseases; the second, the succeeding and more certain signs of the malady. Under the first, he comprehends a heaviness of the body, with a spontaneous lassitude, generally most sensibly felt after exercise; a tightness of the breast, and a weakness of the legs; an itching, redness and pains of the gums; a change of colour in the face to a darkish hue: and observes, that where all these concur, we may foretel an approaching scurvy. But the more immediate and certain signs he enumerates under the second class, viz. a foetid breath, a spungy swelling of the gums, which are apt to bleed, with a loosening of the teeth; an eruption of leaden-coloured, purple or livid spots, on the legs; or of somewhat broader speckled or dark-coloured maculae, sometimes on the face, at other times on the legs. As the disease advances, the patients lose the use of their legs, and are subjected to a difficulty of breathing. [...] There is sometimes observed an aggravation of the symptoms; with some on the fourth or fifth day, in others on the third. Some few have it every day, but without any fever: others become feverish. Preceeding fevers may terminate critically, as it were, in the scurvy: and with that scurvy whole families and monasteries are together infected; which generally end either in a deadly dysentery, or, at other times, in a sudden and mortal faint.*
>
> James Lind, A Treatise of the Scurvy, pp. 262–263

Basically, later descriptions of scorbutic symptoms did not contradict Echtius' observations. However, some interesting points were added by Lord Anson in his report (published in 1748) on his voyage around the world:

> *This disease [...] is surely the most singular and unaccountable of any that affects the human body. Its symptoms are unconstant and innumerable, and its progress and effects extremely irregular: for scarcely any two persons have the same complaints; and where there hath been found some conformity in the symptoms, the order of their appearance has been totally different.*
>
> James Lind, A Treatise of the Scurvy, p. 338

Lord Anson then described the most common symptoms (swelled legs, putrid gums, spots, lassitude). He also added that:

> *This disease is likewise usually attended with a strange dejection of spirits; and with shiverings, tremblings, and a disposition to be seized with the most dreadful terrors, on the slightest accident. [...] A most extraordinary circumstance, and what would be scarcely credible upon any single evidence, is, that the scars of wounds which had been for many years healed, were forced open again by this virulent distemper. [...] Nay, what is still more astonishing, the callous of a broken bone, which had been compleatly formed for a long time, was found to be hereby dissolved; and the fracture seemed as if it had never been consolidated.*
>
> James Lind, A Treatise of the Scurvy, p. 339

In some cases, the therapies used to cure scurvy were even more dreadful than the disease. Still in 1720, 400 Hungarian troops were killed by a remedy for scurvy containing mercury. However, beneficial effects of juices extracted from medicinal herbs (*Cochlearia*, also known as scurvygrass, *Nasturtium*, *Fumaria* and many others) on the course of the illness were known already in the sixteenth century. Concluding evidence that scurvy could be defeated by some principle present in fresh fruit was obtained by Lind, the 'Father of Nautical Medicine', in his well-known experiment on the 20th of May 1747, in which he compared the efficacy of different remedies used to cure 12 patients affected by scurvy:

> *The result of all my experiments was that oranges and lemons were the most effectual remedies for this distemper at sea. I am apt to think oranges preferable to lemons, though it was principally oranges which so speedily and surprisingly recovered Lord Anson's people at the Island of Tinian, of which that noble, brave and experienced commander was so sensible that before he left the island one man was ordered on shore from each mess to lay in a stock of them for their future security.*
>
> James Lind, A Treatise of the Scurvy, p. 150

After such discovery, British sailors received limejuice during long voyages (hence the name of 'limeys'). It took almost two centuries to understand that the 'antiscorbutic factor' present in James Lind's oranges was a mysterious six-carbon sugar initially called by Albert von Szent-Györgyi 'ignose' and 'godnose' (from 'God knows'), then 'hexuronic acid'. Only after the appraisal of its biological properties, 'hexuronic acid' was eventually named ASC by Szent-Györgyi and Haworth.

How can dietary deprivation of a single molecule account for the 'unconstant and innumerable' symptoms of scurvy? It is conceivable to think that, beside ASC, the depletion of other factors present in fruit may contribute to the long list of symptoms reported by Echtius and Lord Anson. Nevertheless, in the last five decades a large number of studies addressing the issue of the specific role of ASC in cell metabolism allowed us to conclude that the molecular mechanism by which ASC prevents scurvy is mainly (if not exclusively) due to its function as an essential co-substrate of a large class of enzymes, known as 2-oxoglutarate-dependent dioxygenases (2-ODDs) (Schofield and Zhang, 1999; Arrigoni and De Tullio, 2000, 2002). In its historical perspective, the identification of the link between ASC and scurvy is a fascinating journey into the birth of molecular medicine.

9.3 Dioxygenases and scurvy

Dioxygenases catalyze the incorporation of O_2 into an organic substrate. 2-ODDs share a catalytic mechanism (*Figure 9.1*) specifically requiring Fe^{2+}, 2-oxoglutarate (α-ketoglutarate) and ASC, although in some cases differences in co-substrate requirement may occur (Arrigoni and De Tullio, 2000)

Incorporation of oxygen into metabolic intermediates is a critical step in a surprisingly large number of pathways. This explains why ASC deprivation results in so many different and apparently nonrelated effects, as Lord Anson noticed 250 years ago.

A list of the main dioxygenases whose activity is possibly related to the prevention of scurvy is reported in *Table 9.1*. Specific dioxygenases are tentatively associated to major scorbutic symptoms although, in some cases, this association is simply correlative and lacks experimental support.

9.3.1 Collagen prolyl and lysyl hydroxylases

Many evident signs of scurvy are clearly related to disorders in collagen structure and function. In 1959, the following statement could be found in an up-to-date review on ASC: 'Probably the most clearly established functional role of L-ASC is in maintaining the normal intercellular material of cartilage, dentin and bone, but the biochemical mechanisms involved are still unknown' (Burns, 1959). Only a few years later, Stone and Meister (1962) reported compelling evidence of ASC

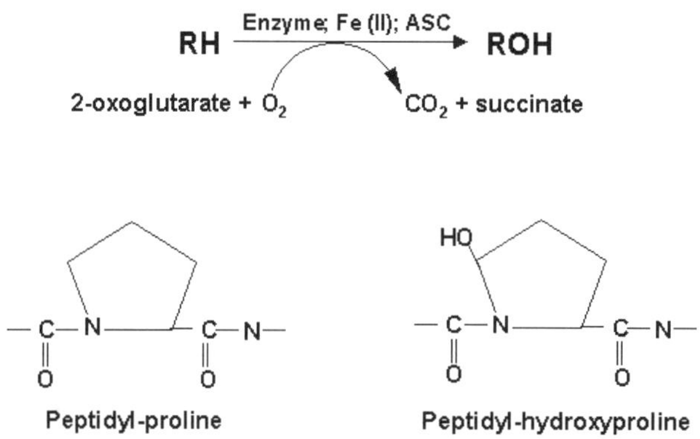

Figure 9.1: The general reaction mechanism of 2-oxoglutarate-dependent dioxygenases and the hydroxylation of peptidyl proline to hydroxyproline.

Table 9.1: Correlation between scurvy symptoms and loss in the activity of specific dioxygenases due to ASC deficiency.

Symptom	Putative target affected	Dioxygenase(s) involved
Hemorrages/spots	Blood vessels/various collagen types (III, IV)	Prolyl/lysyl hydroxylase
Rotten/loose teeth	Dentin/various collagen types (I, III, IV)	Prolyl/lysyl hydroxylase
Rigid tendons	Various collagen types (I, XII, XIV)	Prolyl/lysyl hydroxylase
Bone/cartilage fragility	Various collagen types (I, II, IX, X, XI)	Prolyl/lysyl hydroxylase
Lassitude	Carnitine biosynthesis	Trimethyl-lysine dioxygenase butyrobetaine dioxygenase
Vision problems	Synthesis of vitamin A	β-carotene 15–15′ dioxygenase
Neurological disorders	Noradrenaline / dopamine biosynthesis	Tyrosine hydroxylase dopamine hydroxylase

involvement in the mechanism controlling the incorporation of hydroxyproline residues in collagen.

Among 2-ODDs, collagen prolyl hydroxylase is by far the most studied and best characterized. Proline residues incorporated into the procollagen chains are hydroxylated at carbon 4 by the enzyme in the endoplasmic reticulum. A similar post-translational mechanism occurs for the hydroxylation of lysine residues by lysyl hydroxylase (Kivirikko and Pihlajaniemi, 1998). Human prolyl-hydroxylase is a tetramer made up of 2α- and 2β-chains. The hydroxylation reaction is catalyzed by the α-subunits, whereas the role of the β-subunits is less clear (Kivirikko and Pihlajaniemi, 1998). The α-subunits of collagen prolyl hydroxylase recognize and bind specific target sequences in elongating polypeptides, catalyzing the selective hydroxylation of proline residues.

The ASC-binding site of prolyl-hydroxylase has been identified (Majamaa *et al.*, 1986). The role of ASC in the reaction mechanism of 2-ODDs is now better understood. ASC is not simply required to keep the nonheme Fe^{2+} bound to the enzyme in the reduced state: if so, it could be replaced by other reducing agents. According to detailed studies on the mechanism of prolyl hydroxylase, the first step of the reaction consists in the oxidative decarboxylation of 2-oxoglutarate by one atom of molecular oxygen. This process generates succinate, CO_2 and a highly reactive iron–oxygen–atom complex, the ferryl ion, which subsequently hydroxylates an appropriate proline residue (Hanauske-Abel and Günzler, 1982; De Jong and Kemp, 1984). The generation of the ferryl ion can proceed without subsequent hydroxylation in so-called uncoupled reaction cycles (Myllyla *et al.*, 1984). ASC is utilized as a specific alternative acceptor of the ferryl oxygen in these reaction cycles. In the absence of ASC, prolyl hydroxylase is rapidly inactivated by self-oxidation (Tschank *et al.*, 1994). The uncoupled decarboxylation reaction involves stoichiometric ASC consumption. Since collagen hydroxylases may catalyze occasional uncoupled reaction cycles even in the presence of the peptide substrates, the main function of ASC in these reactions *in vivo* has been suggested to be that of reactivating the enzyme after such uncoupled cycles (Myllyla *et al.*, 1984; Wu *et al.*, 2000).

Collagen is by far the most abundant protein in animals. Mature collagen is present as chains wound in tight triple helices organized into fibrils of great tensile strength and thermal stability (Holmgren *et al.*, 1998). If ASC is not available, proline residues of collagen are not properly hydroxylated and collagen trimers cannot form. Human unhydroxylated collagen (obtained by heterologous expression in transgenic plants) showed increased flexibility as well as a reduced melting temperature, and this recombinant unhydroxylated collagen did not self-assemble into banded fibrils in physiological conditions (Perrett *et al.*, 2001). It is noteworthy that both ASC deficiency (scurvy) and the malfunctioning of prolyl hydroxylase due to genetic defects, as in the Ehlers–Danlos or *osteogenesis imperfecta* syndromes (Myllyharju and Kivirikko, 2001), converge to the same result: collagen underhydroxylation and related disorders. Inactivation of both genes coding for peptidyl-prolyl hydroxylase in *Caenorhabditis elegans* demonstrates that collagen hydroxylation is necessary for viability and development (Friedman *et al.*, 2000).

Up to now, 21 different types of collagen have been described (Chou and Li, 2002). Such different collagen molecules have specific functions in maintaining the integrity of skin, tendons, cartilage, bones, teeth, cornea, muscles, blood vessels and other tissues and organs (Prockop and Kivirikko, 1995; Myllyharju and Kivirikko, 2001). In some cases, the link between ASC deficiency and altered structure of specific collagen types has been directly investigated. For instance, guinea pigs fed an ASC-free diet showed lower expression of type IV collagen and elastin mRNAs in blood vessels, as well as lower amount of hydroxyproline in the protein. This may contribute to defects observed in blood vessels during scurvy (Mahmoodian and Peterkofsky, 1999).

9.3.2 Hydroxylation of peptidyl proline in other proteins

It is now very clear that the hydroxylation of peptidyl proline is necessary for collagen stability. Recent data show that, in some noncollagen proteins, this post-translational modification can have unexpected additional roles in the control of protein degradation and in fundamental signaling pathways. In 2001, two independent groups discovered that hydroxylation of two proline residues present in a specific sequence of hypoxia-inducible factor-1α (HIF-1α) underlies the mechanism of oxygen sensing in mammalian cells (Ivan *et al.*, 2001; Jaakkola *et al.*, 2001). According to the model proposed (*Figure 9.2*), when oxygen is available (normoxic conditions), proline residues of HIF 1α can be properly hydroxylated. Hydroxylation of HIF 1α is necessary for the binding of a multiprotein complex containing the von Hippel–Lindau tumor suppression protein (pVHL). In turn, the formation of this multiprotein complex leads to rapid targeting of HIF 1α to proteasome-mediated degradation (Chun *et al.*, 2002). When less oxygen is available (hypoxia), hydroxylation cannot take place, HIF 1α cannot bind pVHL, and therefore it is not degraded. HIF 1α is a transcription factor that binds a specific sequence (the hypoxia responsive element) in the promoter of an array of hypoxia-induced genes, including those for the synthesis of erythropoietin, transferrin, ceruloplasmin, vascular endothelial growth factor, nitric oxide synthase, carbonic anhydrase 9, insulin-like growth factor, glyceraldehyde-3-phosphate dehydrogenase, and many more (Chun *et al.*, 2002). The activation of these genes

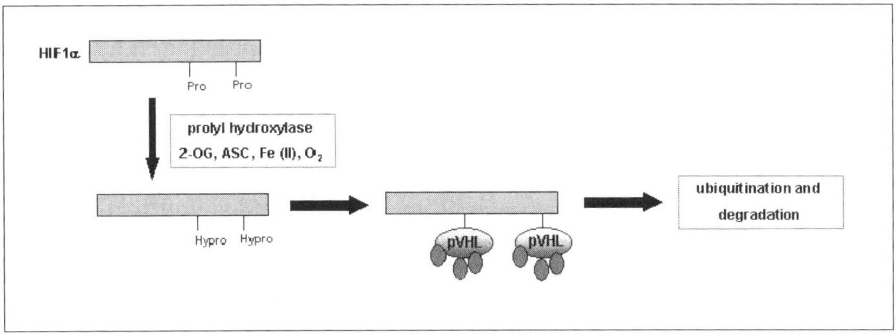

Figure 9.2: Mechanism of the degradation of the transcription factor hypoxia-inducing factor 1α (HIF 1α).
Two proline residues in the protein are hydroxylated by a specific prolyl hydroxylase in the presence of 2-oxoglutarate (2-OG), ASC, ferrous ions and oxygen. The von Hippel–Lindau tumor repressor protein (pVHL) binds hypro residues forming a complex with interacting proteins. This complex targets HIF 1α to ubiquitination and degradation. If hydroxylation does not occur (due to the absence of oxygen and/or other co-substrates required for the reaction), pVHL cannot bind and stable HIF 1α induces transcription of hypoxia-related genes.

is of paramount importance in many important processes, including angiogenesis in cancer (Harris, 2002) and ischemia.

Prolyl hydroxylases that modify HIF share little sequence homology with those acting in collagen hydroxylation, and are apparently located in the cytosol, rather than the endoplasmic reticulum (Bruick and McKnight, 2001). Remarkably, it has been demonstrated that hydroxylation of proline residues in HIF 1α by specific prolyl hydroxylases is strictly dependent upon the amount of oxygen available *in vitro* (Epstein *et al.*, 2001), thus confirming that the mechanism of hydroxylation is an efficient system for sensing oxygen availability.

Although different in sequence and location, prolyl hydroxylases involved in HIF post-translational modification share ASC-dependency with their collagen counterparts *in vitro* (Bruick and McKnight, 2001). At the present, no information is available concerning the effect of ASC deficiency on the activity of these dioxygenases *in vivo*. However, if lower ASC availability yielded limited hydroxylation of HIF 1α, thus avoiding degradation of the protein, we could imagine an apparently paradoxical situation in which lower ASC content could activate molecular responses to hypoxia, thus being beneficial in case of ischemia. This could possibly explain the contradictory results obtained when attempting to correlate plasma ASC concentrations with the risk of myocardial infarction (Padayatty and Levine, 2000). In support of this hypothesis, recent data show that ASC can suppress HIF 1α protein levels and its transcriptional targets in normoxic conditions (Knowles *et al.*, 2003).

The discovery of the role of peptidyl proline hydroxylation in the control of the stability of HIF 1α and in oxygen sensing urged more scientists to investigate if the same mechanism could be involved in the control of the turnover of other proteins. Indeed, Kuznetsova *et al.* (2003) reported that ubiquitination and degradation of the large subunit of RNA polymerase II follows a mechanism requiring hydroxylation of proline residues in the protein, followed by binding of pVHL. In this case, degradation is regulated by UV radiation and not by oxygen availability. A striking difference between the mechanism of ubiquitination of HIF 1α and that of RNA polymerase II is that in the former, degradation is the default process occurring under normal conditions (normoxia), whereas in the latter, hydroxylation, binding of pVHL and ubiquitination occur under stress conditions (UV). Hydroxylation of proline residues represents therefore a dynamic and efficient tool for the control of protein turnover. It can be easily predicted that in the near future many more proteins will be identified, whose degradation is regulated by similar mechanisms.

The presence of hydroxyproline in some prion proteins (Van Rheede *et al.*, 2003) is another interesting point. Proline residues in the N terminus of prion protein are post-translationally hydroxylated. It has been suggested that this modification could be involved in the regulation of the physiology and pathobiology of the cell (Gill *et al.*, 2000).

Although apparently not directly related to scurvy, these findings on the wide range of possible biological roles of ASC-dependent prolyl hydroxylation, further stress the central role of ASC in cell metabolism.

9.3.3 Carnitine biosynthesis

ASC is required for the activity of two enzymes involved in the biosynthesis of carnitine, namely ϵ-trimethyl-lysine dioxygenase (EC 1.14.11.8) and γ-butyrobetaine dioxygenase (EC 1.14.11.1). Carnitine is an essential metabolite having an indispensable role in energy metabolism, since it enables activated fatty acids to enter the mitochondria, where they are broken down via β-oxidation (Vaz and Wanders, 2002). It has been suggested that lassitude and fatigue, which have been indicated among early symptoms of scurvy, could be a consequence of impaired carnitine biosynthesis due to ASC deficiency (Rebouche, 1991).

A striking correlation between ASC availability and carnitine synthesis has been observed. Thoma and Henderson (1984) reported that liver and skeletal muscle carnitine levels were reduced in scorbutic guinea pigs, apparently due to lower activity of γ-butyrobetaine dioxygenase. Injection of ASC reversed the decline in the activity of the enzyme in scorbutic animals. Moreover, Otsuka *et al.* (1999) observed that high doses of ASC in guinea pigs fed high-fat diets contribute to the enhancement of carnitine synthesis and improvement of the triacylglycerol content in the plasma.

9.3.4 Other dioxygenases possibly involved in scorbutic symptoms

The basis of neurological disorders observed in scurvy can be tentatively attributed to low activity of different dioxygenases. The conversion of tyrosine to 3,4-hydroxyphenyl alanine (DOPA) is catalyzed by tyrosine hydroxylase. In turn, DOPA is further modified by dopamine-β-hydroxylase, yielding the neurotransmitter noradrenalin. DOPA and noradrenalin metabolism is apparently correlated with depression (Racagni and Brunello, 1999). Absence of dopamine-β-hydroxylase activity due to a genetic defect is also associated with severe orthostatic hypotension and the lack of sympathetic noradrenergic function (Garland *et al.*, 2002). It is conceivable to hypothesize that also ASC deficiency, by impairing the activity of this ASC-dependent enzyme, could have similar effects.

Indoleamine 2,3-dioxygenase, the first and rate-limiting enzyme in human tryptophan metabolism, has been implicated in the pathogenesis of many diseases (Littlejohn *et al.*, 2000). This enzyme is necessary to form kynurenine (Widner *et al.*, 2000). The kynurenine pathway regulates the metabolism of tryptophan to neuroactive compounds, and also seems to be a key factor in the communication between the nervous and immune systems (Stone and Darlington, 2002).

Synthesis of vitamin A requires the activity of a dioxygenase catalyzing the cleavage of β-carotene into two molecules of retinal (von Lintig and Vogt, 2000). Lack of vitamin A could be well in accordance with visual problems sometimes observed in scorbutic patients.

9.3.5 General remarks on the correlation between dioxygenases and scurvy

A complete list of known and putative ASC-dependent dioxygenases is not the scope of this article. More candidates, including hydroxyphenyl pyruvate dioxygenase, hypusine hydroxylase and enzymes involved in plant metabolism, have been described elsewhere (Arrigoni and De Tullio, 2002). In the absence of sound experimental data, further extending the game of correlating scorbutic symptoms with dioxygenases possibly involved in different metabolic pathways might lead to pure guessing. Nevertheless, this approach could be useful to understand the grounds of ASC involvement in so many physiological processes and, on the other hand, to find an answer to Lord Anson's uncertainty concerning the multiple signs of scurvy.

It should also be considered that we still do not have a clear picture of all ASC can actually do. For instance, Wells and co-workers observed that ASC induced a 2.2-fold increase in the activity of mitochondrial glycerol-3-phosphate dehydrogenase (mGPDH) and played an essential role in glucose-induced insulin release (Wells *et al.*, 2001). Inhibitor studies strongly suggest that ASC elicits mGPDH activity via an unidentified di-iron metalloenzyme. Is there a new class of ASC-dependent enzymes waiting for discovery? This could further broaden the already broad action spectrum of ASC.

A final consideration on scurvy symptoms concerns their order of appearance. According to many clinical descriptions, lassitude generally appears in the early stages of the disease. This could be interpreted as a sign that carnitine biosynthesis (i.e., the activity of ϵ-trimethyl-lysine dioxygenase and/or γ-butyrobetaine dioxygenase) requires a relatively high amount of ASC. It is reasonable to assume that, upon decrease of plasma ASC content in subjects fed an ASC-free diet, ASC-dependent enzymes cannot work below a threshold concentration of ASC. The threshold may be different for different dioxygenases. In principle, the order of appearance of scorbutic symptoms could be obtained by placing the different ASC-dependent enzymes in hierarchical order on the basis of their affinity for

ASC: the ones with lower affinity (higher K_m) are anticipated to sense ASC deficiency first. However this is not necessarily true, since ASC availability may depend on many different factors (transport, subcellular distribution, recycling of oxidized forms, catabolism), thus making any mechanistic forecast quite difficult and inadequate. This could possibly account for Lord Anson's observation that symptoms are very variable in their order of appearance.

9.4 Multi-step control of 2-ODD expression and activity by ASC

Studies on the regulation of collagen hydroxylation and other related phenomena revealed a mechanism operating at different levels.

ASC is specifically required for the expression of osteocalcin, a calcium-binding protein made by osteoblasts (Xiao *et al.*, 1997). ASC treatment increases osteocalcin gene transcription by way of a complex mechanism that is blocked by the inhibition of collagen hydroxylation, and that requires a specific sequence present in the promoter of the osteocalcin gene. As previously reported, ASC deficiency results in lower abundance of type IV collagen mRNA in blood vessels (Mahmoodian and Peterkofsky, 1999).

The picture emerging from these data suggests that ASC controls gene expression by still unclear mechanisms. Early observations suggested that vitamin C could be involved in gene transcription (Price, 1966). More recent reports further substantiate this hypothesis. In some cases, vitamin C seems to stabilize specific mRNAs (Arrigoni and De Tullio, 2002 and references therein). Interestingly, at least some of the genes that have their transcription induced (or stability of the transcript enhanced) by ASC, encode proteins that contain hydroxylated proline residues or that require ASC for their catalytic activity. For example, ASC-dependent stabilization of collagen transcript and destabilization of elastin transcript in smooth muscle cells and skin fibroblasts has been observed (Davidson *et al.*, 1997). Transcription of the gene encoding tyrosine hydroxylase is enhanced by ASC (Seitz *et al.*, 1998).

To date microarray technology has been used in only a few cases to investigate the profile of ASC-dependent gene expression. Catani *et al.* (2001) recently demonstrated that ASC administration induces transcription of the *fra-1* gene, which encodes a transcription factor of the Fos family, and downregulates the activator protein-1 (AP-1)-target genes. Expression profiles of human mesenchymal stem cells during the mineralization process were analyzed in the presence of ASC and other osteogenic supplements (Doi *et al.*, 2002). ASC and other agents were necessary for the mineralization process and for transcriptional stimulation of 55 genes and repression of 82 genes among more than 20 000 examined. Hopefully, more data concerning the effects of ASC (and, conversely, of ASC deficiency) on gene expression will soon be available.

All in all, ASC seems to operate in transcription; mRNA stabilization; post-translational modification of proteins; hydroxylation, epoxidation and desaturation of many different substrates. Although the picture is far from being complete, this is enough to get an idea of how ASC prevents scurvy.

9.5 Conclusions

In the last few decades, much attention has been given to the antioxidant properties of vitamin C. This attention partially eclipsed the importance of the non-antioxidant functions of this fascinating molecule.

Humans (and guinea pigs) fed a diet without ASC die. And they die of scurvy. And scurvy is a disease related to the malfunction of at least a few key dioxygenases.

What can we learn from scurvy? Perhaps that we should take into due consideration the nonantioxidant functions of ASC, whereas its antioxidant activity should not be overestimated. It could be just the tip of the iceberg.

Acknowledgments

I shall always be grateful to Prof. Oreste Arrigoni for having introduced me to the amazing world of ASC, for his contagious enthusiasm, his mind-opening suggestions and his support. I also wish to thank Prof. Hartmut Follmann for his advice on the etymology of the word scurvy.

References

Arrigoni O, De Tullio MC (2000) The role of ascorbic acid in cell metabolism: between gene-directed functions and unpredictable chemical reactions. *J. Plant Physiol.* **157**, 481–488.

Arrigoni O, De Tullio MC (2002) Ascorbic acid: much more than just an antioxidant. *Biochim. Biophys. Acta* **1569**, 1–9.

Bruick RK, McKnight SL (2001) A conserved family of prolyl-4-hydroxylases that modify HIF. *Science* **294**, 1337–1340.

Burns JJ (1959) Biosynthesis of L-ascorbic acid; basic defect in scurvy. *Am. J. Med.* **26**, 740–748.

Catani MV, Rossi A, Costanzo A, Sabatini S, Levrero M, Melino G, Avigliano L. (2001) Induction of gene expression via activator protein-1 in the ascorbate protection against UV-induced damage. *Biochem. J.* **356**, 77–85.

Chou MY, Li HC (2002) Genomic organization and characterization of the human type XXI collagen (COL21A1) gene. *Genomics* **79**, 395–401.

Chun YS, Kim MS, Park JW (2002) Oxygen-dependent and -independent regulation of HIF1α. *J. Kor. Med. Sci.* **17**, 581–588.

Davidson JM, LuValle PA, Zoia O, Quaglino Jr D, Giro M (1997) Ascorbate differentially regulates elastin and collagen biosynthesis in vascular smooth muscle cells and skin fibroblasts by pretranslational mechanisms. *J. Biol. Chem.* **272**, 345–352.

De Jong L, Kemp A (1984) Stoichiometry and kinetics of the prolyl 4-hydroxylase partial reaction. *Biochim. Biophys. Acta* **787**, 105–111.

Doi M, Nagano A, Nakamura Y (2002) Genome-wide screening by cDNA microarray of genes associated with matrix mineralization by human mesenchymal stem cells *in vitro*. *Biochem. Biophys. Res. Commun.* **290**, 381–390.

Epstein AC, Gleadle JM, McNeill LA, Hewitson KS, O'Rourke J, Mole DR, *et al.* (2001) *C. elegans* EGL-9 and mammalian homologs define a family of dioxygenases that regulate HIF by prolyl hydroxylation. *Cell* **107**, 43–54.

Friedman L, Higgin JJ, Moulder G, Barstead R, Raines RT, Kimble J (2000) Prolyl 4-hydroxylase is required for viability and morphogenesis in *Caenorhabditis elegans*. *Proc. Natl Acad. Sci. USA* **97**, 4736–4741.

Garland EM, Hahn MK, Ketch TP, Keller NR, Kim CH, Kim KS, Biaggioni I, Shannon JR, Blakely RD, Robertson D (2002) Genetic basis of clinical catecholamine disorders. *Ann. N. Y. Acad. Sci.* **971**, 506–514.

Gill AC, Ritchie MA, Hunt LG, Steane SE, Davies KG, Bocking SP, Rhie AG, Bennett AD, Hope J (2000) Post-translational hydroxylation at the N-terminus of the prion protein reveals presence of PPII structure *in vivo*. *EMBO J.* **19**, 5324–5331.

Hanauske-Abel HM, Günzler V (1982) A stereochemical concept for the catalytic mechanism of prolyl-hydroxylase: applicability to classification and design of inhibitors. *J. Theor. Biol.* **94**, 421–455.

Harris AL (2002) Hypoxia – a key regulatory factor in tumour growth. *Nature Rev. Cancer* **2**, 38–47.

Holmgren SK, Taylor KM, Bretscher LE, Raines RT (1998) Code for collagen's stability deciphered. *Nature* **392**, 666–667.

Ivan M, Kondo K, Yang H, Kim W, Valiando J, Ohh M, *et al.* (2001) HIF-1α targeted for VHL-mediated destruction by proline hydroxylation: Implications for O_2 sensing. *Science* **292**, 464–468.

Jaakkola P, Mole DR, Tian YM, Wilson MI, Gielbert J, Gaskell SJ, *et al.* (2001) Targeting of HIF-1α to the von Hippel-Lindau ubiquitylation complex by O_2-regulated prolyl hydroxylation. *Science* **292**, 468–472.

Kivirikko KI, Pihlajaniemi T (1998) Collagen hydroxylases and the protein disulfide isomerase subunit of prolyl 4-hydroxylases. *Adv. Enzymol. Relat. Areas Mol. Biol.* **72**, 325–398.

Knowles HJ, Raval RR, Harris AL, Ratcliffe PJ (2003) Effect of ascorbate on the activity of hypoxia-inducible factor in cancer cells. *Cancer Res.* **63**, 1764–1768.

Kuznetsova AV, Meller J, Schnell PO, Nash JA, Ignacak ML, Sanchez Y, Conaway JW, Conaway RC, Czyzyk-Krzeska MF (2003) von Hippel-Lindau protein binds hyperphosphorylated large subunit of RNA polymerase II through a proline hydroxylation motif and targets it for ubiquitination. *Proc. Natl Acad. Sci. USA* **100**, 2706–2711.

Lind J (1753) *A Treatise of the Scurvy in Three Parts.* Containing an inquiry into the Nature, Causes and Cure of that Disease, together with a Critical and Chronological View of what has been published on the subject. A. Millar, London.

Littlejohn TK, Takikawa O, Skylas D, Jamie JF, Walker MJ, Truscott RJ (2000) Expression and purification of recombinant human indoleamine 2,3-dioxygenase. *Protein Expr. Purif.* **19**, 22–29.

Mahmoodian F, Peterkofsky B (1999) Vitamin C deficiency in guinea pigs differentially affects the expression of type IV collagen, laminin, and elastin in blood vessels. *J. Nutr.* **129**, 83–91.

Majamaa K, Günzler V, Hanauske-Abel HM, Myllyla R, Kivirikko KI (1986) Partial identity of the 2-oxoglutarate and ascorbate binding sites of prolyl 4-hydroxylase. *J. Biol. Chem.* **261**, 7819–7823.

Myllyharju J, Kivirikko KI (2001) Collagen and collagen-related diseases. *Ann. Med.* **33**, 7–21.

Myllyla R, Majamaa K, Günzler V, Hanauske-Abel HM, Kivirikko KI (1984) Ascorbate is consumed stoichiometrically in the uncoupled reactions catalyzed by prolyl 4-hydroxylase and lysyl hydroxylase. *J. Biol. Chem.* **259**, 5403–5405.

Nishikimi M, Udenfriend S (1976) Immunologic evidence that the gene for L-gulono-gamma-lactone oxidase is not expressed in animals subject to scurvy. *Proc. Natl Acad. Sci. USA* **73**, 2066–2068.

Nishikimi M, Udenfriend S (1977) Scurvy as an inborn error of ascorbic acid biosynthesis. *Trends Biochem. Sci.* **2**, 111–113.

Otsuka M, Matsuzawa M, Ha TY, Arakawa N (1999) Contribution of a high dose of L-ascorbic acid to carnitine synthesis in guinea pigs fed high-fat diets. *J. Nutr. Sci. Vitaminol. (Tokyo)* **45**, 163–171.

Padayatty SJ, Levine M (2000) Vitamin C and myocardial infarction: the heart of the matter. *Am. J. Clin. Nutr.* **71**: 1027–1028.

Perrett S, Merle C, Bernocco S, Berland P, Garrone R, Hulmes DJS, Theisen M, Ruggiero F (2001) Unhydroxylated triple helical collagen I produced in transgenic plants provides new clues on the role of hydroxyproline in collagen folding and fibril formation. *J. Biol. Chem.* **276**, 43693–43698.

Price CE (1966) Ascorbate stimulation of RNA synthesis. *Nature* **212**, 1481.

Prockop DJ, Kivirikko KI (1995) Collagens: molecular biology, diseases, and potentials for therapy. *Annu. Rev. Biochem.* **64**, 403–434.

Racagni G, Brunello N (1999) Physiology to functionality: the brain and neurotransmitter activity. *Int. Clin. Psychopharmacol.* **14**, S3–S7.

Rebouche CJ (1991) Ascorbic acid and carnitine biosynthesis. *Am. J. Clin. Nutr.* **54**, 1147S–1152S.

Schofield CJ, Zhang Z (1999) Structural and mechanistic studies on 2-oxoglutarate-dependent oxygenases and related enzymes. *Curr. Opin. Struct. Biol.* **9**, 722–731.

Seitz G, Gebhardt S, Beck JF, Bohm W, Lode HN, Niethammer D, Bruchelt G (1998) Ascorbic acid stimulates DOPA synthesis and tyrosine hydroxylase gene expression in the human neuroblastoma cell line SK-N-SH. *Neurosci. Lett.* **244**, 33–36.

Stone N, Meister A (1962) Function of ascorbic acid in the conversion of proline to collagen hydroxyproline. *Nature* **194**, 555–557.

Stone TW, Darlington LG (2002) Endogenous kynurenines as targets for drug discovery and development. *Nature Rev. Drug Discov.* **1**, 609–620.

Thoma WJ, Henderson LM (1984) Effect of vitamin C deficiency on hydroxylation of trimethylaminobutyrate to carnitine in the guinea pig. *Biochim. Biophys. Acta* **797**, 136–139.

Tschank G, Sanders J, Baringhaus KH, Dallacker F, Kivirikko K, Günzler V (1994) Structural requirements for the utilisation of ascorbate analogues in the prolyl 4-hydroxylase reaction. *Biochem. J.* **300**, 75–79.

Van Rheede T, Smolenaars MM, Madsen O, De Jong WW (2003) Molecular evolution of the Mammalian prion protein. *Mol. Biol. Evol.* **20**, 111–121.

Vaz FM, Wanders RJA (2002) Carnitine biosynthesis in mammals. *Biochem. J.* **361**, 417–429.

von Lintig J, Vogt K (2000) Filling the gap in vitamin A research. Molecular identification of an enzyme cleaving beta-carotene to retinal. *J. Biol. Chem.* **275**, 11915–11920.

Xiao G, Cui Y, Ducy P, Karsenty G, Franceschi RT (1997) Ascorbic acid-dependent activation of the osteocalcin promoter in MC3T3-E1 preosteoblasts: requirement for collagen matrix synthesis and the presence of an intact OSE2 sequence. *Mol. Endocrinol.* **11**, 1103–1113.

Wells WW, Xu DP, Washburn MP, Cirrito HK, Olson LK (2001) Polyhydroxybenzoates inhibit ascorbic acid activation of mitochondrial glycerol-3-phosphate dehydrogenase. Implications for glucose metabolism and insulin secretion. *J. Biol. Chem.* **276**, 2404–2410.

Widner B, Ledochowski M, Fuchs D (2000) Interferon-gamma-induced tryptophan degradation: neuropsychiatric and immunological consequences. *Curr. Drug Metab.* **1**, 193–204.

Wu M, Moon HS, Pirskanen A, Myllyharju J, Kivirikko KI, Begley TP (2000) Mechanistic studies on prolyl-4-hydroxylase: the vitamin C requiring uncoupled oxidation. *Bioorg. Med. Chem. Lett.* **10**, 1511–1514.

10

Ascorbate as an antioxidant

Garry R. Buettner and Freya Q. Schafer

10.1 Basic chemistry of ascorbate

Ascorbate (ASC) is a ubiquitous nutrient found in nearly all fresh foods; it is essential for human health. Most mammals make their own ascorbate, using liver-enzymes to convert glucose to ascorbate. However, during human evolution this ability was lost and thus for humans ascorbate is a vitamin; it must be obtained from outside sources (see Chapter 3 by Running *et al.*). When deficient in ASC, diseases such as scurvy can occur (see chapter by De Tullio). We are dependent on adequate intake of vitamin C not only for the prevention of scurvy, but for overall well being (Davies *et al.*, 1991; Stone, 1974; Lewin, 1976).

10.1.1 Ascorbate structure

Ascorbic acid ($AscH_2$) is a diacid with a pK_1 of 4.25 and a pK_2 of 11.8. The first proton exceeds the acidity of typical carboxylic acids (pK_a of ~ 4.7) (Weast, 1987). Ascorbic acid is unique in that pK_2 (11.8) of its diol is far removed from pK_1, $\Delta pK = 7.6$. As a result, at physiological pH values, 99.9% of ASC will be present as the monoanion ($AscH^-$); only ~ 0.06% will be present as the diacid ($AscH_2$); ~ 0.01% as the ASC dianion (Asc^{2-}). Thus, the chemistry, biochemistry, and biology of vitamin C revolve around the ASC monoanion and not ascorbic acid.

ASC ($AscH^-$) has an unusual structure in that it has at its core a five-membered lactone ring[1]; on this ring is a bifunctional ene-diol group (carbons 2 and 3) with

[1]A lactone is an intramolecular, organic anhydride formed from a hydroxyacid by the loss of water between an –OH and a carboxylic acid group resulting in an ester. The reaction below shows this process. Many steps in the biosynthesis of ASC from glucose are not shown.

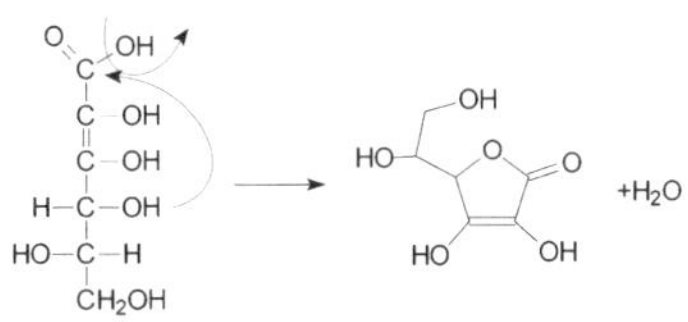

an adjacent carbonyl group (*Figure 10.1*). This conjugated structure allows ASC to serve as an excellent reducing agent and thereby an effective donor antioxidant. The loss of one electron results in the semidehydroascorbate radical (also known as: monodehydroascorbate, ascorbate radical, and $Asc^{\bullet-}$), a very stable π-radical with the unpaired electron delocalized over the highly conjugated tricarbonyl system. Loss of a second electron yields dehydroascorbic (DHA), which upon hydrolysis is converted to dehydroascorbic acid.[2] DHA can be reduced back to ASC and thus ASC is recycled. However, DHA is unstable at physiological pH and if it is not reduced to $AscH^-$ it will undergo rupture of the ring to form 2, -diketo-1-gulonic acid, and ASC will be lost (*Figure 10.2*) as the products formed after ring-rupture are not recycled back to $AscH^-$.

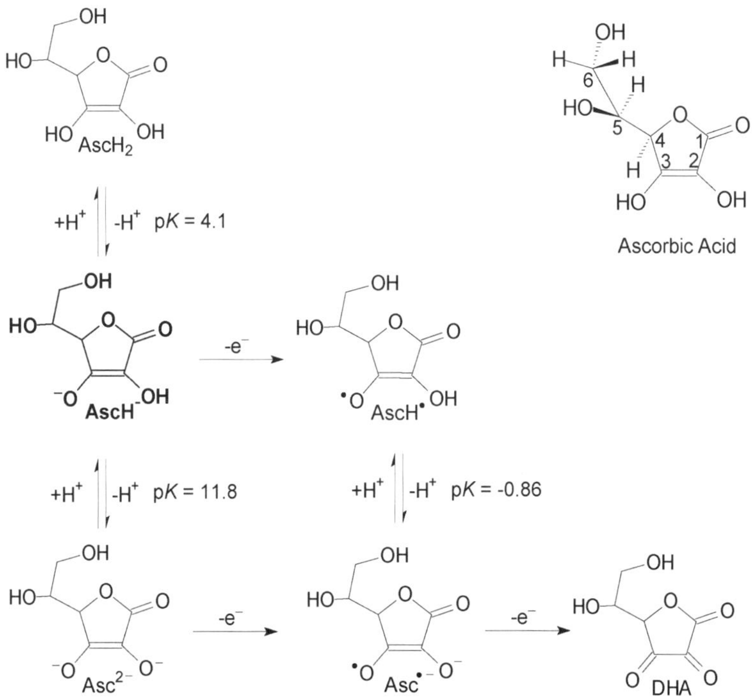

Figure 10.1: All the forms of 'ASC'.
The structure in the upper right shows the stereochemistry of the ascorbic acid molecule. The schematic shows all the forms that 'ASC' can take as it undergoes oxidation/reduction and protonation/deprotonation. The abbreviations used are: $AscH_2$, ascorbic acid; $AscH^-$, ASC monoanion; Asc^{2-}, ASC dianion; AscH, ascorbyl free radical; $Asc^{\bullet-}$, ascorbate free radical or semidehydroascorbate; DHA, dehydroascorbic.

[2]Notes on nomenclature: (i) $Asc^{\bullet-}$ is usually referred to in brief as the ASC free radical. The ending 'ate' being used because it is a charged species. The short name ascorbyl radical would be used for $AscH^{\bullet}$, the neutral protonated form of $Asc^{\bullet-}$. The ending 'yl' being used for neutral species. (ii) DHA is dehydroascorbic; it has no acidic hydrogens and therefore is not an acid. However, once it hydrolyzes there are acid hydrogens available and thus the term dehydroascorbic acid (DHAA) is appropriate.

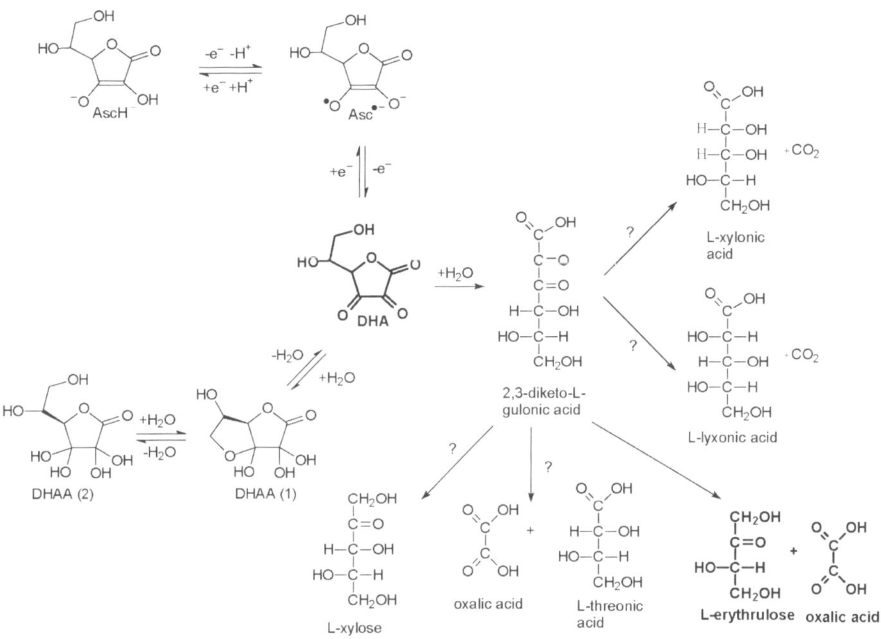

Figure 10.2: Oxidation of ASC.
ASC readily undergoes sequential one-electron oxidations to form the ascorbate free radical and then dehydroascorbic. Upon hydrolysis the ring of DHA can open, and ASC is essentially lost. DHA can be reduced by glutathione as well as enzymatic systems thereby recycling ASC (see Chapter 8 by May and Asard).

The ASC free radical, $Asc^{\bullet -}$, is a strong acid having a pK_a of –0.86 (Davis *et al.*, 1986). It will be present as a monoanion over the entire biological pH range.

10.1.2 Ascorbate oxidation products

The five-membered ring of the ASC molecule is somewhat fragile. Upon oxidation of $AscH^-$ to DHA the strain in the bonds of the ring will increase. Because the O–C_1 bond in the ring is an ester bond, it will readily hydrolyze to 2,3-diketo-L-gulonic acid, a straight chain molecule (*Figure 10.2*). The diketone moiety of 2,3-diketo-L-gulonic acid is also subject to hydrolysis.

10.1.3 Dismutation of $Asc^{\bullet -}$

Because the unpaired electron of $Asc^{\bullet -}$ is delocalized in the pi-system of orbitals, it decays relatively slowly compared to highly oxidizing radicals such as hydroxyl, alkoxyl, or peroxyl radicals. In typical, near neutral buffered solutions, the dominant mode of decay is disproportionation. A detailed study by Bielski *et al.* (1981) indicates that the stability of $Asc^{\bullet -}$ is also due to obligatory dimer formation as an intermediate during the disproportionation reaction.

$$2\,\mathrm{Asc}^{\cdot-} \underset{k_{-1}}{\overset{k_1}{\rightleftharpoons}} (\mathrm{Asc})_2^{2-} \tag{1}$$

$$(\mathrm{Asc})_2^{2-} + \mathrm{H}^+ \xrightarrow[k_3]{k_2} \mathrm{AscH}^- + \mathrm{DHA} \tag{2}$$

$$(\mathrm{Asc})_2^{2-} + \mathrm{H_2O} \xrightarrow[k_{obs}]{} \mathrm{AscH}^- + \mathrm{DHA} + \mathrm{OH}^- \tag{3}$$

Net Reaction: $$2\,\mathrm{Asc}^{\cdot-} + \mathrm{H}^+ \longrightarrow \mathrm{AscH}^- + \mathrm{DHA} \tag{4}$$

Thus, we have

$$-\mathrm{d}\,[\mathrm{Asc}^{\cdot-}]/\mathrm{dt} = -2\,k_{\mathrm{obs}}\,[\mathrm{Asc}^{\cdot-}]^2 \tag{5}$$

where $$k_{\mathrm{obs}} = k_1 / [1 + (k_{-1} / (k_2\,[\mathrm{H}^+] + k_3))] \tag{6}$$

An increase in ionic strength has been observed to both increase and decrease the rate of dismutation of $\mathrm{Asc}^{\cdot-}$ (Wieczorek *et al.* , 1987). However, a most important observation is that phosphate buffer accelerates this dismutation (Bielski *et al.*, 1981). For example in 50 mM phosphate buffer, the k_{obs} for reaction (4) at pH 7.0 increases from 2×10^5 to $2 \times 10^6 \cdot \mathrm{M}^{-1} \cdot \mathrm{s}^{-1}$. Because of the wide use of phosphate buffers, this must be considered when interpreting data.

10.1.4 Equilibrium reactions of AscH⁻ and DHA

The ascorbate free radical will be present in solutions due to autoxidation of AscH^- and Asc^{2-}, metal catalyzed oxidation, as well as comproportionation of AscH^- and DHA (Foerster *et al.* , 1965).

$$\mathrm{AscH}^- + \mathrm{DHA} \rightleftharpoons 2\,\mathrm{Asc}^{\cdot-} + \mathrm{H}^+ \tag{7}$$

$$K = \frac{[\mathrm{Asc}^{\cdot-}]^2}{[\mathrm{AscH}^-]\,[\mathrm{DHA}]} \tag{8}$$

The equilibrium constant for reaction 7 is pH dependent; K varies from 5.6×10^{-12} at pH 4.0 to 5.1×10^{-9} at pH 6.4. When the acid–base properties of ascorbic acid and ascorbate free radical were understood, it was possible to develop an expression for K at any pH value (Bielski, 1982):

$$K = \frac{[\mathrm{Asc}^{\cdot-}]^2\,[\mathrm{H}^+]+(1+[\mathrm{H}^+]/10^{-\mathrm{p}K1})}{[\mathrm{DHA}]\,[\mathrm{AscH_2}]_{\mathrm{total}}} = 2.0 \times 10^{-15} \cdot \mathrm{M}^{-2} \tag{9}$$

where pK_1 is the first ionization constant of ascorbic acid (4.25) and $[AscH_2]_{total}$ is the analytical concentration of $AscH_2$ i.e., $[AscH_2]_{total} = [AscH_2] + [AscH^-]$. From this we can see that at near-neutral pH $Asc^{\cdot-}$ can readily arise from comproportionation, but this is less likely to happen as the pH decreases.

10.2 ASC as a donor antioxidant

10.2.1 What is a donor antioxidant?

There are two general groups of antioxidants, preventive antioxidants and chain-breaking antioxidants. Preventive antioxidants intercept oxidizing species before damage can be done. Chain-breaking antioxidants slow or stop oxidative processes after they begin, by intercepting the chain-carrying radicals. Chain-breaking antioxidants include donor antioxidants and sacrificial antioxidants. A donor antioxidant provides a hydrogen atom or an electron to an oxidizing species thereby stopping the cycle of chain reactions. ASC is considered to be a water-soluble, donor, chain-breaking antioxidant (reaction 10).

$$AscH^- + X^{\cdot} \longrightarrow Asc^{\cdot-} + XH \qquad (10)$$

Both $AscH^-$ and $Asc^{\cdot-}$ are relatively unreactive; $Asc^{\cdot-}$ decays to harmless products; $Asc^{\cdot-}$ does not add O_2 to make a peroxyl radical and the resulting oxidation products can be recycled. Thus, ASC has all the characteristics of a good chain-breaking antioxidant.

10.2.2 Kinetics of AscH⁻ reactions

In general chain-breaking antioxidants act by reacting with radicals that will initiate chain reactions (e.g. hydroxyl, $HO^{\cdot}$, or alkoxyl radicals, $RO^{\cdot}$), but most importantly with the chain-carrying radicals, (e.g. peroxyl radicals, $ROO^{\cdot}$). As seen in *Table 10.1*, ASC reacts readily with highly reactive oxidizing free radicals such as $HO^{\cdot}$, $RO^{\cdot}$, and $ROO^{\cdot}$. However, ASC also reacts efficiently with less oxidizing radicals such as the tocopheroxyl ($TO^{\cdot}$) and urate radicals ($UH^{\cdot-}$). That these reactions are relatively fast suggests that ASC may have a significant role in the recycling of other small molecule antioxidants. This ability to react with the free radicals formed by antioxidants and thereby recycle these antioxidants is consistent with the thermodynamic properties of the $Asc^{\cdot-}$, $H^+/AscH^-$ couple.

10.2.3 Thermodynamics of ascorbate reactions/pecking order

If a substance is to function as an effective donor, chain-breaking antioxidant, then the redox couple for that substance must have favorable thermodynamic characteristics, i.e., the $E^{\circ\prime}$ should be such that neither half of the couple is too oxidizing or too reducing. The thermodynamic pecking order describes the hierarchy of free radical electron/hydrogen atom transfer reactions using one-electron reduction potentials. Antioxidants have reduction potentials that place them in the middle of the thermodynamic pecking order (Buettner, 1993). This location in

Table 10.1: Rate constants for the reaction of the equilibrium mixture of $AscH_2/AscH^-/Asc^{2-}$. All measurements were made at pH 7.4 unless noted otherwise.

Radical	k_{obs} ($\cdot M^{-1}\cdot s^{-1}$; pH 7.4)
$HO^{\cdot}$	1.1×10^{10}
$RO^{\cdot}$ (*tert*-butyl alkoxyl radical)	1.6×10^{9}
$ROO^{\cdot}$ (alkyl peroxyl radical, *e.g.* $CH_3OO^{\cdot}$)	$1\text{–}2 \times 10^{6}$
$Cl_3COO^{\cdot}$	1.8×10^{8}
$GS^{\cdot}$ (glutathiyl radical)	6×10^{8} (5.6)
$UH^{\cdot-}$ (urate radical)	1×10^{6}
$TO^{\cdot}$ (tocopheroxyl radical)	2×10^{5} [a]
$Asc^{\cdot-}$ (dismutation)	2×10^{5} [b]
$CPZ^{\cdot+}$ (chlorpromazine radical cation)	1.4×10^{9} (5.9)
Fe(III)EDTA/ Fe(II)EDTA	$\sim 10^{2}$
$O_2^{\cdot-}/HO_2^{\cdot}$	2.7×10^{5}
Fe(III)Desferal®/ Fe(II)Desferal®	Very slow

[a]Estimated k_{obs} for $TO^{\cdot}$ when in a biological membrane. [b]k is pH dependent; this is k_{obs} at pH 7.4.

the pecking order provides antioxidants with enough reducing power to react with reactive oxidizing species. At the same time, they are too weak to initiate readily damaging reductive reactions.

One-electron reduction potentials, such as tabulated in *Table 10.2*, can be used to predict the direction of free radical reactions. For example, the ascorbate radical has a one-electron reduction potential of +280 mV while the tocopherol radical ($TO^{\cdot}$) has a reduction potential of +480 mV. Thus, ASC can react with $TO^{\cdot}$ and thereby recycle vitamin E.

$$AscH^- + TO^{\cdot} \longrightarrow Asc^{\cdot-} + TOH \quad \Delta E^{\circ\prime} = +200 \text{ mV} \tag{11}$$

Recall, that a positive value for $\Delta E^{\circ\prime}$ indicates a thermodynamically favorable process.

It is our experience that when freshly harvested biological fluids or tissues are examined by electron paramagnetic resonance spectroscopy (EPR), $Asc^{\cdot-}$ will most likely be observed (Buettner *et al.*, 1987; Jurkiewicz and Buettner, 1994; Sharma *et al.* , 1994). This is consistent with the proposal that a role for ASC is as a terminal small-molecule antioxidant for one-electron processes (Buettner and Jurkiewicz, 1993).

In these one-electron reactions, e.g., reaction 10, ASC is oxidized to a free radical. A potentially very damaging radical ($X^{\cdot}$, such as $HO^{\cdot}$, $RO^{\cdot}$ or $ROO^{\cdot}$) is replaced by the more domesticated $Asc^{\cdot-}$. The ASC radical does not react by an addition reaction with O_2 to form dangerous peroxyl radicals. ASC (probably Asc^{2-}) and/or $Asc^{\cdot-}$ appear to produce very low levels of superoxide (Scarpa *et al.*, 1993; Williams and Yandell, 1982). But by removing $O_2^{\cdot-}$, superoxide dismutase provides protection from this possibility (Winterbourn, 1993; Koppenol, 1993). Thus, the biological organism is protected from further free radical-mediated oxidations. In addition, $Asc^{\cdot-}$ as well as DHA can be reduced back to ASC by enzyme systems (see chapter by May and Asard). Thus, ASC is recycled.

Table 10.2: Standard one-electron reduction potentials at pH 7.0 for selected radical couples. Adapted from Buettner, 1993. An extensive tabulation of one-electron reduction potentials can be found in Wardman, 1989.

Redox couple	$E^{\circ\prime}$ (mV)
$HO^{\cdot}$, H^+/H_2O	+2310
$RO^{\cdot}$, H^+/ROH (aliphatic alkoxyl radical)	+1600
$ROO^{\cdot}$, $H^+/ROOH$ (alkyl peroxyl radical)	+1000
$GS^{\cdot}/GS^-$ (glutathione)	+920
$PUFA^{\cdot}$, H^+/PUFA-H) (*bis*-allylic-H)	+600
$HU^{\cdot-}$, H^+/UH^{2-} (urate)	+590
$TO^{\cdot}$, H^+/TOH (tocopherol)	+480
H_2O_2, H^+/H_2O, $HO^{\cdot}$	+320
$Asc^{\cdot-}$, $H^+/AscH^-$[a]	**+282**
Fe(III)EDTA/Fe(II)EDTA	+120
DHA/$Asc^{\cdot-}$[a]	**–174**
$O_2/O_2^{\cdot-}$	–330
$Paraquat^{\cdot+}/paraquat^{2+}$	–448
Fe(III)Desferal®/Fe(II)Desferal®"	–450
$RSSR/RSSR^{\cdot-}$ (e.g., GSSG)	–1500
H_2O/e_{aq}^-	–2870

[a]The two-electron couple of DHA, $H^+/AscH^-$ has a reduction potential of +54 mV. GSSG, Glutathione disulfide; PUFA, polyunsaturated fatty acid.

10.2.4 C and E as co-antioxidants

The first experimental evidence that ASC enhances the antioxidant action of vitamin E was gathered at the University of Iowa in the 1940s in a study of the antioxidant action of tocopherol on oils (Golumbic and Mattill, 1941). ASC is water-soluble and vitamin E is lipid soluble. Thus, they are clearly localized in physically different biological/cellular structures/compartments. At first glance it would appear that in a biological setting there would be little or no possible interaction between these species. However, the phenol group of tocopherol, which is the basis of its antioxidant action, appears to be located at the water–membrane interface of a biological membrane (Van Acker *et al.* , 1993; Burton and Ingold, 1986; Fukuzawa *et al.*, 1993) (*Figure 10.3*).

This physical arrangement allows easy reaction between the tocopheroxyl free radical, $TO^{\cdot}$, and $AscH^-$:

$$TO^{\cdot} + AscH^- \xrightarrow{k = 2 \times 10^5 \cdot M^{-1} \cdot s^{-1}} TOH + Asc^{\cdot-} \qquad (11)$$

Although there are many measurements of the rate constant for this reaction in a variety of conditions and solvent systems, we believe that $k = 2 \times 10^5 \cdot M^{-1} \cdot s^{-1}$ is an excellent estimate for this reaction when TOH is in a biological membrane (Scarpa *et al.*, 1984; Bisby and Parker, 1995; Buettner, 1993).

Both the thermodynamics and kinetics of the $TO^{\cdot} + AscH^-$ reaction are consistent with the position of the $Asc^{\cdot-}$,$H^+/AscH^-$ and $TO^{\cdot}$,H^+/TOH couples in the thermodynamic pecking order for free radical reactions. The reaction of $TO^{\cdot}$

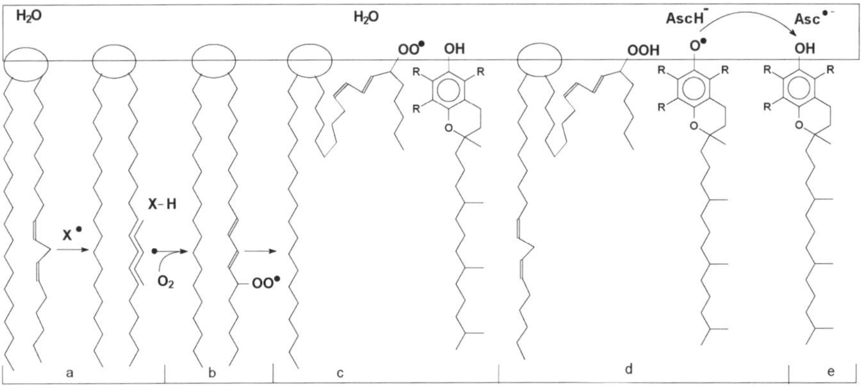

Figure 10.3: Vitamin C and vitamin E as co-antioxidants in lipid peroxidation. The cartoon depicts the role of ASC and vitamin E as co-antioxidants in membrane lipid peroxidation. One leaflet of the bilayer is represented. (a) Initiation of the peroxidation process by an oxidizing radical, X·, by abstraction of a *bis*-allylic hydrogen yielding a pentadienyl radical. (b) Oxygenation to form a peroxyl radical and a conjugated diene; the addition of oxygen introduces a large dipole moment, making this fatty acid chain more water soluble. (c) The peroxyl radical moiety partitions to the water-membrane interface, where it is poised for repair by tocopherol. (d) The peroxyl radical is repaired by tocopherol (TOH) converting it to a lipid hydroperoxide; (e) TO· is recycled by ASC. Asc·⁻ can be recycled by enzyme systems. This cartoon cannot show the dynamic aspects of this process. TOH in the membrane will undoubtedly be 'bobbing up and down' so that the position of the 'OH' on tocopherol in the membrane (or LDL) is changing. In addition, TOH and TO· may have somewhat different positions at the interface. The importance of this process is being debated for cell membranes, but is probably of great importance in the protection of LDL against oxidation.

with AscH⁻ results in the export of oxidative free radicals from the membrane or low density lipoprotein (LDL). Thus, TOH protects the membrane/LDL by stopping the propagation reactions of lipid peroxyl radicals, while ASC protects the membrane/LDL against the possible pro-oxidant reactions of TO· (Bowry and Stocker, 1993; Mukai *et al.* , 1993), while recycling vitamin E and protecting the organism. This action is an example of ASC and tocopherol acting in concert, i.e., as co-antioxidants.

Evidence that reaction 11 will occur in a biological setting was provided by a study that examined the temporal changes in the levels of Asc·⁻ and TO· in freshly collected human plasma (*Figure 10.4*; Sharma and Buettner, 1993). Upon introduction of an oxidative stress there was an immediate increase in the level of Asc·⁻ as would be expected if AscH⁻ indeed functions as the terminal small-molecule antioxidant. However, in plasma there are no enzyme systems available to recycle ASC as it oxidizes. Thus, ASC is soon depleted. Upon its depletion, as indicated by the diminishing levels of Asc·⁻, TO· is no longer being reduced to TOH (reaction 11). TO· can then be detected by EPR. The steady-state concentration of TO· increases and because it is not being recycled by AscH⁻, TOH is also depleted and TO· can no longer be detected.

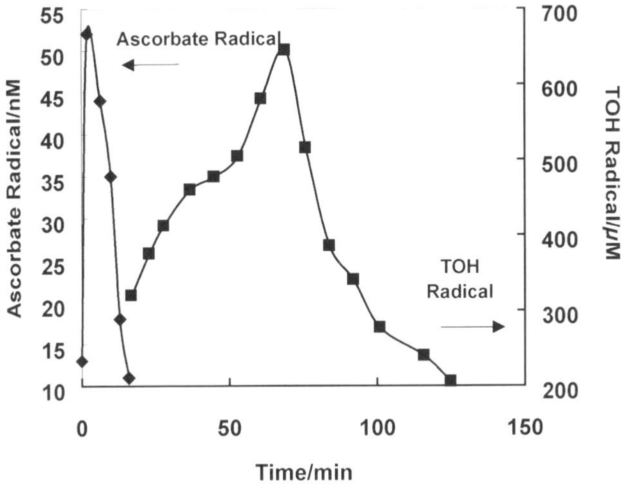

Figure 10.4: ASC and tocopheroxyl free radical concentrations over time in human blood plasma subjected to oxidative insult.
Time zero corresponds to the intensity observed at baseline in the plasma. After collection of this point hypoxanthine and xanthine oxidase were introduced into the sample. The increase in the level of ascorbate radical with the second data point demonstrates increased oxidative stress in the sample. The endogenous levels of vitamins C and E in this plasma sample were 63 and 48 μM, respectively. EPR spectroscopy was used to monitor the level of ascorbate radical in the plasma. The freshly harvested plasma sample was placed in an aqueous flat cell for determination of ascorbate radical concentration. See Sharma and Buettner (1993) for additional details.

10.2.5 Reactions of ascorbate with reactive nitrogen species

ASC reacts readily with many one-electron oxidants (*Table 10.1*); it might be expected that ASC would react with peroxynitrite. This reaction has been studied in detail and has been found to be pH dependent (Bartlett *et al.*, 1995). At pH 7.4 the reaction is quite slow with a second-order rate constant of $k_{obs} = 47 \cdot M^{-1} \cdot s^{-1}$; compare this to $k_{obs} = 4800 \cdot M^{-1} \cdot s^{-1}$ for the reaction of cysteine with peroxynitrite at this same pH (Radi *et al.* , 1991) and $k_{obs} = 4.8 \cdot 10^7 \cdot M^{-1} \cdot s^{-1}$ for the reaction of ASC with ozone (O_3) (Kanofsky and Sima, 1995). Thus, ASC appears to be a poor antioxidant when acting against peroxynitrite.

There appears to be no appreciable reaction of ASC with nitric oxide ($NO^{\cdot}$) as ASC seems not to influence the lifetime of $NO^{\cdot}$ in buffered solutions (G.R. Buettner and F.Q. Schafer, unpublished results). However, there is indirect evidence suggesting that $NO^{\cdot}$ reacts with $Asc^{\cdot-}$ (Eiserich *et al.*, 2002).

The nitrogen dioxide radical, $NO_2^{\cdot}$, is quite oxidizing; $E^{\circ\prime} = +990$ mV for the $NO_2^{\cdot}/NO_2^-$ couple (Koppenol *et al.* , 1992). Because it is a one-electron oxidant it reacts readily with ASC with $k_{obs} = 2 \times 10^7 \cdot M^{-1} \cdot s^{-1}$ at near neutral pH (Alfassi *et al.*, 1990; Forni *et al.* , 1986).

10.2.6 Reaction of ascorbate with singlet oxygen

Singlet oxygen ($^1O_2$) undergoes an addition reaction with ascorbic acid (diacid) at carbons 2 and 3 in cold methanol (–80°C) forming hydroperoxide products (Kwon and Foote, 1988). However, in room temperature/neutral aqueous solutions, singlet oxygen oxidizes ASC to produce H_2O_2; one half of the oxygen consumed appears as H_2O_2 (Buettner and Need, 1985). Thus, even though ASC efficiently quenches $^1O_2$ ($k = 8 \times 10^6 \cdot M^{-1} \cdot s^{-1}$; Rooney, 1983), so this chemical quenching brings about the formation of another reactive oxygen species, H_2O_2.

$$AscH^- + {}^1O_2 + H^+ \longrightarrow DHA + H_2O_2 \qquad (12)$$

The presence of DHA has not been demonstrated, but the two-electron oxidation of $AscH^-$ is consistent with the two-electron reduction of O_2 to form H_2O_2.

10.3 $Asc^{\cdot-}$ as a marker of oxidative stress

The ascorbate free radical is naturally detectable by EPR at low steady-state levels in biological samples. Examples are: leaves (Stegmann and Schuler, 1993), plasma (Sharma and Buettner, 1993; Minetti *et al.*, 1992; Pagan-Carlo *et al.*, 1999), synovial fluid (Buettner and Chamulitrat, 1990), skin (Buettner *et al.*, 1987; Jurkiewicz and Buettner, 1994) and milk (Friel *et al.*, 2002; Raghuveer *et al.*, 2002). As oxidative stress increases in a system, the steady-state $Asc^{\cdot-}$ concentration increases. These findings are consistent with a role for ASC as the terminal small-molecule antioxidant (*Table 10.2*). Because of the antioxidant reaction of ASC to form the ascorbate free radical, the EPR detection of the latter (*Figure 10.5*) can be used as a real-time indicator of oxidative stress (Buettner and Jurkiewicz, 1993). It is an excellent indicator of ongoing oxidative stress, but consideration must be given to pH, ASC concentration, temperature, oxygen and catalytic metal concentrations and in some settings, incident light. In the most successful applications of this approach the sample often serves as its own control. For example in the skin/UV light studies, EPR was used to determine the level of $Asc^{\cdot-}$ before and during exposure to the light making it straightforward to assess changes (Buettner *et al.*, 1987; Jurkiewicz and Buettner, 1994).

10.4 Ascorbate as a pro-oxidant

ASC is an outstanding antioxidant. However, it is also widely used as a pro-oxidant (Girotti *et al.*, 1985; Buettner and Jurkiewicz, 1996). This paradoxical behavior results because it is an excellent reducing agent. As a reducing agent it is able to reduce catalytic metals such as Fe^{3+} to Fe^{2+}. The redox cycling of these metals is essential to the oxidation of the vast majority of singlet state organic molecules. For the majority of organic molecules, true autoxidation is negligible because of the minimum 23 $kcal \cdot mol^{-1}$ activation energy required to overcome the spin restriction of dioxygen. However, redox active metals can overcome this spin

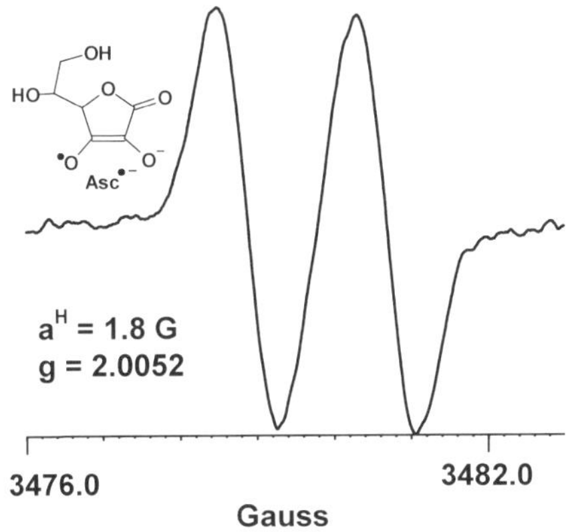

Figure 10.5: EPR spectrum of ascorbate radical.
The ascorbate radical is usually observed as a doublet with a peak-to-peak linewidth of about 0.7 gauss. This narrow linewidth and the relatively long lifetime of $Asc^{\bullet -}$ allow it to be detected readily by EPR. With modern EPR instrumentation it is possible to easily detect on the order of 1 nM $Asc^{\bullet -}$. This ability to detect low levels of $Asc^{\bullet -}$ contributes to its usefulness as a real-time indicator of oxidative stress. See Buettner and Jurkiewicz (1993) for details.

restriction and thereby catalyze oxidations of otherwise stable organics. In the presence of oxygen the reduced metal can initiate these oxidation processes (Miller *et al.*, 1990; Qian and Buettner, 1999). Thus, ASC can be a pro-oxidant when in the presence of redox active, catalytic metals.

10.5 Autoxidation of ascorbate and working with ascorbate solutions

Before beginning this discussion it must be understood that we use the term autoxidation to mean oxidation in the absence of metal catalysts. The term oxidation is used more broadly and includes all oxidations, with or without catalysts (Miller *et al.*, 1990).

ASC is readily oxidized. However, the rate of this oxidation is dependent upon pH (Buettner and Jurkiewicz, 1993) and the presence of catalytic metals (Buettner, 1986, 1988). The diacid is very slow to oxidize. Consequently, at low pH, i.e., less than 2 or 3, ascorbic acid solutions are quite stable, assuming catalytic transition metal ions are not introduced. However, as the pH is raised above pK_1, $AscH^-$ becomes dominant and Asc^{2-} will appear and the stability of the ASC in solution decreases (*Figure 10.6*). This loss of stability is usually the result of the presence of adventitious catalytic metals in the buffers and salts that are typically used in studies at near neutral pH. Typical buffers employed in biochemical and biological research have on the order of 1 μM iron and < 0.1 μM copper. Copper is ~ 80 times more efficient as a catalyst for ASC oxidation than iron; therefore, in typical buffers it is often the adventitious copper that is the biggest culprit in catalyzing ASC oxidation (Buettner, 1986, 1988).

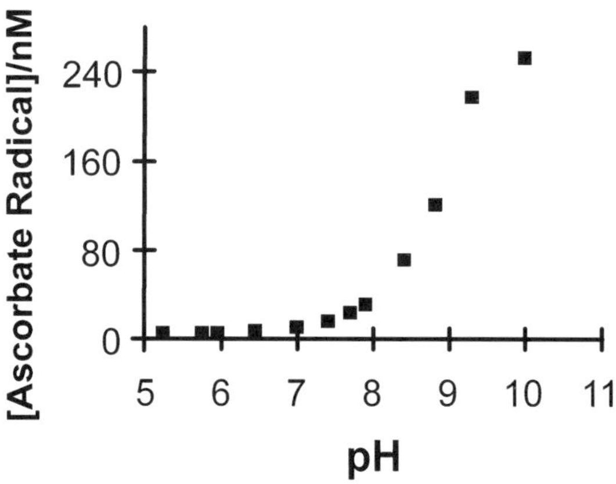

Figure 10.6: Oxidation of ASC vs. pH.
The steady-state level of $Asc^{\cdot-}$ can be used as a measure of the rate of oxidation of ASC in solution. Here, each solution of ASC (500 μM) was made with 50 mM phosphate buffer that had been treated with chelating resin using the batch method. In addition 50 μM deferrioxamine was present to insure that that any metals from the glassware and sample-handling apparatus did not contribute to the oxidation of ASC. To these solutions ASC was added and the EPR spectra were collected. Note that at acidic pH values, little oxidation occurs, but at higher pH values significant oxidation occurs. Trace levels of adventitious, redox active metals will greatly increase the rate of ASC oxidation. If no precautions are taken to remove these metals, ASC will be completely oxidized in a few minutes in near-neutral buffers (Buettner, 1990).

Adventitious metals can be removed from buffers using ion-exchange resins such as Chelex 100® or chelating resin (Buettner, 1988). ASC is an excellent tool to determine if significant adventitious catalytic metals are present in solutions or the effectiveness of efforts to remove these adventitious catalytic metals near-neutral buffer systems. In this method the loss of ASC is followed due to oxidation by monitoring its change in absorbance at 265 nm. In the standard test we add ~ 3.5 μl of 0.100 M ascorbic acid solution to 3.00 ml of the buffer in a standard 1-cm quartz cuvette. This results in an initial absorbance of 1.8. The loss of ASC is followed for 15 min. A loss of more than ~ 0.5% of the $AscH^-$ in this time indicates significant metal contamination. (If using a diode array spectrometer, interrogate the solution only a few times as the UV radiation near 200 nm will itself initiate ASC photo-oxidation.)

As one might expect, the quality of a stock solution of ASC will determine the quality and reproducibility of the results. We prepare ASC stock solutions using only the diacid. It is prepared as a 0.100 M stock solution (typically 10 ml) using high purity water. This solution is colorless, having a pH of ~ 2. It is stored in a volumetric flask with a tight-fitting plastic stopper, thus oxygen is kept from the solution during long-term storage. As the solubility of oxygen in air-saturated water is ~ 0.25 mM, the solution will become anaerobic with loss of < 1% of the original ASC. If the flask is indeed clean, we have found that the solution can be kept for several weeks without significant loss of ASC due to the low pH and lack of oxygen. The appearance of a yellow color is an indication of ASC deterioration. We avoid the use of sodium ASC as it invariably contains substantial quantities of oxidation products as evidenced by the yellow color of the solid and of the solution.

Chelators can be excellent tools to modulate the metal catalysis of ASC oxidation. Because the principal adventitious metals of concern are copper and iron, to slow or stop oxidative catalysis by adventitious metals we typically use diethylenetriaminepentaacetic acid (DTPA alias DETAPAC) and Desferal® as chelators of these metals. Desferal® is outstanding at quenching the catalysis of ASC oxidation by 'loosely bond' iron, but its quenching of copper catalysis of ASC oxidation, although good, is not complete (Buettner, 1986, 1988). Another consideration is that the hydroxamic acid moieties of Desferal® can become involved in free radical oxidations; in addition DETAPAC can serve as a reducing agent. The actual choice of chelating agent is determined by what is surmised to be the most troubling adventitious metal in a particular study. The next question is: how much? We have learned that millimolar or greater amounts of chelating agent are unnecessary and in many instances troublesome. We typically use only 50–250 μM chelating agent, usually at the lower end of this range. Both of these chelating agents are hexadentate, thus stoichiometry with copper or iron ions is 1:1. Adventitious levels of these metals in buffers are typically in the range 0.1–10 μM (Buettner, 1988). Thus, 50 μM chelator in principle should catch all of these metals, but each experimental setting needs to be considered. Other chelating agents may be of value in some settings and should be considered as appropriate. For example, exogenously added lactoferrin has been demonstrated to be of great value in protecting components of milk from oxidation when nutritional iron is added (Raghuveer *et al.*, 2002).

10.6. Human health and ascorbate

If the thermodynamic and kinetic considerations of ASC and tocopherol are important, then one might hypothesize that a certain ratio of these antioxidants would provide opportunity for better health. Gey (1998) tested this hypothesis by re-examining data from epidemiological studies. He concluded in cardiovascular disease that vitamin E acts as first risk discriminator, vitamin C as second one; optimal health requires sufficient levels of each vitamin. In addition he concluded that plasma values desirable for primary prevention are ≥ 30 μM lipid-standardized vitamin E and ≥ 50 μM vitamin C aiming at [vitamin C]/[vitamin E] > 1.3–1.5. This analysis in conjunction with the basic thermodynamic and kinetic properties of these two vitamins provides a new way to think about how these co-nutrients may contribute to optimal health. This type of study combined with new biochemical approaches to study nutritional needs of vitamins (Levine *et al.*, 1996) may bring about new thinking on optimal nutrition for antioxidants.

Acknowledgments

This work was supported in part by NIH Grants CA66081, CA81090, CA84462.

References

Alfassi ZB, Huie RE, Neta P, Shoute LCT (1990) Temperature-dependence of the rate constants for the reaction of inorganic radicals with organic reductants. *J. Phys. Chem.* **94**, 8800–8805.

Bartlett D, Church DF, Bounds PL, Koppenol WH (1995) The kinetics of the oxidation of L-ascorbic acid by peroxynitrite. *Free Radic. Biol. Med.* **18**, 85–92.

Bielski BHJ (1982) Chemistry of ascorbic acid radicals. In: *Ascorbic Acid: Chemistry, Metabolism, and Uses* (eds PA Seib, BM Tolbert). American Chemical Society, Washington DC, pp. 81–100.

Bielski BHJ, Allen AO, Schwarz HA (1981) Mechanism of disproportionation of ascorbate radicals. *J. Am. Chem. Soc.* **103**, 3516–3518.

Bisby RH, Parker AW (1995) Reaction of ascorbate with the α-tocopheroxyl radical in micellar and bilayer membrane systems. *Arch. Biochem. Biophys.* **317**, 170–178.

Bowry VW, Stocker R (1993) Tocopherol-mediated peroxidation. The prooxidant effect of vitamin E on the radical-initiated oxidation of human low-density lipoprotein. *J. Am. Chem. Soc.* **115**, 6029–6044.

Buettner GR (1986) Ascorbate autoxidation in the presence of iron and copper chelates. *Free Rad. Res. Commun.* **1**, 349–353.

Buettner GR (1988) In the absence of catalytic metals ascorbate does not autoxidize at pH 7, Ascorbate as a test for catalytic metals. *J. Biochem. Biophys. Methods* **16**, 27–40.

Buettner GR (1990) Ascorbate oxidation: UV absorbance of ascorbate and ESR spectroscopy of the ascorbate radical as assays for iron. *Free Radic. Res. Commun.* **10**, 5–9.

Buettner GR (1993) The pecking order of free radicals and antioxidants: Lipid peroxidation, α-tocopherol, and ascorbate. *Arch. Biochem. Biophys.* **300**, 535–543.

Buettner GR, Chamulitrat W (1990) The catalytic activity of iron in synovial fluid as monitored by the ascorbate free radical. *Free Radic. Biol. Med.* **8**, 55–56.

Buettner GR, Jurkiewicz BA (1993) Ascorbate free radical as a marker of oxidative stress: An EPR study. *Free Radic. Biol. Med.* **14**, 49–55.

Buettner GR, Jurkiewicz BA (1996) Catalytic metals, ascorbate, and free radicals: combinations to avoid. *Radic. Research.* **145**, 532–541.

Buettner GR, Need MJ (1985) Hydrogen peroxide and hydroxyl free radical production by hematoporphyrin derivative, ascorbate and light. *Cancer Lett.* **25**, 297–304.

Buettner GR, Motten AG, Hall RD, Chignell CF (1987) ESR detection of endogenous ascorbate free radical in mouse skin: Enhancement of radical production during UV irradiation following topical application of chlorpromazine. *Photochem. Photobiol.* **46**, 161–164.

Burton GW, Ingold KU (1986) Vitamin E: Application of the principles of physical organic chemistry to the exploration of its structure and function. *Acc. Chem. Res.* **19**, 194–201.

Davies MB, Austin J, Partridge DA (1991) *Vitamin C: Its Chemistry and Biochemistry.* The Royal Society of Cambridge, Cambridge.

Davis HF, McManus HJ, Fessenden RW (1986) An ESR study of free-radical protonation equilibria in strongly acid media. *J. Phys. Chem.* **90**, 6400–6404.

Eiserich JP, Baldus S, Brennan ML, Ma W, Zhang C, Tousson A, *et al.* (2002) Myeloperoxidase, a leukocyte-derived vascular NO oxidase. *Science* **296**, 2391–2394.

Foerster G, Weis W, Staudinger H (1965) Messung der elektronenspinresonanz an semidehydroascorbinsaeure. *Annalen. der Chemie.* **690**, 166–169.

Forni LG, Mora-Arellano VO, Packer JE, Willson RL (1986) Nitrogen dioxide and related free radicals: Electron transfer reactions with organic compounds in solutions containing nitrite and nitrate. *J. Chem. Soc. Perkin Trans.* **2**, 1–6.

Friel JK, Martin SM, Langdon M, Herzberg GR, Buettner GR (2002) Milk from mothers of both premature and full-term infants provides better antioxidant protection than does infant formula. *Pediatr. Res.* **51**, 612–618.

Fukuzawa K, Ikebata W, Sohmi K (1993) Location, antioxidant and recycling dynamics of α-tocopherol in liposome membranes. *J. Nutr. Sci. Vitaminol.* **39**, S9–S22.

Gey KF (1998) Vitamins E plus C and interacting co-nutrients required for optimal health. A critical and constructive review of epidemiology and supplementation data regarding cardiovascular disease and cancer. *Biofactors* **7**, 113–174.

Girotti AW, Thomas JP, Jordan JE (1985) Prooxidant and antioxidant effects of ascorbate on photosensitized peroxidation of lipids in erythrocyte membranes. *Photochem. Photobiol.* **41**, 267–276.

Golumbic C, Mattill HA (1941) Antioxidants and the autoxidation of fats XIII: The anti-oxygenic action of ascorbic acid in association with tocopherols, hydroquinones, and related compounds. *J. Am. Chem. Soc.* **63**, 1279–1280.

Jurkiewicz BA, Buettner GR (1994) Ultraviolet light-induced free radical formation in skin: An electron paramagnetic resonance study. *Photochem. Photobiol.* **59**, 1–4.

Kanofsky JR, Sima PD (1995) Reactive absorption of ozone by aqueous biomolecule solutions: Implications for the role of sulfhydryl compounds as targets for ozone. *Arch. Biochem. Biophys.* **316**, 52–62.

Koppenol WH (1993) A thermodynamic appraisal of the radical sink hypothesis. *Free Radic. Biol. Med.* **14**, 91–94.

Koppenol WH, Moreno JJ, Pryor WA, Ischiropoulos H, Beckman JS (1992) Peroxynitrite, a cloaked oxidant formed by nitric oxide and superoxide. *Chem. Res. Toxicol.* 5, 834–842.

Kwon BM, Foote CS (1988) Chemistry of singlet oxygen 50, Hydroperoxide intermediates in the photooxygenation of ascorbic acid. *J. Am. Chem. Soc.* **110**, 6582–6583.

Levine M, Conry-Cantilena C, Wang Y, Welch RW, Washko PW, Dhariwal KR, Park JB, Lazarev A, Graumlich JF, King J *et al.* (1996) Vitamin C pharmacokinetics in healthy volunteers: Evidence for a recommended dietary allowance. *Proc. Natl Acad. Sci. USA* **93**, 3704–3709.

Lewin S (1976) *Vitamin C: Its Molecular Biology and Medical Potential.* Academic Press, New York.

Miller DM, Buettner GR, Aust SD (1990) Transition metals as catalysts of 'autoxidation' reactions. *Free Radic. Biol. Med.* **8**, 95–108.

Minetti M, Forte T, Soriani M, Quaresima V, Menditoo A, Ferrari M (1992) Iron-induced ascorbate oxidation in plasma as monitored by ascorbate free radical formation. *Biochem. J.* **282**, 459–465.

Mukai K, Sawada K, Kohno Y, Terao J (1993) Kinetic study of the pro-oxidant effect of tocopherol: Hydrogen abstraction from lipid hydroperoxides by tocopheroxyls. *Lipids* **28**, 747–752.

Pagan-Carlo LA, Garcia LA, Buettner GR, Kerber RE (1999) Captopril lowers coronary venous free radical concentration after direct current cardiac shocks. *Chest* **116**, 484–487.

Qian SY, Buettner GR (1999) Iron and dioxygen chemistry is an important route to initiation of biological free radical oxidations: An electron paramagnetic resonance spin trapping study. *Free Radic. Biol. Med.* **26**, 1447–1456.

Radi R, Beckman JS, Bush KM, Freeman BA (1991) Peroxynitrite oxidation of sulfhydryls. The cytotoxic potential of superoxide and nitric oxide. *J. Biol. Chem.* **266**, 4244–4245.

Raghuveer T, McGuire EM, Martin SM, Wagner BA, Rebouche CJ, Buettner GR, Widness JA (2002) Lactoferrin attenuates iron-induced oxidation products in preterm infants' diet. *Pediatr. Res.* 52, 964–972.

Rooney ML (1983) Ascorbic acid as a photooxidation inhibitor. *Photochem. Photobiol.* **38**, 619–621.

Scarpa M, Rigo A, Maiorino M, Ursini F, Gregolin C (1984) Formation of α-tocopherol radical and recycling of α-tocopherol by ascorbate during peroxidation of phosphatidylcholine liposomes. An EPR study. *Biochim. Biophys. Acta* **801**, 215–219.

Scarpa M, Stevanato R, Viglino P, Rigo A (1993) Superoxide ion as active intermediate in the autoxidation of ascorbate by molecular oxygen. *J. Biol. Chem.* **258**, 6695–6697.

Sharma MK, Buettner GR (1993) Interaction of vitamin C and vitamin E during free radical stress in plasma: An ESR study. *Free Radic. Biol. Med.* **14**, 649–653.

Sharma MK, Buettner GR, Spencer K, Kerber RE (1994) Ascorbyl free radical as a real-time marker of free radical generation during myocardial reperfusion: An electron paramagnetic resonance study. *Circ. Res.* **74**, 650–658.

Stegmann HB, Schuler P (1993) Oxidative stress of crops monitored by EPR. *Z. Naturforsch.* **48c**, 766–772.

Stone I (1974) *The Healing Factor: Vitamin C Against Disease.* Grosset and Dunlap, New York.

Van Acker SABE, Koymans LMH, Bast A (1993) Molecular pharmacology of Vitamin E: Structural aspects of antioxidant activity. *Free Radic. Biol. Med.* **15**, 311–328.

Wardman P (1989) Reduction potentials of one-electron couples involving free radicals in aqueous solutions. *J. Phys. Chem. Ref. Data.* **18**, 1637–1755.

Weast RC, ed. (1987) *Handbook of Chemistry and Physics*. 67th Edn. CRC Press, Boca Raton, FL.
Wieczorek P, Ogonski T, Machoy Z (1987) Interaction of Na, Li Cs, K ions with ascorbyl radicals. *Z. Naturforsch.* **42c**, 215–216.
Williams NH, Yandell JK (1982) Outer-sphere electron-transfer reaction of ascorbate anions. *Aust. J. Chem.* **35**, 1133–1144.
Winterbourn CC (1993) Superoxide as an intracellular radical sink. *Free Radic. Biol. Med.* **14**, 85–90.

11

Vitamin C and oxidative damage to DNA

Henrik E. Poulsen, Peter Møller, Jens Lykkesfeldt, Allan Weimann and Steffen Loft

11.1 Introduction

Numerous epidemiological studies (for citations and review, see Ames, 2001; Halliwell 2001), have pointed at a close correlation between vitamin C intake on one side and arteriosclerosis and cancer incidence on the other side. Clearly vitamin C is important for humans: we cannot synthesize it; deficient intake results in scurvy; there are efficient carrier proteins for its uptake (Levine, 1999); it is a necessary cofactor for several enzymes and metabolic roles (Halliwell, 2001); and it is the most important water soluble reductant. It is present in high levels in sperm fluid (see section 11.4) and may also be important as an inhibitor of nitrosamine formation in the stomach (Mirvish *et al.*, 1998).

Among all the functions of vitamin C there has been a considerable focus on its function as an antioxidant in the prevention of various diseases such as arteriosclerosis, cancer, degenerative cerebral diseases and aging. Vitamin C is present in high concentrations in cells and body fluids and it is straightforward to assume that it could prevent oxidation of important macromolecules such as DNA. Surprisingly, in spite of huge, long-lasting and costly trials with antioxidant vitamins, none of these have tested out vitamin C as a single compound, and only very few have included vitamin C in a cocktail together with β-carotene and vitamin E.

In this review we will therefore focus on the scientific data available about the relationship between oxidative DNA damage and vitamin C, including whether vitamin C can act as an antioxidant as well as a prooxidant *in vivo*.

11.2 Measurements of DNA oxidation

There is a generally accepted concept that changes in DNA are necessary but not sufficient for cancer development. There also are suggestions that DNA changes are of importance for arteriosclerosis development, however low density

Vitamin C, edited by Han Asard, James M May and Nicholas Smirnoff

lipoprotein (LDL) oxidation is still considered the prime initial event. The attack of free radical and reactive oxygen species (ROS) formed by cellular metabolism and external factors modifies DNA mainly in the base region and these modifications are mutagenic in simple *in vitro* systems. However only a small portion of the induced modifications are assumed to induce mutations *in vivo* because of efficient repair mechanisms (Ames, 2001; Bohr, 2002; Cooke *et al.*, 2002; Halliwell 2001; Jackson and Loeb, 2001; Poulsen *et al.*, 1998c).

The methods available to measure DNA oxidation include measurement of the levels of oxidized lesions in tissue or cell samples, measurement of the urinary excretion of oxidized bases that most likely is the result of DNA repair, and measurement on single cells using the Comet assay.

Measurement of the level of oxidative DNA modifications in tissues or cells, and the excretion into urine of modified bases or nucleosides, is most often done by either chemical measurement, e.g., HPLC with mass spectrometry or electrochemical detection, details on these methods can be found elsewhere (ESCODD 2000, 2002a, 2002b; Hartmann *et al.*, 2003; Lunec, 1998; Riis and ESCODD, 2002). ELISA methods are also available but are still not sufficiently specific for urinary analysis. Other approaches such as ligase-mediated PCR and mutation analysis are outside the scope of this review, as there are no data relating to vitamin C. The interpretation of these measurements in tissue or cells on the one hand and of the urinary excretion on the other hand is quite different although the methodology of the chemical estimation is very similar (Poulsen *et al.*, 1998a; Poulsen and Loft, 1998). Assuming that the individual under investigation is in steady state, e.g., not acutely but chronically exposed, the urinary excretion e.g., in 24-h urine is interpreted as the rate of damage to DNA by oxidative stress. It is important to stress that although the excretion of the 8-hydroxylated form of guanine (8-oxodG), or other indicators, in contrast to tissue concentration measurement, is based on the DNA repair mechanism; the measure is independent of DNA repair in the sense that a change in repair will not influence the measurement in steady state. Also it should be noted that the urinary excretion is an average of the modification from oxidative stress in the body and cannot be related to a single organ. Measurement of the tissue level, on the other hand, results from a balance between the rate of formation (oxidative stress) and the rate of removal (repair). If one or both of these factors change the level will change, e.g., a decreased repair will give increased levels, so will increased oxidative stress. By measuring tissue levels it is not possible to distinguish whether repair or oxidative stress is behind the change.

We recently assessed the rate of damage related to the presently known major DNA modification 8-oxodG to a minimum of about 500 hits per cell per day in humans (Poulsen *et al.*, 1998b), which would result in a doubling time for this particular modification of about 60–190 days, meaning that after a lifespan of 8 years 1% of DNA would be oxidized, a situation not compatible with life. This situation is avoided by specific repair mechanisms that keep the levels down to about 1 per million bases.

Estimations of the tissue levels of 8-oxodG in normal human tissues and cells are very controversial, but there is agreement that the levels as estimated by different methods are about 1 per million guanines (ESCODD, 2000, 2002a; Lunec, 1998; Riis and ESCODD, 2002). The levels of 8-oxodG in mitochondrial

DNA reported in the literature vary to an extreme degree (Beckman and Ames 1999) and there is no final conclusion about the true mitochondrial level except that it seems to be higher than in genomic DNA.

Although 8-oxodG is the only lesion studied extensively, about 100 different modifications in DNA from oxidation have been demonstrated (Dizdaroglu, 1992; Dizdaroglu *et al.*, 2002) of which maybe 20 are relevant *in vivo*.

Measurement of genomic and mitochondrial DNA modifications from oxidation requires isolation, extraction and digestion of DNA before subsequent analysis. The methods available are HPLC-ECD, GC-MS, LC-MS, LC-MS/MS (ESCODD, 2000, 2002a, 2002b; Lunec, 1998; Riis and ESCODD, 2002). A major problem in the analysis is the possibility of artificial oxidation of guanine/deoxyguanosine during the work-up procedures, in nuclear as well as in mitochondrial DNA. The ratio of nonoxidized guanine/deoxyguanosine to the oxidized ones is 1 : 1 000 000. Slight oxidation, i.e., just 0.1 or 0.01%, during the preparative and the work-up procedures therefore will result in very high measured levels and will obscure most changes from increased or decreased oxidative stress. In the literature there are many reports of very high levels of 8-oxodG in tissue or cells; these data should be interpreted with great care and in this review we have avoided citing papers with such levels. Measurement of 8-oxodG or the corresponding oxidized base in urine is generally accepted as accurate, since the levels of unoxidized guanine in urine are low. Thus, the problem of artefacts due to *post hoc* oxidation is much less severe, if it exists at all.

The Comet Assay in principle detects strand breaks as movement of DNA fragments in agarose gels. In the nucleus the DNA is organized on the nuclear scaffold as loops of supercoiled helices that unwind by introduction of strand breaks (Collins *et al.*, 1997). The procedure for making this visible consists of several steps. First the cells are lysed in solutions containing detergents and high salt concentrations, thus producing a nucleus-like structure that is virtually devoid of proteins (referred to as a nucleoid). A subsequent alkaline electrophoresis forces the DNA to migrate towards the anode. Upon staining with a fluorescent dye the migration can be visualized in a microscope as images that resemble the tail of a comet. In an enzyme-modified version of comet assay, nucleoids are incubated with DNA repair enzymes that excise specific types of oxidatively damaged DNA from the strand. The most frequently used repair enzymes have been formamidopyrimidine DNA glycosylase (FPG) and endonuclease III (ENDOIII) (Collins *et al.*, 1997). The FPG protein recognizes mainly oxidized purines, e.g., 8-oxodG and formamidopyridine derivates of guanine and adenine, whereas ENDOIII recognizes a diverse array of oxidized pyrimidines, e.g., thymidine glycol, urea residues, deoxyuridineglycol, 5-hydroxydeoxycytosine and 5-hydroxydeoxyuridine.

Quantification of the Comet Assay can be done by visual scoring or by automated image analysis systems. A detailed discussion of the quantification methods is not included here.

11.3 Antioxidant and prooxidant properties of vitamin C

Vitamin C has been called the most important water soluble antioxidant in plasma (Frei *et al.*, 1988) and its antioxidant properties both *in vitro* and *in vivo* have been documented extensively (Carr and Frei, 1999). However, the compound also has the ability to reduce transition metal ions such as copper and iron and can therefore potentially act as a prooxidant (Halliwell, 1996). While Fenton type systems employing vitamin C as prooxidant have been frequently used *in vitro* (Wills, 1969), it has been subject to much debate whether similar reactions occur *in vivo* and if so what the biological implications are.

Vitamin C, or ascorbic acid (ASC), fulfils the criteria of an effective antioxidant. The oxidation of ASC is outlined in *Figure 11.1*. ASC and even the MDHA have low reduction potentials making it possible for them to serve as electron donors for a variety of potentially damaging radicals and oxidants. The MDHA itself has low reactivity due to its highly resonance-stabilized structure (*Figure 11.1*). Furthermore, two MDHAs have been shown to readily dismutate into one molecule of ASC and one of DHA thereby lowering their energy state even further (Buettner and Jurkiewicz, 1996). DHA – the two electron oxidation product of ASC – is most likely formed via dismutation or further oxidation of the MDHA. DHA rapidly breaks down nonenzymatically to 2,3-diketogulonate as well as oxalate, threonate and other oxidation products with a half-life of only a few minutes at physiological pH (Bode *et al.*, 1990).

Consequently, efficient ways of regenerating ASC have evolved. ASC can be regenerated from MDHA by means of an NADH-dependent semidehydroascorbate reductase (Wells and Jung, 1997). However as reported by Meister (1992), regeneration of ASC from the MDHA was not sufficient to protect GSH-deficient animals from oxidative damage, indicating that this route is probably not of major importance (Lykkesfeldt, 2002).

DHA is readily reduced back to ASC by cellular antioxidants such as glutathione and lipoic acid (Wells and Jung, 1997; Wills, 1969, see Chapter 8 by May and Asard). The efficient chemical reduction of DHA under physiological conditions (e.g., via glutathione) has previously led to the conclusion that ASC recycling occurs non-enzymatically in vivo (Wills, 1969; Winkler, 1992; Winkler *et al.*, 1994). However, more recently ASC regeneration from DHA has been shown to occur intracellularly via enzymatic as well as non-enzymatic pathways. Thus, several enzymes are now known to possess DHA reductase activity including the NADPH-dependent thioredoxin reductase (May *et al.*, 1997) and the glutathion-dependent enzymes glutaredoxin and protein disulfide isomerase (Park and Levine, 1996; Wells *et al.*, 1990). Moreover, enzymatic regeneration of ASC *in vivo* was recently demonstrated in guinea pigs in a vitamin C deficiency model (Lykkesfeldt, 2002).

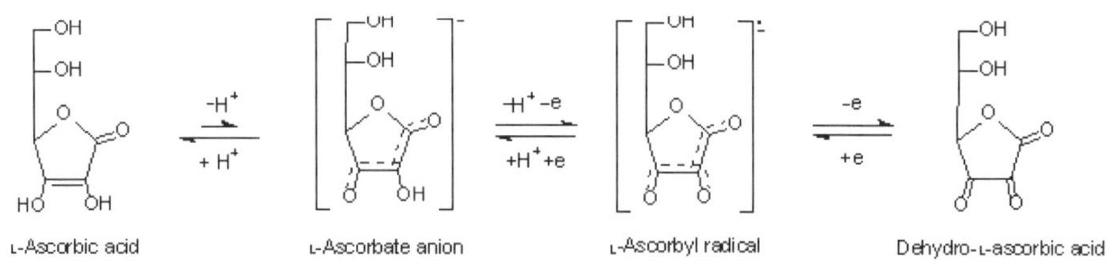

Figure 11.1: Oxidation of ASC by means of two successive one-electron oxidation steps. Direct formation of DHA by two-electron oxidation is also possible. Furthermore, two molecules of MDHA can dismutate into one molecule of ASC and one of DHA.

The classical system of hydroxyl radical formation *in vitro* involves ASC in the role of prooxidant:

$$Fe^{3+} + ASC \rightarrow Fe^{2+} + MDHA \qquad \text{(Scheme 1)}$$

$$Fe^{2+} + H_2O_2 \rightarrow Fe^{3+} + OH^- + OH^\cdot \qquad \text{(Scheme 2)}$$

In fact, ASC acts as an antioxidant in reducing Fe^{3+} while itself getting oxidized (Scheme 1). However, due to the ability of free Fe^{2+} to participate in reactions generating hydroxyl radicals (Scheme 2), it is overall regarded as a prooxidant function.

From a chemical point of view, it is clear that depending on the circumstances, vitamin C can act as an antioxidant or as a prooxidant. In the *in vivo* situation it remains a matter of controversy whether Fenton type reactions occur and if so to what extent they play a role in oxidative damage to cellular macromolecules, e.g., DNA. In the organism there are efficient systems to control free transition metals. Prime examples of this are the hemoglobin molecule and the cytochrome P450 enzyme superfamily, where the special affinity between iron ions and oxygen is utilized in oxygen transport and in controlled oxidation reactions of foreign compounds. In healthy humans, iron and copper ions are largely bound in complexes making them unavailable for free radical reactions (Halliwell and Gutteridge 1986, 1990). Presently there is no evidence that demonstrate an *in vivo* uncontrolled prooxidant mechanism of vitamin C, and if existing it may be biologically well controlled and without any importance in the intact organism. More relevant are perhaps disease conditions, in which elevated levels of transition metal ions are found *in vivo* either as a result of traumatic tissue injury, iron overload as seen in hemochromatosis or in severe infectious conditions with massive production of nitric oxide, e.g., septic shock. However, data concerning DNA oxidation in these conditions are scarce and inconclusive.

11.4 Vitamin C and oxidative DNA damage in sperm

Oxidative damage to DNA is abundant in sperm. Moreover, the male fertility appears to be closely related to the level of 8-oxodG in sperm DNA, indicating important effects of oxidative stress (Loft *et al.*, 2003; Ni *et al.*, 1997; Sharma and Agarwal, 1996; Shen *et al.*, 1999; Shen and Ong, 2000). Similarly, vitamin C levels are high in sperm plasma and may influence both the extent of oxidative DNA damage and fertility. In a US study the sperm count was significantly lower in samples with low vitamin C concentration than in samples with high vitamin C (Fraga *et al.*, 1991; Shen and Ong, 2000). Moreover, the 8-oxodG level in sperm DNA was highly dependent on the intake and seminal plasma concentration of vitamin C as shown by a significant correlation among 24 men and a depletion and replenishment study in eight men (Fraga *et al.*, 1991; Jacob *et al.*, 1991). In a small study in infertile men, administration of a combination of vitamin C, α-tocopherol and glutathione for 2 months decreased the levels of 8-oxodG in sperm DNA to the levels of fertile men (Kodama *et al.*, 1997). Furthermore, supplementation with vitamin C, 200 or 1000 mg per day, increased sperm concentrations and viability in heavy smokers (Dawson *et al.*, 1992). *In vitro* addition of ascorbate (ASC) to sperm samples protects from induction of DNA damage by hydrogen peroxide or irradiation, whereas addition of ASC together with tocopherol and/or glutathione may enhance damage (Donnelly *et al.*, 1999a, 1999b; Hughes *et al.*, 1998).

Smokers have decreased plasma levels of vitamin C (Lykkesfeldt *et al.*, 1997) due to both increased expenditure and lower intake as compared to nonsmokers (Lykkesfeldt *et al.*, 1996; Zondervan *et al.*, 1996). In a study on 8-oxodG in sperm from US men, the intake of vitamin C assessed by questionnaire in a subset was 26% lower in smokers than in nonsmokers although the level of vitamin C in seminal plasma was not significantly different between the groups (Fraga *et al.*, 1996). Nevertheless, the level of α-tocopherol was significantly lower in seminal plasma from smoking as compared to nonsmoking US men (Fraga *et al.*, 1996). It would thus be expected that smokers would have higher levels of 8-oxodG in sperm. This was in fact found in studies of men from Argentina, California and China (Fraga *et al.*, 1996; Shen *et al.*, 1997). Incubation of sperm cells from nonsmokers in sperm plasma from smokers leads to decreased viability of the cells (Zavos *et al.*, 1998). However, in a large Danish study (Loft *et al.*, 2003; Ni *et al.*, 1997; Sharma and Agarwal, 1996; Shen *et al.*, 1999; Shen and Ong, 2000) there was no difference between smokers and nonsmokers with respect to 8-oxodG in sperm DNA. Another study (Sergerie *et al.*, 2000) found no association between smoking and DNA fragmentation in sperm. It is possible that the apparent discrepancies could be related to differences in genetic background, sperm counts and diet, including vitamin C intake between the populations.

11.5 Vitamin C intervention trials relating to oxidative DNA damage and related disease

Searching the PubMed for ASC or DHA gives 23 072 hits, restriction to 'human' gives 11 474 hits, restricting to controlled trials gives 164 hits (search performed February 2003). A more refined search may give more hits but will not be able to change the pattern of many chemical and biological (e.g., cell) studies and very few controlled intervention trials with vitamin C. When restricted to cancer trials with hard end-points or to surrogate markers for cancer development, the number of trials falls dramatically to those listed in *Tables 11.1 and 11.2*. Three controlled trials, two with a large number of subjects, have been conducted (Harris *et al.*, 1989; Heart Protection Study Collaborative Group, 2002) with diseases as endpoints, however, none of them used vitamin C as a single substance and all of them included beta-carotene or vitamin E as well as other antioxidant vitamins or substances. All three provided negative results in the sense that vitamin C did not show any positive or negative effects (*Table 11.1*). *Table 11.1* also includes the studies we have been able to identify that use a global biomarker for oxidative stress, all these results are also negative (*Table 11.1*).

Negative studies also dominate *Table 11.2*, which lists the 15 studies where all except one have used lymphocytes as a surrogate tissue on a relatively small number of subjects, range 8–86. The positive studies indicate that the level of

Table 11.1: Clinical trials with mortality/morbidity or global biomarker measurement.

Subjects (n)	Vitamin C dose	Plasma vitamin C	Marker/tissue	Effect	Other vitamins	Duration	Reference
20 536	250 mg·day^{-1}	33% up	Mortality, morbidity, hospitalization	None	600 mg vitamin E, 20 mg β carotene	5 years	Harris *et al.*, 1989; Heart Protection Study Collaborative Group 2002
18 000 (approx)	120 mg·day^{-1}	Increased by 240% (relative to placebo)	Death by cancer and crebrovascular disease	None	Factorial with retinol, zinc, riboflavin, niacin, molybdenum, β-carotene, selenium, vitamin E		Blot *et al.*, 1993
205	1000 mg·day^{-1}	n.d.	Colorectal adenoma	None	Factorial with vitamin E 400 mg	4 years	Greenberg *et al.*, 1994
184	500 mg·day^{-1}	Increased by 30%	8-oxodG in urine (ELISA)	None	Factorial with vitamin 400 mg	60 days	Huang *et al.*, 2000
142	500 mg·day^{-1}	Increased by 70%	8-oxodG in urine	None vitamin E	Factorial with 182 mg	60 days	Prieme *et al.*, 1997
48	500 mg·day^{-1}	30%	8-oxodG in urine	None	Alone or with vitamin E 182 mg	36 months	Porkkala-Sarataho *et al.*, 2000
23	1000 mg·day^{-1}	n.d.	8-oxodG in urine	Decreased in HIV-positive after zidovudine treatment	600 mg vitamin E	1 month	Asuncion *et al.*, 1996
11	1000 mg·day^{-1}	Increased by 30%	8-oxodG in urine	None	553 mg vitamin E, 10 mg β carotene	1 month	Witt *et al.*, 1992

ELISA, Enzyme linked immunosorbent assay; n.d., not determined.

Table 11.2: Clinical trials with vitamin C using biomarkers on surrogate tissue.

Subjects (n)	Vitamin C dose	Plasma vitamin C	Marker/tissue	Effect	Other vitamins	Duration	Reference
80	100 mg·day^{-1}		Comet/ lymphocytes	Decrease ENDOIII sites and sensitivity to H_2O_2	Vitamin E 280 IU, β carotene 25 mg	20 weeks	Duthie *et al.*, 1996
72	1000 mg·day^{-1}	Increased by 66%	Comet/ esophagus biopsy cells	None	Vitamin E 200 mg	12 weeks	White *et al.*, 2002
63	500 mg·day^{-1}	Increased by 40%	8-oxodG / lymphocytes	None	400 IU vitamin E, 12 mg β carotene	6 month	Jacobson *et al.*, 2000
48	60–6000 mg·day^{-1}	Increased by 23–80%	Comet/ lymphocytes	None	None	2 weeks	Anderson *et al.*, 1997
38	60–260 mg·day^{-1}	Unchanged, n = 20 Increased by 47%, n = 12	8-oxodG/ lymphocytes 12 other DNA modifications	Decrease in n = 12 with low initial plasma vitamin C	14 mg Fe	12 weeks	Rehman *et al.*, 1998
24	1000 mg·day^{-1}	Increased by 15–110%	Micronuclei/ lymphocytes	Reduction, particular in smokers	None	7 days	Schneider *et al.*, 2001
21	350 mg·day^{-1}		8-oxodG/ lymphocytes	None	250 IU vitamin E, 60 mg β carotene, 80 μg Se	4 weeks	Welch *et al.*, 1999
20	500 mg·day^{-1}	Increased by 50%	Comet/ lymphocytes	None	200 IU vitamin E, 9 mg β carotene, 1.8 g red ginseng	4 weeks	Lee *et al.*, 1998
20	260 mg·day^{-1}	Increased by 15%	8-oxodG/ lymphocytes 10 other DNA modifications	None	14 mg Fe	6 weeks	Proteggente *et al.*, 2000

20	500–5000 mg·day^{-1}	Increased by 50%	8-oxodG, NK cell activity, cell cycle progression, apoptosis	None	None	2 weeks	Vojdani 2001
19	60 mg·day^{-1}	Increased by 26%	8-oxodG/ lymphocytes	Decreased after supplementing	3 mg vitamin B, 30 IU vitamin E, 40 mg Zn, 2 mg Cu	60 days	Howard *et al.*, 1998
15	1000 mg·day^{-1}	n.d.*	Micronuclei/ lymphocytes	Reduced micronuclei	None	1 month	Perminova *et al.*, 2001
14	1000 mg·day^{-1}	Increased by 50%	Comet/ lymphocytes	No effect to H_2O_2	None	42 days	Brennan *et al.*, 2000
12	1000 mg·day^{-1}	n.d.	Comet/ lymphocytes	Sensitivity to H_2O_2 down	None	Single dose	Panayiotidis and Collins, 1997)
6	35 mg·day^{-1}	n.d.	Comet/ lymphocytes	Sensitivity to ionizing radiation *ex vivo* down	None	Single dose	Green *et al.*, 1994

n.d., Not determined; IU, international units – 100 IU vitamin E corresponds to 91 mg vitamin E; NK, natural killer.

oxidative DNA modification is not changed by vitamin C, however, some protection may be provided against increased oxidative stress.

By means of the Comet assay it has been possible to study effects of vitamin C supplementation on DNA damage in human subjects. These studies are listed in *Table 11.2* together with studies using other endpoints, mainly measurement of 8-oxodG and micronuclei. In general a preventive effect is apparent within a few hours after intake, whereas continued administration has failed to yield significant effects. In studies with administration of a single high dose of vitamin C, the findings include decreased levels of strand breaks and alkali labile sites, and lower sensitivity to *ex vivo* challenge with H_2O_2 and ionizing radiation in leukocytes (Green *et al.*, 1994; Panayiotidis and Collins, 1997). The effect of vitamin C appeared to peak early, within 2–8 h after consumption (Panayiotidis and Collins, 1997). In contrast, in three studies the steady state level of strand breaks and alkali labile sites were unaffected by vitamin C supplementation (Anderson *et al.*, 1997; Brennan *et al.*, 2000; Welch *et al.*, 1999). Since strand breaks are unaltered in most antioxidant intervention studies, the unaltered strand break levels in long-term vitamin C studies may also reflect the inability of this particular endpoint to reveal antioxidant effects (Moller and Loft, 2002). In *ex vivo* lymphocyte challenge studies, two investigations found no effect of either X-ray or hydrogen peroxide (Anderson *et al.*, 1997; Welch *et al.*, 1999), whereas one study shows decreased sensitivity to hydrogen peroxide after 42 days of 1000 mg·day^{-1} vitamin C consumption (Brennan *et al.*, 2000). So far, no study has examined the effect of prolonged administration of vitamin C alone on oxidation of DNA bases assessed by the comet assay. In a study with administration of cocktail of vitamin C (100 mg·day^{-1}), vitamin E (280 mg·day^{-1}), and β-carotene (25 mg·day^{-1}) for 20 weeks, the level of ENDOIII sensitive sites were decreased in smokers (Duthie *et al.*, 1996).

11.6 Conclusions

There is a plethora of evidence for antioxidant functions of vitamin C *in vitro* and in cell systems and a large number of epidemiological studies consistently showing relations between vitamin C intake and cancer.

Although vitamin C can act as a prooxidant in chemical systems, there is no substantial evidence that such reactions occur in humans *in vivo*. Large scale clinical intervention trials are scarce and the three trials with hard end-points, two with thousands of volunteers, have not been able to identify a cancer preventive effect of vitamin C. Studies using surrogate measures for global oxidative stress to DNA or using surrogate tissue (lymphocytes) for measuring the number of oxidative modifications of DNA (8-hydroxylation of guanine) generally have not been able to identify a long lasting protective effect of vitamin C. Vitamin C seems to have a particular function in protecting sperm DNA, however there are negative studies as well.

We conclude that there are surprisingly few intervention trials with vitamin C with regard to prevention of cancer or its surrogate markers. Consequently, and in view of the abundant *in vitro* and epidemiological evidence, there still is a real need for well designed clinical intervention trials using both hard end-points and surrogate end-points on surrogate tissue. We also conclude that the methodology for using surrogate endpoints and surrogate tissue is still in its infancy and that much development is needed.

References

Ames BN (2001) DNA damage from micronutrient deficiencies is likely to be a major cause of cancer. *Mutat. Res.* **475**, 7–20.

Anderson D, Phillips BJ, Yu TW, Edwards AJ, Ayesh R, Butterworth KR (1997) The effects of vitamin C supplementation on biomarkers of oxygen radical generated damage in human volunteers with "low" or "high" cholesterol levels. *Environ. Mol. Mutagenesis* **30**, 161–174.

Asuncion JDL, Millan A, Pla R, Bruseghini L, Esteras A, Pallardo FV, Sastre J, Vina J (1996) Mitochondrial glutathione oxidation correlates with age-associated oxidative damage to mitochondrial DNA. *FASEB J.* **10**, 333–338.

Beckman KB, Ames BN (1999) Endogenous oxidative damage of mtDNA. *Mutat. Res.* **424**, 51–58.

Blot WJ, Li JY, Taylor PR, Guo W, Dawsey S, Wang GQ, Yang CS, Zheng SF, Gail M, Li GY (1993) Nutrition intervention trials in Linxian, China: supplementation with specific vitamin/mineral combinations, cancer incidence, and disease-specific mortality in the general population. *J. Natl Cancer Inst.* **85**, 1483–1492.

Bode AM, Cunningham L, Rose RC (1990) Spontaneous decay of oxidized ascorbic acid (dehydro-L-ascorbic acid) evaluated by high-pressure liquid chromatography. *Clin. Chem.* **36**, 1807–1809.

Bohr VA (2002) Repair of oxidative DNA damage in nuclear and mitochondrial DNA, and some changes with aging in mammalian cells. *Free Radic. Biol. Med.* **32**, 804–812.

Brennan LA, Morris GM, Wasson GR, Hannigan BM, Barnett YA (2000) The effect of vitamin C or vitamin E supplementation on basal and H_2O_2-induced DNA damage in human lymphocytes. *Br. J. Nutr.* **84**, 195–202.

Buettner GR, Jurkiewicz BA (1996) Catalytic metals, ascorbate and free radicals: combinations to avoid. *Radiat. Res.* **145**, 532–541.

Carr A, Frei B (1999) Does vitamin C act as a pro-oxidant under physiological conditions? *FASEB J.* **13**, 1007–1024.

Collins AR, Dobson VL, Dusinska M, Kennedy G, Stetina R (1997) The comet assay: what can it really tell us? *Mutat. Res.* **375**, 183–193.

Cooke MS, Lunec J, Evans MD (2002) Progress in the analysis of urinary oxidative DNA damage. *Free Radic. Biol. Med.* **33**, 1601–1614.

Dawson EB, Harris WA, Teter MC, Powell LC (1992) Effect of ascorbic acid supplementation on the sperm quality of smokers. *Fertil. Steril.* **58,** 1034–1039.

Dizdaroglu M (1992) Oxidative damage to DNA in mammalian chromatin. *Mutat. Res.* **275**, 331–342.

Dizdaroglu M, Jaruga P, Birincioglu M, Rodriguez H (2002) Free radical-induced damage to DNA: mechanisms and measurement. *Free Radic. Biol. Med.* **32**, 1102–1115.

Donnelly ET, McClure N, Lewis SE (1999a) Antioxidant supplementation in vitro does not improve human sperm motility. *Fertil. Steril.* **72,** 484–495.

Donnelly ET, McClure N, Lewis SE (1999b) The effect of ascorbate and alpha-tocopherol supplementation in vitro on DNA integrity and hydrogen peroxide-induced DNA damage in human spermatozoa. *Mutagenesis* **14,** 505–512.

Duthie SJ, Ma A, Ross MA, Collins AR (1996) Antioxidant supplementation decreases oxidative DNA damage in human lymphocytes. *Cancer Res.* **56,** 1291–1295.

ESCODD (2000) Comparison of different methods of measuring 8-oxoguanine as a marker of oxidative DNA damage. *Free Rad. Res.* **32**, 333–341.

ESCODD (2002a) Comparative analysis of baseline 8-oxo-7,8-dihydroguanine in mammalian cell DNA, by different methods in differents laboratories: An approach to concensus. *Carcinogenesis* **23**, 2129–2133.

ESCODD (2002b) Inter-laboratory validation of procedures for measuring 8-oxo-7,8-dihydroguanine/8-oxo-7,8-dihydro-2′-deoxyguanosine in DNA. *Free Rad. Res.* **36**, 239–245.

Fraga CG, Motchnik PA, Shigenaga MK, Helbock HJ, Jacob RA, Ames BN (1991) Ascorbic acid protects against endogenous oxidative DNA damage in human sperm. *Proc. Natl Acad. Sci. USA* . **88**, 11003–11006.

Fraga CG, Motchnik PA, Wyrobek AJ, Rempel DM, Ames BN (1996) Smoking and low antioxidant levels increase oxidative damage to sperm DNA. *Mutat. Res.* **351**, 199–203.

Frei B, Stocker R, Ames BN (1988) Antioxidant defenses and lipid peroxidation in human blood plasma. *Proc. Natl Acad. Sci. USA* **85**, 9748–9752.

Green MH, Lowe JE, Waugh AP, Aldridge KE, Cole J, Arlett CF (1994) Effect of diet and vitamin C on DNA strand breakage in freshly-isolated human white blood cells. *Mutat. Res.* **316**, 91–102.

Greenberg ER, Baron JA, Tosteson TD, Freeman DH, Beck GJ, Bond JH, *et al.* (1994) A clinical trial of antioxidant vitamins to prevent colorectal adenoma. *N. Engl J. Med.* **331**, 141–147.

Halliwell B (1996) Vitamin C: antioxidant or pro-oxidant in vivo? *Free Rad. Res.* **25**, 439–454.

Halliwell B (2001) Vitamin C and genomic stability. *Mutat. Res.* **475**, 29–35.

Halliwell B, Gutteridge JM (1986) Oxygen free radicals and iron in relation to biology and medicine: some problems and concepts. *Arch. Biochem. Biophys.* **246**, 501–514.

Halliwell B, Gutteridge JM (1990) The antioxidants of human extracellular fluids. *Arch. Biochem. Biophys.* **280**, 1–8.

Hartmann A, Agurell E, Beevers C, Brendler-Schwaab S, Burlinson B, Clay P, *et al.* (2003) Recommendations for conducting the in vivo alkaline Comet assay. *Mutagenesis* **18**, 45–51.

Heart Protection Study Collaborative Group (2002) MRC/BHF Heart Protection Study of antioxidant vitamin supplementation in 20 536 high-risk individuals: a randomised placebo-controlled trial. *Lancet* **23**.

Howard DJ, Ota RB, Briggs LA, Hampton M, Pritsos CA (1998) Oxidative stress induced by environmental tobacco smoke in the workplace is mitigated by antioxidant supplementation. *Cancer Epidemiol Biomarkers Prev.* **7**, 981–988.

Huang HY, Helzlsouer KJ, Appel LJ (2000) The effects of vitamin C and vitamin E on oxidative DNA damage: results from a randomized controlled trial. *Cancer Epidemiol. Biomarkers Prev.* **9**, 647–652.

Hughes CM, Lewis SE, McKelvey-Martin VJ, Thompson W (1998) The effects of antioxidant supplementation during Percoll preparation on human sperm DNA integrity. *Hum. Reprod.* **13**, 1240–1247.

Jackson AL, Loeb LA (2001) The contribution of endogenous sources of DNA damage to the multiple mutations in cancer. *Mutat. Res.* **477**, 7–21.

Jacob RA, Kelley DS, Pianalto FS, Swendseid ME, Henning SM, Zhang JZ, Ames BN, Fraga CG, Peters JH (1991) Immunocompetence and oxidant defence during ascorbate depletion of healthy men. *Am. J. Clin. Nutr.* **54** (suppl 6), 1302S–1309S.

Jacobson JS, Begg MD, Wang LW, Wang Q, Agarwal M, Norkus E, Singh VN, Young TL, Yang D, Santella RM (2000) Effects of a 6-month vitamin intervention on DNA damage in heavy smokers. *Cancer Epidemiol. Biomarkers Prev.* **9**, 1303–1311.

Kodama H, Yamaguchi R, Fukuda J, Kasai H, Tanaka T (1997) Increased oxidative deoxyribonucleic acid damage in the spermatozoa of infertile male patients. *Fertil. Steril.* **68**, 519–524.

Lee BM, Lee SK, Kim HS (1998) Inhibition of oxidative DNA damage, 8-OHdG, and carbonyl contents in smokers treated with antioxidants (vitamin E, vitamin C, beta-carotene and red ginseng) *Cancer Lett.* **132**, 219–227.

Levine M (1999) Criteria and recommendations for vitamin C intake. *JAMA* 281: 1415–1423.

Loft S, Jensen TK, Hjørlund NH, Giwercman A, Gyllemborg J, Olsen EE, Scheike T, Poulsen HE, Bonde JP (2003) Oxidative DNA damage in human sperm influences time to pregnancy. *Hum. Reproduction* **18**, 1265–1272.

Lunec J (1998) ESCODD: European Standards Committee on Oxidative DNA Damage. *Free Rad. Res.* **29**, 601–608.

Lykkesfeldt J (2002) Increased oxidative damage in vitamin C deficiency is accompanied by induction of ascorbic acid recycling capacity in young but not mature guinea pigs. *Free Radic. Res.* **36**, 567–574.

Lykkesfeldt J, Loft S, Nielsen JB, Poulsen HE (1997) Ascorbic acid and dehydroascorbic acid as biomarkers of oxidative stress caused by smoking. *Am. J. Clin. Nutr.* **65**, 959–963.

Lykkesfeldt J, Prieme H, Loft S, Poulsen HE (1996) Effect of smoking cessation on plasma ascorbic acid concentration. *Br. Med. J.* **313**, 91.

May JM, Mendiratta S, Hill KE, Burk RF (1997) Reduction of dehydroascorbate to ascorbate by the selenoenzyme thioredoxin reductase. *J. Biol. Chem.* **272**, 22607–22610.

Meister A (1992) Commentary: On the antioxidant effects of ascorbic acid and glutathione. *Biochem. Pharmacol.* **44**, 1905–1915.

Mirvish SS, Grandjean AC, Reimers KJ, Connelly BJ, Chen SC, Morris CR, Wang X, Haorah J, Lyden ER (1998) Effect of ascorbic acid dose taken with a meal on nitrosoproline excretion in subjects ingesting nitrate and proline. *Nutr. Cancer* **31**, 106–110.

Moller P, Loft S (2002) Oxidative DNA damage in human white blood cells in dietary antioxidant intervention studies. *Am. J. Clin. Nutr.* **76**, 303–310.

Ni ZY, Liu YQ, Shen HM, Chia SE, Ong CN (1997) Does the increase of 8-hydroxydeoxyguanosine lead to poor sperm quality? *Mutat. Res.* **381**, 77–82.

Panayiotidis M, Collins AR (1997) Ex vivo assessment of lymphocyte antioxidant status using the comet assay. *Free Rad. Res.* **27**, 533–537.

Park JB, Levine M (1996) Purification, cloning and expression of dehydroascorbic acid-reducing activity from human neutrophils: identification as glutaredoxin. *Biochem. J.* **315**, 931–938.

Perminova IN, Sinel'shchikova TA, Alekhina NI, Perminova EV, Zasukhina GD (2001) Individual sensitivity to genotoxic effects of nickel and antimutagenic activity of ascorbic acid. *Bull Exp. Biol. Med.* **131**, 367–370.

Porkkala-Sarataho E, Salonen JT, Nyyssonen K, Kaikkonen J, Salonen R, Ristonmaa U, Diczfalusy U, Brigelius-Flohe R, Loft S, Poulsen HE (2000) Long-term effects of vitamin E, vitamin C, and combined supplementation on urinary 7-hydro-8-oxo-2′-deoxyguanosine, serum cholesterol oxidation products, and oxidation resistance of lipids in nondepleted men. *Arterioscler. Thromb. Vasc. Biol.* **20**, 2087–2093.

Poulsen HE, Loft S (1998) Interpretation of oxidative DNA modification: Relation between tissue levels, excretion of urinary repair products and single cell gel electrophoresis (comet assay). In: *DNA and Free Radicals: Techniques, Mechanisms and Applications* (eds Aruoma OI and Halliwell B) OICA International, London, pp. 261–270.

Poulsen HE, Loft S, Prieme H, Vistisen K, Lykkesfeldt J, Nyyssonen K, Salonen JT (1998a) Oxidative DNA damage in vivo: relationship to age, plasma antioxidants, drug metabolism, glutathione-S-transferase activity and urinary creatinine excretion. *Free Rad. Res.* **29**, 565–571.

Poulsen HE, Prieme H, Loft S (1998b) Role of oxidative DNA damage in cancer initiation and promotion. *Eur J. Cancer Prev.* **7**, 9–16.

Poulsen HE, Prieme H, Loft S (1998c) Role of oxidative DNA damage in cancer initiation and promotion. *Eur. J. Cancer Prev.* **7**, 9–16.

Prieme H, Loft S, Nyyssonen K, Salonen JT, Poulsen HE (1997) No effect of supplementation with vitamin E, ascorbic acid, or coenzyme Q10 on oxidative DNA damage estimated by 8-oxo-7,8-dihydro-2′-deoxyguanosine excretion in smokers. *Am. J. Clin. Nutr.* **65**, 503–507.

Proteggente AR, Rehman A, Halliwell B, Rice-Evans CA (2000) Potential problems of ascorbate and iron supplementation: pro-oxidant effect in vivo? *Biochem. Biophys. Res. Commun.* **277**, 535–540.

Rehman A, Collis CS, Yang M, Kelly M, Diplock AT, Halliwell B, Rice-Evans C (1998) The effects of iron and vitamin C co-supplementation on oxidative damage to DNA in healthy volunteers. *Biochem. Biophys. Res. Commun.* **246**, 293–298.

Riis B, ESCODD (2002) Comparison of results from different laboratories in measuring 8-oxo-2′-deoxyguanosine in synthetic oligonucleotides. *Free Rad. Res.* **36**, 649–659.

Schneider M, Diemer K, Engelhart K, Zankl H, Trommer WE, Biesalski HK (2001) Protective effects of vitamins C and E on the number of micronuclei in lymphocytes in smokers and their role in ascorbate free radical formation in plasma. *Free Radic. Res.* **34**, 209–219.

Sergerie M, Ouhilal S, Bissonnette F, Brodeur J, Bleau G (2000) Lack of association between smoking and DNA fragmentation in the spermatozoa of normal men. *Hum. Reprod.* **15**, 1314–1321.

Sharma RK, Agarwal A (1996) Role of reactive oxygen species in male infertility. *Urology* **48**, 835–850.

Shen HM, Chia S-E, Ni Z-Y, New A-L, Lee B-L, Ong C-N (1997) Detection of oxidative DNA damage in human sperm and the association with cigarette smoking. *Reproduct. Toxicol.* **11**, 675–680.

Shen HM, Chia SE, Ong CN (1999) Evaluation of oxidative DNA damage in human sperm and its association with male infertility. *J. Androl.* **20**, 718–723.

Shen HM, Ong C-N (2000) Detection of oxidative DNA damage in human sperm and its association with sperm function and male infertility. *Free Radic. Biol. Med.* **28**, 529–536.

Vojdani A (2000) New evidence for antioxidant properties of vitamin C (vol 24, pg 508 2000) *Cancer Detection and Prevention* **24**, 508–523.

Welch RW, Turley E, Sweetman SF, Kennedy G, Collins AR, Dunne A, Livingstone MB, McKenna PG, McKelvey-Martin VJ, Strain JJ (1999) Dietary antioxidant supplementation and DNA damage in smokers and nonsmokers. *Nutr. Cancer* **34**, 167–172.

Wells WW, Jung CH (1997) Regeneration of Vitamin C. In: *Vitamin C in Health and Disease.* (ed. Packer FL and Fuch). Marcel Deldeer, New York, pp. 109–121.

Wells WW, Xu DP, Yang YF, Rocque PA (1990) Mammalian thioltransferase (glutaredoxin) and protein disulfide isomerase have dehydroascorbate reductase activity. *J. Biol. Chem.* **265**, 15361–15364.

White KL, Chalmers DM, Martin IG, Everett SM, Neville PM, Naylor G, Sutcliffe AE, Dixon MF, Turner PC, Schorah CJ (2002) Dietary antioxidants and DNA damage in patients on long-term acid-suppression therapy: a randomized controlled study. *Br. J. Nutr.* **88**, 265–271.

Wills ED (1969) Lipid peroxide formation in microsomes – role of non-haem iron. *Biochem. J.* **113**, 325–332.

Winkler BS (1992) Unequivocal evidence in support of the nonenzymatic redox coupling between glutathione/glutathione disulfide and ascorbic acid/dehydroascorbic acid. *Biochim. Biophys. Acta* **1117**, 287–290.

Winkler BS, Orselli SM, Rex TS (1994) The redox couple between glutathione and ascorbic acid: a chemical and physiological perspective. *Free Radic. Biol. Med.* **17**, 333–349.

Witt EH, Reznick AZ, Viguie CA, Starke-Reed P, Packer L (1992) Exercise, oxidative damage and effects of antioxidant manipulation. *J. Nutr.* **122**, 766–773.

Zavos PM, Correa JR, Antypas S, Zarmakoupis-Zavos PN, Zarmakoupis CN (1998) Effects of seminal plasma from cigarette smokers on sperm viability and longevity. *Fertil. Steril.* **69**, 425–429.

Zondervan KT, Ocke MC, Smit HA, Seidell JC (1996) Do dietary and supplementary intakes of antioxidants differ with smoking status? *Int. J. Epidemiol.* **25**, 70–79.

12

Vitamin C status declines with age

Alexander J. Michels and Tory M. Hagen

12.1 Introduction

Aging is a natural process that is marked by the inability to adapt to external stresses (Hagen *et al.*, 2000; Hall *et al.*, 2000; Kregel *et al.*, 1995; Lorens *et al.*, 1990; Seals *et al.*, 1994; Suh *et al.*, 2002). Although many of the mechanisms behind these changes are not yet known and are undoubtedly multifactorial, significant evidence suggests that oxidative stress plays a key role in the development of the aging phenotype (Beal, 2002; Beckman and Ames, 1998; Berlett and Stadtman, 1997; Hagen *et al.*, 1999; Muscari *et al.*, 1996; Sohal and Weindruch, 1996; Sohal, 2002; Sohal *et al.*, 2002; Suh *et al.*, 2001). With the growing elderly population, there is increasing interest in dietary antioxidants and antioxidant mechanisms as a means to modulate age-related increases in oxidative stress. Vitamin C is regarded as one of the most useful antioxidants to combat cellular oxidative stress (Arrigoni and De Tullio, 2002; Bendich and Langseth, 1995; Carr and Frei, 1999a, 1999b; Frei *et al.*, 1989). However vitamin C levels appear to decline in many tissues with age, suggesting that the elderly may require increased dietary intake for maintenance of normal cellular levels of this vitamin. The mechanisms behind the age-related decline of vitamin C are unknown but may be due to a variety of factors, including increased utilization and turnover, poor dietary absorption, or reduced uptake from the plasma. This chapter will review the available knowledge regarding vitamin C status in the elderly and the current theories behind the loss of vitamin C during the aging process.

12.2 Aging, antioxidants, and oxidative stress

The connection between aging and oxidant production was first proposed by Denham Harman (Harman, 1956). This so-called 'free radical theory of aging' postulates that mammalian cells constantly produce reactive oxygen and nitrogen species (ROS/RNS) as a consequence of living in an aerobic environment. Both

Vitamin C, edited by Han Asard, James M May and Nicholas Smirnoff

endogenous and exogenous sources of ROS/RNS contribute to a cellular oxidant pool, which is not necessarily detrimental and actually contributes to normal cell proliferation, cell turnover and maintenance of a stress response (Finkel and Holbrook, 2000). Thus, a critical balance exists between oxidant production and host defenses for normal cell function. When this balance is upset, such as in aging, there is an accumulation of oxidative damage and the induction of aberrant cell signaling. Over time this leads to widespread cellular degeneration.

Numerous studies support the hypothesis that aging results in a prooxidant cellular environment (Beal 2002; Beckman and Ames 1998; Berlett and Stadtman 1997; Sohal and Brunk, 1992; Sohal and Orr, 1992; Sohal, 2002; Sohal *et al.*, 2002). While this altered redox state is attributed to increased oxidant flux, there are also losses in antioxidant defense capacity (Ames *et al.*, 1993; Shigenaga *et al.*, 1994; Sohal and Orr, 1992; Sohal *et al.*, 2002). Aerobic species protect against oxidative damage with an elaborate and overlapping array of antioxidant defenses. Cellular enzymatic and nonenzymatic antioxidants maintain a normal cellular redox balance but decline with age leading to heightened ROS/RNS appearance and increased oxidative damage. However, age-related loss of antioxidant capacity is not systemic; rather, certain antioxidant systems appear to be affected more markedly than others. For example, some antioxidant enzymes decline while other enzyme levels are unchanged or increase during aging (Carrillo *et al.*, 1992; Ji *et al.*, 1990; Kasapoglu and Ozben, 2001; Pansarasa *et al.*, 1999; Sahoo and Chainy, 1997; Tian *et al.*, 1998; Vertechy *et al.*, 1989). Despite the equivocal and often inconsistent reports regarding antioxidant enzyme activities, there is little doubt that an age-related loss of certain low molecular weight antioxidants occurs. Steady-state glutathione levels, for example, decline 25–50% in many tissues with age (Arockia Rani and Panneerselvam, 2001; Azhar *et al.*, 1995; Droge, 2002; Erden-Inal *et al.*, 2002; Hagen *et al.*, 1999, 2000; Head *et al.*, 2002; Rikans and Moore, 1988). Additionally, the ratio of oxidized to reduced glutathione can increase up to 300% (Azhar *et al.*, 1995; Droge, 2002; Erden-Inal *et al.*, 2002), which supports the contention that aged tissues exhibit a more prooxidant environment.

Vitamin C also declines markedly in many tissues with age (see below), which may have profound implications for cell function due to the diverse and vital roles that this vitamin plays in the cell. Vitamin C is a potent antioxidant, able to react with a wide variety of biological oxidants (Carr and Frei, 1999b), and thus inhibits the generation of lipid and protein peroxidation (Carr and Frei, 1999a; Hunt *et al.*, 1992) and consequent DNA damage (Fraga *et al.*, 1991). Diets high in vitamin C decrease oxidative damage (McCall and Frei, 1999) and reduce the frequency of age-related diseases and other pathological conditions (Barja, 1996). Additionally, vitamin C may be interconnected with other antioxidants, preserving or recycling glutathione or vitamin E in a so called 'antioxidant network' (Packer, 1997). Thus, lower vitamin C levels could lead to the loss of other cellular antioxidants, precipitating age-related increases in oxidative stress.

In addition to its role as a potent cellular antioxidant, vitamin C participates in the synthesis of collagen, carnitine and neurotransmitters (Arrigoni and De Tullio, 2002; Gershoff, 1993; Levine, 1986). Vitamin C also contributes to nitric oxide production in vascular endothelial cells, thus contributing to vasorelaxation (Smith *et al.*, 2002). Therefore, significant alterations in cellular vitamin C status may have serious consequences for human health (Bendich and Langseth, 1995), especially in the elderly. Indeed, the role of vitamin C in aging may be of much greater importance and complexity than current research suggests.

12.3 Evidence for age-related decline of vitamin C

The loss of vitamin C appears to be a general hallmark of aging that occurs across species, however, as reviewed below, there is no consensus yet as to the extent and precise nature of this altered vitamin C status or its consequences for cellular function.

Most of the available data related to vitamin C and aging come from animals. However, application of these results to humans is problematic because most animal models, except primates and guinea pigs, synthesize vitamin C (see chapter by Smirnoff *et al.*), which may obscure the primary cause(s) of vitamin C loss with age. Despite this problem, it is remarkable how age-associated declines in vitamin C status in animals closely parallels that measured in humans.

12.3.1 Animal studies

Vitamin C levels tend to decline with age, but the extent of this loss in general, as well as specific age-related changes in particular organs, varies widely depending on the study (*Table 12.1*). This variability may be due to species or strain differences or to the small population size. Despite such variability, hepatic vitamin C becomes consistently lower in both rats and mice with age, regardless of the particular strain studied (Azhar *et al.*, 1995; De and Darad, 1991; Leibovitz and Siegel 1980; Lykkesfeldt *et al.*, 1998; Michels *et al.*, 2003; Rikans *et al.*, 1996). For example, hepatic vitamin C levels in Fischer 344 (F344) rats declined from 1.76 nmol·mg^{-1} protein to 0.53 nmol·mg^{-1} protein in cells from young and old animals, respectively (Michels *et al.*, 2003). In this model, sex also may play a role in age-related losses of vitamin C, because only age-related declines in males (Rikans *et al.*, 1996) were evident.

In rodents, other organs also show a consistent loss of vitamin C with age. The adrenal gland and eye lens of rats normally have very high concentrations of vitamin C (Jacob, 1996), but both exhibit significant age-related declines in this vitamin (Azhar *et al.*, 1995; Rikans and Moore, 1988). In the eye lens, Rikans and Moore observed that vitamin C levels are depressed nearly 62% in older animals (Rikans and Moore, 1988). Additional reports show a loss of vitamin C in the thymus and lung of both aging rats and mice (Leibovitz and Siegel, 1980; Patnaik and Kanungo, 1966; Rikans and Moore, 1988). Vitamin C in the rat brain is normally at high levels (Spector and Lorenzo 1973; Spector 1977), but declines up to 25% in both the tissue and extracellular fluid with age (Arockia Rani and Panneerselvam, 2001; Kanungo and Patnaik, 1964; Kume-Kick and Rice, 1998; O'Donnell and Lynch, 1998; Svensson *et al.*, 1993). This decline is moderate compared to other organs. Age-related declines of vitamin C have been demonstrated in rat heart, sometimes by as much as 50% (Suh *et al.*, 2001). In contrast, there is sparse and inconsistent data on age-related changes in kidney vitamin C

Table 12.1: Vitamin C levels in animals.

Samples	Reference	Population	Effect of age on vitamin C levels
Liver	Lykkesfeldt *et al.*, 1998	Male Fischer 344 rats	Significant decline
	De and Darad, 1991	Male Wistar rats	No change
	Azhar *et al.*, 1995	Male Sprague–Dawley rats	Significant decline
	Rikans *et al.*, 1988	Male Fischer 344 rats	Significant decline
	Siegel and Leibovitz, 1980	Male and female BALB/c, DBA/2 and NZB mice	Significant decline
	Berger at al., 1989	Male guinea pigs	No change
	Rikans *et al.*, 1996	Male and female Fischer 344 rats	Male: significant decline. Female: no change
	Kanungo and Patnik, 1964	Male Wistar rats	Significant decline
Plasma	Azhar *et al.*, 1995	Male Sprague–Dawley rats	Significant decline
	Rikans *et al.*, 1988	Male Fischer 344 rats	Nonsignificant decline
	Scrofano *et al.*, 1998	Male and female Emory mice	Significant decline
	Berger at al., 1989	Male guinea pigs	Significant decline
	Preston *et al.*, 2001	Male and female rhesus monkeys	Significant decline
Serum	De and Darad, 1991	Male Wistar rats	Significant decline
	Wartanowicz and Ziemlanski, 1984	Male Wistar rats	Significant decline
	Siegel and Leibovitz, 1980	Male and female BALB/c, DBA/2 and NZB mice	Significant decline
Macrophages	de la Fuente *et al.*, 1995	Male guinea pigs, male BALB/c mice	Significant decline
Adrenal	Azhar *et al.*, 1995	Male Sprague–Dawley rats	Significant decline
	Hughes and Jones, 1971	Male and female guinea pigs	Significant decline
Lens	Rikans *et al.*, 1988	Male Fischer 344 rats	Significant decline
	Berger at al., 1989	Male guinea pigs	Significant decline
	Hughes and Jones, 1971	Male and female guinea pigs	Significant decline
Lung	Rikans *et al.*, 1988	Male Fischer 344 rats	Significant decline
	Siegel and Leibovitz, 1980	Male and female BALB/c, DBA/2 and NZB mice	Significant decline
	Patnik and Kanungo 1966	Male Wistar Rats	Significant decline

Thymus	Siegel and Leibovitz, 1980	Male and female BALB/c, DBA/2 and NZB mice	Significant decline
	Patnik and Kanungo, 1966	Female Wistar rats	Significant decline
Brain	Rikans *et al.*, 1988	Male Fischer 344 rats	No change
	Hughes and Jones, 1971	Male and female guinea pigs	Significant increase
	Kume-Kick and Rice, 1998	Male and female Long–Evans rats	Significant decline
	Sahoo and Chainy, 1997	Male Wistar rats	Significant increase
	Arockia Rani and Panneerselvam, 2001	Male Wistar rats	Significant decline
	Zhang *et al.*, 1993	Male Mongolian gerbils	No change
	O'Donnell and Lynch, 1998	Male Wistar rats	Significant decline
	Kanungo and Patnik, 1964	Male Wistar rats	Significant decline
	Svensson *et al.*, 1993	Male Sprague–Dawley rats	Significant decline
Heart	Rikans *et al.*, 1988	Male Fischer 344 rats	No change
	Suh *et al.*, 2001	Male Fischer 344 rats	Significant decline
	Patnik and Kanungo, 1966	Male and Female Wistar rats	Significant decline
Kidney	Rikans *et al.*, 1988	Male Fischer 344 male rats	Nonsignificant decline
	Berger *et al.*, 1989	Male Guinea pigs	No change
	Kanungo and Patnik, 1964	Male Wistar rats	Significant decline
Spleen	Siegel and Leibovitz, 1980	Male and female BALB/c, DBA/2 and NZB mice	Significant increase
	Hughes and Jones, 1971	Male and female guinea pigs	Significant decline

levels (Kanungo and Patnaik, 1964; Rikans and Moore, 1988). For most organs, there is abundant evidence indicating a clear loss of tissue vitamin C regardless of the animal model examined (*Table 12.1*).

Consistent with changes in the vitamin C content of tissues, there is also a notable change in plasma vitamin C levels with age (Azhar *et al.*, 1995; De and Darad, 1991; Leibovitz and Siegel, 1980; Preston *et al.*, 2001; Scrofano *et al.*, 1998; Wartanowicz and Ziemlanski, 1984). Some studies have reported that plasma vitamin C declines in both old rats and mice by as much as 50% (De and Darad, 1991; Leibovitz and Siegel, 1980). As anticipated by altered plasma status, macrophage vitamin C levels are consistently lower in both aging mice and guinea pigs (De La Fuente *et al.*, 1995), mirroring some of the data seen in human lymphocytes (see below).

Guinea pigs and rhesus monkeys, two species that require dietary vitamin C, also show a trend toward impaired vitamin C status with age. In the guinea pig, no age-related change was observed in hepatic or renal vitamin C (Berger *et al.*, 1989), but other organs, such as the lens, spleen, and adrenal glands as well as the blood plasma, are markedly affected (Hughes *et al.*, 1980; Preston *et al.*, 2001). For rhesus monkeys supplemented with vitamin C, plasma vitamin C in young animals was 0.70 $mg \cdot dl^{-1}$ compared to 0.56 $mg \cdot dl^{-1}$ in old animals. Moreover, Berger *et al.*, observed that plasma and tissue vitamin C in old guinea pigs reached saturation at 2 mg of vitamin C per gram of diet, which was twice the amount needed for saturation in young animals (Berger *et al.*, 1989). These results underscore the increasing need for adequate vitamin C intake in old animals.

In conclusion, a preponderance of the data suggests diminished vitamin C levels in many tissues and plasma in experimental animal models during the aging process.

Contribution of vitamin C synthesis

In mammals, including the mouse and rat, vitamin C synthesis occurs in the liver (Nandi *et al.*, 1997). Primates and guinea pigs cannot synthesize vitamin C because of a mutation in the enzyme catalyzing the terminal step in vitamin C synthesis (Nishikimi and Yagi, 1991; see Chapter 1 by Smirnoff *et al.*). Except for the liver and perhaps kidney in animals, every tissue in both humans and animals is completely dependent on vitamin C uptake from the plasma. Thus, the study of vitamin C uptake in most organs, regardless of the animal model, is directly applicable to humans.

Several lines of evidence indicate that the age-related decline in vitamin C status is not due to the endogenous synthetic capacity of the liver. Lykkesfeldt *et al.* (1998) examined the synthetic capacity of hepatocytes isolated from young and old F344 rats and observed no age-related changes in vitamin C synthesis. Vitamin C levels in the liver and plasma of rats, which would be directly impacted by changes in vitamin C synthetic capacity, are also the most adversely impacted with age (Azhar *et al.*, 1995; De and Darad, 1991; Leibovitz and Siegel, 1980). Since there are no apparent age-related changes in vitamin C synthesis, other mechanism(s) must contribute to the loss of vitamin C in aging animals. Lastly, animals that do not have endogenous hepatic vitamin C synthesis, such as the guinea pig or monkey, also show declines in plasma and tissue vitamin C levels despite constant supplementation with this vitamin in the diet (Hughes and Jones, 1971; Preston *et al.*, 2001). These results imply that the mechanism(s) behind the age-associated loss in vitamin C are not due to synthesis but may be related to dietary or plasma uptake of the vitamin, regardless of its source.

12.3.2 Human studies

'Baseline' vitamin C status in human populations relies on analysis of plasma or blood cells from volunteers. In depth analysis of age-related changes in particular organs and tissues in healthy subjects are usually impractical. Moreover, environmental factors that may affect vitamin C levels are difficult to control. For instance, poor diet (Bidlack and Smith, 1988; Schmuck *et al.*, 1996; Van Der Wielen *et al.*, 1995), socio-economic status (Bianchetti *et al.*, 1990; Stitt *et al.*, 1995), and physical disabilities (Bidlack and Smith, 1988; Schmuck *et al.*, 1996) all contribute to declines in vitamin C in the elderly. Studies on vitamin C status in individuals in nursing homes found that some subjects exhibit vitamin C levels bordering on scurvy (Birlouez-Aragon *et al.*, 2001; Harrill and Cervone, 1977; Loh and Wilson, 1971; Newton *et al.*, 1985). However, supplementation of these individuals with vitamin C increased their plasma concentrations markedly, suggesting no biological basis for loss of vitamin C with age (Birlouez-Aragon *et al.*, 1995; Monget *et al.*, 1996; Newton *et al.*, 1985). Furthermore, subjects exhibiting clinical pathologies, such as diabetes mellitus (Birlouez-Aragon *et al.*, 2001; Cunningham 1998a) and Alzheimer's disease (Martin *et al.*, 2002), tend to have lower plasma vitamin C. These effects can be further exacerbated by smoking, which also adversely affects plasma vitamin C concentration in and of itself (Birlouez-Aragon *et al.*, 2001; Heseker and Schneider, 1994). Although vitamin C status declines in the elderly, it is difficult to ascertain the extent that this loss is due to aging *per se* versus life-style and disease state.

Plasma vitamin C levels in older individuals vary widely from study to study, most likely due to the aforementioned lifestyle variables as well as to differences in analytical methods used to assay vitamin C (Brubacher *et al.*, 2000). Unfortunately, many of these studies do not directly relate vitamin C levels in the elderly to baseline values in younger individuals using identical analytical procedures. Additionally, some studies use depletion and/or repletion of vitamin C and report no apparent age-related changes after administration of a high dose of vitamin C (Blanchard *et al.*, 1989, 1990; Blanchard 1991; Newton *et al.*, 1985; Vanderjagt *et al.*, 1987). However, the pharmacokinetics of vitamin C absorption in the elderly are unknown and comparisons using only a single dose are difficult (see below). Thus, only studies where baseline comparisons between young and old were made will be reviewed.

Plasma vitamin C

Plasma vitamin C status compares well with tissue vitamin C content, based on laboratory animal studies and the correlation between plasma and lymphocyte vitamin C concentrations (Loh and Wilson, 1971; Lykkesfeldt *et al.*, 1997; McClean *et al.*, 1976). The average saturated plasma concentration of vitamin C for young adults is approximately 60–70 μM (Levine *et al.*, 1996, 2001). Consistent with the loss of plasma vitamin C in old animals, most reports show a similar loss of vitamin C in the elderly (*Table 12.2*). For example, Paolisso *et al.*, observed a significant 20% decline in vitamin C levels in adults aged between 75 and 99 years compared to those under the age of 50 years (Paolisso *et al.*, 1998). Similar to published animal studies, most reports in humans have compared young adults (under the age of 60 years) and elderly (above 60 years) as separate groups.

Increasingly, a number of reports also examine the effects of 'advanced age' on plasma vitamin C, comparing elderly under the age of 70 years to older individuals (Birlouez-Aragon *et al.*, 2001; Burr *et al.*, 1974; Jacob *et al.*, 1988; Mecocci *et al.*, 2000; Paolisso *et al.*, 1998). As shown previously, most of these studies show a further vitamin C loss for every decade of life up to 80 years. In addition, there are some reports indicating even greater losses of vitamin C beyond the age of 80 years compared to subjects between the ages of 60 and 80 (Birlouez-Aragon *et al.*, 2001; Burr *et al.*, 1974; Mecocci *et al.*, 2000). In contrast, other reports show either an increase (Paolisso *et al.*, 1998) or no further declines (Jacob *et al.*, 1988) in plasma

Table 12.2: Vitamin C levels in human plasma and serum.

Samples	Reference	Study population	Effects of age on vitamin C levels
Plasma	Birlouez-Aragon *et al.*, 2001	1133 men, 1451 women (ages over 60 years)	Significant decline in men. No change in women
	Meccoci *et al.*, 2000	35 men, 62 women (ages under 61 to over 100 years)	Significant decline
	Paolisso *et al.*, 1998	82 men and women (ages under 50 to over 100 years)	Significant decline
	Lykkesfeldt *et al.*, 1997	206 men and women (ages 20–90 years)	Nonsignificant decline
	Menzzetti *et al.*, 1996	191 men and women (under 70 to over 80 years)	Significant decline
	Heseker and Schneider 1994	2006 men and women (ages 18–88 years)	Significant decline
	Blanchard *et al.*, 1989	16 women (ages 20–29 and 65–71 years)	No change
	Jacob *et al.*, 1988	235 men, 442 women (ages over 60 years)	No change
	Sasaki *et al.*, 1983	95 men, 105 women (ages 12–96 years)	Significant decline
	Schorah *et al.*, 1979	88 men and women (ages 19–97 years)	Significant decline
	McClean *et al.*, 1976	178 men (ages 17–68 years)	Significant decline
	Burr *et al.*, 1974	295 men, 529 women (ages over 65 years)	Significant decline
Serum	Kasapoglu and Öben, 2001	50 men, 50 women (ages 20–69 years)	No change

vitamin C levels of the 'oldest old'. In some cases, a significant loss was sex-related in advanced age. For example, while vitamin C levels were 8% lower in men over 80 years compared to those between 60 and 64 years, no significant loss was detected in similar groups of women (Birlouez-Aragon *et al.*, 2001).

Sex differences in plasma vitamin C concentrations have been extensively reported, revealing a higher vitamin C level in women (Birlouez-Aragon *et al.*, 2001; Burr *et al.*, 1974; Heseker and Schneider, 1994; Jacob *et al.*, 1988; Kasapoglu and Ozben 2001; Lykkesfeldt *et al.*, 1997). In fact, it does not appear that women suffer any significant age-related declines in plasma vitamin C levels (Birlouez-Aragon *et al.*, 2001; Blanchard *et al.*, 1989; Jacob *et al.*, 1988; Kasapoglu and Ozben, 2001; Lykkesfeldt *et al.*, 1997). Even in the few studies that do indicate a slight change for elderly women, these losses are less severe than in men (Heseker and Schneider, 1994). In contrast, plasma vitamin C concentrations are consistently and significantly lower in elderly men compared to young men (Heseker and Schneider, 1994; McClean *et al.*, 1976) and further declines are noted with advanced age (Birlouez-Aragon *et al.*, 2001; Burr *et al.*, 1974). This suggests that sex is a significant factor in vitamin C status in the elderly. However, when both young men (Levine *et al.*, 1996) and women (Levine *et al.*, 2001) were depleted of vitamin C and then replenished, there was no difference in plasma vitamin C levels between men and women. Additionally, when differences in renal excretion of vitamin C were examined, there was no difference between young adults or the elderly of either sex (Oreopoulos *et al.*, 1993). These observations suggest that either a sex-related difference in diet plays a very important role in aging or that size or body composition differences between men and women are also important factors in determining vitamin C status.

12.3.3 Cellular vitamin C

Most studies examining cellular vitamin C use white blood cells for a representative measure of vitamin C uptake into tissues. Although leukocyte vitamin C content is often used to determine nutritional status in younger adults (Levine *et al.*, 1996, 2001), very few studies report the levels in healthy elderly subjects. Despite this limitation, a trend for lower leukocyte concentrations of vitamin C with age is evident (*Table 12.3*). Studies using a mixed population of men and women report significant age-related declines in leukocyte vitamin C levels (Attwood *et al.*, 1978; Burr *et al.*, 1974; Schorah *et al.*, 1979), and this is also seen when only lymphocytes were analyzed (Lenton *et al.*, 2000). As an example, Attwood *et al.* showed that vitamin C declines by 50% in lymphocytes from subjects aged over 85 years when compared to those under the age of 60 years (Attwood *et al.*, 1978). Like plasma studies, sex differences in leukocyte vitamin C are evident, with younger women having significantly higher vitamin C levels than men of the same age (Attwood *et al.*, 1978; Burr *et al.*, 1974; Milne *et al.*, 1971). One study using exclusively women found no differences in leukocyte vitamin C with age (Blanchard *et al.*, 1989). Additionally, it is equally unclear whether leukocyte vitamin C status declines with 'advanced' age. While one study reported no further changes above the age of 65 years (Attwood *et al.*, 1978), Burr *et al.* observed a significant 23% loss of vitamin C in men and women above the age of 85 years when compared to those between the ages of 65 and 69 years (Burr *et al.*, 1974). Another group also reported leukocyte vitamin C declines with increasing age, but this only reached significance in women (Milne *et al.*, 1971).

Thus, while there appears to be age-related decrements of vitamin C in human plasma and leukocytes similar to what has been observed in aging animals, the data on vitamin C status after the age of 65 years are conflicting. Furthermore, it appears that vitamin C levels in aging men decline more precipitously than in women.

12.3.4 Tissue vitamin C

Due to the difficulty in obtaining tissue samples from living subjects, there is virtually no information regarding alterations in tissue vitamin C in the elderly. Samples can be obtained readily at autopsy, although they may not be representative

Table 12.3: Vitamin C levels in human leukocytes.

Samples	Reference	Study population	Effects of age on vitamin C levels
Leukocytes	Lenton *et al.*, 2000	240 men and women (ages under 59 and over 60 years)	Significant decline
	Blanchard *et al.*, 1989	16 women (ages 20–29 and 65–71 years)	No change
	Schorah *et al.*, 1979	88 men and women (ages 19–97 years)	Significant decline
	Attwood *et al.*, 1978	434 men and women (ages 16–53 and over 65 years)	Significant decline
	McClean *et al.*, 1976	178 men (ages 17–68 years)	Nonsignificant decline
	Burr *et al.*, 1974	295 men, 529 women (ages over 65 years)	Significant decline
	Milne *et al.*, 1971	451 men and women (ages 62–94 years)	No decline in men. Significant decline in women

of living tissue. In one study on cadavers, vitamin C levels in tissues taken at the time of autopsy showed significant declines in the heart, brain and skeletal muscle in old compared to young adults (Schaus, 1957). Vitamin C levels in the cerebral cortex, for example, were 77% lower in individuals older than 80 years than in those under 50 years (Schaus, 1957). Regardless of potential artifacts introduced by delays in sampling and by differences in sample handling, this single study suggests that an age-related loss in tissue vitamin C, similar to that observed in animal studies, occurs in humans.

12.3.5 Vitamin C pharmacokinetics in the elderly

Age-related declines in plasma and tissue vitamin C levels implicate diminished vitamin C bioavailability or inadequate dietary intake. Declining tissue levels may reflect changes in vitamin C intake and/or absorptive capacity with age. Since healthy elderly populations consume more dietary vitamin C than the general population (Garry *et al.*, 1982a, 1982b; Vanderjagt *et al.*, 1987), this would argue that the correlation between intake and plasma vitamin C no longer holds with respect to the elderly. Pharmacokinetic studies of vitamin C absorption in young men and women (Levine *et al.*, 1996, 2001) have influenced the recommended daily allowance (RDA) for vitamin C. To date, no similar studies with geriatric subjects have been published. Thus, optimal doses of vitamin C for individuals over the age of 65 years to maintain or maximize plasma vitamin C levels are not known.

Absorptive capacity of the gastrointestinal tract for a number of key macromolecules and micronutrients, possibly including vitamin C, declines with age. Supplementation studies have revealed that an amount of vitamin C many times greater than the RDA is necessary to reverse lower plasma levels in the elderly

(Blanchard *et al.*, 1989, 1990; Blanchard 1991; Newton *et al.*, 1985; Vanderjagt *et al.*, 1987). A recent meta-analysis of 30 publications on vitamin C intake during aging found that the relationship between intake and plasma vitamin C concentrations in people aged 60–96 years was remarkably different from that in younger adults (Brubacher *et al.*, 2000). Subjects over 65 years of age required twice the amount of vitamin C in their diet to achieve the same plasma concentration seen in young adults. Also, at any given dose of dietary vitamin C, corresponding plasma levels in the elderly were consistently lower, suggesting that the lower age-dependent plasma vitamin C levels may be due to altered absorption from the diet. Thus, higher intake of vitamin C in the elderly may normalize plasma levels, but the optimal intake remains unclear.

12.3.6 Consequences of vitamin C loss

Although vitamin C levels decline with age, the metabolic consequence of this loss on the aging individual is not clear. Certainly, the lower vitamin C levels would impair antioxidant defenses against both endogenously and exogenously produced oxidants. Additionally, lower vitamin C may adversely affect the relationship between this vitamin and other low molecular weight antioxidants, which could impair this network of antioxidant protection. Thus, insufficient vitamin C likely contributes to the formation of a prooxidant environment in cells with age.

The role vitamin C plays in human health stretches far beyond its antioxidant capacity. Vitamin C functions as an electron donor and necessary cofactor for hepatic carnitine production (Rebouche, 1991), an amino acid derivative used in the transport of fatty acids into the mitochondria for ATP generation. Carnitine levels decline significantly with age, reflecting changes in vitamin C status (Hagen *et al.*, 1998; Hansford and Castro, 1982; Paradies *et al.*, 1992). Thus, lower vitamin C levels, especially in the liver, may markedly affect mitochondrial β-oxidation. This will also be true for organs (e.g., heart, kidney cortex) that use fatty acids as a primary fuel source.

Collagen production also requires vitamin C for both regulation of the collagen genes and post-translational protein maturation (Barnes, 1975; Chung *et al.*, 1997; Murad *et al.*, 1981). Again, lower vitamin C may be partly responsible for the loss of age-related collagen production (Chung *et al.*, 1997). It is also known that endothelial-derived nitric oxide production declines in aging rodents (Marin and Rodriguez-Martinez, 1999). Vitamin C has recently been found to enhance and participate in nitric oxide production in vascular endothelial cells and thus contributes to vascular tone (Smith *et al.*, 2002). Additionally, vitamin C has roles in prostaglandin synthesis (Horrobin 1996), immune surveillance (Jariwalla and Harakeh, 1996), gene expression (Arrigoni and De Tullio, 2002; Hitomi and Tsukagoshi, 1996) and cell proliferation (Brigelius-Flohe and Flohe, 1996), all of which appear to be negatively affected in the elderly.

Perhaps the strongest evidence suggesting an adverse consequence of vitamin C deficiency is provided by the striking correlation between high vitamin C intake and disease prevention. Indeed, high vitamin C consumption is directly associated with a decreased overall mortality (Block, 1992b; Khaw *et al.*, 2001; Sahyoun *et al.*, 1996; Wilson, 1989). Age-associated diseases, including cardiovascular disease and cancer, also exhibit an inverse correlation with increased vitamin C intake (Bendich and Langseth, 1995; Block, 1992a; Carr and Frei, 1999a, 1999b; Enstrom *et al.*, 1992; Howe *et al.*, 1990) although some of this evidence is controversial (see chapter by Stocker *et al.*). Vitamin C consumption is positively correlated with the prevention of macular degeneration and cataracts (Robertson *et al.*, 1991; Valero *et al.*, 2002). Evidence also suggests that increased vitamin C intake helps prevent complications due to diabetes (Cunningham, 1998b) and neurodegenerative disease (Martin *et al.*, 2002).

We are only beginning to understand the manifold ways that vitamin C or aging affects cellular function. However, the consequences of low vitamin C levels or aging on human health have some unmistakable similarities. Establishing these correlations will have a profound impact on gerontology and the maintenance of heath.

12.4 Age-dependent changes in vitamin C transport

To be effective, dietary antioxidants, including vitamin C, must be readily absorbed by the gastrointestinal tract and transported to the plasma, cells and subcellular compartments. Loss of vitamin C absorptive capacity, as already suggested, may be due to age-dependent changes in cellular transporters responsible for the uptake of this vitamin. We will now discuss the available evidence for such age-associated changes in vitamin C transport and augment this discussion with new evidence from our laboratory indicating that one particular vitamin C transport system, the so-called SVCT, is adversely affected with age.

12.4.1 Vitamin C transport

There are two distinct transport mechanisms (*Figure 12.1*) responsible for vitamin C uptake in cells (Liang *et al.*, 2001; see also Chapter 6 by Wilson). Dehydroascorbic acid (DHA), the oxidized form of vitamin C, is taken up by glucose transport (GLUT) proteins (Rumsey *et al.*, 1997, 2000; Vera *et al.*, 2001). DHA is rapidly reduced intracellularly by reduced glutathione (GSH)- and NADPH-dependent mechanisms to ascorbic acid (ASC). A sodium-dependent vitamin C transport (SVCT) protein catalyzes the direct uptake of ASC (Malo and Wilson, 2000; Tsukaguchi *et al.*, 1999). Although just recently discovered, the SVCT system is widely distributed (Berger and Hediger, 2000; Fujita *et al.*, 2001; Korcok *et al.*, 2000; Rajan *et al.*, 1999; Tsukaguchi *et al.*, 1999; Wang *et al.*, 1999, 2000). It is not yet clear which transport system is responsible for the majority of vitamin C uptake *in vivo*.

12.4.2 GLUT-dependent dehydroascorbic acid uptake

DHA is structurally similar to glucose and is recognized by specific glucose transport proteins (Vera *et al.*, 2001). DHA uptake is by facilitated transport. Specifically, GLUT1 and GLUT3, when expressed in *Xenopus* oocytes, transport DHA with an affinity similar to that of glucose (Rumsey *et al.*, 1997). GLUT4 also catalyzes DHA uptake, but with a lower capacity than GLUT1 and 3 (Rumsey *et al.*,

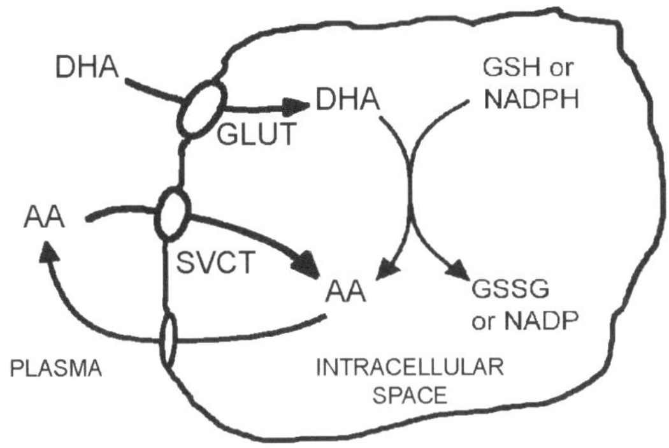

Figure 12.1: Mechanisms of vitamin C transport.
Vitamin C accumulation occurs via dehydroascorbic acid (DHA) uptake through glucose transport (GLUT) or direct ascorbic acid (ASC) uptake through sodium-dependent vitamin C transporter (SVCT). It is controversial as to which transport mechanism is more physiologically relevant for overall ASC uptake.

2000). No other extracellular glucose transporters have DHA transport activity, including the predominant sodium-dependent glucose transporter, SGLT1 (Rumsey *et al.*, 1997). The regulation and functional expression of the GLUT system is under the control of the insulin-signaling pathway (Klip and Douen, 1989; Taha and Klip, 1999), and DHA uptake can be stimulated by extracellular insulin (Rumsey *et al.*, 2000). GLUT4 is the major insulin-responsive glucose transporter, and levels of this protein increase on the cell surface of insulin-responsive tissues after insulin exposure (James *et al.*, 1989). GLUT1 expression is not directly under the control of insulin, but its transcription is affected by insulin exposure (Taha *et al.*, 1999).

The physiological relevance of DHA uptake has been questioned for a number of reasons. DHA is a very minor component in plasma (≤ 0.1 μM) and tissue (≤ 10 μM), while vitamin C levels are approximately 60–100 μM in plasma (Schorah *et al.*, 1996). Cellular vitamin C levels reach approximately 1–10 mM, depending on tissue type (Jacob, 1996). DHA is very unstable in aqueous systems and has a half-life of only 7 min. Although auto-oxidation of vitamin C is possible *in vitro,* vitamin C primarily acts as an antioxidant physiologically and thus is required to be in a reduced form. Additionally, circulating glucose levels are over 1000-fold higher than those of DHA, suggesting that significant competitive inhibition of DHA uptake would occur (Rumsey *et al.*, 1997). However, periods of heightened oxidant production may produce, at least in cellular microenvironments, appreciable amounts of extracellular DHA. Thus, the relevance of DHA uptake, particularly related to altered age-related vitamin C levels, is still unknown.

12.4.3 SVCT-mediated ascorbate uptake

Reduced vitamin C uptake occurs via a sodium-dependent transport process (Liang *et al.*, 2001; Tsukaguchi *et al.*, 1999; see also Chapter 6 by Wilson). Two isoforms of SVCT with distinct kinetic properties (Wang *et al.*, 2000) are found in both rat (Tsukaguchi *et al.*, 1999) and human tissues (Daruwala *et al.*, 1999; Wang *et al.*, 1999). SVCT1 is a low-affinity carrier found in the kidney, small intestine, colon, prostate and ovary (Tsukaguchi *et al.*, 1999; Wang *et al.*, 1999, 2000). SVCT2 is found in epithelial tissue, heart, placenta, brain, eye, liver, and kidney and is responsible for low-capacity, high-affinity ASC transport (Rajan *et al.*, 1999; Tsukaguchi *et al.*, 1999; Wang *et al.*, 1999). Nonfunctional splice variants of SVCT1 are detected in colon cell lines, and putative SVCT2 mRNA splice variants exist in some tissues (Wang *et al.*, 1999). However, the function or expression of these variants is not known. Since specific antibodies to the two SVCT protein isoforms have only recently become available, current studies are limited to monitoring SVCT transcription or transient transfection of cell types. Whether SVCT RNA levels correspond to actual protein content remains to be seen.

SVCT proteins specifically require sodium, and transport will not occur if sodium is replaced with any other cation (Tsukaguchi *et al.*, 1999). The process is electrogenic, as two sodium ions are required for every molecule of ascorbate. Thus, SVCT activity is driven by sodium gradients and membrane potential (Tsukaguchi *et al.*, 1999). Recent studies show that the expression of SVCT2 is required for prenatal development, as knockout mice do not live past birth (Sotiriou *et al.*, 2002). However, little is known about the functional or regulatory signals that control SVCT expression, activity, or potential age-related alterations in these parameters.

12.4.4 Age-related changes in the glucose transport system

The various isoforms of glucose transport proteins and the tissue-specific expression make the determination of age-associated changes in glucose transport difficult to determine. Although GLUTs are able to transport vitamin C, no studies have correlated age-related changes in glucose transport and vitamin C levels in animal models. Since there is limited data on glucose transport activity with age, it is necessary to extrapolate transport activity from the published data on GLUT expression.

In the brain, some groups report age-related declines in GLUT1 levels (Gschanes *et al.*, 2000; Mooradian *et al.*, 1997) that are ascribed to higher turnover of the protein (Mooradian and Shah, 1997). However, this loss of GLUT1 has not always been detected (Oka *et al.*, 1992). GLUT3 levels decline with age in specific areas of the brain (Fattoretti *et al.*, 2001), and GLUT4 levels fall appreciably in adipose tissue, heart, and skeletal muscle with age (Armoni *et al.*, 1995; Cartee, 1993; Gulve *et al.*, 1993; Houmard *et al.*, 1995; Larkin *et al.*, 2001; Mooradian *et al.*, 1999; Oka *et al.*, 1992). However, GLUT4 loss in these tissues may be due to increased insulin resistance generated by an *ad libitum* diet (Hardy *et al.*, 2002) and the lack of exercise (Cartee, 1994; Youngren and Barnard, 1995). On the other hand, myocardial GLUT4 levels decline in older animals even when exercised, which suggests the involvement of a true aging component (Hall *et al.*, 1994). No changes in GLUT1 levels are evident in heart, skeletal muscle or adipose tissue with respect to aging (Mooradian *et al.*, 1999; Wang *et al.*, 1997). In the intestine, no age-related changes are detected in any of the transport proteins, including GLUT2, GLUT5 and

SGLT1 (Casirola *et al.*, 1997), which suggests that these transporters may not be responsible for the lower absorption of vitamin C in the elderly. No studies have related GLUT protein expression in the liver or kidney to aging.

To determine if there are any age-related changes in DHA uptake, we incubated freshly isolated hepatocytes from young and old Fischer 344 rats with 100 μM DHA. Cells from both young and old animals rapidly accumulated intracellular ASC as a result of DHA uptake (*Figure 12.2*). Cells from both young and old rats showed a similar increase in intracellular ASC levels for approximately 9 min following DHA addition, when a new steady-state level of ASC was reached. Additionally, there were no apparent changes in the capacity for DHA uptake in this model system. This suggests that there are no age-related changes in the ability of the liver to take up DHA from the extracellular media, which confirms other studies reporting no age-related changes in GLUT levels.

Thus, GLUT transport in general does not change in the aging animal and by extrapolation, cannot account for the declines in vitamin C absorptive capacity. The lack of data for most organs makes comparisons difficult, although it appears that small changes in GLUT transport activity do occur in the brain and heart. On the other hand, age-related differences in GLUT transporter expression do not necessarily result in altered GLUT activity (Armoni *et al.*, 1995). Without specific data on the contribution of GLUT-dependent vitamin C transport in these organs, the relevance to vitamin C status is unknown.

12.4.5 Changes in SVCT transport

Sodium-dependent vitamin C transport was observed in a number of different studies before the discovery of the SVCT proteins (Goldenberg and Schweinzer, 1994; Helbig *et al.*, 1989; Padh and Aleo, 1987; Prasad *et al.*, 1998; Wilson, 1989; Wilson and Dixon, 1989; Welch *et al.*, 1995; Zreik *et al.*, 1999). Unfortunately,

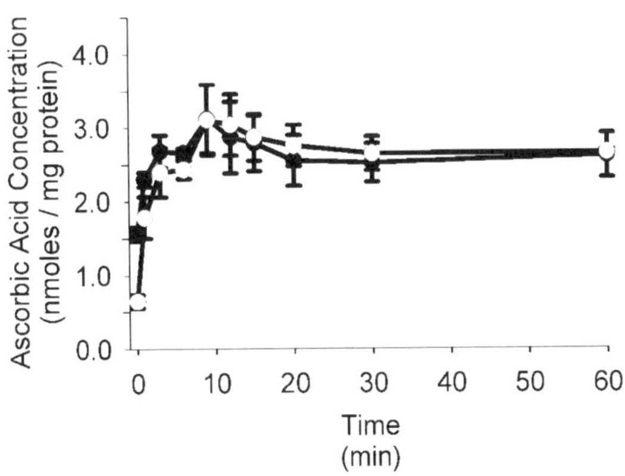

Figure 12.2: No age-related changes in hepatocellular DHA uptake.
Hepatocytes isolated from young (●) and old (○) rats were exposed to 100 μM DHA. Intracellular ASC was measured by HLPC coupled electrochemical detection. No significant changes in ASC accumulation were detected.

SVCT proteins have not been isolated or functionally reconstituted. The mRNA sequence encoding SVCT proteins has been elucidated, but few studies have investigated SVCT mRNA expression (Rajan *et al.*, 1999; Tsukaguchi *et al.*, 1999; Wang *et al.*, 1999, 2000). Only our recent study (Michels *et al.*, 2003) examined age-associated differences in sodium-dependent vitamin C uptake and the corresponding changes in SVCT expression.

Our work was designed to determine if the loss of tissue vitamin C was due, in part, to an impaired ability to transport ASC from the exogenous milieu. Freshly isolated hepatocytes from young (3–5 months) and old (24–26 months) male Fischer 344 rats were incubated with 100 μM ASC and uptake was monitored over time. In cells from young rats, ASC increased at a linear rate for up to 20 min, when a steady-state level was reached (*Figure 12.3*). A similar mode of ASC uptake was seen in cells from old rats, in which a linear rate of ASC uptake was observed for 15 min after ASC addition (*Figure 12.3*). However, two major differences in transport characteristics were observed between cells from young and old rats. First, a significant age-related decline in the rate of ASC uptake was observed. Second, steady-state ASC levels were markedly lower in hepatocytes from old rats. Removal of sodium from the extracellular media of the cells resulted in a large decline in ASC uptake from both young and old cells (Michels *et al.*, 2003). Additionally, when a competitive inhibitor to GLUT transport function, 2-deoxyglucose, was added in excess relative to vitamin C, there was no appreciable difference in either the rate or accumulation of vitamin C (Michels *et al.*, 2003). This suggests that transporter-driven vitamin C uptake was primarily through SVCT, independent of activity of the GLUT system.

Unfortunately, available analytical methods to assess the presence or activity of SVCT *in vivo* are very limited. A kinetic analysis of vitamin C uptake shows an age-related change to a lower capacity yet higher affinity system (Michels *et al.*,

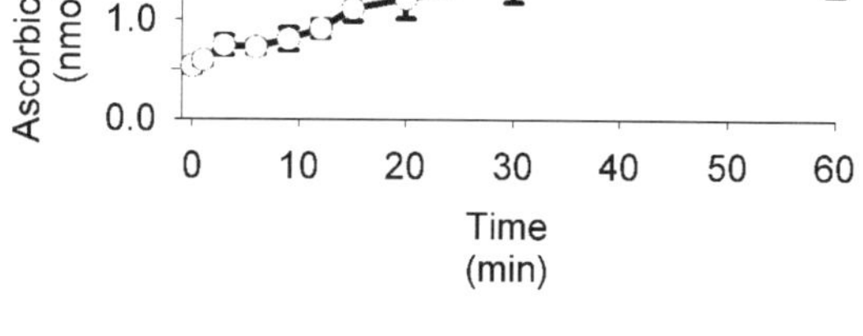

Figure 12.3: Hepatocellular ASC uptake declines with age.
Hepatocytes isolated from young (●) and old (○) rats were exposed to 100 μM ASC. The initial rate of ASC uptake into hepatocytes from old rats (n = 6) was significantly ($P < 0.005$) lower than that seen in cells from young rats (n = 6). Reprinted from Michels *et al.*, 2003 with permission from Elsevier.

2003). These studies were performed in isolated hepatocytes, which express both SVCT1 and 2 and have different kinetic properties (Michels *et al.*, 2003). A preferential age-related loss of SVCT1 levels could result in the observed kinetic changes. Reverse transcriptase–PCR analysis on liver mRNA showed that only one of the SVCT proteins declines with age. While SVCT1 levels fall approximately 45% in tissue from old animals when compared to young, SVCT2 levels remain the same (Michels *et al.*, 2003).

Cell types other than the liver develop age-related declines in vitamin C status, although not to the same degree. However, most other organs, including the heart, brain, and eye, do not express SVCT1 and express only SVCT2. No data on age-related changes in SVCT2 levels in these organs are presently available, and the regulatory systems for SVCT are speculative (Michels *et al.*, 2003). Since vitamin C levels decline in other organs, further work must be done to elucidate the role of aging on the SVCT transport system.

12.5 Conclusions and future directions

Vitamin C status is impaired during the aging process. The data in animal models indicate that the decline in vitamin C is not due to an age-related change in synthetic capacity. Even in animals that do not synthesize the vitamin, the loss of vitamin C in tissues is notable even when the vitamin is constantly supplied in the diet. Human data indicate that similar losses in vitamin C occur in plasma and tissues in the elderly, despite increased levels of the vitamin in their diets. Furthermore, when studies of vitamin C intake were correlated with plasma vitamin C status, the elderly showed greater alterations in the amount of vitamin C absorbed, suggesting that increased intake, possibly by supplementation, in the elderly may be necessary. Indeed, supplementation with doses of vitamin C many times the amount required to saturate the plasma in younger individuals is necessary to increase plasma vitamin C levels in elderly subjects. While reversal of age-related losses in plasma status is possible, true pharmacokinetic studies of vitamin C uptake for those over 65 years of age have yet to be performed. Since plasma and tissue vitamin C levels are dependent on uptake, transport may play a pivotal role in the loss of vitamin C.

Vitamin C transport occurs through one of two transport systems, each specific for one redox state of the molecule. The oxidized form of vitamin C, DHA, is taken up via GLUT proteins. As reviewed here, there is no conclusive evidence that GLUT expression or activity changes with age, especially in organs where vitamin C levels decline. Further studies are required to correlate changes in GLUT expression or activity with physiological vitamin C transport. ASC, the reduced, active form of vitamin C, is taken up through a newly discovered SVCT protein. Corresponding to the loss of vitamin C in the liver, it appears that SVCT1 mRNA levels decline with age, although no apparent changes are evident in SVCT2 (Michels *et al.*, 2003). This suggests that SVCT function is required for normal vitamin C status in the liver, although no other tissues have been examined for age-related changes in SVCT levels or activity. Loss of SVCT1 in the intestine, where a majority of vitamin C uptake occurs, may explain much of the age-related changes in absorptive capacity. Although loss of vitamin C is observed in the heart and the brain, both of these tissues exhibit only SVCT2, not SVCT1, and age-related changes in SVCT2 levels have not been elucidated.

Some of these observations may be explained by a differential regulation of SVCT1 versus SVCT2 and expression may also be affected by tissue-specific factors. Nothing is known about the transcriptional regulation of SVCT1 or SVCT2 in the rat, and only limited work in the human genome is available (Rajan *et al.*, 1999; Wang *et al.*, 1999, 2000). No specific transcription factors are positively linked to changes in either SVCT1 or SVCT2 levels for any species, primarily due to the lack of sequence data on the 5′-untranslated regions of SVCT1 and SVCT2. Future work in this area has broad implications on the age-related changes in the regulation of SVCT and should be focused on reversing these declines in SVCT1 and, depending on the tissue, SVCT2.

The exploration of the effect of aging on vitamin C is only beginning. Since vitamin C has multi-faceted and diverse biochemical and physiological roles, suboptimum levels could have numerous serious ramifications on health. Although the protection against oxidative damage is one function under extensive research, other functions of vitamin C that change with age have yet to be elucidated. Carnitine synthesis declines significantly with age and may be predicated on the loss of vitamin C. Vitamin C-dependent loss of carnitine would affect cellular energy production by the mitochondria, especially in those organs that use fatty acids as their primary fuel, such as the heart or the kidney cortex.

Normal collagen synthesis, which also requires vitamin C, declines with age, and aberrant increases in crosslinking of collagen are evident, resulting in thinning skin, and effects on normal joint, blood vessel and myocardial function. Vitamin C also plays a role in nitric oxide synthase activity and vessel relaxation. Age-related changes in blood vessel relaxation may be largely due to inadequate vitamin C. Similarly, prostaglandin synthesis, immune surveillance, gene expression and cell proliferation, to name a few, are influenced by vitamin C and also appear to be negatively affected with age.

Vitamin C status and aging is a critical topic that is only beginning to be explored, yet will have important consequences on public health. Future research on vitamin C transport will no doubt bring great advances in our understanding of age-related loss in vitamin C status, as well as guide recommendations for proper dietary intake of this key vitamin in the elderly.

References

Ames BN, Shigenaga MK, Hagen TM (1993) Oxidants, antioxidants and digenerative diseases of aging. *Proc. Natl Acad. Sci. USA* **90**, 7915–7922.

Armoni M, Harel C, Burvin R, Karnieli E (1995) Modulation of the activity of glucose transporters (GLUT) in the aged/obese rat adipocyte: suppressed function, but enhanced intrinsic activity of GLUT. *Endocrinology* **136**, 3292–3298.

Arockia Rani PJ, Panneerselvam C (2001) Carnitine as a free radical scavenger in aging. *Exp. Gerontol.* **36**, 1713–1726.

Arrigoni O, De Tullio MC (2002) Ascorbic acid: much more than just an antioxidant. *Biochim. Biophys. Acta* **1569**, 1–9.

Attwood E C, Robey E, Kramer JJ, Ovenden N, Snape S, Ross J, Bradley F (1978) A survey of the haematological, nutritional and biochemical state of the rural elderly with particular reference to vitamin C. *Age Ageing* 7(1): 46–56.

Azhar S, Cao L, Reaven E (1995) Alteration of the adrenal antioxidant defense system during aging in rats. *J. Clin. Invest.* **96**: 1414–1424.

Barja G (1996) Ascorbic acid and aging. *Subcell. Biochem.* **25**, 157–188.

Barnes MJ (1975) Function of ascorbic acid in collagen metabolism. *Ann. NY Acad. Sci.* **258**, 264, 277.

Beal MF (2002) Oxidatively modified proteins in aging and disease. *Free Radic. Biol. Med.* **32**, 797–803.

Beckman KB, Ames BN (1998) The free radical theory of aging matures. *Physiol. Rev.* **78**, 547–581.

Bendich A, Langseth L (1995) The health effects of vitamin C supplementation: a review. *J. Am. Coll. Nutr.* **14**, 124–136.

Berger J, Shepard D, Morrow F, Taylor A (1989) Relationship between dietary intake and tissue levels of reduced and total vitamin C in the nonscorbutic guinea pig. *J. Nutr.* **119**, 734–740.

Berger UV, Hediger MA (2000) The vitamin C transporter SVCT2 is expressed by astrocytes in culture but not in situ. *NeuroReport* **11**, 1395–1399.

Berlett BS, Stadtman ER (1997) Protein oxidation in aging, disease, and oxidative stress. *J. Biol. Chem.* **272**, 20313–20316.

Bianchetti A, Rozzini R, Carabellese C, Zanetti O, Trabucchi M (1990) Nutritional intake, socioeconomic conditions, and health status in a large elderly population. *J. Am. Geriatr. Soc.* **38**, 521–526.

Bidlack WR, Smith CH (1988) Nutritional requirements of the aged. *Crit. Rev. Food Sci. Nutr.* **27**, 189–218.

Birlouez-Aragon I, Girard F, Ravelontseheno L, Bourgeois C, Belliot JP, Abitbol G (1995) Comparison of two levels of vitamin C supplementation on antioxidant vitamin status in elderly institutionalized subjects. *Int. J. Vitamin Nutr. Res.* **65**, 261–266.

Birlouez-Aragon I, Delcourt C, Tessier F, Papoz L (2001) Associations of age, smoking habits and diabetes with plasma vitamin C of elderly of the POLA study. *Int. J. Vitamin Nutr. Res.* **71**, 53–59.

Blanchard J, Conrad KA, Watson RR, Garry PJ, Crawley JD (1989) Comparison of plasma, mononuclear and polymorphonuclear leucocyte vitamin C levels in young and elderly women during depletion and supplementation. *Eur. J. Clin. Nutr.* **43**, 97–106.

Blanchard J, Conrad KA, Mead RA, Garry PJ (1990) Vitamin C disposition in young and elderly men. *Am. J. Clin. Nutr.* **51**, 837–845.

Blanchard J (1991) Depletion and repletion kinetics of vitamin C in humans. *J. Nutr.* **121**, 170–176.

Block G (1992a) Vitamin C status and cancer. Epidemiologic evidence of reduced risk. *Ann. N Y Acad. Sci.* **669**, 280–290; discussion 290–282.

Block G (1992b) Vitamin C and reduced mortality. *Epidemiology* **3**, 189–191.

Brigelius-Flohe R, Flohe L (1996) Ascorbic acid, cell proliferation, and cell differentiation in culture. *Subcell. Biochem.* **25**, 83–107.

Brubacher D, Moser U, Jordan P (2000) Vitamin C concentrations in plasma as a function of intake: a meta-analysis. *Int. J. Vitamin Nutr. Res.* **70**, 226–237.

Burr ML, Elwood PC, Hole DJ, Hurley RJ, Hughes RE (1974) Plasma and leukocyte ascorbic acid levels in the elderly. *Am. J. Clin. Nutr.* **27**, 144–151.

Carr A, Frei B (1999a) Does vitamin C act as a pro-oxidant under physiological conditions? *FASEB J.* **13**, 1007–1024.

Carr AC, Frei B (1999b) Toward a new recommended dietary allowance for vitamin C based on antioxidant and health effects in humans. *Am. J. Clin. Nutr.* **69**, 1086–1107.

Carrillo MC, Kanai S, Sato Y, Kitani K (1992) Age-related changes in antioxidant enzyme activities are region and organ, as well as sex, selective in the rat. *Mech. Ageing Dev.* **65**, 187–198.

Cartee GD (1993) Myocardial GLUT-4 glucose transporter protein levels of rats decline with advancing age. *J. Gerontol.* **48**, B168–B170.

Cartee GD (1994) Influence of age on skeletal muscle glucose transport and glycogen metabolism. *Med. Sci. Sports Exerc.* **26**, 577–585.

Casirola DM, Lan Y, Ferraris RP (1997) Effects of changes in calorie intake on intestinal nutrient uptake and transporter mRNA levels in aged mice. *J. Gerontol. A Biol. Sci. Med. Sci.* **52**, B300–B310.

Chung JH, Youn SH, Kwon OS, Cho KH, Youn JI, Eun HC (1997) Regulations of collagen synthesis by ascorbic acid, transforming growth factor-beta and interferon-gamma in human dermal fibroblasts cultured in three-dimensional collagen gel are photoaging- and aging-independent. *J. Dermatol. Sci.* **15**, 188–200.

Cunningham JJ (1998a) Micronutrients as nutriceutical interventions in diabetes mellitus. *J. Am. Coll. Nutr.* **17**, 7–10.

Cunningham JJ (1998b) The glucose/insulin system and vitamin C: implications in insulin-dependent diabetes mellitus. *J. Am. Coll. Nutr.* **17**, 105–108.

Daruwala R, Song J, Koh WS, Rumsey SC, Levine M (1999) Cloning and functional characterization of the human sodium-dependent vitamin C transporters hSVCT1 and hSVCT2. *FEBS Lett.* **460**, 480–484.

De AK, Darad R (1991) Age-associated changes in antioxidants and antioxidative enzymes in rats. *Mech. Ageing Dev.* **59**, 123–128.

De La Fuente M, Hernanz A, Collazos ME, Barriga C, Ortega E (1995) Effects of physical exercise and aging on ascorbic acid and superoxide anion levels in peritoneal macrophages from mice and guinea pigs. *J. Comp. Physiol.* **165**, 315–319.

Droge W (2002) Aging-related changes in the thiol/disulfide redox state: implications for the use of thiol antioxidants. *Exp. Gerontol.* **37**, 1333–1345.

Enstrom JE, Kanim LE, Klein MA (1992) Vitamin C intake and mortality among a sample of the United States population. *Epidemiology* **3**, 194–202.

Erden-Inal M, Sunal E, Kanbak G (2002) Age-related changes in the glutathione redox system. *Cell Biochem. Funct.* **20**, 61–66.

Fattoretti P, Bertoni-Freddari C, Di Stefano G, Casoli T, Gracciotti N, Solazzi M, Pompei P (2001) Quantitative immunohistochemistry of glucose transport protein (Glut3) expression in the rat hippocampus during aging. *J. Histochem. Cytochem.* **49**, 671–672.

Finkel T, Holbrook NJ (2000) Oxidants, oxidative stress and the biology of ageing. *Nature* **408**, 239–247.

Fraga CG, Motchnik PA, Shigenaga MK, Helbock HJ, Jacob RA, Ames BN (1991) Ascorbic acid protects against endogenous oxidative DNA damage in human sperm. *Proc. Natl Acad. Sci. USA* **88**, 11003–11006.

Frei B, England L, Ames BN (1989) Ascorbate is an outstanding antioxidant in human blood plasma. *Proc. Natl Acad. Sci. USA* **86**, 6377–6381.

Fujita I, Hirano J, Itoh N, Nakanishi T, Tanaka K (2001) Dexamethasone induces sodium-dependent vitamin C transporter in a mouse osteoblastic cell line MC3T3-E1. *Br. J. Nutr.* **86**, 145–149.

Garry PJ, Goodwin JS, Hunt WC, Gilbert BA (1982a) Nutritional status in a healthy elderly population: vitamin C. *Am. J. Clin. Nutr.* **36**, 332–339.

Garry PJ, Goodwin JS, Hunt WC, Hooper EM, Leonard AG (1982b) Nutritional status in a healthy elderly population: dietary and supplemental intakes. *Am. J. Clin. Nutr.* **36**, 319–331.

Gershoff SN (1993) Vitamin C (ascorbic acid): new roles, new requirements? *Nutr. Rev.* **51**, 313–326.

Goldenberg H, Schweinzer E (1994) Transport of vitamin C in animal and human cells. *J. Bioenerg. Biomembr.* **26**, 359–367.

Gschanes A, Boado R, Sametz W, Windisch M (2000) The drug cerebrolysin and its peptide fraction E021 increase the abundance of the blood–brain barrier GLUT1 glucose transporter in brains of young and old rats. *Histochem. J.* **32**, 71–77.

Gulve EA, Henriksen EJ, Rodnick KJ, Youn JH, Holloszy JO (1993) Glucose transporters and glucose transport in skeletal muscles of 1- to 25-mo-old rats. *Am. J. Physiol.* **264**, E319–E327.

Hagen TM, Ingersoll RT, Wehr CM, Lykkesfeldt J, Vinarsky V, Bartholomew JC, Song MH, Ames BN (1998) Acetyl-L-carnitine fed to old rats partially restores mitochondrial function and ambulatory activity. *Proc. Natl Acad. Sci. USA* **95**, 9562–9566.

Hagen TM, Ingersoll RT, Lykkesfeldt J, Liu J, Wehr CM, Vinarsky V, Bartholomew JC, Ames AB (1999) (R)-alpha-lipoic acid-supplemented old rats have improved mitochondrial function, decreased oxidative damage, and increased metabolic rate. *FASEB J.* **13**, 411–418.

Hagen TM, Vinarsky V, Wehr CM, Ames BN (2000) (R)-alpha-lipoic acid reverses the age-associated increase in susceptibility of hepatocytes to tert-butylhydroperoxide both in vitro and in vivo. *Antioxid. Redox Signal.* **2**, 473–483.

Hall DM, Xu L, Drake VJ, Oberley LW, Oberley TD, Moseley PL, Kregel KC (2000) Aging reduces adaptive capacity and stress protein expression in the liver after heat stress. *J. Appl. Physiol.* **89**, 749–759.

Hall JL, Mazzeo RS, Podolin DA, Cartee GD, Stanley WC (1994) Exercise training does not compensate for age-related decrease in myocardial GLUT-4 content. *J. Appl. Physiol.* **76,** 328–332.

Hansford RG, Castro F (1982) Age-linked changes in the activity of enzymes of the tricarboxylate cycle and lipid oxidation, and of carnitine content, in muscles of the rat. *Mech. Ageing Dev.* **19**, 191–200.

Hardy RW, Meckling-Gill KA, Williford J, Desmond RA, Wei H (2002) Energy restriction reduces long-chain saturated fatty acids associated with plasma lipids in aging male rats. *J. Nutr.* **132**, 3172–3177.

Harman D (1956) Theory based on free radical and radiation chemistry. *J. Gerontol.* **11,** 298–300.

Harrill I, Cervone N (1977) Vitamin status of older women. *Am. J. Clin. Nutr.* **30,** 431–440.

Head E, Liu J, Hagen TM, Muggenburg BA, Milgram NW, Ames BN, Cotman CW (2002) Oxidative damage increases with age in a canine model of human brain aging. *J. Neurochem.* **82**, 375–381.

Helbig H, Korbmacher C, Wohlfarth J, Berweck S, Kuhner D, Wiederholt M (1989) Electrogenic Na+-ascorbate cotransport in cultured bovine pigmented ciliary epithelial cells. *Am. J. Physiol.* **256**, C44–49.

Heseker H, Schneider R (1994) Requirement and supply of vitamin C, E and beta-carotene for elderly men and women. *Eur. J. Clin. Nutr.* **48,** 118–127.

Hitomi K, Tsukagoshi N (1996) Role of ascorbic acid in modulation of gene expression. *Subcell. Biochem.* **25**, 41–56.

Horrobin DF (1996) Ascorbic acid and prostaglandin synthesis. *Subcell. Biochem.* **25**, 109–115.

Houmard JA, Weidner MD, Dolan PL, Leggett-Frazier N, Gavigan KE, Hickey MS, *et al.* (1995) Skeletal muscle GLUT4 protein concentration and aging in humans. *Diabetes* **44**, 555–560.

Howe GR, Hirohata T, Hislop TG, Iscovich JM, Yuan JM, Katsouyanni A, *et al.* (1990) Dietary factors and risk of breast cancer: combined analysis of 12 case–control studies. *J. Natl Cancer Inst.* **82**, 561–569.

Hughes RE, Jones PR (1971) The influence of sex and age on the deposition of L-xyloascorbic acid in tissues of guinea-pigs. *Br. J. Nutr.* **25**, 77–83.

Hughes RE, Hurley RJ, Jones E (1980) Dietary ascorbic acid and muscle carnitine (beta-OH-gamma-(trimethylamino) butyric acid) in guinea-pigs. *Br. J. Nutr.* **43**, 385–387.

Hunt JV, Bottoms MA, Mitchinson MJ (1992) Ascorbic acid oxidation: a potential cause of the elevated severity of atherosclerosis in diabetes mellitus? *FEBS Lett.* **311**, 161–164.

Jacob RA, Otradovec CL, Russell RM, Munro HN, Hartz SC, McGandy RB, Morrow FD, Sadowski JA (1988) Vitamin C status and nutrient interactions in a healthy elderly population. *Am. J. Clin. Nutr.* **48**, 1436–1442.

Jacob RA (1996) Three eras of vitamin C discovery. *Subcell. Biochem.* **25**, 1–16.

James DE, Strube M, Mueckler M (1989) Molecular cloning and characterization of an insulin-regulatable glucose transporter. *Nature* **338**, 83–87.

Jariwalla RJ, Harakeh S (1996) Antiviral and immunomodulatory activities of ascorbic acid. *Subcell. Biochem.* **25**, 213–231.

Ji LL, Dillon D, Wu E (1990) Alteration of antioxidant enzymes with aging in rat skeletal muscle and liver. *Am. J. Physiol.* **258**, R918–R923.

Kanungo MS, Patnaik BK (1964) Ascorbic acid and aging in the rat. Uptake of ascorbic acid by skin and bone marrow, and its concentration in various organs. *Biochem. J.* **90,** 637–640.

Kasapoglu M, Ozben T (2001) Alterations of antioxidant enzymes and oxidative stress markers in aging. *Exp. Gerontol.* **36**, 209–220.

Khaw KT, Bingham S, Welch A, Luben R, Wareham N, Oakes S, Day N (2001) Relation between plasma ascorbic acid and mortality in men and women in EPIC-Norfolk prospective study: a prospective population study. European Prospective Investigation into Cancer and Nutrition. *Lancet* **357**, 657–663.

Klip A, Douen AG (1989) Role of kinases in insulin stimulation of glucose transport. *J. Membr. Biol.* **111**, 1–23.

Korcok J, Yan R, Siushansian R, Dixon SJ, Wilson JX (2000) Sodium-ascorbate cotransport controls intracellular ascorbate concentration in primary astrocyte cultures expressing the SVCT2 transporter. *Brain Res.* **881**, 144–151.

Kregel KC, Moseley PL, Skidmore R, Gutierrez JA, Guerriero V Jr (1995) HSP70 accumulation in tissues of heat-stressed rats is blunted with advancing age. *J. Appl. Physiol.* **79**, 1673–1678.

Kume-Kick J, Rice ME (1998) Estrogen-dependent modulation of rat brain ascorbate levels and ischemia-induced ascorbate loss. *Brain Res.* **803**, 105–113.

Larkin LM, Reynolds TH, Supiano MA, Kahn BB, Halter JB (2001) Effect of aging and obesity on insulin responsiveness and glut-4 glucose transporter content in skeletal muscle of Fischer 344 × Brown Norway rats. *J. Gerontol. A Biol. Sci. Med. Sci.* **56**, B486–B492.

Leibovitz BE, Siegel BV (1980) Aspects of free radical reactions in biological systems: aging. *J. Gerontol.* **35**, 45–56.

Lenton KJ, Therriault H, Cantin AM, Fulop T, Payette H, Wagner JR (2000) Direct correlation of glutathione and ascorbate and their dependence on age and season in human lymphocytes. *Am. J. Clin. Nutr.* **71**, 1194–1200.

Levine M (1986) New concepts in the biology and biochemistry of ascorbic acid. *N. Engl. J. Med.* **314**, 892–902.

Levine M, Conry-Cantilena C, Wang Y, Welch RW, Washko PW, Dhariwal KR, *et al.* (1996) Vitamin C pharmacokinetics in healthy volunteers: evidence for a recommended dietary allowance. *Proc. Natl Acad. Sci. USA* **93**, 3704–3709.

Levine M, Wang Y, Padayatty SJ, Morrow J (2001) A new recommended dietary allowance of vitamin C for healthy young women. *Proc. Natl Acad. Sci. USA* **98**, 9842–9846.

Liang WJ, Johnson D, Jarvis SM (2001) Vitamin C transport systems of mammalian cells. *Mol. Membr. Biol.* **18**, 87–95.

Loh HS, Wilson CW (1971) Relationship between leucocyte and plasma ascorbic acid concentrations. *Br. Med. J.* **3**, 733–735.

Lorens SA, Hata N, Handa RJ, Van De Kar LD, Guschwan M, Goral J, *et al*, (1990) Neurochemical, endocrine and immunological responses to stress in young and old Fischer 344 male rats. *Neurobiol. Aging* **11**, 139–150.

Lykkesfeldt J, Loft S, Nielsen JB, Poulsen HE (1997) Ascorbic acid and dehydroascorbic acid as biomarkers of oxidative stress caused by smoking. *Am. J. Clin. Nutr.* **65**, 959–963.

Lykkesfeldt J, Hagen TM, Vinarsky V, Ames BN (1998) Age-associated decline in ascorbic acid concentration, recycling, and biosynthesis in rat hepatocytes – reversal with (R)-alpha-lipoic acid supplementation. *FASEB J.* **12**, 1183–1189.

Malo C, Wilson JX (2000) Glucose modulates vitamin C transport in adult human small intestinal brush border membrane vesicles. *J. Nutr.* **130**, 63–69.

Marin J, Rodriguez-Martinez MA (1999) Age-related changes in vascular responses. *Exp. Gerontol.* **34**, 503–512.

Martin A, Youdim K, Szprengiel A, Shukitt-Hale B, Joseph J (2002) Roles of vitamins E and C on neurodegenerative diseases and cognitive performance. *Nutr. Rev.* **60**, 308–326.

McCall MR, Frei B (1999) Can antioxidant vitamins materially reduce oxidative damage in humans? *Free Radic. Biol. Med.* **26**, 1034–1053.

McClean HE, Dodds PM, Abernethy MH, Stewart AW, Beaven DW (1976) Vitamin C concentration in plasma and leucocytes of men related to age and smoking habit. *N Z Med. J.* **83**, 226–229.

Mecocci P, Polidori MC, Troiano L, Cherubini A, Cecchetti R, Pini G, *et al.* (2000) Plasma antioxidants and longevity: a study on healthy centenarians. *Free Radic. Biol. Med.* **28**, 1243–1248.

Michels AJ, Joisher N, Hagen TM (2003) Age-related decline of sodium-dependent ascorbic acid transport in isolated rat hepatocytes. *Arch. Biochem. Biophys.* **410**, 112–120.

Milne JS, Lonergan ME, Williamson J, Moore FM, McMaster R, Percy N (1971) Leucocyte ascorbic acid levels and vitamin C intake in older people. *Br. Med. J.* **4**, 383–386.

Monget AL, Galan P, Preziosi P, Keller H, Bourgeois C, Arnaud J, Favier A, Hercberg S (1996) Micronutrient status in elderly people. Geriatrie/Min. Vit. Aux Network. *Int. J. Vitamin Nutr. Res.* **66**, 71–76.

Mooradian AD, Girgis W, Shah GN (1997) Thyroid hormone-induced GLUT-1 expression in rat cerebral tissue: effect of age. *Brain Res.* **747**, 144–146.

Mooradian AD, Shah GN (1997) Age-related changes in glucose transporter-one mRNA structure and function. *Proc. Soc. Exp. Biol. Med.* **216**, 380–385.

Mooradian AD, Chehade JM, Kim J (1999) Age-related changes in thyroid hormone effects on glucose transporter isoforms of rat heart. *Life Sci.* **65**, 981–989.

Murad S, Grove D, Lindberg KA, Reynolds G, Sivarajah A, Pinnell SR (1981) Regulation of collagen synthesis by ascorbic acid. *Proc. Natl Acad. Sci. USA* **78**, 2879–2882.

Muscari C, Giaccari A, Giordano E, Clo C, Guarnieri C, Caldarera CM (1996) Role of reactive oxygen species in cardiovascular aging. *Mol. Cell. Biochem.* **160/161**, 159–166.

Nandi A, Mukhopadhyay CK, Ghosh MK, Chattopadhyay DJ, Chatterjee IB (1997) Evolutionary significance of vitamin C biosynthesis in terrestrial vertebrates. *Free Radic. Biol. Med.* **22**, 1047–1054.

Newton HM, Schorah CJ, Habibzadeh N, Morgan DB, Hullin RP (1985) The cause and correction of low blood vitamin C concentrations in the elderly. *Am. J. Clin. Nutr.* **42**, 656–659.

Nishikimi M, Yagi K (1991) Molecular basis for the deficiency in humans of gulonolactone oxidase, a key enzyme for ascorbic acid biosynthesis. *Am. J. Clin. Nutr.* **54** (6 Suppl), 1203S–1208S.

O'Donnell E, Lynch MA (1998) Dietary antioxidant supplementation reverses age-related neuronal changes. *Neurobiol. Aging* **19**, 461–467.

Oka Y, Asano T, Lin JL, Tsukuda K, Katagiri H, Ishihara H, Inukai K, Yazaki Y (1992) Expression of glucose transporter isoforms with aging. *Gerontology* **38** (Suppl 1), 3–9.

Oreopoulos DG, Lindeman RD, Vanderjagt DJ, Tzamaloukas AH, Bhagavan HN, Garry PJ (1993) Renal excretion of ascorbic acid: effect of age and sex. *J. Am. Coll. Nutr.* **12**, 537–542.

Packer L (1997) Oxidants, antioxidant nutrients and the athlete. *J. Sports Sci.* **15**, 353–363.

Padh H, Aleo JJ (1987) Activation of serum complement leads to inhibition of ascorbic acid transport. *Proc. Soc. Exp. Biol. Med.* **185**, 153–157.

Pansarasa O, Bertorelli L, Vecchiet J, Felzani G, Marzatico F (1999) Age-dependent changes of antioxidant activities and markers of free radical damage in human skeletal muscle. *Free Radic. Biol. Med.* **27**, 617–622.

Paolisso G, Tagliamonte MR, Rizzo MR, Manzella D, Gambardella A, Varricchio M (1998) Oxidative stress and advancing age: results in healthy centenarians. *J. Am. Geriatr. Soc.* **46**, 833–838.

Paradies G, Ruggiero FM, Gadaleta MN, Quagliariello E (1992) The effect of aging and acetyl-L-carnitine on the activity of the phosphate carrier and on the phospholipid composition in rat heart mitochondria. *Biochim. Biophys. Acta* **1103**, 324–326.

Patnaik BK, Kanungo MS (1966) Ascorbic acid and aging in the rat. Uptake of ascorbic acid by teeth and concentration of various forms of ascorbic acid in different organs. *Biochem. J.* **100**, 59–62.

Prasad PD, Huang W, Wang H, Leibach FH, Ganapathy V (1998) Transport mechanisms for vitamin C in the JAR human placental choriocarcinoma cell line. *Biochim. Biophys. Acta* **1369**, 141–151.

Preston AM, Bercovitch FB, Rodriguez CA, Lebron MR, Rivera CE (2001) Plasma ascorbic acid concentrations in a population of rhesus monkeys (*Macaca mulatta*). *Contemp. Top. Lab. Anim. Sci.* **40**, 30–32.

Rajan DP, Huang W, Dutta B, Devoe LD, Leibach FH, Ganapathy V, Prasad PD (1999) Human placental sodium-dependent vitamin C transporter (SVCT2): molecular cloning and transport function. *Biochem. Biophys. Res. Commun.* **262**, 762–768.

Rebouche CJ (1991) Ascorbic acid and carnitine biosynthesis. *Am. J. Clin. Nutr.* **54** (6 Suppl), 1147S–1152S.

Rikans LE, Moore DR (1988) Effect of aging on aqueous-phase antioxidants in tissues of male Fischer rats. *Biochim. Biophys. Acta* **966**, 269–275.

Rikans LE, Lopez TR, Hornbrook KR (1996) Age and gender differences in hepatic ascorbic acid concentrations and NADPH-dehydroascorbatic acid reductase activity. *Mech. Ageing Dev.* **91**, 165–169.

Robertson JM, Donner AP, Trevithick JR (1991) A possible role for vitamins C and E in cataract prevention. *Am. J. Clin. Nutr.* **53** (1 Suppl), 346S–351S.

Rumsey SC, Kwon O, Xu GW, Burant CF, Simpson I, Levine M (1997) Glucose transporter isoforms GLUT1 and GLUT3 transport dehydroascorbic acid. *J. Biol. Chem.* **272**, 18982–18989.

Rumsey SC, Daruwala R, Al-Hasani H, Zarnowski MJ, Simpson IA, Levine M (2000) Dehydroascorbic acid transport by GLUT4 in *Xenopus* oocytes and isolated rat adipocytes. *J. Biol. Chem.* **275**, 28246–28253.

Sahoo A, Chainy GB (1997) Alterations in the activities of cerebral antioxidant enzymes of rat are related to aging. *Int. J. Dev. Neurosci.* **15**, 939–948.

Sahyoun NR, Jacques PF, Russell RM (1996) Carotenoids, vitamins C and E, and mortality in an elderly population. *Am. J. Epidemiol.* **144**, 501–511.

Schaus R (1957) The ascorbic acid content of human pituitary, cerebral cortex, heart and skeletal muscle and its relation to age. *Am. J. Clin. Nutr.* **5**, 39–41.

Schmuck A, Ravel A, Coudray C, Alary J, Franco A, Roussel AM (1996) Antioxidant vitamins in hospitalized elderly patients: analysed dietary intakes and biochemical status. *Eur. J. Clin. Nutr.* **50**, 473–478.

Schorah CJ, Newill A, Scott DL, Morgan DB (1979) Clinical effects of vitamin C in elderly inpatients with low blood-vitamin-C levels. *Lancet* **1**(8113), 403–405.

Schorah CJ, Dowing C, Piripitsi A, Gallivan L, Al-Hazaa AH, Sanderson MJ, Bodenham A (1996) Total vitamin C, ascorbic acid, and dehydroascorbic acid concentrations in plasma of critically ill patients. *Am. J. Clin. Nutr.* **63**, 760–765.

Scrofano MM, Jahngen-Hodge J, Nowell TR Jr, Gong X, Smith DE, Perrone G, *et al.* (1998) The effects of aging and calorie restriction on plasma nutrient levels in male and female Emory mice. *Mech. Ageing Dev.* **105**, 31–44.

Seals DR, Taylor JA, Ng AV, Esler MD (1994) Exercise and aging: autonomic control of the circulation. *Med. Sci. Sports Exerc.* **26**, 568–576.

Shigenaga MK, Hagen TM, Ames BN (1994) Oxidative damage and mitochondrial decay in aging. *Proc. Natl Acad. Sci. USA* **91**, 10771–10778.

Smith AR, Visioli F, Hagen TM (2002) Vitamin C matters: increased oxidative stress in cultured human aortic endothelial cells without supplemental ascorbic acid. *FASEB J.* **16**, 1102–1104.

Sohal RS, Brunk UT (1992) Mitochondrial production of pro-oxidants and cellular senescence. *Mutat. Res.* **275**, 295–304.

Sohal RS, Orr WC (1992) Relationship between antioxidants, prooxidants, and the aging process. *Ann. N Y Acad. Sci.* **663**, 74–84.

Sohal RS, Weindruch R (1996) Oxidative stress, caloric restriction, and aging. *Science* **273**, 59–63.

Sohal RS (2002) Role of oxidative stress and protein oxidation in the aging process. *Free Radic. Biol. Med.* **33**, 37–44.

Sohal RS, Mockett RJ, Orr WC (2002) Mechanisms of aging: an appraisal of the oxidative stress hypothesis. *Free Radic. Biol. Med.* **33**, 575–586.

Sotiriou S, Gispert S, Cheng J, Wang Y, Chen A, Hoogstraten-Miller S, *et al* (2002) Ascorbic acid transporter Slc23a1 is essential for vitamin C transport into the brain and for perinatal survival. *Nature Med.* **8**, 514–517.

Spector R, Lorenzo AV (1973) Ascorbic acid homeostasis in the central nervous system. *Am. J. Physiol.* **225**, 757–763.

Spector R (1977) Vitamin homeostasis in the central nervous system. *N. Engl J. Med.* **296**, 1393–1398.

Stitt S, O'Connell C, Grant D (1995) Old, poor & malnourished. *Nutr. Health* **10**, 135–154.

Suh JH, Shigeno ET, Morrow JD, Cox B, Rocha AE, Frei B, Hagen TM (2001) Oxidative stress in the aging rat heart is reversed by dietary supplementation with (R)-(alpha)-lipoic acid. *FASEB J* **15**, 700–706.

Suh Y, Lee KA, Kim WH, Han BG, Vijg J, Park SC (2002) Aging alters the apoptotic response to genotoxic stress. *Nature Med.* **8**, 3–4.

Svensson L, Wu C, Hulthe P, Johannessen K, Engel JA (1993) Effect of ageing on extracellular ascorbate concentration in rat brain. *Brain Res.* **609**, 36–40.

Taha C, Klip A (1999) The insulin signaling pathway. *J. Membr. Biol.* **169**, 1–12.

Taha C, Liu Z, Jin J, Al-Hasani H, Sonenberg N, Klip A (1999) Opposite translational control of GLUT1 and GLUT4 glucose transporter mRNAs in response to insulin. Role of mammalian target of rapamycin, protein kinase b, and phosphatidylinositol 3-kinase in GLUT1 mRNA translation. *J. Biol. Chem* **274**, 33085–33091.

Tian L, Cai Q, Wei H (1998) Alterations of antioxidant enzymes and oxidative damage to macromolecules in different organs of rats during aging. *Free Radic. Biol. Med.* **24**, 1477–1484.

Tsukaguchi H, Tokui T, Mackenzie B, Berger UV, Chen XZ, Wang Y, Brubaker RF, Hediger MA (1999) A family of mammalian Na+-dependent L-ascorbic acid transporters. *Nature* **399**, 70–75.

Valero MP, Fletcher AE, De Stavola BL, Vioque J, Alepuz VC (2002) Vitamin C is associated with reduced risk of cataract in a Mediterranean population. *J. Nutr.* **132**, 1299–1306.

Van Der Wielen RP, Van Heereveld HA, De Groot CP, Van Staveren WA (1995) Nutritional status of elderly female nursing home residents; the effect of supplementation with a physiological dose of water-soluble vitamins. *Eur. J. Clin. Nutr.* **49**, 665–674.

Vanderjagt DJ, Garry PJ, Bhagavan HN (1987) Ascorbic acid intake and plasma levels in healthy elderly people. *Am. J. Clin. Nutr.* **46**, 290–294.

Vera JC, Reyes AM, Velasquez FV, Rivas CI, Zhang RH, Strobel P, Slebe JC, Nunez-Alarcon J, Golde DW (2001) Direct inhibition of the hexose transporter GLUT1 by tyrosine kinase inhibitors. *Biochemistry* **40**, 777–790.

Vertechy M, Cooper MB, Ghirardi O, Ramacci MT (1989) Antioxidant enzyme activities in heart and skeletal muscle of rats of different ages. *Exp. Gerontol.* **24**, 211–218.

Wang H, Dutta B, Huang W, Devoe LD, Leibach FH, Ganapathy V, Prasad PD (1999) Human Na(+)-dependent vitamin C transporter 1 (hSVCT1): primary structure, functional characteristics and evidence for a non-functional splice variant. *Biochim. Biophys. Acta* **1461**, 1–9.

Wang Y, Russo TA, Kwon O, Chanock S, Rumsey SC, Levine M (1997) Ascorbate recycling in human neutrophils: induction by bacteria. *Proc. Natl Acad. Sci. USA* **94**, 13816–13819.

Wang Y, Mackenzie B, Tsukaguchi H, Weremowicz S, Morton CC, Hediger MA (2000) Human vitamin C (L-ascorbic acid) transporter SVCT1. *Biochem. Biophys. Res. Commun.* **267**, 488–494.

Wartanowicz M, Ziemlanski S (1984) Effect of ageing process on the content of certain vitamins in rat serum and liver. *Acta Physiol. Pol.* **35**, 338–345.

Wilson JX (1989) Ascorbic acid uptake by a high-affinity sodium-dependent mechanism in cultured rat astrocytes. *J Neurochem* **53**, 1064–1071.

Wilson JX, Dixon SJ (1989) High-affinity sodium-dependent uptake of ascorbic acid by rat osteoblasts. *J. Membr. Biol.* **111**, 83–91.

Welch RW, Wang Y, Crossman AJ, Park JB, Krik KL, Levine M (1995) Accumulation of vitamin C (ascorbate) and its oxidized metabolite dehydroascorbic acid occurs by separate mechanisms. *J. Biol. Chem.* **270**, 12584–12592.

Youngren JF, Barnard RJ (1995) Effects of acute and chronic exercise on skeletal muscle glucose transport in aged rats. *J. Appl. Physiol.* **78**, 1750–1756.

Zreik TG, Kodaman PH, Jones EE, Olive DL, Behrman H (1999) Identification and characterization of an ascorbic acid transporter in human granulosa-lutein cells. *Mol. Hum. Reprod.* **5**, 299–302.

13

Ascorbic acid in the central nervous system: uptake, distribution and functions

Russel J. Reiter

13.1 Introduction

Despite the large amount of research on ascorbic acid (ASC, vitamin C) generally, and its high concentrations in certain brain regions, definitive knowledge of its specific functions in the central nervous system (CNS) is unexpectedly sparse. Although best known as a reducing agent, i.e., as an antioxidant due to its ability to quench several reactive oxygen species (ROS) (Rose and Bode, 1993), it reportedly has a variety of other effects in the brain. For example, ASC may be involved in neurotransmitter metabolism and as a cofactor for several enzymes found in neurons and/or glia. Even though neurodegenerative diseases of the aged are believed to benefit from antioxidant intake, it has been difficult to show any definitive relationship between these conditions and the consumption of vitamin C.

Given that humans are incapable of synthesizing their own ASC, the human CNS, as well as other organs, is dependent on the dietary intake of this constituent for the maintenance of optimal function. Calculating the specific amounts of vitamin C consumed daily has been difficult because of the wide variations of ASC concentrations in different foodstuffs and the fact that ASC is readily destroyed by heat, by an alkaline environment and upon exposure to air; as a result the storage and cooking of foods significantly alters their ASC levels. Even when ASC-rich foods are consumed, different efficiencies of absorption among individuals makes calculations of how much gets into the blood and tissues uncertain.

One feature that characterizes much of the literature summarized below is the fact that many of the publications appeared more than 10 years ago and progress in the field of research relating to ASC–brain interactions, with few exceptions, has been rather slow. This is surprising considering the widespread public interest

Vitamin C, edited by Han Asard, James M May and Nicholas Smirnoff

in antioxidant ingestion and the potential importance of ASC in a variety of essential physiological processes and possibly in aging (Ely and Krone, 2002).

The current review considers the relationship of ASC to the brain. Herein, the findings related to the route by which ASC enters the mammalian CNS and its differential distribution in different brain regions are summarized. Furthermore, the multiple, albeit sometimes ill-defined, actions of ASC in the CNS are described.

13.2 Transport of ASC into the brain

ASC levels in different fluids/tissues are clearly not in equilibrium within mammals. Some regions of the brain have ASC levels that rank high among the tissues that contain this molecule. ASC does not enter the brain directly but rather is initially transferred from the blood into the cerebrospinal fluid (CSF) and then into the interstitial fluid of the brain from where it is taken up by neurons and glia. The most complete early studies that described the transport of ASC into the CNS are those of Hammarstrom (1966), Spector and Lorenzo (1973; 1974) and Spector (1981).

The report of Hammarstrom (1966) showed that peripherally administered ^{14}C-labelled ASC given to guinea pigs eventually localized in the brain, but the route was not a direct one. Rather, based on the autoradiographic distribution of the labeled product over time, ASC first penetrated the choroid plexus to enter the CSF from where it diffused into the surrounding neural tissue. The author presumed, on the basis of the results obtained, that the blood–brain barrier limited the direct transfer of ASC from the blood into the brain.

Considering these results and those showing that CSF concentrations of ASC are sometimes greater than those in the plasma (Tatsumi *et al.*, 1935; Booij, 1940), Spector and Lorenzo (1973) tested whether there was an active, saturable carrier for ASC in the choroid plexus of the rabbit. When the *in vitro* choroid plexus was incubated with artificial CSF to which radioactive ASC was added, the cells of the plexus accumulated the radioactivity against a concentration gradient; the process was found to be active and saturable. The uptake system was consistent with the Michaelis–Menten transport model. In the associated *in vivo* studies, performed in the unanesthetized rabbits, these workers found that ASC is transferred to the CSF from the blood via the choroid plexus; this is followed by the uptake of ASC by the brain; again, the process was found to be saturable. This uptake process maintains CSF and brain ASC concentrations within narrower limits than would be predicted on the basis of the wide fluctuations in plasma ASC concentrations. The transport of ASC across the cells of the choroid plexus is now known to be accomplished by Na^+-dependent transporters, SVCT1 and SVCT2 (Tsukaguchi *et al.*, 1999; see also Chapter 6 by Wilson).

The uptake of ASC by the choroid plexus also was shown to be stereospecific both in *in vitro* and *in vivo* studies (Spector and Lorenzo, 1974). *In vitro*, ASC, isoascorbate and D-glucoascorbate interfered with the intracellular accumulation of ^{14}C-labelled product in the choroid plexus. *In vivo*, studies again performed using unanesthetized rabbits, the transport of ASC into the CSF and brain was more readily saturated by high plasma concentrations of ASC than by increases in isoascorbate in the blood. In follow-up studies, Spector (1981) found that increasing concentrations ^{14}C-ASC in the neural tissues, after its intraventricular injection, correlated with its disappearance from the CSF. Likewise, brain tissue in close proximity to the ventricles initially contained the greatest amounts of radioactive ASC; with increasing time the ASC penetrated further into the neural tissue.

It is currently accepted that ASC is less likely to enter neural tissue directly from the blood due to the presence of a restrictive blood–brain barrier (Lam and Daniel, 1986; Agus *et al.*, 1997). Rather, from the plasma ASC is shunted, via the choroid plexi, into the ventricular CSF; from the CSF ASC passively diffuses into the parenchymal interstitial fluid which surrounds blood vessels and cells of the CNS; this likely accounts in part for the high concentrations of ASC in the extracellular fluid (ECF; see below) since they equilibrate with those of the CSF. From the ECF ASC is taken into neurons via the SVCT2 transporter (Tsukaguchi *et al.*, 1999). Glia also contain ASC, although in much lower concentrations (roughly 10-fold lower levels) than neurons (where they are in the millimolar range). The reasons for this difference remain to be determined. On the basis of its diffusion into the ECF from the CSF, it was initially anticipated that brain nuclei nearest the ventricular walls may contain the highest final concentrations of ASC, an assumption not verified, however, when specific measurements were made. The transport and compartmentalization of ASC in the brain is summarized in *Figure 13.1*.

Besides ASC diffusing into the ECF of the brain from the ventricular CSF, it also is discharged from cellular elements in the CNS. This process is a dynamic one and is controlled by complex and, for the most part, unknown mechanisms. The efflux of ASC from neurons/glia seems to vary with time of day, with the level of neuronal activity and the levels change markedly in response to psychoactive drugs. Because of this dynamic regulation of ECF ASC levels, the vitamin was presumed to have neuromodulatory functions.

In general, behavioral activation in animals is typically associated with increased extracellular concentrations of ASC with the levels in the striatum of rats differing between rest and activity by as much as 90% (O'Neill *et al.*, 1982; Boutelle *et al.*, 1989). Giving anti-anxiolytic drugs to stressed rats lowers the normal rise in ASC in the ECF (Boutelle *et al.*, 1990). Finally, drugs that promote behavioral activation cause an elevation of neostriatal ASC release (Fillenz *et al.*, 1986). In reference to this latter point, behavioral activation of rats following the peripheral administration of amphetamine increases neural ECF ASC levels (O'Neill and Fillenz, 1985; Zetterström *et al.*, 1992), at least in the striatum but not necessarily in other brain regions (Yount *et al.*, 1991). The regional variability as well as the wide individual variations in the response of ECF ASC concentrations in these studies make determinations of their significance difficult.

As noted above, the mechanisms governing the release of ASC from brain cells appear complex, but one process seems to involve high affinity transporters for the amino acid neurotransmitters. Thus, during the re-uptake of neurotransmitters, the carrier proteins may also transport either a neurotransmitter (homoexchange) or another substance (heteroexchange) in the outward direction (Bernath, 1992). That

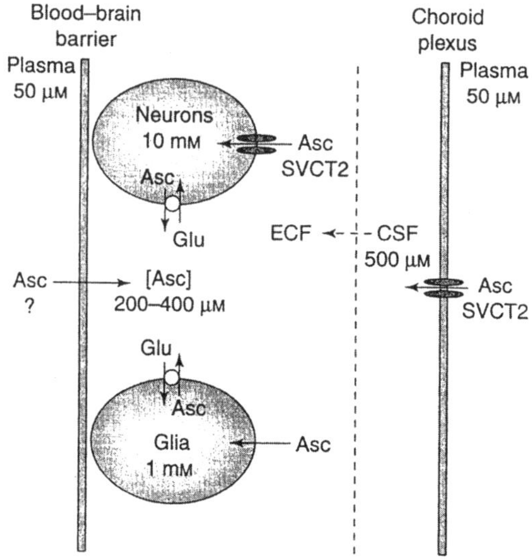

Figure 13.1: Diagrammatic representation of the processes involved in the transfer of ascorbate (ASC) into the CNS.
Typically, plasma ASC levels are around 50 μM although these values vary widely. Initially, the vitamin is actively transported across the choroid plexus where it concentrates in the ventricular cerebrospinal fluid (CSF); concentrations of ASC in the CSF can be as high as 500 mM. ASC in the CSF is in equilibrium with that in the extracellular fluid (ECF) of the brain; from this compartment ASC is transported into neurons via the SVCT2 transporter and, less actively, into the glia by mechanisms which have not been clarified. The entrance of vitamin C into the brain from blood capillaries and across the blood–brain barrier seems greatly limited. Also illustrated in this figure is the release of ASC from neurons and glia via a heteroexchange mechanism on the glutamate (GLU) transporter. From Rice (2000). Reprinted from Rice, M.E. (2000) with permission from Elsevier.

ASC is released via a heteroexchange mechanism via amino acid transporters is supported by a variety of studies (Bigelow *et al.*, 1984; Grünewald and Fillenz, 1984; Cammack *et al.*, 1991). This process of ASC release may apply to both neurons and glia (Rebec and Pierce, 1994). While both glutamate and β-aminobutyric acid influence the discharge of ASC into the ECF in the striatum, experimental evidence shows that dopaminergic activity in this brain region is essentially inconsequential in terms of vitamin C release (Kuhr *et al.*, 1987).

13.3 Distribution of ASC in the brain

In general, the brain contains rather high concentrations of ASC although these levels vary widely according to the brain region in which the constituent is measured and the species from which the neural tissue is obtained.

A detailed study of the CNS of the Sprague–Dawley rat revealed the nucleus accumbens contains the highest levels [464 ± 38 (SEM) $\mu g \cdot g^{-1}$ wet tissue weight] of ASC with the substantia nigra having the lowest mean (190 ± 11 $\mu g \cdot g^{-1}$) concentrations (Milby *et al.*, 1982). Other nuclear groups or regions in which ASC levels were estimated fell between these extremes (*Table 13.1*). In this study the thalamus, caudate and hippocampus (exclusive of the dentate gyrus) also were divided along their anterior–posterior axes and measurements of ASC were made. In each of these structures the samples collected anteriorly had higher concentrations of ASC, i.e., for each tissue there was an anterior–posterior gradient. The physiological significance, if any, of these anterior–posterior gradients remains unknown. In the hippocampal formation, the dentate gyrus had roughly 50% higher ASC concentrations than did the remaining hippocampal tissue. In this study, the tissue samples were collected using the punch technique after the brain was sliced coronally and the measurement of ASC was done utilizing liquid chromatography with electrochemical detection, a method that unequivocally verifies the presence of ASC.

The results obtained by Milby and colleagues (1982) are generally consistent with earlier, less detailed studies wherein ASC levels were biochemically measured in much larger regions of the rat brain (Rajalakshmi *et al.*, 1967; Hornig, 1975). According to the former report, placing rats on a low protein diet reduced ASC synthesis in the rat liver and likewise lowered regional brain concentrations of ASC. During early development, ASC levels in whole rat brain change substantially in the first 4 weeks of life. According to Adlard *et al.* (1973) ASC levels are 80% higher in the CNS of 5-day-old rats compared to prepubertal rats at 39 days of age. Food restriction for 16 h during the neonatal period was without influence on brain ASC levels.

Kuo *et al.* (1978) preliminarily examined the subcellular distribution of ASC in the rat brain. In this study, synaptosomal cytoplasm was found to have the highest concentrations of ASC. Conversely, little ASC was associated with the particulate fractions. More detailed studies on the intracellular localization of this vitamin are needed.

ECF levels of ASC have been estimated by Schenk *et al.* (1982) using rat brain slides and voltammetric measurements. Accordingly, they reported that the ECF concentrations of ASC in the caudate nucleus are on the order of 284 μM. In an

Table 13.1: Mean (± SEM) ASC concentrations in the brain of adult Sprague–Dawley rats. The brains of both males and females are included (n = 5). Data from Milby *et al.* (1982).

Nuclear group/brain region	ASC concentration ($\mu g \cdot g^{-1}$ wet tissue)
Nucleus accumbens	464 ± 38
Hippocampus	394 ± 15
Neocortex	380 ± 15
Amygdala	370 ± 52
Hypothalamus	349 ± 44
Septum	317 ± 28
Caudate-putamen	306 ± 8
Globus pallidus	235 ± 18
Thalamus	220 ± 17
Reticular formation	217 ± 18
Substantia nigra	190 ± 11

ancillary *in vivo* study, ECF levels of ASC in the caudate nucleus were found to be 227 μM, a value statistically indistinguishable from that obtained in the *in vitro* experiments. Schenk and co-workers (1982) also showed that the ECF ASC concentrations are often maintained at the expense of total tissue levels.

Other species in which neural concentrations of ASC have been routinely measured include the guinea pig (Chinoy, 1972; Hornig, 1975; Kratzing *et al.*, 1982) and human (Schaus, 1957; Hornig, 1975, Mefford *et al.*, 1981; Oke *et al.*, 1987). These species are similar in as much as neither has the capability of synthesizing the vitamin. In the guinea pig brain the distribution and concentrations of ASC are not markedly different than those in the rat. When the brain of the guinea pig was grossly dissected into large regional sections, each contained levels of ASC similar to those in the rat brain (Chinoy, 1972). The cerebral cortex, cerebellar cortex and the brain stem (pons/medulla) of the 2-day-old guinea pig brain contained larger amounts of ASC than did the brain of older guinea pigs, i.e., 4-month-old or 10–12-month-old animals (Kratzing *et al.*, 1982). Particularly in the study of Chinoy (1972), the function of ASC in the neural tissue was presumed to be that of a free radical scavenger although no evidence was presented to support this assumption. Also, given that merely rinsing the dissected brain regions with 37°C saline removed much of the ASC, Kratzing and co-workers (1982) surmised that in the intact brain much of the ASC is in the ECF rather than intracellular. The percentage of ASC removed from the brain by rinsing was even greater when the animals were rendered scorbutic.

The human brain has been dissected into a number of nuclear groups in which the ASC concentrations have been measured. Among these nuclear groups, the septal nuclei and the hypothalamus seem to contain the highest levels of ASC (roughly 200 $mg \cdot g^{-1}$ wet tissue), although there are no major differences between the various areas (Mefford *et al.*, 1981). In the most thorough examination, Oke *et al.* (1987) claimed an increasing dorsal to ventral gradient of ASC concentrations in the human brain; however, this was not particularly uniform from one brain to another and was not apparent in all brains. Certainly, the study of Oke and colleagues (1987) provides the most detailed information on the regional distribution of ASC in the human brain but the functional significance of the measured concentrations remains, for the most part, unknown.

Increasing age reportedly is associated with a substantial reduction in brain ASC levels, at least in the cerebral cortex (Schaus, 1957). The supposed age-dependent changes are, however, in need of further study.

13.4 Actions of ASC in the brain

The number of published reports relating to the functional implications of ASC in the CNS is certainly not voluminous and the reports that do exist, in many cases, are isolated findings. Even though the widespread importance of ASC as a significant reductive force in many tissues is well known (Frei *et al.*, 1990; May, 1999), this function has not been experimentally exploited to a major degree in reference to the brain.

13.4.1 Antioxidant actions of ASC in ischemia/reperfusion injury

Ischemia/reperfusion (I/R) injury, i.e., stroke, is known to cause damage to the brain via processes that involve free radicals (Nagel *et al.*, 1997; Cheung, 2003) as well as by other means (Takizawa *et al.*, 1996). When Stamford *et al.* (1999) tested the ability of ASC to combat cellular death in the brain after I/R-induced injury, the vitamin was found to be only marginally protective. Using a complex scheme of high-dose ASC administration, given before and after I/R induction, ASC only reduced neuronal loss in the striatum (measured at 5 days after 5 min bilateral carotid artery occlusion in gerbils) at the highest dose given; in the hippocampus, ASC afforded no protection from the I/R insult. In a follow-up study, the authors (Stamford *et al.*, 1999) showed that ASC inhibited dopamine (DA) (measured by voltammetry in an *in vitro* model) in the striatum. Given that DA is normally released during I/R injury (Phebus *et al.*, 1986) and under these conditions it has neurotoxic actions (Weinberger *et al.*, 1985), it was surmised that ASC was protective in the *in vivo* gerbil brain against I/R due to its ability to limit DA release. The authors also deduced that the antioxidant properties of ASC had no role in protecting the hypoxic-reoxygenated brain under the conditions of this study.

A novel phosphate ester of vitamin C, i.e., L-ascorbic acid 2-[3,4-dihydro-2,5,7,8-tetramethyl-2-(4,8,12-trimethyltridecly)-2-H-1-benzopyran-6-yl-hydrogen phosphate] potassium salt (EPC-K1) has been tested as to its ability to reduce neural I/R injury after transient middle cerebral artery occlusion (MCAO) in rats (Zhang *et al.*, 2001). This newly synthesized molecule is soluble in both aqueous and lipid media. Presumably via its antioxidative properties, EPC-K1 reduced DNA damage, i.e., 8-hydroxy-2-deoxyguanosine levels, apoptosis (as assessed by the Terminal UTP-nick end labeling (TUNEL) assay) and, to a lesser degree, caspase-3 staining particularly in the marginal area of the infarct. In this study, EPC-K1 was given at a dose of 20 $mg \cdot kg^{-1}$, intravenously, at ischemia onset and 3 h later (2 h after reperfusion onset).

Evidence has been presented that ASC protects rat brain mitochondrial respiration from the disruptive effects of I/R injury (Sciamanna and Lee, 1993). A 30-min period of ischemia caused roughly a 60% reduction in state 3 respiratory rates with both succinate and NAD-linked substrates and also in energy-linked Ca^{2+} transport; conversely, state 4 respiratory rates were not changed under these conditions. These ischemic responses were changed only slightly by ASC administration. However, ASC, given 5 min in advance of the ischemic episode (induced by seven-vessel occlusion of neural arteries in the rat), increased dose dependently state 3 and reduced state 4 respiratory rates of mitochondria collected from ischemic, reperfused brains; maximal protection was achieved with a ASC dose of 0.8 $mg \cdot kg^{-1}$ body weight. A time course study demonstrated that ASC was most protective against I/R-induced mitochondrial alterations when administered prior to ischemia and only partially so when it was given after reperfusion was re-established. While the authors surmised that ASC was protective because of its free radical scavenging activity, no direct evidence for this was presented.

ASC was prophylactically used to reduce infarct size in a primate model of focal cerebral ischemia (Ranjan *et al.*, 1993). In this study, adult monkeys (*Macaca radiata*) were pre-treated intraperitoneally for 6 days with 1 g ASC daily after which half were subjected to MCAO for 2 h. For the occlusive procedure the monkeys were anesthetized with pentobarbitone. During the 2-h period of ischemia ASC levels in the basal ganglia decreased relative to those in the sham operated (without the imposition of ischemia) monkeys. At 24 h after the induced ischemia, macroscopic infarct size (estimated from three brain slices) was assessed using triphenyl tetrazolium chloride-stained brains; this revealed the area of infarction was reduced in size in the ASC supplemented monkeys. There was, however, significant variation in terms of infarct size and the degree of protection provided by ASC among the different animals.

Partial functional restoration after I/R injury of the spinal cord of rabbits was noted following the intravenous administration of a hypothermic solution containing 30 mg·kg^{-1} methyl prednisolone and 10 mg·kg^{-1} each of vitamins C and E (Tetik *et al.*, 2001). The rabbits were functionally evaluated for the degree of spastic paraplegia at 24 and 48 h after I/R. When given individually methyl prednisolone, vitamin C, and vitamin E were less effective in terms of functional restoration than when they were given in combination. In this case the protective actions of vitamins C and E were presumed to involve their antioxidative properties while methyl prednisolone was likely protective for other reasons (Tetik *et al.*, 2001). Since the solutions were cooled before administration, their protective actions in these studies may have been in part a consequence of the reduced temperature of the I/R tissue.

While extracellular striatal ASC levels reportedly rise during the early stages of I/R (Yasu, 2000), it is clear from the data discussed above that the role of this vitamin in protecting against transient hypoxia and reoxygenation is still somewhat ambiguous. Also, in a review of the literature regarding epidemiological risk factors for neural damage, Wolf (1997) concluded that the consumption of vitamin C has not been shown to reduce the incidence of stroke.

The relationship of ASC and glutathione (GSH), both low molecular weight antioxidants in the mammalian brain, is of special interest. Total tissue concentrations for both ASC and GSH are on the order of 2–3 mM, values much higher than those of other antioxidants in the brain. The differential properties and distribution of these molecules suggest they may have complementary roles in the CNS. ASC is more highly concentrated in neurons (Lyrer *et al.*, 1991) while GSH is preferentially localized in glia (Rice and Russo-Menna, 1998). Their complementary actions are indicated by the fact that depletion of either one is compensated for by the other and thus they act as a redox couple (Miester, 1994; Winkler *et al.*, 1994). However, the use of these agents to potentially alter I/R-induced injury to the brain seems not to have been performed.

13.4.2 Prooxidant actions of ASC in neural tissue

While the significance of ASC as a major protective antioxidant of the brain after I/R injury is still unclear, there are findings to suggest that in some of these conditions ASC may actually have prooxidant actions in the CNS as in other organs (Carr and Frei, 1999). The prooxidant actions of ASC in the brain in some cases likely relate to its ability to interact with hemoglobin-derived iron to eventually generate the highly toxic hydroxyl radical ($OH^{\cdot}$) (*Figure 13.2*). Particularly in the case of hemorrhage into the brain, hemoglobin is liberated into the interstitium. Oxyhemoglobin quite readily loses its heme via spontaneous dissociation (Hebbel *et al.*, 1988). Once released, heme, a hydrophobic molecule, is likely inserted into the lipid, especially important in the CNS given the high concentrations of phospholipids in this tissue. At this site, heme is eventually broken down probably in a reaction with partially oxidized fatty acids; in this process free iron is released (Sadrzadeh *et al.*, 1987). ASC, which as noted above is particularly abundant in the CNS, then functions in the reduction of Fe^{3+} to Fe^{2+} (*Figure 13.2*). Via the Fenton reaction Fe^{2+} may then react with available H_2O_2 to generate the $OH^{\cdot}$ which readily destroys any molecule it encounters (Sadrzadeh and Eaton, 1988). Additionally, there are other potential actions of Fe^{2+}, e.g., homolytic decomposition of fatty acid hydroperoxides, which may contribute to the damaging effects of reducing agents such as ASC in the brain.

Recently, an alternative mechanism has been proposed to explain the toxic reactions of ASC in the CNS. Since the previously discussed findings, as summarized in *Figure 13.2*, were demonstrated exclusively with cellular homogenates in protein-free medium, Song and co-workers (1999, 2001) have proposed another scheme to explain ASC-mediated damage to intact brain cells. They formulated their hypothesis after observing that ASC functioned as a reducing agent in brain tissue homogenates but had an oxidizing effect in brain slides which contained intact cells. They provide evidence that ASC-mediated oxidative damage is not a result of the interaction of ASC with iron (Song *et al.*, 1999). Rather, they propose the following interactions (Si *et al.*, 1998; Song *et al.*, 1999). After ASC is oxidized to dehydroascorbate, it is transported into cells via a glucose transporter; in the cytosol dehydroascorbate is reduced back to ASC and, as a result of this reductive reaction, other essential cellular components are oxidized (Song *et al.*, 2001).

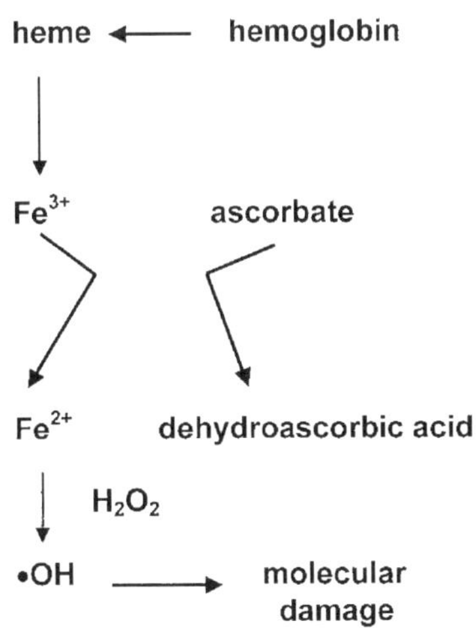

Figure 13.2: A simplified scheme summarizing one of the potential means whereby ASC promotes molecular damage via free radical mechanisms in the CNS. The molecular damage in the brain most often involves lipids; however, the integrity of proteins and DNA can be destroyed by this means as well.

The commonly used drug, 6-hydroxydopamine (6-OHDA), is known to be neurotoxic with the mechanism of molecular damage involving free radicals. During the auto-oxidation of 6-OHDA *in vitro*, ASC induced the continuous generation of $OH^{\cdot}$ (Mendez-Alvarez *et al.*, 2001). The *in vivo* injection of 6-OHDA in combination with ASC caused large striated lesions as judged by the loss of immunocytochemically-detected tyrosine hydroxylase. According to Mendez-Alvarez and co-workers (2001), these findings establish that ASC is a critical element in the neurotoxicity of 6-OHDA. This finding is consistent with those of Choi and co-workers (2000) who also reported that ASC markedly potentiated the toxicity of 6-OHDA in pheochromocytoma-derived PC12 cells. While the pro-oxidant actions of ASC have been documented under a variety of experimental conditions, under what conditions ASC functions in this capacity *in vivo* remain unknown. Perhaps its interaction with Fe^{2+} is potentially the most damaging.

13.4.3 ASC and models of neurodegeneration

While the epidemiological evidence linking vitamin C intake to the onset and progression of Alzheimer disease (AD) is not compelling (Foley and White, 2002), there is a modicum of data indicating that ASC may reduce the toxicity of amyloid β-peptide (Aβ) under experimental conditions (Prehn *et al.*, 1996). Aβ, a prominent constituent in the brain of AD patients, is widely accepted as being toxic to neurons due to its ability to generate ROS. As a result, many antioxidants have been tested for their ability to reduce neuronal death after exposure to Aβ; this has also been done for ASC. Using rat hippocampal neurons grown in culture, Prehn *et al.* (1996) found that the addition of ASC (300 mM) to the incubation medium of neurons treated with either $A\beta_{1-40}$ or $A\beta_{25-35}$ (the oxidatively active fragment), both at concentrations of 1 μM, significantly reduced neuronal death after 5 days; a lipophilic antioxidant, 2-mercaptoethanol (10 μM) had a similar protective action. The authors in this case did not, however, establish that the beneficial actions of ASC were directly related to its antioxidative capabilities.

The ability of ASC to modulate DA oxidation-mediated neuronal toxicity was investigated by Choi *et al.* (2000). During treatment of PC12 cells with DA, ASC was added to the incubation medium at various intervals after DA addition; not unexpectedly, the earlier ASC was added, the more effective it was in reducing DA-mediated death as estimated by measuring acid phosphatase activity and lactate dehydrogenase leakage. Combining ASC with a sulfhydryl-rich antioxidant, i.e., glutathione, eliminated the toxicity of DA indicating, according to the authors, that the use of reducing agents may be a beneficial strategy in protecting against DA-mediated neuronal cytotoxicity such as occurs in Parkinsonism.

ASC has also been implicated as a beneficial agent in an experimental model of Huntington disease (HD), although the actual data remain sparse. As already established, the ECF of the brain contains high levels of ASC. Considering this, Rebec *et al.* (2002) evaluated, using slow scan voltammetry, the extracellular dynamic changes in ASC concentrations in a mouse strain (R6/2) engineered to express the gene for HD. The measures were done in the striatum and the mice were anesthetized for the base-line measures. During the period of anesthesia, concentrations of ASC were the same in striatal ECF from controls and HD mice.

During behavioral activation, however, ASC levels increased in the ECF of the control mice while decreasing in the HD mice. The authors interpreted this difference to mean that striatal ASC release is disrupted in HD, correlating with the abnormal behavior that is apparent at this time. The findings have potential significance in as much as HD is theoretically related to a deficiency of antioxidant protection (Coyle and Puttfarcken, 1993).

Kok (1997) has proposed that the age-related decline in ASC triggers amyotrophic lateral sclerosis (ALS). The reduction in ASC levels is followed by a lowering of ASC in neurons, normally at millimolar concentrations, to greatly diminished levels which then precipitate upper, and to a greater degree, lower motoneuron loss. However, there is no epidemiological evidence to suggest such an association in humans.

13.4.4 ASC and hypothalamic neuropeptide release

Thyrotropin-releasing hormone (TRH; pyro-Glu-His-Pro-NH_2) is one of a family of COOH-terminal α-amidated peptides. TRH, although best known for its release of thyroid stimulating hormone from the anterior pituitary after its synthesis in the hypothalamus, actually is rather widely distributed in the CNS. The α-amidation of TRH by the enzyme peptidyl-α-amidating monooxygenase (PAM) utilizes ASC, along with copper, and molecular oxygen, as a cofactor (Glembotski *et al.*, 1986). In the study in question, primary hypothalamic neuroendocrine cells were treated with ASC and TRH levels were estimated after treatment of the cells with an antiserum to the tripeptide. The presence of ASC in the medium markedly increased (two- to threefold) cellular TRH levels with the response being dose dependent. In this study the concentration of ASC that caused a 50% maximal response, i.e., the EC_{50}, was 20 mM. D-isoascorbic acid, an ASC isomer also shown to be a cofactor for PAM *in vitro*, augmented cellular immunoreactive TRH levels as well. The authors concluded that ASC, along with copper and molecular oxygen, is an essential cofactor for PAM and, therefore, for TRH synthesis (Glembotski, 1987).

Another hypothalamic releasing hormone, luteinizing hormone releasing hormone (LHRH), is reportedly influenced by vitamin C (Miller and Cicero, 1987). ASC (0.5 mM) caused a calcium-dependent discharge of LHRH from rat medial basal hypothalamic fragments *in vitro*. This response was blocked both by the α-adrenergic receptor antagonist, phentolamine, and by the mu receptor opiate agonist, sufentanil. On the other hand, propranolol, a β-receptor blocker, was without effect on the ability of ASC to modulate LHRH release. These findings are consistent with the hypothesis that ASC augments LHRH release via an enhancement of norepinephrine activity at the level of α1-adrenergic receptors. The ability of ASC to alter LHRH release *in vivo* has yet to be demonstrated.

13.4.5 ASC and synaptic transmission

ASC is often used as a reducing agent in the medium of brain homogenates when the binding of ligands to their receptors are investigated. Under these conditions, ASC frequently has been found to lower ligand binding, most likely as a consequence of ASC-induced lipid peroxidation. Examples of receptors whose binding by ligands have been lowered in the presence of ASC include opiate receptors, serotonin receptors, α-adrenergic and β-adrenergic receptors, and dopamine receptors (Heikkila and Monzino, 1987). Given these findings, the authors caution about using an appropriate concentration of ASC in these experimental settings, e.g., a concentration that prevents ligand decomposition but one that does not promote significant lipid breakdown. Alternatively, investigators could combine ASC with a chelating agent or an inhibitor of lipid peroxidation or, finally, use another reducing agent that limits ligand decomposition but does not induce the peroxidation of membrane lipids. Under the best of circumstances, since ASC functions as a 'double edged sword', it must be used cautiously under the experimental conditions described in these reports.

ASC may have some actions in reference to synaptic modulation. Certainly, this seems to be the case for the actions of DA. A number of laboratories have reported that concentrations of ASC ranging from 0.1 to 1.5 mM reduced the binding of radiolabeled agonists to the DA receptor in brain homogenates (Kayaalp and Neff, 1980; Arana *et al.*, 1982; Tolbert *et al.*, 1992); there has, however, also been a failure in reproducing these finding (Bacopoulos, 1982). Overall, while there seems to be an effect of ASC in modulating the ability of the dopaminergic receptor to interact with agonists, there is no universal agreement among the published reports and documenting this effect *in vivo* is yet to be reported.

Besides reducing agonist binding to the DA receptor, the addition of ASC to neural homogenates also inhibited specific binding of DA antagonists (Kayaalp and Neff, 1980; Chan *et al.*, 1982; Dorris, 1987; Wiener *et al.*, 1989). While ASC reduces agonist binding in a dose-dependent manner, it blocks antagonist binding according to an inverted U-shaped function with maximal inhibition occurring at intermediate concentrations of the vitamin (Heikkila *et al.*, 1981, 1982). Another difference is that whereas ASC competitively lowers binding of DA agonists, its action in reference to the antagonists is noncompetitive (Heikkila *et al.*, 1982). Since the inhibition of antagonist binding by ASC results in an increased level of lipid peroxidation products (Ebersole and Molinoff, 1991), and that the inhibition is prevented by the addition of other antioxidants to the homogenates (Dunlap *et al.*, 1979; Muakkassah-Kelly *et al.*, 1981), the prooxidant actions of ASC seem to be involved in its ability to hinder DA agonist and antagonist binding to the receptor. While there may be disagreements as to the mechanisms involved in the inhibitory actions of ASC at the level of the DA receptor, it does oppose the action of the monoamine in the striatum. Given that ASC is released from glutaminergic terminals in the neostriatum, a role for this vitamin at this location seems likely.

The initial documentation that ASC influenced glutaminergic neurotransmission in the striatum was indirect; in these investigations it was shown that the systemic administration of ASC increased electrically recorded neural activity (Ewing *et al.*, 1982); in this study it was shown that the rise in neuronal firing rate correlated with the arrival of vitamin C at the appropriate sites. The study obviously could not determine whether the action of ASC on glutaminergic transmission was either presynaptic or postsynaptic. Subsequent investigations showed that ASC reduces the binding of glutamate to the N-methyl-D-aspartate (NMDA) receptor (Majewska *et al.*, 1990). This action was believed to relate to the antioxidant effects of the vitamin since other antioxidants, although not all, had a similar effect. The other antioxidants actually promoted NMDA receptor activity. As a result of the differential actions of the antioxidants used, Majewska and co-workers (1990) suggested that ASC acts on the regulatory site of the NMDA receptor that alters its electric charge, whereas the disulfide antioxidants (which enhanced NMDA receptor physiology) act on the regulatory sulfhydryl site.

If the actions of ASC on glutaminergic neurotransmission are valid, *in vivo* the vitamin should alter the behavioral effects of drugs that act on glutamate receptors. When tested, ASC did in fact enhance the cataleptic effects of haloperidol (Rebec *et al.*, 1985; Dorris and Dill, 1986), documenting in fact that ASC potentiates glutamate neurotransmission *in vivo*.

While a number of actions of ASC on the NMDA receptor and influences on glutamate have been observed, the underlying mechanisms are far from established. A carrier-mediated heteroexchange of ASC at the glutamate transporter (described above) suggests a presynaptic site of action of ASC; however, ASC could potentiate the postsynaptic effects of glutamate by a direct interaction at NMDA receptors as well.

13.4.6 Relative efficacies of ASC and other antioxidants in neural tissue

The efficacy of ASC in protecting the brain from processes which lead to tissue loss via free radical mechanisms has been compared with that of other antioxidants. Using a model of phenylketonuria in rat pups, Martinez-Cruz and colleagues (2002) examined the ability of three antioxidants, i.e., ASC, vitamin E and melatonin in reducing oxidative damage in the CNS. The daily doses of antioxidants used were 30 $mg \cdot kg^{-1}$ for both ASC and vitamin E and 20 $mg \cdot kg^{-1}$ for melatonin. The oxidative endpoints in this study were Ehrlich adducts, levels of products of lipid peroxidation, reduced and oxidized GSH levels, and activities of glutathione peroxidase and reductase in the cerebrum and cerebellum of rats. While each of the antioxidants reduced some parameters of free radical damage, the protective actions of ASC were somewhat less pervasive than those of melatonin or vitamin E.

In another study, Martin *et al.* (2000) compared the ability of ASC, vitamin E and melatonin to maintain GSH homeostasis in isolated brain mitochondria oxidatively stressed using *t*-butyl hydroperoxide (*t*-BHP). When mitochondria were incubated with 100 mM *t*-BHP, the majority of the GSH was converted to the oxidized form (GSSG). While the antioxidant melatonin (100 nM) counteracted this effect of *t*-BHP, neither vitamin antioxidant was capable of reversing the oxidation of GSH. Other parameters yielded similar results with only melatonin being capable of reversing the oxidative actions of *t*-BHP in the mitochondria.

DA autoxidation, a process that can lead to degeneration of the nigrostriatal dopaminergic system and consequentially to signs of Parkinson disease, has been used as a model system to compare the antioxidative properties of ASC, trolox (water soluble vitamin E), deprenyl and melatonin (Khaldy *et al.*, 2000). In this study the rate of accumulation of 2,3-dihydroxybenzoate (2,3-DHBA) in the presence of salicylate, an $OH^{\cdot}$ scavenger, during the autoxidation of DA indicated that ASC was ineffective in neutralizing the OH· while the other agents did so dose dependently. The authors thus concluded that under the conditions of this experiment, ASC was incapable of reducing the autoxidation of DA.

Vitamin C has also been compared with the antioxidant melatonin in lowering oxidative stress and restoring antioxidant enzyme activity in a model of AD (Montilla-Lopez *et al.*, 2002). The model of AD used in this case was the treatment of cultured neuroblastoma cells with okadaic acid. After 2 h incubation with 50 nM okadaic acid, the neuroblastoma cells exhibited increased lipid peroxidation and depressed activities of glutathione transferase, glutathione reductase and catalase. Both ASC (0.5 $\mu g \cdot ml^{-1}$) and melatonin (10^{-5} M) prevented the oxidative changes. Although no dose–response relationships were determined, the available data indicates that vitamin C was less effective than melatonin in reducing the oxidative damage in these cells.

In the studies summarized herein ASC generally was less protective of brain tissue from free radical generating processes than were the other antioxidants with which it was compared. The bulk of the studies, however, were performed *in vitro* and how these findings relate to the *in vivo* situation remains unknown; as a result, it would not be justified to draw conclusions concerning how effective these agents would be under conditions where factors such as absorption, uptake, distribution and metabolism are so different than in cultured cells or homogenized tissues.

13.5 Concluding remarks

Some of the main points summarized in this resumé include the uptake routes of ASC into the brain, its uneven distribution in the CNS, its differential release mechanisms into the ECF, its actions during I/R injury and in experimental models of neurodegeneration, and its ability to modulate synaptic processes. In many of these areas, especially in terms of the functions of ASC in the CNS, the literature does not provide a particularly clear picture of specific actions of the vitamin in the brain. This seems somewhat incongruous given that investigations of ASC have a very long experimental history and its high levels in the brain certainly portend critical functions in this tissue. Interestingly, progress in defining the actions of ASC in the brain have also been relatively slow in recent years. Thus, we seem to be far from understanding all the actions of this important factor in brain neurochemistry and physiology.

References

Adlard BPF, de Souza SW, Moon S (1973) The effect of age, growth retardation and asphyxia on ASCorbic acid concentrations in developing brain. *J. Neurochem.* **21**, 877–881.

Agus DB, Gambhir SS, Pardridge WM, Spielholz C, Baselga J, Vera JC, Golde DW (1997) Vitamin C crosses the blood–brain barrier in the oxidized form through glucose transporters. *J. Clin. Invest.* **1000**, 2842–2848.

Arana GW, Baldessarini JR, Kula NS (1982) Differential effects of ascorbate and EDTA on high affinity binding of ^{3}H-apomorphine and ^{3}H-ADTN to calf caudate membranes. *Neuropharmacology* **21**, 601–604.

Bacopoulos NG (1982) Dopamine receptors in rat brain regions – optimal conditions for ^{3}H-agonist binding, pH dependency and lack of inhibition by ascorbic acid. *Biochem. Pharmacol.* **31**, 3085–3091.
Bernath S (1992) Calcium-independent release of amino acid neurotransmitters – fact or artifact. *Prog. Neurobiol.* **38**, 57–91.
Bigelow JC, Brown DS, Wightman KM (1984) β-Aminobutyric acid stimulates the release of endogenous ascorbic acid from rat striatal tissue. *J. Neurochem.* **42**, 412–419.
Booij J (1940) The vitamin-C content of blood and cerebrospinal fluid. *Rec. Trav. Chim.* **59**, 713–719.
Boutelle MG, Svensson L, Fillenz M (1989) Rapid changes in striatal ascorbate in response to tail-pinch monitored by constant potential voltammetry. *Neuroscience* **30**, 11–17.
Boutelle MG, Svensson L, Fillenz M (1990) Effect of diazepam on behavior and associated changes in ascorbate concentration in rat brain areas: striatum, n. accumbens and hippocampus. *Psychopharmacology* **100**, 230–236.
Cammack J, Ghasemzadeh B, Adams RN (1991) The pharmacological profile of glutamate-evoked ascorbic acid efflux measured by in vivo electrochemistry. *Brain Res.* **565**, 17–22.
Carr A, Frei B (1999) Does vitamin C act as a pro-oxidant under physiological conditions. *FASEB J.* **13**, 1007–1024.
Chan B, Seeman P, Davis A, Madras BK (1982) Ascorbate injury and EDTA (or manganese) protection of D-2 dopamine receptors. *Eur. J. Pharmacol.* **81**, 111–116.
Cheung RTF (2003) The utility of melatonin in reducing cerebral damage resulting from ischemia and reperfusion. *J. Pineal Res.* **34**, 153–160.
Chinoy NJ (1972) Ascorbic acid levels in mammalian tissues and its metabolic significance. *Comp. Biochem. Physiol.* **42A**, 945–952.
Choi HY, Song JH, Park DK, Ross GM (2000) The effects of ascorbic acid on dopamine-induced death of PC12 cells are dependent on exposure kinetics. *Neurosci. Lett.* **296**, 81–84.
Coyle JT, Puttfarcken P (1993) Oxidative stress, glutamate, and neurodegenerative disorders. *Science* **262**, 689–695.
Dorris RL (1987) Ascorbic acid reduces accumulation of [^{3}H]spiperone in mouse striatum in vivo. *Proc. Soc. Exp. Biol. Med.* **186**, 13–16.
Dorris RL, Dill RE (1986) Potentiation of haloperidol-induced catalepsy by ascorbic acid in rats and nonhuman primates. *Pharmacol. Biochem. Behav.* **24**, 781–783.
Dunlap CE III, Leslie FM, Rado M, Cox BM (1979) Ascorbate destruction of opiate stereospecific binding in guinea pig brain homogenate. *Mol. Pharmacol.* **16**, 105–119.
Ebersole BJ, Molinoff PB (1991) Inhibition of binding of [^{3}H]PN200-110 to membranes from rat brain and heart by ascorbate is mediated by lipid peroxidation. *J. Pharmacol. Exp. Ther.* **259**, 337–344.
Ely JT, Krone CA (2002) Aging: predictions of a new perspective on old data. *Exp. Biol. Med.* **227**, 939–942.
Ewing AG, Wightman RM, Dayton MA (1982) In vivo voltammetry with electrodes that discriminate between dopamine and ascorbate. *Brain Res.* **249**, 361–370.
Fillenz M, O'Neill RD, Grünewald RA (1986) Changes in extracellular fluid ascorbate concentration as an index of excitatory amino acid release. In: *Monitoring Neurotransmitter Release During Behavior* (eds. MH Joseph, M Fillenz, IA MacDonald, CA Marsden). Ellis Harwood, Chichester, pp. 144–153.
Foley DJ, White LR (2002) Dietary intake and of antioxidants and risk of Alzheimer's disease. *J. Am. Med. Assoc.* **287**, 3261–3263.
Frei B, Stocker R, Engl C, Ames BN (1990) Ascorbate: the most effective antioxidant in human blood plasma. *Adv. Exp. Med. Biol.* **264**, 155–163.
Glembotski CC (1987) The role of ascorbic acid in the biosynthesis of neuroendocrine peptides α-MSH and TRH. *Ann. N.Y. Acad. Sci.* **498**, 54–62.
Glembotski CC, Menaker S, Winokur A, Gibson TR (1986) Ascorbic acid increases thyprotropin-releasing hormone content of hypothalamic cell cultures. *J. Neurosci.* **6**, 1796–1802.
Grünewald RA, Fillenz M (1984) Release of ascorbate from symptosomal fraction of rat brain. *Neurochem. Int.* **6**, 491–500.
Hammarstrom L (1966) Autoradiographic studies on the distribution of C^{14}-labelled ascorbic acid and dehydroascorbic acid. *Acta Physiol. Scand. Suppl.* **289**, 1–70.
Hebbel RP, Morgan WT, Estin JW, Hedlund BE (1988) Accelerated autoxidation and heme loss due to instability of sickle hemoglobin. *Proc. Natl. Acad. Sci. USA* **85**, 237–241.

Heikkila RE, Manzino L (1987) Ascorbic acid, redox cycling, lipid peroxidation, and the binding of dopamine receptor antagonists. *Ann. N.Y. Acad. Sci.* **498**, 63–73.

Heikkila RE, Cabbat FS, Manziano L (1981) Differential inhibitory effects of ascorbic acid on the binding of dopamine agonists and antagonists to neostriatal membrane preparations: correlations with behavioral effects. *Res. Commun. Chem. Pathol. Pharmacol.* **34**, 409–421.

Heikkila RE, Cabbat FS, Manzino L (1982) Inhibitory effects of ascorbic acid on the binding of ^{3}H-dopamine antagonists to neostriatal membrane preparations: relationship to lipid peroxidation. *J. Neurochem.* **38**, 1000–1006.

Hornig D (1975) Distribution of ascorbic acid, metabolites and analogues in man and animals. *Ann. N.Y. Acad. Sci.* **258**, 103–118.

Kayaalp SO, Neff NH (1980) Differentiation by ascorbic acid of dopamine agonist and antagonist binding sites in striatum. *Life Sci.* **26**, 1837–1841.

Khaldy H, Escames G, Leon J, Vives F, Luna JD, Acuña-Castroviejo D (2000) Comparative effects of melatonin, L-deprenyl, Trolox and ascorbate in the suppression of hydroxyl radical formation during dopamine autoxidation in vitro. *J. Pineal Res.* **29**, 100–107.

Kok AB (1997) Ascorbate availability and neurodegeneration disorders in amyotrophic lateral sclerosis. *Med. Hypothesis* **48**, 281–296.

Kratzing CC, Kelly JD, Obelisks BA (1982) Ascorbic acid in neural tissues. *J. Neurochem.* **59**, 625–627.

Kuhr WG, Wightman RM, Rebec G (1987) Dopaminergic neurons: simultaneous measurements of dopamine release and single unit activity during stimulation of the medial forebrain bundle. *Brain Res.* **418**, 122–128.

Kuo CH, Yonehara N, Hata F, Yoshida H (1978) Subcellular distribution of ascorbic acid in the rat brain. *Jap. J. Pharmacol.* **28**, 789–790.

Lam DKC, Daniel PM (1986) The influx of ascorbic acid into the rat's brain. *Q. J. Exp. Physiol.* **71**, 483–489.

Lyrer P, Lolt H, Kabiersch A, Langemann H, Kaeser A (1991) Levels of low molecular weight scavengers in the rat brain during focal ischemia. *Stroke* **22**, 1548–1553.

Majewska MD, Bell JA, London ED (1990) Regulation of the NMDA receptor by redox phenomena – inhibitory role of ascorbate. *Brain Res.* **537**, 328–332.

Martin M, Macias M, Escames G, Leon J, Acuña-Castroviejo D (2000) Melatonin but not vitamins C and E maintains glutathione homeostasis in t-butyl hydroperoxide-induced mitochondrial oxidative stress. *FASEB J.* **14**, 1677–1679.

Martinez-Cruz F, Pozo D, Osuna C, Espinar A, Merchante C, Guerrero JM (2002) Oxidative stress induced by phenylketonuria in the rat: prevention by melatonin, vitamin E, and vitamin C. *J. Neurosci. Res.* **69**, 550–558.

May JM (1999) Is ascorbic acid an antioxidant for the plasma membrane? *FASEB J.* **13**, 995–1006.

Mefford IN, Oke AF, Adams RN (1981) Regional distribution in human brain. *Brain Res.* **212**, 223–226.

Meister A (1994) Glutathione-ascorbic acid antioxidant system in animals. *J. Biol. Chem.* **269**, 9397–9400.

Mendez-Alvarez E, Soto-Otero R, Hermida-Ameijeiras A, Lopez-Martin ME, Labeira-Garcia JL (2001) Effect of iron and magnesium on hydroxyl radical production by 6-hydroxydopamine: mediation by antioxidants. *Free Rad. Biol. Med.* **31**, 986–998.

Milby K, Oke A, Adams RN (1982) Detailed mapping of ascorbate distribution in rat brain. *Neurosci. Lett.* **28**, 15–20.

Miller BT, Cicero TJ (1987) Ascorbate-induced release of LHRH: noradrenergic and opioid modulation. *Brain Res. Bull.* **19**, 95–99.

Montilla-Lopez P, Munoz-Agueda MC, Lopez MF, Munoz-Castaneda JR, Bujalance-Arenas I, Tunez-Finana I (2002) Comparison of melatonin versus vitamin C on oxidative stress and antioxidative enzyme activity in Alzheimer's disease induced by okadaic acid in neuroblastoma cells. *Eur. J. Pharmacol.* **451**, 237–243.

Muakkassah-Kelly SF, Andersen JW, Shih JC, Hochstein P (1981) Dual effects of ascorbate on serotonin and spiperone binding in rat cortical membranes. *J. Neurochem.* **41**, 1429–1439.

Nagel E, Meyer zu Vilsendorf A, Bartels M, Pichlmayr R (1997) Antioxidative vitamins in prevention of ischemia/reperfusion injury. *Int. J. Vit. Nutr. Res.* **67**, 298–306.

O'Neill RD, Fillenz M (1985) Circadian changes in extracellular ascorbate in rat cortex, accumbus, striatum and hippocampus – correlations with motor activity. *Neurosci. Lett.* **60**, 331–336.

O'Neill RD, Grünewald RA, Fillenz M, Albery WJ (1982) Linear sweep voltammetry with carbon paste electrodes in rat striatum. *Neuroscience* **7**, 1945–1954.

Oke AF, May L, Adams RN (1987) Ascorbic acid distribution patterns in human brain. *Ann. N.Y. Acad. Sci.* **498**, 1–12.

Phebus LA, Perry KW, Clemens JA, Fuller RW (1986) Brain anoxia releases striatal dopamine in rats. *Life Sci.* **38**, 2447–2453.

Prehn JH, Bindokas VP, Jordan J, Galindo MF, Ghadge GD, Roos RP, *et al.* (1996) Protective effect of transforming factor beta-1 on beta-amyloid neurotoxicity in rat hippocampal neurons. *Mol. Pharmacol.* **49**, 319–328.

Rajalakshmi R, Malathy J, Ramakrishnan CV (1967) Effect of dietary protein content on regional distribution of ascorbic acid in the rat brain. *J. Neurochem.* **14**, 161–167.

Ranjan A, Theodore D, Haran RP, Chy MJ (1993) Ascorbic acid and focal cerebral ischemia in a primate model. *Acta Neurochir. (Wien)* **123**, 87–91.

Rebec GV (1994) A vitamin as neuromodulatory: ascorbate release into the extracellular fluid of the brain regulates dopaminergic and glutaminergic transmission. *Prog. Neurobiol.* **43**, 537–565.

Rebec GV, Centore JM, White LK, Alloway KD (1985) Ascorbic acid and the behavioral response to haloperidol: implications for the action of antipsychotic drugs. *Science* **227**, 438–440.

Rebec GV, Barton SJ, Ennis MD (2002) Dysregulation of ascorbate release in the striatum of behaving mice expressing the Huntington's disease gene. *J. Neurosci.* **22**, 1–5.

Rice ME (2000) Ascorbate regulation and its neuroprotective role in the brain. *Trends Neurosci.* **23**, 209–216.

Rice ME, Russo-Menna I (1998) Differential compartmentalization of brain ascorbate and glutathione between neurons and glia. *Neuroscience* **82**, 1213–1223.

Rose RC, Bode AM (1993) Biology of free radical scavengers: an evaluation of ascorbate. *FASEB J.* **7**, 1135–1142.

Sadrzadeh SMH, Eaton JW (1988) Hemoglobin mediated oxidant damage to the central nervous system requires endogenous ascorbate. *J. Clin. Invest.* **82**, 1510–1515.

Sadrzadeh SMH, Anderson DK, Panter SS, Halloway PE, Eaton JW (1987) Hemoglobin potentiates central nervous system damage. *J. Clin. Invest.* **79**, 662–670.

Schaus R (1957) The ascorbic acid content of human pituitary, cerebral cortex, heart and skeletal muscle and its relation to age. *Am. J. Clin. Nutr.* **5**, 39–41.

Schenk JO, Miller E, Gaddis R, Adams RN (1982) Homeostatic control of ascorbate concentration in CNS extracellular fluid. *Brain Res.* **253**, 353–356.

Sciamanna MA, Lee CP (1993) Ischemia/reperfusion-induced injury of forebrain mitochondria and protection by ascorbate. *Arch. Biochem. Biophys.* **305**, 215–224.

Si F, Ross GM, Shin SH (1998) Glutathione protects PC12 cells from ascorbate and dopamine-induced apoptosis. *Exp. Brain Res.* **123**, 263–268.

Song JH, Shin SH, Ross GM (1999) Prooxidant effects of ascorbate in rat brain slices. *J. Neurosci. Res.* **58**, 328–336.

Song JH, Shin SH, Ross GM (2001) Oxidative stress induced by ascorbate causes neuronal damage in an in vitro system. *Brain Res.* **895**, 66–72.

Spector R (1981) Penetration of ascorbic acid from the cerebrospinal fluid into brain. *Exp. Neurol.* **72**, 645–653.

Spector R, Lorenzo AV (1973) Ascorbic acid homeostasis in the central nervous system. *Am. J. Physiol.* **225**, 757–763.

Spector R, Lorenzo AV (1974) Specificity of ascorbic acid transport system in the central nervous system. *Am. J. Physiol.* **226**, 1468–1473.

Stamford JA, Isaac D, Hicks CA, Ward MA, Osbourne DJ, O'Neill MJ (1999) Ascorbic acid is neuroprotective against global ischemia in striatum but not hippocampus: histological and voltammetric data. *Brain Res.* **835**, 229–240.

Takizawa S, Matsushima K, Fujita H, Ogawa S, Nakazawa H, Shinohara Y (1996) Neuroprotection against ischemic injury through inhibition of glutamate release: N-type calcium channel antagonist and Cu, Zn-superoxide dismutase. In: *Free Radicals in Brain Physiology and Disorders* (eds L Packer, M Hiramatsu, T Yoshikawa). Academic Press, San Diego, pp. 265–278.

Tatsumi M, Nagao Y, Okamura K, Gamo J (1935) Uber der Ubergang von Vitamin C in den Liquor und das augen Kammerwasser. *Klin. Wochschr.* **14,** 1007.

Tetik Ö, Islamoglu F, Yagdi T, Calkavur T, Ozbek C, Canpolat L, Büket S, Vüksel M (2001) An intraaortic solution to prevent spinal cord injury in a rabbit model. *Eur. J. Endovasc. Surg.* **22**, 175–179.

Tolbert LC, Morris PE, Spollen JJ, Ashe SC (1992) Stereospecific effects of ascorbic acid on d1 and d2 agonist binding. *Life Sci.* **51,** 921–930.

Tsukaguchi H, Tokui T, MacKenzie B, Berger UV, Hediger MA (1999) A family of mammalian Na^+-dependent L-ascorbic acid transporters. *Nature* **399,** 70–75.

Weinberger J, Nieves-Rosa J, Cohen G (1985) Nerve terminal damage in cerebral ischemia: protective effect of alpha-methyl-para-tyrosine. *Stroke* **16,** 864–870.

Wiener HL, Lajtha A, Sershen A (1989) Ascorbic acid inhibits [^{3}H]SCH-23390 binding to striatal dopamine D1 receptors. *J. Recept. Res.* **9,** 331–339.

Winkler BS, Orselli SM, Rex TS (1994) The redox couple between glutathione and ascorbic acid: a chemical and physiological perspective. *Free Rad. Biol. Med.* **17,** 333–349.

Wolf PA (1997) Epidemiology and risk factor management. In: *Primer on Cerebrovascular Diseases* (eds KMA Welch, LR Caplan, DJ Reis, BK Siesjo, B Weir). Academic Press, San Diego, pp. 751–757.

Yasu T (2000) Continuous real-time measurement of extracellular ascorbate release in the rat striatum in vivo following forebrain ischemia-reperfusion. *Neurosci. Lett.* **293,** 123–126.

Yount SE, Kraft MEE, Rebec G (1991) Acute and long-term amphetamine treatments alter extracellular ascorbate in neostriatum but not nucleus accumbens of freely moving rats. *Life Sci.* **49,** 1237–1244.

Zetterström T, Wheeler DB, Bautelle MG, Fillenz M (1992) Striatal ascorbate and its relationship to dopamine receptor stimulation and motor activity. *Eur. J. Neurosci* **3,** 940–946.

Zhang WR, Hayashi T, Sasaki C, Sato K, Nagano I, Manabe Y, Abe K (2001) Attenuation of oxidative DNA damage with a novel antioxidant EPC-K1 in rat brain neuronal cells after transient middle cerebral artery occlusion. *Neurol. Res.* **23,** 676–680.

14

Physiology of vitamin C in neutrophils during inflammation

Jay W. Heinecke

14.1 Introduction

The cellular hallmark of the acute inflammatory response is the appearance of neutrophils at sites of tissue injury. Neutrophils are phagocytic white blood cells that play a critical role in host defense by ingesting and killing invading pathogens. One important component of their antimicrobial armamentarium is a membrane-associated NADPH oxidase that generates reactive intermediates from oxygen. Humans and animals deficient in this oxidase suffer from recurrent bacterial and fungal infections, indicating that oxidant production by neutrophils is clinically important. However, production of reactive intermediates is also potentially dangerous to the host, and oxidative damage during inflammation is implicated in tissue injury and the pathogenesis of diseases ranging from ischemia-reperfusion injury to atherosclerosis and arthritis.

Many lines of evidence indicate that reactive species generated by neutrophils and other white blood cells can oxidatively modify proteins, lipids, and nucleic acids during acute and chronic inflammation. These observations suggest that antioxidants might play important roles in protecting host tissues against oxidant-mediated injury. Because *in vitro* studies demonstrate that vitamin C is a potent scavenger of reactive intermediates and that neutrophils accumulate high levels of this compound, the vitamin might be of central importance in maintaining normal neutrophil function and preventing tissue damage.

Remarkably, the exact function of vitamin C in neutrophils is unknown, despite intense study. In this chapter, I review its uptake by neutrophils, its proposed roles in protecting neutrophils and tissue from oxidative damage, and the evidence that it affects neutrophil function and inhibits oxidative tissue injury during inflammation in humans.

Vitamin C, edited by Han Asard, James M May and Nicholas Smirnoff

14.2 Oxidant production by neutrophils

The use of potentially lethal oxidants by professional phagocytic white blood cells – neutrophils, monocytes, macrophages, and eosinophils – suggests that the generation of reactive intermediates must be carefully controlled (Babior *et al.*, 2002). The mechanisms that regulate oxidant production can be studied in neutrophils, which are readily isolated in large quantities from plasma.

Circulating neutrophils are quiescent but are quickly and synchronously activated by a wide array of agonists. One important agonist *in vivo* is a component of opsonized bacteria that binds to receptors on the neutrophil membrane. This binding leads to the activation of NADPH oxidase and also triggers the phagocytosis of particulate bacteria (Winterbourn *et al.*, 2000; Babior *et al.*, 2002).

A wide variety of soluble and particulate agents can activate the NADPH oxidase *in vitro* (Winterbourn *et al.*, 2000; Babior *et al.*, 2002). Commonly used soluble activators include phorbol ester, calcium ionophore, complement peptide C5a, tumor necrosis factor, and *N*-formylated peptides (such as fMetLeuPhe) that mimic bacterial cell wall proteins. Particulate material such as opsonized bacteria and yeast cell-wall preparations are also potent agonists.

When activated, the NADPH oxidase of phagocytes reduces molecular oxygen to superoxide ($O_2^{\cdot-}$) by a pathway that is insensitive to cyanide (Winterbourn *et al.*, 2000; Babior *et al.*, 2002). The enzyme uses NADPH, but not NADH, as a cofactor (Reaction 1). The required NADPH is produced by activating the hexose monophosphate shunt.

$$\text{NADPH} + 2O_2 \rightarrow 2O_2^{\cdot-} + \text{NADP}^+ + \text{H}^+ \qquad \text{(Reaction 1)}$$

The major electron transport element of the neutrophilic NADPH oxidase is a membrane-bound b-type cytochrome, b_{558} (Babior *et al.*, 2002). This protein contains two heme groups, a noncovalently associated flavin group, a large β subunit (gp91-*phox*), and a small α subunit (p22-*phox*; gp for glycoprotein, p for protein, and *phox* for *ph*agocyte *ox*idase). There are probably equal amounts of the two subunits, and there is good evidence that gp91-*phox* contains one heme group; a second may be interposed between the heavy and light subunits of the flavocytochrome (Babior *et al.*, 2002).

Active NADPH oxidase is assembled from both membrane and cytosolic components. Assembly requires translocation of an activation complex from the cytosol to membrane-bound flavocytochrome b_{558} (Babior *et al.*, 2002). Genetic studies have identified two critical cytosolic components, p47-*phox* and p67-*phox*. A small cytosolic GTP-binding protein, Racl or Rac2, may also be required for activation.

The superoxide made by NADPH oxidase is secreted into the phagolysosome containing the bacterium, and it dismutates into hydrogen peroxide (H_2O_2). Neutrophil granules containing an array of microbicidal proteins also are secreted into the phagocytic vacuole, further enhancing the toxic environment and bathing the microbe in high local concentrations of oxidants (Babior *et al.*, 2002; Winterbourn *et al.*, 2000). A major component of these granules is the heme protein myeloperoxidase, which converts H_2O_2 into a variety of more potent oxidants (Winterbourn *et al.*, 2000). Phagocytosis also stimulates neutrophils to secrete superoxide, H_2O_2, and myeloperoxidase into the extracellular milieu, as do soluble agonists that trigger oxidant production.

14.3 Physiology of vitamin C

Vitamin C, also known as ascorbic acid or ascorbate (ASC), is an essential human vitamin that is normally present at 20–100 μM concentrations in plasma (May, 1999; Padayatty and Levine, 2001). Most mammals synthesize vitamin C in the liver, but humans, nonhuman primates, and guinea pigs have lost the enzyme gulonolactone oxidase, which catalyzes the last step in vitamin C synthesis. Thus, humans are dependent upon dietary vitamin C, and inadequate intake results in scurvy, which is clinically characterized by fatigue, weakness, skin hemorrhage, and bleeding gums.

Vitamin C is an essential electron donor for enzymes involved in collagen biosynthesis, catecholamine synthesis, tyrosine metabolism, and amidation of peptide hormones (Padayatty and Levine, 2001). The clinical manifestations of scurvy are thought to reflect disruption of collagen synthesis (see chapter by De Tullio). Vitamin C also has nonenzymatic functions, including the ability to scavenge reactive intermediates (May, 1999; Padayatty and Levine, 2001). This has led to the proposal that the vitamin is of central importance in protecting host tissue from oxidative stress.

Vitamin C bears a negative charge at neutral pH (the ASC anion). It has a high reduction potential and is readily oxidized with the loss of one or two electrons and protons (formally represented as a hydrogen atom, $H^{\cdot}$). It can be oxidized by a wide range of reactive intermediates (Reaction 2) into ASC radical (semi-dehydroascorbic acid, MDHA) and dehydroascorbate (DHA) in concert with reduction of the oxidant (Reaction 3).

$$\text{ascorbate anion} \leftrightarrows \text{MDHA} + H^{\cdot} \leftrightarrows \text{DHA} + H^{\cdot} \qquad \text{(Reaction 2)}$$

$$\text{oxidizing intermediate} + H^{\cdot} \rightarrow \text{reduced intermediate} \qquad \text{(Reaction 3)}$$

DHA is relatively unstable at physiological pH, undergoing spontaneous hydrolysis to diketogulonic acid (Reaction 4) (see Chapter 2 by Bánhegyi and Loewus).

$$\text{DHA} + H_2O \rightarrow \text{diketogulonic acid} \qquad \text{(Reaction 4)}$$

Formation of diketogulonic acid is likely to be irreversible *in vivo* (Padayatty and Levine, 2001). In contrast, DHA and MDHA are readily converted back to ASC anion. Vitamin C circulates almost exclusively as ASC anion in human plasma, which contains relatively little DHA under normal circumstances.

14.4 Mechanisms for vitamin C uptake in neutrophils

The intracellular concentration of vitamin C in resting neutrophils is ~ 1.5 mM, which is 10- to 50-fold higher than in plasma (Washko *et al.*, 1989). This discrepancy strongly implies that neutrophils have specialized transport mechanisms for taking up the vitamin. Studies with cultured cells have demonstrated at least two pathways. One involves the GLUT transporters, a family of proteins that originally were described as facilitators of glucose diffusion across the plasma membrane (Vera *et al.*, 1993; Rumsey *et al.*, 1997). Because vitamin C is also a reducing sugar, it is not surprising that its derivatives can interact with the GLUT transporters as well. However, these membrane-associated proteins do not recognize vitamin C itself; instead, they interact with DHA. Studies of transfected cells demonstrate that GLUT1 and GLUT3 mediate facilitated diffusion of DHA by a pathway that does not require sodium ions (Vera *et al.*, 1993; Rumsey *et al.*, 1997). Once inside cells, DHA is reduced rapidly back to vitamin C, which cannot escape through the plasma membrane of neutrophils (Vera *et al.*, 1993; Rumsey *et al.*, 1997, see chapter by Wilson). Because the concentration of DHA inside cells is extremely low, this diffusion-controlled uptake assists the cells in accumulating vitamin C against a concentration gradient. Many cell types (including red blood cells) possess GLUT transporters, which are thought to be partly responsible for plasma's very low ratio of DHA to reduced vitamin C (May *et al.*, 2001).

Many lines of evidence support the proposal that DHA uptake by GLUT transporters is one of the major pathways that allow neutrophils to accumulate vitamin C (Vera *et al.*, 1993; Rumsey *et al.*, 1997; Welch *et al.*, 1995). For example, glucose inhibits the uptake of dehydroascorbic acid by both GLUT transporters and neutrophils. In both systems, DHA uptake is independent of sodium and inhibited by glucose analogs. Moreover, neutrophils express GLUT1 (Vera *et al.*, 1993).

Many cells also have uptake mechanisms that recognize vitamin C rather than DHA (Welch *et al.*, 1995; Tsukaguchi *et al.*, 1999). The sodium-dependent vitamin C transporters SVCT1 and SVCT2, which are found in many cell types, provide one mechanism for taking in the reduced vitamin (Tsukaguchi *et al.*, 1999). Because they couple vitamin C uptake with the cotransport of sodium down the ion's concentration gradient, they enable cells to accumulate the vitamin even when its intracellular concentration is higher than its concentration in plasma. Isolated neutrophils have a sodium-dependent pathway that is specific for the reduced form of vitamin C, indicating that transporters in the SVCT family likely contribute to vitamin C uptake by neutrophils *in vivo* (Welch *et al.*, 1995).

14.5 Vitamin C recycling in human neutrophils

Activated neutrophils generate high local concentrations of oxidants, which react rapidly with vitamin C to yield MDHA and DHA (Washko *et al.*, 1993; Carr *et al.*, 2000). GLUT transporters on neutrophils and bystander cells rapidly take up DHA, raising the possibility that generation of DHA by reactive intermediates might be a mechanism for increasing vitamin C uptake into cells exposed to oxidative stress (Washko *et al.*, 1993; May *et al.*, 2001). The processes that oxidize vitamin C to DHA and usher the oxidized vitamin into cells have been termed 'ASC recycling'.

Elegant *in vitro* studies demonstrate that neutrophils are central to ASC recycling (Washko *et al.*, 1993). Activated neutrophils produce superoxide, H_2O_2, hypochlorous acid, nitrogen dioxide radical and other species, resulting in the conversion of extracellular vitamin C to DHA. The concentration of vitamin C inside cells increases 10-fold when oxidant production is underway. Superoxide dismutase and catalase, two scavengers of reactive intermediates, inhibit extracellular DHA production and the increase in intracellular vitamin C (Washko *et al.*, 1993). Moreover, neutrophils isolated from patients suffering from chronic granulomatous disease, a genetic deficiency in components of NADPH oxidase, fail to oxidize vitamin C or to accumulate high levels of vitamin C (Washko *et al.*, 1993; Wang *et al.*, 1997). These results strongly support the hypothesis that oxidant production by the NADPH oxidase of activated neutrophils is needed for the oxidation of vitamin C to DHA, which is then taken into cells and reduced back to vitamin C.

Bacteria and fungi stimulate neutrophils to generate oxidants (Babior *et al.*, 2002), suggesting that one pathway for increasing intracellular vitamin C levels in humans might involve microbial pathogens. Indeed, neutrophils exposed to bacteria and vitamin C accumulate high concentrations of the vitamin. Induction of ASC recycling in neutrophils by bacteria requires functional NADPH oxidase and involves dehydroascorbic acid production (Washko *et al.*, 1993). These observations suggest that sepsis and other forms of acute infection in humans might trigger oxidant production by neutrophils, which in turn might raise intracellular levels of vitamin C. The observation that plasma levels of vitamin C decrease when humans develop systemic bacterial infections is consistent with this hypothesis (Galley *et al.*, 1996).

ASC recycling might also protect other cells from oxidative stress. For example, red blood cells take up DHA by a GLUT transporter-dependent pathway and reduce the compound back to vitamin C intracellularly (May *et al.*, 2001; see also Chapter 8 by May and Asard). In contrast to neutrophils, however, they slowly release the reduced vitamin back into plasma. This release might be critical for keeping DHA at a low level in plasma.

14.6 Reduction of DHA

Based on *in vitro* studies, a number of mechanisms for reducing DHA to vitamin C have been proposed (Padayyatty *et al.*, 2001; May *et al.*, 2001; see chapter by May and Asard). One nonenzymatic mechanism involves the direct reaction of glutathione (GSH) with the oxidized form of the vitamin to yield glutathione disulfide (GSSG) and reduced vitamin C (Reaction 5). The intracellular concentration of glutathione, a tripeptide containing a cysteine residue, is typically in the millimolar range, suggesting that direct reduction of DHA by glutathione's thiol group (SH) could be physiologically important. This reaction has been demonstrated in dialyzed red blood cell lysates (May *et al.*, 2001), but its relevance to the intracellular reduction of DHA in neutrophils is uncertain.

$$2\text{GSH} + \text{DHA} \rightarrow \text{GSSG} + \text{vitamin C} \qquad \text{(Reaction 5)}$$

Glutathione disulfide would then be reduced back to glutathione by the action of glutathione reductase, an NADPH-dependent enzyme.

A number of *in vitro* enzymatic pathways for reducing DHA have also been described, including those requiring glutaredoxin, thioredoxin, or protein disulfide isomerase (Padayatty and Levine, 2001; May *et al.*, 2001; Park and Levine, 1996). Using activity assays, glutaredoxin, a thioltransferase, was purified and identified as one mechanism for reducing DHA in lysates of human neutrophils (Park and Levine, 1996). The mechanism involves the oxidation of the protein-bound thiol in glutaredoxin in concert with the reduction of DHA to vitamin C. The rate of protein-mediated reduction of DHA by glutaredoxin was at least 10-fold greater than when glutathione alone mediated the reduction. Other *in vitro* studies estimated that the protein was responsible for at least 50% of the DHA-reducing activity of neutrophil lysates, suggesting that glutaredoxin is a major contributor to the intracellular reduction of oxidized vitamin C (Park and Levine, 1996).

These observations suggest that NADPH production by the hexose monophosphate shunt of activated neutrophils might play multiple roles when neutrophils recycle ASC (Babior *et al.*, 2002; Vera *et al.*, 1993; Washko *et al.*, 1993; Park and Levine, 1996). NADPH allows the phagocyte NADPH oxidase to generate oxidants, leading to the conversion of vitamin C to DHA. The cells take up the DHA and reduce it to vitamin C using glutaredoxin, which is simultaneously converted to its oxidized form. Its reduced form is regenerated as the reductase activity of glutaredoxin oxidizes glutathione to glutathione disulfide. NADPH from the hexose monophosphate shunt provides the reducing equivalents needed for glutathione reductase to convert glutathione disulfide back to glutathione. Thus, NADPH plays a critical role both in generating DHA (via the production of oxidants) and in reducing DHA back to vitamin C (via glutaredoxin and perhaps by direct chemical reduction of DHA by glutathione).

14.7 Vitamin C scavenges oxidants *in vitro*

The NADPH oxidase of neutrophils and other phagocytic white blood cells generates superoxide by directly reducing molecular oxygen. Because superoxide rapidly dismutates (Reaction 6), the cells also generate hydrogen peroxide (H_2O_2) (Babior *et al.*, 2002; Winterbourn *et al.*, 2000).

$$2O_2^{\cdot -} + 2H^+ \rightarrow H_2O_2 + O_2 \qquad \text{(Reaction 6)}$$

The myeloperoxidase that phagocytes secrete interacts with hydrogen peroxide to generate antimicrobial toxins (Babior *et al.*, 2002). In contrast to many *in vitro* oxidation reactions, this enzyme does not require free metal ions. One of its major actions is to convert chloride ion to hypochlorous acid (HOCl; Reaction 7).

$$H_2O_2 + Cl^- + H^+ \rightarrow HOCl + H_2O \qquad \text{(Reaction 7)}$$

Hypochlorous acid is a potent bactericidal oxidant that may inadvertently damage host proteins at sites of inflammation *in vivo*. We have shown that it chlorinates tyrosine to 3-chlorotyrosine, a highly specific marker for myeloperoxidase-initiated protein oxidation, and that levels of 3-chlorotyrosine are elevated in inflamed tissue (Gaut *et al.*, 2001; Hazen *et al.*, 1996).

The phenolic amino acid L-tyrosine is also a substrate for oxidation by myeloperoxidase (Heinecke, 2002). The reaction involves direct one-electron oxidation of tyrosine by compound I, a complex of myeloperoxidase and H_2O_2. Tyrosyl radical formed by myeloperoxidase can damage proteins, resulting in the formation of protein-bound dityrosine.

$$\text{myeloperoxidase} + H_2O_2 \rightarrow \text{compound I} + H_2O \qquad \text{(Reaction 8)}$$

$$\text{compound I} + \text{L-tyrosine} \rightarrow \text{compound II} + \text{tyrosyl radical} \qquad \text{(Reaction 9)}$$

Recent studies indicate that myeloperoxidase also uses nitrite, a decomposition product of nitric oxide, to generate nitrogen dioxide radical ($NO_2^{\cdot}$), a potent nitrating intermediate (Reaction 8; Eiserich *et al.*, 1998; Gaut *et al.*, 2002). The reaction involves direct one-electron oxidation of nitrite (NO_2^-) by compound I of myeloperoxidase (Reaction 10).

$$NO_2^- + \text{compound I} + H^+ \rightarrow NO_2^{\cdot} + H_2O + \text{compound II} \qquad \text{(Reaction 10)}$$

These observations raise the possibility that the myeloperoxidase pathway both nitrates and chlorinates host tissues *in vivo*.

Superoxide, H_2O_2, HOCl, tyrosyl radical, nitrogen dioxide radical, and other reactive intermediates produced by neutrophils have been implicated in damage to proteins and DNA during inflammation (Winterbourn *et al.*, 2000; Heinecke, 1999). Because plasma contains high concentrations of vitamin C, which reacts rapidly with HOCl, tyrosyl radical and nitrogen dioxide radical *in vitro*, the vitamin might be a physiologically important scavenger of these species in the aqueous phase (Winterbourn *et al.*, 2000; Carr and Frei, 2002; Carr *et al.*, 2000). It reacts much more slowly with superoxide, suggesting that other enzymatic or nonenzymatic mechanisms might destroy this reactive intermediate.

Lipid peroxidation is thought to be involved in many forms of oxidant-mediated tissue injury. Biochemical studies and electron paramagnetic resonance spectroscopy demonstrate that peroxidation of polyunsaturated lipids can be initiated by tyrosyl radical, which myeloperoxidase generates from the phenolic amino acid L-tyrosine (Heinecke, 2002). This process bears remarkable biochemical similarities to the enzymatic peroxidation of polyunsaturated fatty acids by cyclooxygenase, a reaction that also is thought to involve tyrosyl radical. Nitrogen dioxide radical generated by myeloperoxidase will also promote lipid peroxidation (Carr *et al.*, 2000). Thus, reactive intermediates produced by neutrophils could initiate or promote lipid oxidation at sites of inflammation.

Vitamin C is highly water soluble and therefore cannot enter phospholipid–cholesterol membranes to inhibit lipid peroxidation by directly scavenging lipid peroxyl radicals. In contrast, vitamin E resides at the interface between the lipid and aqueous environment, and it reacts rapidly with peroxyl radicals, terminating lipid peroxidation as it converts into the tocopheroxyl (vitamin E) radical (Carr *et al.*, 2000; Stocker, 1999). In turn, tocopheroxyl radical rapidly oxidizes vitamin C to MDHA, converting back to vitamin E. Repair of tocopheroxyl radical by vitamin C has been proposed to be critical to inhibition of lipid peroxidation. *In vitro* studies demonstrate that a mixture of vitamin C and vitamin E inhibits lipid peroxidation more potently than either vitamin alone, suggesting that the water- and lipid-soluble vitamins collaborate to prevent oxidative damage.

Activation of neutrophils in plasma leads to lipid peroxidation (Frei *et al.*, 1988, 1989). However, peroxidation is not observed until neutrophil-derived oxidants have consumed all of the vitamin C. After that, vitamin E appears to offer little protection against lipid peroxidation. Similar observations have been reported for the initiation of lipid peroxidation by tyrosyl radical generated by myeloperoxidase or neutrophils (Savenkova *et al.*, 1994). These observations suggest that the major mechanism by which vitamin C inhibits lipid peroxidation involves the scavenging of water-soluble reactive intermediates generated by neutrophils.

14.8 Does vitamin C scavenge oxidants *in vivo*?

Oxidative stress is thought to contribute to disorders ranging from cigarette smoke-induced tissue damage to atherosclerosis and cancer (Heinecke, 1999; Ames *et al.*, 1993). Cigarette smoking is of particular interest because smokers are known to have low levels of vitamin C, cigarette smoke is a potent source of reactive intermediates, and smoking is a major risk factor for atherosclerotic disease (Ames *et al.*, 1993; Loft and Poulsen, 1996). Despite the availability of genetically altered guinea pigs and mice that require dietary supplementation to maintain normal plasma and tissue levels of vitamin C, there is a dearth of well-controlled studies on the vitamin's effects on oxidative stress *in vivo*. However, careful investigations of guinea pigs have shown that cigarette smoke makes white blood cells adhere to the vascular wall and aggregate with platelets (Lehr *et al.*, 1994, 1997). Significantly, dietary supplementation with vitamin C appears to block or reverse these effects. In one model, the biological effects of cigarette smoke appeared to be mediated in part by lipid derivatives (Lehr *et al.*, 1997), which activated platelets. Because cigarette smoke promotes lipid peroxidation and because lipid oxidation generates molecules that activate platelets, these observations support the possibility that vitamin C is a physiologically relevant scavenger of oxidants *in vivo*.

Almost nothing is known about vitamin C's ability to inhibit oxidative tissue damage in humans. In a remarkable clinical study, however, healthy women were placed on a vitamin C-deficient diet to lower their plasma and white blood cell levels of the vitamin (Levine *et al.*, 2001). They then were given increasing levels of dietary vitamin C, ranging from 30 $mg{\cdot}day^{-1}$ to 2500 $mg{\cdot}day^{-1}$. The different regimens produced levels of vitamin C that varied eightfold in plasma (from ~10 μM to ~80 μM) and threefold in neutrophils (from ~0.5 mM to 1.4 mM). Surprisingly, neither vitamin C deficiency nor supplementation with the vitamin affected plasma or urinary levels of F2-isoprostanes (Levine *et al.*, 2001), which are specific and sensitive markers of lipid peroxidation *in vivo*. These observations suggest that vitamin C does not inhibit lipid peroxidation in healthy women.

Another interesting aspect of this (Levine *et al.*, 2001) and earlier studies (Levine *et al.*, 1996) was the observation that vitamin C levels in both plasma and white blood cells plateau when humans supplement their diets with ~200 mg of vitamin C each day. Thus, dietary supplementation with higher amounts of the vitamin does not boost levels of the vitamin in either plasma or neutrophils, suggesting that high-dose vitamin C therapy is not likely to offer any benefits over low-dose therapy.

Epidemiological studies regarding the role of vitamin C in the pathogenesis of heart disease have reached conflicting results (Tribble, 1999). However, three prospective, randomized clinical trials – the best test of a proposed antioxidant's actions – have examined vitamin C's ability to prevent artery disease.

The Heart Protection Study randomly assigned more than 20 000 patients with established heart disease, other occlusive vascular disease, or diabetes to either placebo or combination 'antioxidant' therapy (600 mg vitamin E, 250 mg vitamin C, and 20 mg β-carotene per day). Analysis of plasma confirmed that the level of vitamin E doubled and that the level of vitamin C increased by ~33% (MRC/BHF Heart Protection Study, 2002). After a 5-year follow-up, however, there was no significant reduction in either mortality or the incidence of cardiovascular disease in the supplemented subjects (MRC/BHF Heart Protection Study, 2002). Thus, vitamin C does not appear to help vitamin E prevent clinical events in patients at high risk for heart disease. These observations are important given the synergistic effect of the two vitamins on lipid peroxidation *in vitro*.

Statins are widely used to lower cholesterol in patients with known atherosclerotic vascular disease. However, there is little information on their potential interactions with vitamin E and other proposed antioxidants. Brown and colleagues recently examined this important issue in patients with coronary artery disease, low levels of high density lipoprotein (the good form of cholesterol), and normal levels of low density lipoprotein (the bad form of cholesterol). Aggressive lipid-lowering therapy with a statin in combination with niacin markedly reduced clinical events (Brown *et al.*, 2001). It also completely blocked the progression of atherosclerosis as assessed by quantitative coronary angiography. In contrast, combination therapy with vitamin E, vitamin C, selenium, and β-carotene failed to impact the risk for cardiovascular disease. Moreover, the beneficial effects of the lipid-lowering regimen on both clinical events and progressive atherosclerosis were significantly blunted by the additional 'antioxidant' regimen (Brown *et al.*, 2001). Careful analyses of the lipoproteins suggested that these deleterious effects resulted from a reduction in a high density lipoprotein subfraction thought to be cardioprotective (Cheung *et al.*, 2001). These provocative observations raise the possibility that proposed antioxidants might be detrimental to lipid-lowering therapy.

In contrast to these negative results, a recent trial of combination antioxidant therapy in heart transplant patients was more encouraging (Fang *et al.*, 2002). Cardiac transplantation associates strongly with oxidative stress and accelerated coronary atherosclerosis. To determine whether antioxidant therapy slows disease progression, 40 transplanted patients received a statin to lower their lipid levels and were randomly assigned to receive either additional treatment with vitamin C (550 mg twice daily) plus vitamin E (400 IU twice daily) or placebo. The treated subjects more than doubled their plasma levels of the two vitamins. Intravascular ultrasonography revealed that the ratio of intimal to medial thickening, a measure of atherosclerosis, increased by 8% in the control group but did not change in the group treated with antioxidants (Fang *et al.*, 2002). These results suggest that combination therapy with vitamin C and vitamin E might have the potential to inhibit oxidative stress and disease progression in humans. However, it is important to note that this study was small and that the investigators used a surrogate endpoint for monitoring the progression of atherosclerosis. In future studies, it will be important to determine whether this intervention has a significant impact on clinically relevant endpoints.

14.9 Vitamin supplementation in patients at increased risk of infection

Neutrophils play a critical role in defending humans against invading bacterial and fungal pathogens (Babior *et al.*, 2002; Winterbourn *et al.*, 2000). Sepsis – a systemic bacterial or fungal infection – is a leading cause of death in patients hospitalized in the intensive care unit. Recent studies demonstrate that sepsis results in oxidant production by neutrophils in patients (Bhattacharjee *et al.*, 2001), and it has long been known that plasma levels of vitamin C are low in hospitalized and septic patients.

Two recent studies have examined the influence of supplementation with a regimen that included vitamin C on clinical outcomes in acutely ill and diabetic patients. In a randomized, prospective trial of surgical patients suffering from various forms of acute trauma, early treatment with high intravenous doses of vitamin C and vitamin E resulted in a significant reduction in the risk of multiorgan failure (Nathens *et al.*, 2002). The patients also required less time of hospitalization in the intensive care unit. There was no significant impact of vitamin supplementation on the risk of pneumonia or overall mortality, but the study examined a relatively small number of patients and lacked statistical power to detect such effects. These observations raise the possibility that vitamin C and/or vitamin E plays a previously unrecognized role in protecting acutely ill humans from multiorgan failure, perhaps in part by affecting host defense mechanisms.

Diabetics suffer from a greatly increased risk of infection with bacteria and fungal organisms (Engelich *et al.*, 2001; Rayfield *et al.*, 1982). The underlying mechanisms are unclear, but neutrophils isolated from diabetic patients exhibit a wide range of functional defects *in vitro*. The metabolic hallmark of diabetes is hyperglycemia, and glucose competes for uptake of DHA by the GLUT transporters. Moreover, the production of reactive intermediates may be increased in diabetics, and plasma levels of vitamin C are low in these patients, suggesting that one component of neutrophil dysfunction in diabetes may involve oxidative stress and impaired vitamin C function (Baynes and Thorpe, 2000; Price *et al.*, 2001). It is thus of interest that dietary supplementation of diabetic patients with a multivitamin that contained vitamin C resulted in dramatic decline in the risk of infection (Barringer *et al.*, 2003). It will be of great interest to determine whether this observation can be confirmed in larger studies, and to use animal models of infection and vitamin C deficiency to explore the role of the vitamin in neutrophil function and host defense mechanisms.

14.10 Prospects

Despite intense interest, we still do not understand the role of vitamin C in neutrophil physiology and inflammation. Although one popular hypothesis suggests that vitamin C scavenges reactive intermediates, there is no compelling evidence that the vitamin augments antioxidant defense mechanisms in humans. Indeed, the doses of vitamin C used in clinical trials have not convincingly inhibited lipid, protein, or DNA oxidation. Thus, these trials have revealed little about the role of oxidative stress in the pathogenesis of disease.

To determine the clinical relevance of vitamin C in protecting tissues and neutrophils against reactive intermediates, it will be critical to identify subjects who display evidence of increased oxidative stress. Establishing the optimal regimen for inhibiting lipid and protein peroxidation also needs to be determined so that vitamin C, perhaps in conjunction with other compounds, can be administered effectively.

One powerful strategy for identifying physiologically relevant antioxidants is to quantify specific oxidation products in tissues and plasma, using sensitive and specific methods such as mass spectrometry (Heinecke, 1999; Baynes and Thorpe, 2000). Because the utility of an antioxidant depends critically on the nature of the oxidant that inflicts tissue damage, interventions found to specifically inhibit physiologically relevant pathways, such as oxidant production by neutrophils, would be logical candidates for clinical trials.

References

Ames BN, Shigenaga, MK, Hagen TM (1993) Oxidants, antioxidants, and the degenerative diseases of aging. *Proc. Natl Acad. Sci. USA* **90**, 7915–7922.

Babior BM, Lambeth JD, Nauseef W (2002) The neutrophil NADPH oxidase. *Arch. Biochem. Biophys.* **397**, 342–344.

Barringer TA, Kirk JK, Santaniello AC, Foley KL, Michielutte R (2003) Effect of a multivitamin and mineral supplement on infection and quality of life. A randomized, double-blind, placebo-controlled trial. *Ann Intern Med.* **138**, 365–371.

Baynes JW, Thorpe SR (2000) Glycoxidation and lipoxidation in atherogenesis. *Free Radic. Biol. Med.* **28**, 1708–1716.

Bhattacharjee S, Pennathur S, Byun J, Crowley J, Mueller D, Gischler J, Hotchkiss RS, Heinecke JW (2001) NADPH oxidase of neutrophils elevates o,o′-dityrosine cross-links in proteins and urine during inflammation. *Arch. Biochem. Biophys.* **395**, 69–77.

Brown BG, Zhao XQ, Chait A, Fisher LD, Cheung MC, Morse JS, *et al.* (2001) Simvastatin and niacin, antioxidant vitamins, or the combination for the prevention of coronary disease. *N. Engl J. Med.* **345**, 1583–92.

Carr AC, Frei B (2002) Human neutrophils oxidize low-density lipoprotein by a hypochlorous acid-dependent mechanism: the role of vitamin C. *Biol Chem.* **383**, 627–636.

Carr AC, McCall MR, Frei B (2000) Oxidation of LDL by myeloperoxidase and reactive nitrogen species: reaction pathways and antioxidant protection. *Arterioscler. Thromb. Vasc. Biol.* **20**, 1716–1723.

Cheung MC, Zhao XQ, Chait A, Albers JJ, Brown BG (2001) Antioxidant supplements block the response of HDL to simvastatin-niacin therapy in patients with coronary artery disease and low HDL. *Arterioscler. Thromb. Vasc. Biol.* **21**, 1320–1326.

Eiserich JP, Hristova M, Cross CE, Jones AD, Freeman BA, Halliwell B, van der Vliet A (1998) Formation of nitric oxide-derived inflammatory oxidants by myeloperoxidase in neutrophils. *Nature* **391**, 393–397.

Engelich G, Wright DG, Hartshorn KL (2001) Acquired disorders of phagocyte function complicating medical and surgical illnesses. *Clin. Infect. Dis.* **33**, 2040–2048.

Fang JC, Kinlay S, Beltrame J, Hikiti H, Wainstein M, Behrendt D, *et al.* (2002) Effect of vitamins C and E on progression of transplant-associated arteriosclerosis: a randomised trial. *Lancet* **359**, 1108–1113.

Frei B, England L, Ames BN (1989) Ascorbate is an outstanding antioxidant in human blood plasma. *Proc. Natl Acad. Sci. USA* **86**, 6377–6381.

Frei B, Stocker R, Ames BN (1988) Antioxidant defenses and lipid peroxidation in human blood plasma. *Proc. Natl Acad. Sci. USA* **85**, 9748–9752.

Galley HF, Davies MJ, Webster NR (1996) Ascorbyl radical formation in patients with sepsis: effect of ascorbate loading. *Free Radic. Biol. Med.* **20**, 139–143.

Gaut JP, Byun J, Tran HD, Lauber WM, Carroll JA, Hotchkiss RS, Belaaouaj A, Heinecke JW (2002) Myeloperoxidase produces nitrating oxidants in vivo. *J Clin Invest.* **109**, 1311–1319.

Gaut JP, Yeh GC, Tran HD, Byun J, Henderson JP, Richter GM, *et al.* (2001) Neutrophils employ the myeloperoxidase system to generate antimicrobial brominating and chlorinating oxidants during sepsis. *Proc. Natl Acad. Sci. USA* **98**, 11961–11966.

Hazen SL, Hsu FF, Mueller DM, Crowley JR, Heinecke JW (1996) Human neutrophils employ chlorine gas as an oxidant during phagocytosis. *J Clin Invest.* **98**, 1283–1289.

Heinecke JW (1999) Mass spectrometric quantification of amino acid oxidation products in proteins: insights into pathways that promote LDL oxidation in the human artery wall. *FASEB J.* **13**, 1113–1120.

Heinecke JW (2002) Tyrosyl radical production by myeloperoxidase: a phagocyte pathway for lipid peroxidation and dityrosine cross-linking of proteins. *Toxicology* **177**, 11–22.

Lehr HA, Frei B, Arfors KE (1994) Vitamin C prevents cigarette smoke-induced leukocyte aggregation and adhesion to endothelium in vivo. *Proc. Natl Acad. Sci. USA* **91**, 7688–7692.

Lehr HA, Weyrich AS, Saetzler RK, Jurek A, Arfors KE, Zimmerman GA, Prescott SM, McIntyre TM (1997) Vitamin C blocks inflammatory platelet-activating factor mimetics created by cigarette smoking. *J. Clin. Invest.* **99**, 2358–2364.

Levine M, Conry-Cantilena C, Wang Y, Welch RW, Washko PW, Dhariwal KR, *et al.* (1996) Vitamin C pharmacokinetics in healthy volunteers: evidence for a recommended dietary allowance. *Proc. Natl Acad. Sci. USA* **93**, 3704–3709.

Levine M, Wang Y, Padayatty SJ, Morrow J (2001) A new recommended dietary allowance of vitamin C for healthy young women. *Proc. Natl Acad. Sci. USA* **98**, 9842–9846.

Loft S, Poulsen HE (1996) Cancer risk and oxidative DNA damage in man. *J. Mol. Med.* **74**, 297–312.

May JM (1999) Is ascorbic acid an antioxidant for the plasma membrane? *FASEB J.* **13**, 995–1006.

May JM, Qu Z, Cobb CE (2001) Recycling of the ascorbate free radical by human erythrocyte membranes. *Free Radic. Biol. Med.* **31**, 117–124.

MRC/BHF Heart Protection Study of antioxidant vitamin supplementation in 20,536 high-risk individuals: a randomised placebo-controlled trial. (2002) *Lancet.* **360**, 23–33.

Nathens AB, Neff MJ, Jurkovich GJ, Klotz P, Farver K, Ruzinski JT, Radella F, Garcia I, Maier RV (2002) Randomized, prospective trial of antioxidant supplementation in critically ill surgical patients. *Ann. Surg.* **236,** 814–822.

Padayatty SJ, Levine M (2001) New insights into the physiology and pharmacology of vitamin C. *CMAJ.* **164**, 353–355.

Park JB, Levine M (1996) Purification, cloning and expression of dehydroascorbic acid-reducing activity from human neutrophils: identification as glutaredoxin. *Biochem. J.* **315**, 931–938.

Price KD, Price CS, Reynolds RD (2001) Hyperglycemia-induced ascorbic acid deficiency promotes endothelial dysfunction and the development of atherosclerosis. *Atherosclerosis* **158**, 1–12.

Rayfield EJ, Ault MJ, Keusch GT, Brothers MJ, Nechemias C, Smith H (1982) Infection and diabetes: the case for glucose control. *Am. J. Med.* 72, 439–450.

Rumsey SC, Kwon O, Xu GW, Burant CF, Simpson I, Levine M (1997) Glucose transporter isoforms GLUT1 and GLUT3 transport dehydroascorbic acid. *J. Biol. Chem.* **272**, 18982–18989.

Savenkova ML, Mueller DM, Heinecke JW (1994) Tyrosyl radical generated by myeloperoxidase is a physiological catalyst for the initiation of lipid peroxidation in low density lipoprotein. *J. Biol. Chem.* **269,** 20394–20400.

Stocker R (1999) The ambivalence of vitamin E in atherogenesis. *Trends Biochem Sci.* **24,** 219–223.

Tribble DL (1999) AHA Science Advisory. Antioxidant consumption and risk of coronary heart disease: emphasis on vitamin C, vitamin E, and beta-carotene: A statement for healthcare professionals from the American Heart Association. *Circulation* **99**, 591–595.

Tsukaguchi H, Tokui T, Mackenzie B, Berger UV, Chen XZ, Wang Y, Brubaker, RF, Hediger MA (1999) A family of mammalian Na+-dependent L-ascorbic acid transporters. *Nature* **399**, 70–75.

Vera JC, Rivas CI, Fischbarg J, Golde DW (1993) Mammalian facilitative hexose transporters mediate the transport of dehydroascorbic acid. *Nature* **364**, 79–82.

Wang Y, Russo TA, Kwon O, Chanock S, Rumsey SC, Levine M (1997) Ascorbate recycling in human neutrophils: induction by bacteria. *Proc. Natl Acad. Sci. USA* **94**, 13816–13819.

Washko P, Rotrosen D, Levine M (1989) Ascorbic acid transport and accumulation in human neutrophils. *J. Biol. Chem.* **264**, 18996–19002.

Washko PW, Wang Y, Levine M (1993) Ascorbic acid recycling in human neutrophils. *J. Biol. Chem.* **268**, 15531–15535.

Welch RW, Wang Y, Crossman A Jr, Park JB, Kirk KL, Levine M (1995) Accumulation of vitamin C (ascorbate) and its oxidized metabolite dehydroascorbic acid occurs by separate mechanisms. *J. Biol. Chem.* **270**, 12584–12592.

Winterbourn CC, Vissers MC, Kettle AJ (2000) Myeloperoxidase. *Curr. Opin. Hematol.* **7**, 53–58.

15

Ascorbic acid in atherosclerosis

Paul K. Witting and Roland Stocker

15.1 Introduction

Atherosclerosis and its clinical sequelae such as myocardial infarct and stroke represent a major source of morbidity and mortality in the developed world. A wealth of data now links serum cholesterol, in particular low density lipoprotein (LDL) cholesterol, to the development of atherosclerosis and advances in the treatment of this disease have included the use of cholesterol-lowering drugs. However, despite the success of lipid-lowering therapy atherosclerosis remains a significant problem in Western societies consuming large proportions of health budgets and health related costs. Therefore, approaches in addition to cholesterol lowering are necessary to combat this disease. One potential approach is increased dietary intake of natural and synthetic agents such as micronutrients and antioxidants that may provide long-term protection against atherosclerosis.

Vitamin C is ubiquitous in biological fluids and, at physiological pH, present as the mono-hydro conjugate base, ascorbate (ASC). ASC is labile due largely to its redox potential ($E^{\circ}_{1/2} \sim +282$ mV, pH 7; Buettner, 1993), and generally considered to function as a reducing agent capable of enhancing enzymatic activity through maintenance of the iron center in the active, ferrous state. This inherent metal-reducing activity is also the basis for the prooxidant activity of ASC *in vitro*, attributed to the effective redox cycling of iron and the production of hydroxyl radicals through Fenton chemistry (Halliwell, 1996). There is compelling evidence, however, that this prooxidant activity is not observed *in vivo* where ASC generally protects biological molecules from oxidation (Carr and Frei, 1999a) and even after iron supplementation (Berger *et al.*, 1997). Indeed, ASC is a quantitatively important antioxidant (Wayner *et al.*, 1987); qualitatively it also provides the first line of defense against oxidative damage in human plasma (Frei *et al.*, 1988, 1989), where it is considered the most effective water-soluble antioxidant (Frei *et al.*, 1990; Frei, 1991) (*Table 15.1*).

Vitamin C, edited by Han Asard, James M May and Nicholas Smirnoff

Table 15.1: Nonproteinaceous and proteinaceous plasma antioxidants. Data from (Stocker, 2000).

Antioxidants	Plasma
Water-soluble (nmol·mg protein^{-1})	
Ascorbate	0.11–1.88
Urate	2.0–5.6
Lipid-soluble (mmol·mol cholesterol^{-1})	
α-Tocopherol	8.9–23.7
α-Tocopherol[a]	8.9–16.3
γ-Tocopherol	0.5–1.1
Ubiquinone-10	0.3–0.8
Lycopene	0.3–0.6
β-Carotene	0.2–0.5
Glutathione-related (mU·mg protein^{-1})	
GSH Peroxidase	5
GSH Reductase	0.38
GST Peroxidase	0
Total GST	0.06
Superoxide dismutase-related (mU·mg protein^{-1})	
EC-SOD	0.06–0.25
Cu, Zn-SOD	ND
Mn-SOD	ND

[a]Expressed as mmol·mol^{-1} cholesteryl linoleate, the most prevalent oxidizable lipid present in human LDL. GST-peroxidase, selenium independent peroxidase; EC-SOD, extracellular superoxide dismutase; Cu,Zn-SOD, copper, zinc superoxide dismutase; Mn-SOD; manganese superoxide dismutase. ND, not detectable.

15.1.1 Oxidative events and atherosclerosis

Atherosclerosis is considered a chronic inflammatory disease of blood vessels characterized by endothelial dysfunction and the accumulation of lipid in the vessel wall. Oxidation of LDL lipid in the subendothelial space is thought to play a key role in the initiation and progression of the disease (Steinberg *et al.*, 1989; Chisolm and Steinberg, 2000). A large body of literature supports this 'oxidation theory of atherosclerosis' (for reviews see Steinberg *et al.*, 1989; Berliner and Heinecke, 1996; Diaz *et al.*, 1997; Witztum and Steinberg, 2001; Steinberg, 2002), and there is now general agreement that LDL undergoes oxidation and that oxidized LDL is present in arterial lesions. However, while the theory has gained acceptance, several key assumptions and questions remain unanswered. Specifically, it is still not known where, how, and to what extent LDL becomes oxidized *in vivo*, nor which of the many biological effects demonstrable for 'oxidized LDL' *in vitro* (*Figure 15.1*) are relevant to atherogenesis *in vivo* and whether LDL oxidation is a cause or consequence of the disease.

The 'oxidation theory' implies that compounds that effectively inhibit LDL oxidation *in vitro* are potentially anti-atherogenic. Accordingly, several structurally different synthetic antioxidants, such as probucol (4,4′-(isopropylidene dithio)*bis*(2,6-di-*tert.*butyl)phenol) (Ferns *et al.*, 1992), butylated hydroxytoluene (Freyschuss *et al.*, 1993), *N,N′*-diphenyl-phenylene-diamine (Tangirala *et al.*, 1995) and BO-653 (2,3-dihydro-5-hydroxy-2,2-dipentyl-4,6-di-*tert*-butylbenzofuran)

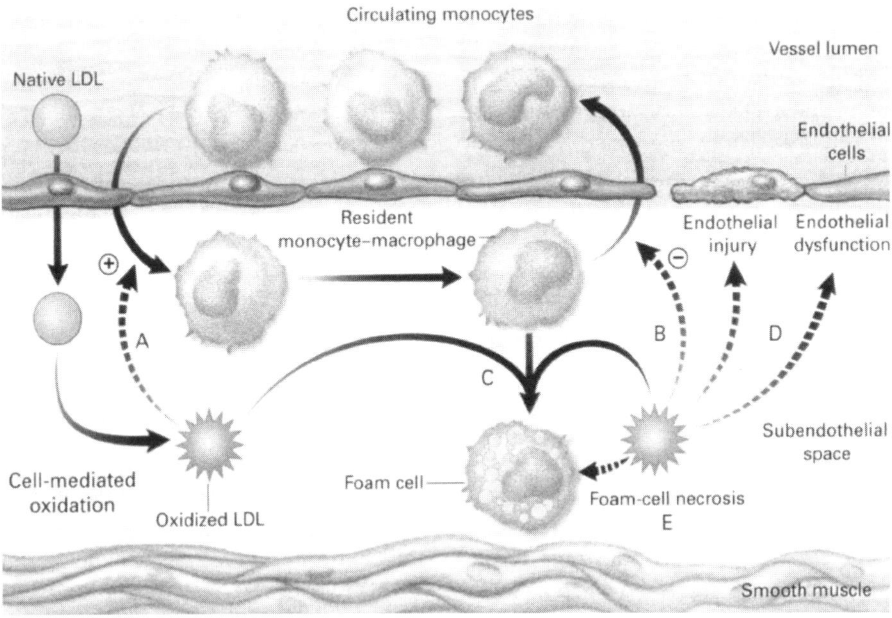

Figure 15.1: The oxidative hypothesis implies that oxidized LDL is a key and early step in the progression of atherosclerosis.
Circulating LDL migrates through the endothelium into the subendothelial space where it becomes oxidized. Oxidized LDL activates a series of events including: expression of adhesion molecules on the surface of endothelial cells (A) that facilitates monocyte adhesion and infiltration into the subendothelial space and (B) impairs monocyte re-entry into the lumen of the vessel. Resident monocytes mature to macrophages that (C) actively take-up oxidized LDL to yield lipid-laden foam cells: an early 'hall-mark' of atherogenesis. Oxidized LDL also promotes endothelial cell injury (D) that leads to vascular or endothelial dysfunction. Therefore, to reduce the progression of atherosclerosis a requirement would be an increased resistance of LDL lipids to oxidation. However, several important additional mechanisms may underlie the role of antioxidants in preventing the clinical manifestations of CAD. For example, there is evidence that plaque stability, and vasomotor function show dependence on antioxidants (Diaz *et al.*, 1997). Also, cellular antioxidants inhibit monocyte adhesion, protect against the cytotoxic effects of oxidized LDL, and inhibit platelet activation. Furthermore, cellular antioxidants protect against the endothelial dysfunction associated with atherosclerosis by preserving endothelium-derived ·NO activity. These mechanisms may have an important role in the benefits of antioxidants in addition to inhibition of LDL oxidation. Reproduced with permission from the National Academy of Sciences (Quinn *et al.*, 1987).

(Cynshi *et al.*, 1998) inhibit lesion development in various animal models of the disease. However, not all synthetic antioxidants inhibit atherosclerosis in animals even if they effectively inhibit LDL oxidation *in vitro* (Fruebis *et al.*, 1994; Witting *et al.*, 1999a). Also, a particular antioxidant (e.g., 3,3′,5,5′-tetrabutyl 1,1′-biphenyl 4,4′-diol) may be protective in one (Witting *et al.*, 1999b) but not another animal model of atherosclerosis (Witting *et al.*, 1999a), while other antioxidants (e.g., probucol) inhibit disease at one site but promote atherosclerosis at another site of the aortic tree at the same time and in the same animal (Witting *et al.*, 2000c). Moreover, effective inhibition of lipid oxidation in the artery wall by antioxidant supplements does not always inhibit atherosclerosis (Pratico *et al.*, 1998; Witting *et al.*, 1999a, 2000c). Finally, controlled, prospective clinical trials with natural antioxidants in humans have overall yielded controversial results (Kritharides and Stocker, 2002 for a recent review). Therefore, the precise role of LDL oxidation in and the impact of antioxidants on atherosclerosis remain unclear.

15.1.2 LDL lipid oxidation

The redox active transition metals copper and iron are generally required for cell-mediated LDL lipid (per)oxidation *in vitro* and LDL oxidation *in vitro* is commonly achieved by exposing the lipoprotein to high molar concentrations of Cu^{2+} (Esterbauer *et al.*, 1987, 1989). Under these conditions, the endogenous antioxidants of LDL including α-tocopherol (α-TOH, biologically the most active form of vitamin E), are consumed rapidly and a 'high uptake' form of LDL is generated due to modification of apolipoprotein B-100 by reactive aldehydes derived from the breakdown of primary lipid hydroperoxides (Esterbauer *et al.*, 1992). However, the presence and relevance of 'free' transition metals for LDL oxidation *in vivo* is debatable (Rae *et al.*, 1999). Formation of lipid aldehydes occurs primarily after complete depletion of α-TOH in LDL (Esterbauer *et al.*, 1988). However, human carotid endarterectomy samples, which represent the most advanced atherosclerotic lesions, still contain apparently normal amounts of α-TOH (Suarna *et al.*, 1995) associated with lipoprotein fractions (Niu *et al.*, 1999), and there is biochemical evidence that most lipid hydro(pero)xides detected in human lesions are formed in the presence of α-TOH (see below). A lack of participation of transition metals is also suggested by the absence in early atherosclerotic lesions of *o*- and *m*-tyrosine (Leeuwenburgh *et al.*, 1997b), previously used as marker molecules indicative of the presence of metals (Halliwell, 1996).

In vitro LDL oxidation can also be achieved by enzymes such as lipoxygenases (Parthasarathy *et al.*, 1989; Kuhn *et al.*, 1994; Upston *et al.*, 1997; Heydeck *et al.*, 2001) and peroxidases (Wieland *et al.*, 1993; Braun and von Bruchhausen, 1994; Witting *et al.*, 1997), although their role *in vivo* is questionable. In contrast, there is strong evidence for a role of reactive chlorine and nitrogen species for *in vivo* LDL oxidation (Mashima *et al.*, 2001).

A common feature of many oxidizing conditions is the generation of α-tocopheroxyl radical (α-TO$^{\cdot}$) from α-TOH, and this contributes to the prooxidant activity of vitamin E in LDL according to the model of tocopherol-mediated peroxidation (TMP) (Bowry *et al.*, 1992; Bowry and Stocker, 1993; Upston *et al.*, 1999) (*Figure 15.2*). In the TMP model of LDL lipid peroxidation, α-TOH acts as a phase transfer agent, drawing aqueous radicals into individual LDL particles thereby increasing the proportion of lipoprotein particles that contain an α-TO$^{\cdot}$ radical (Reaction 1). In turn, α-TO$^{\cdot}$ may initiate a radical chain reaction via abstraction of a hydrogen atom from a bisallylic hydrogen-containing lipid molecule (mostly esterified linoleic and arachidonic acid) (LH), resulting in α-TOH and a carbon-centered lipid radical (L$^{\cdot}$) (Reaction 2). The latter rapidly adds to molecular oxygen dissolved in LDL to form a lipid peroxyl radical (LOO$^{\cdot}$)

(Reaction 3) that is scavenged by α-TOH, producing a molecule of lipid hydroperoxide (LOOH) and regenerating α-TO· (Reaction 4). In this model, the process of lipid peroxidation occurs within discrete lipid particles separated by an aqueous phase and with α-TO· as the 'chain-carrying radical' unable to diffuse between particles. The rate-limiting step is Reaction 2, the rate of which is two to three orders of magnitude smaller than that of LOO· with LH. Therefore, the TMP model (*Figure 15.2*) represents 'retarded' peroxidation when compared with that seen in the absence of α-TOH, yet the presence of the vitamin can still increase the overall oxidizability of LDL (see Bowry and Stocker, 1993 for a detailed description). Reaction 2 can be inhibited readily, for example by co-antioxidants (Bowry *et al.*, 1995; Witting *et al.*, 1996) that eliminate α-TO· (see below).

There is indirect support for TMP being important for LDL oxidation *in vivo*. In human lesions the predominant lipid oxidation products are those of cholesteryl-linoleate (C18 : 2) contained in lipoproteins, with the hydroxy derivatives (C18 : 2-OH) being the most abundant (Suarna *et al.*, 1995, 1997; Waddington *et al.*, 2003). Four discrete positional and configurational isomers (i.e., 9-*trans/cis*, 13-*cis/trans*, 9-*trans/trans* and 13-*trans/trans* C18 : 2-OH) may be formed of each of the two diasteromers (*R* and *S*), with the presence of α-TOH affecting their relative distribution: in the presence of the vitamin, the kinetically favored *trans/cis* and *cis/trans* isomers accumulate preferentially, whereas in the absence of α-TOH the thermodynamically favored *trans/trans* isomers predominate (Porter *et al.*, 1980; Porter and Wujek, 1984; Kenar *et al.*, 1996; Upston *et al.*, 1997; Heydeck *et al.*, 2001). Concordant with the finding that large amounts of oxidized lipid co-exist with normal levels of α-TOH (Suarna *et al.*, 1995; Upston *et al.*, 2002a), *trans/cis* and *cis/trans* predominate over

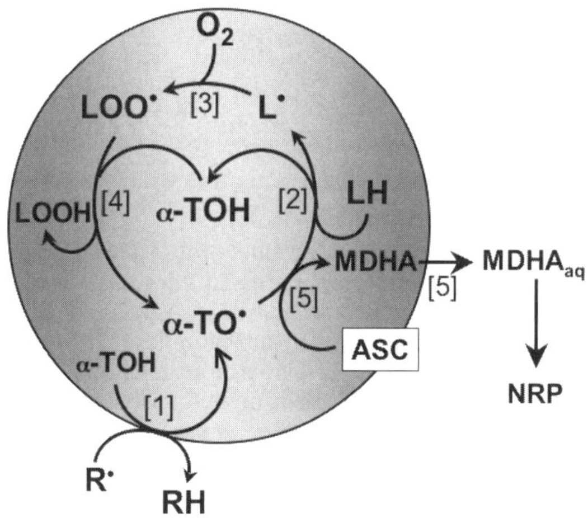

Figure 15.2: Prooxidant activity of vitamin E in lipoprotein particles: lipid peroxidation via tocopherol-mediated peroxidation.
Schematic of an oxidizing LDL particle illustrates the phase- and chain-transfer activities of α-TOH in the process yielding lipid hydroperoxides (LOOH). The co-antioxidant ASC effectively eliminates the chain carrying α-TO· via export of the radical from the particle (Reaction 5) thereby inhibiting tocopherol-mediated peroxidation of LDL lipids (LH).

trans/trans isomers of C18 : 2-OH in human carotid and aortic lesions and in lipoproteins isolated from them (Upston *et al.*, 2002b) (*Table 15.2*). Further, dietary supplementation with vitamin E significantly increases aortic concentrations of α-TOH and *trans/cis* and *cis/trans* isomers in rabbits after arterial injury (Upston *et al.*, 2001). Finally, supplementing animals with a synthetic co-antioxidant whose ability to scavenge peroxyl radicals is inferior to that of α-TOH, such that the tissue concentrations of the co-antioxidant approaches that of endogenous α-TOH, inhibits lipoprotein lipid oxidation *in vivo* (Witting *et al.*, 1999a, 1999b).

Together, these data indicate that in atherosclerotic lesions, lipoproteins still contain α-TOH and their lipids oxidize predominantly in the presence, not absence, of vitamin E via TMP. Therefore, effective inhibition of lipoprotein lipid oxidation can be achieved by retardation of TMP, and this represents a likely major mechanism by which vitamin C may attenuate formation and accumulation of oxidized LDL.

The following reviews activities of ASC that may provide protection against atherosclerosis, with a focus on potential molecular actions and a summary of available data on the effect of vitamin C supplements on vascular disease in humans.

15.2 Protective activities of vitamin C in atherosclerosis

A separate chapter of this volume is dedicated to the function of ASC as an antioxidant (see Chapter 10 by Buettner and Schafer). The following text is therefore limited to potential protective activities of vitamin C related to atherosclerosis, a topic reviewed earlier by others (Carr *et al.*, 2000c).

15.2.1 Direct oxidant scavenging by vitamin C

Among the water-soluble biological antioxidants, ASC is generally regarded as a primary, first line protective agent that repairs or nullifies free radicals ($R^{\cdot}$ or one-electron oxidants) that otherwise cause damage. This is achieved by ASC donating

Table 15.2: *Cis/trans*-isomers dominate the product profile of C18 : 2-OH in aortic lesions[a]. Endogenous C18 : 2-OH was isolated from supernatants derived from aortic homogenates of various stage human lesions. Configurational and stereo-isomers of C18 : 2 were isolated by reversed-phase HPLC and further analyzed by normal-phase HPLC using a chiral column to identify the products. Total C18:2-OH isolated from fatty streak lesions (nine separate tissue samples) were pooled to yield a single sample for isomer analyses. For subsequent groups the data represent mean ± SD, expressed as a percentage of the total C18 : 2-OH (*cis/trans*)-isomers or 13- *cis/trans*-C18 : 2-OH (*S* isomers). [a]Data from Upston *et al.* (2002b) with permission from Portland Press.

Lesion stage	n	*Cis/trans*-C18:2-OH (%)	*Cis/trans*-C18 : 2-OH 13-*S* isomer (%)
Fatty streak	9	87	50
Fibro-fatty	7	85 ± 2	50 ± 23
Ulcerated/complex	8	79 ± 6*	50 ± 1

* Significantly different to the corresponding value from the fibro-fatty lesions, $P < 0.05$.

a single electron followed by a proton to yield a *chemically* reduced, nonradical product (RH) and ascorbyl radical (MDHA) that is less reactive toward oxidizable targets than $R^{\cdot}$ (Reaction 6).

$$R^{\cdot} + ASC \rightarrow RH + MDHA \qquad \text{(Reaction 6)}$$

The inert MDHA readily dismutates to ASC and dehydroascorbic acid. Both MDHA (Buettner, 1993; Alcain *et al.*, 1991) and dehydroascorbic acid (Nardai *et al.*, 2001; May *et al.*, 1998a; Mendiratta *et al.*, 1998) can be reduced by enzymatic systems utilizing NADPH or NADH as reducing cofactors, effectively recycling the pool of bioavailable vitamin C (see chapter by May and Asard).

There is a significant body of evidence from *in vitro* studies to support the idea that physiological concentrations of ASC are able to inhibit LDL lipid (per)oxidation initiated by vascular cells (Martin and Frei, 1997), activated neutrophils (Stocker *et al.*, 1991) and cell-free systems (Sato *et al.*, 1990). This is achieved by ASC scavenging aqueous oxidants and acting as a co-antioxidant for LDL's α-TOH (Frei *et al.*, 1990; Witting *et al.*, 1997). For example, ASC efficiently scavenges aqueous peroxyl (Barclay *et al.*, 1985, 1989) and protein radicals (Deterding *et al.*, 1998; Witting *et al.*, 1999c, 2000a) that can initiate LDL lipid peroxidation.

Ascorbate also effectively inhibits LDL and plasma protein oxidation induced by the two-electron oxidant hypochlorous acid (HOCl) (Hazell *et al.*, 1994; Hawkins and Davies, 1999; Carr *et al.*, 2000b; Carr and Frei, 2002). HOCl, produced by myeloperoxidase (MPO) from chloride anion and hydrogen peroxide, readily transforms LDL into a high uptake form *in vitro* (Hazell and Stocker, 1993), and there is increasing evidence for MPO as an oxidant in atherosclerosis (for a review see Heinecke, 1999). For example, circulating concentrations of MPO are associated with the risk of coronary artery disease (Zhang *et al.*, 2001), and active MPO (Daugherty *et al.*, 1994) and oxidation makers indicative of MPO/HOCl are present in human lesions (Hazell *et al.*, 1996; Hazen and Heinecke, 1997; Hazen *et al.*, 2000). Furthermore, MPO can bind to LDL (Carr *et al.*, 2000a) such that MPO may associate with and directly oxidize LDL in the subendothelial space.

In contrast, ASC appears unlikely to be a physiologically important scavenger of peroxynitrite ($ONOO^-$), as indicated by the low rate constant for the reaction between ASC and $ONOO^-$ at physiological pH (Bartlett *et al.*, 1995). A powerful oxidant capable of oxidizing LDL (Jessup *et al.*, 1992; Thomas *et al.*, 1998) and proteins (Radi *et al.*, 1991) *in vitro*, $ONOO^-$ is generated from nitric oxide ($^{\cdot}NO$) and superoxide anion radical ($O_2^{\cdot -}$) (Ischiropoulos *et al.*, 1992) and has been implicated in oxidative reactions in atherosclerosis (Beckman *et al.*, 1994; Leeuwenburgh *et al.*, 1997a; Zou *et al.*, 2002).

In spite of its potent antioxidant activity *in vitro* and its presence at relatively high concentration (Suarna *et al.*, 1995), ASC appears unable, however, to prevent the accumulation of lipid oxidation products (Suarna *et al.*, 1995) and HOCl-damaged proteins (Hazell *et al.*, 1996; Leeuwenburgh *et al.*, 1997a) during atherogenesis.

15.2.2 Synergistic actions of vitamin C

The ability of ASC to reduce α-TO˙ to α-TOH was recognized first by Packer *et al.* (1979). This redox cycling process is now well established as a mechanism that spares vitamin E from oxidation in isolated biological membranes (Kagan *et al.*, 1990; May *et al.*, 1998b) or complex emulsions including human plasma *in vitro* (Barclay *et al.*, 1985; Sharma and Buettner, 1993). Despite these data however, the importance of this interaction for *in vivo* maintenance of vitamin E and inhibition of lipid oxidation remains to be established (Burton *et al.*, 1990; Huang *et al.*, 2002).

With regards to inhibition of LDL lipid peroxidation, it is important to recognize that it is the 'trapped' α-TO˙ in an oxidizing LDL particle that is responsible for the prooxidant activity of vitamin E (Bowry and Stocker, 1993). Therefore, TMP is prevented by elimination of α-TO˙, and co-antioxidants that engage in Reaction 5 (*Figure 15.2*), are efficient anti-TMP agents (Thomas *et al.*, 1995). ASC efficiently quenches α-TO˙ in micelles (Bisby and Parker, 1991; Witting *et al.*, 1996), and it completely inhibits LDL lipid oxidation initiated by aqueous and lipophilic peroxyl radicals (Bowry *et al.*, 1995; Witting *et al.*, 1996), horseradish peroxidase (Witting *et al.*, 1997) and lipoxygenase (Upston *et al.*, 1997) *in vitro*. At least in the case of horseradish peroxidase, addition of ASC results in immediate disappearance of α-TO˙ and cessation of lipid peroxidation, with concomitant formation of MDHA (Witting *et al.*, 1997) (*Figure 15.3*).

Another example of synergistic activity of ASC is the maintenance of cellular glutathione (GSH), an important reductant involved in intracellular redox

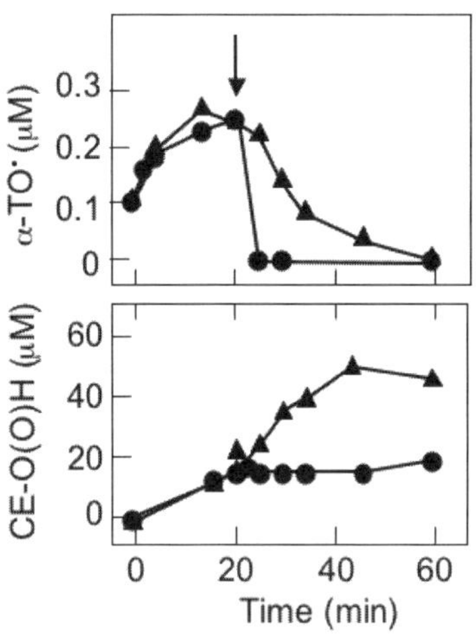

Figure 15.3: Ascorbate is an effective co-antioxidant for LDL lipids.
Data show the time-dependent changes in α-TO˙ concentration and consumption of α-TOH (top panel) and accumulation of hydroperoxides and hydroxides of cholesterylesters (bottom panel) in LDL undergoing oxidation initiated with horseradish peroxidase and treated sequentially with catalase followed by either physiologic urate (100 μM), a poor co-antioxidant or ascorbate (50 μM), a highly effective co-antioxidant. Experimental conditions are as detailed in (Witting *et al.*, 1997). Reproduced with permission from the American Chemical Society.

homeostasis (Winterbourn and Munday, 1990). GSH is readily oxidized by one-electron oxidants to yield the corresponding glutathione thiyl radical ($GS^{\cdot}$) that is able to damage a wide range of molecules (Eling *et al.*, 1986). Reduction of $GS^{\cdot}$ by ASC is thermodynamically favored and kinetically fast (ΔG ~60 $kJ{\cdot}mol^{-1}$, k ~$6 \times 10^{8}{\cdot}M^{-1}{\cdot}s^{-1}$) (Wardman, 1995; Tamba and O'Neill, 1991), so that ASC is a preferred radical sink in cells (Sturgeon *et al.*, 1998) maintaining GSH in the antioxidant active, reduced form.

15.2.3 Other protective activities of ASC

In addition to oxidant scavenging, ASC has a number of potentially protective activities related to cardiovascular disease (CVD) that will be referred to briefly in the following.

Vitamin C and endothelial function

The presence of heightened oxidative stress associated with atherosclerosis is not limited to lipoproteins in the vascular wall. One important consequence of oxidative stress is the impairment of vascular homeostasis resulting, in part, from impaired endothelial function and reduced bioactivity of endothelium-derived $^{\cdot}NO$. $^{\cdot}NO$, synthesized from L-arginine by endothelial nitric oxide synthase (eNOS), stimulates the relaxation of vascular smooth muscle cells and plays an important role in the maintenance of vascular tone. $^{\cdot}NO$ is highly reactive and can be inactivated rapidly in the vessel lumen and wall by reaction with a plethora of molecules. Endothelial function of coronary and peripheral arteries is impaired in atherosclerosis and a variety of related conditions including hypertension, smoking and diabetes. It manifests as a loss of endothelium-dependent vasodilation and is typically associated with decreased bioavailability of $^{\cdot}NO$. Conversely, increasing the bioavailability of $^{\cdot}NO$ is considered a useful strategy for the prevention or therapy of CVD (Bult *et al.*, 1999), particularly as endothelial dysfunction predicts the occurrence of clinical events in patients with atherosclerosis.

Patients with unstable coronary syndromes (Vita *et al.*, 1998) or hypertension have low plasma concentration of ASC, and vitamin C supplements have the potential to attenuate defective endothelium-dependent vasodilation. For example, when given acutely, ASC reverses endothelial dysfunction in humans with established coronary artery disease (Levine *et al.*, 1996; Gokce *et al.*, 1999), and chronic treatment with ascorbic acid lowers blood pressure in hypertensive patients (Duffy *et al.*, 1999). In addition, a recent study reported preservation of $^{\cdot}NO$-dependent coronary vasomotor function in cardiac transplant recipients to relate to plasma concentrations of vitamin C, but not plasma F_2-isoprostanes (a marker of lipid oxidation; Fang *et al.*, 2000). However, a short-term improvement with vitamin C supplementation does not always extend to a sustained benefit, as seen in smokers for endothelium-dependent, flow-mediated dilation (FMD) of the brachial artery (Raitakari *et al.*, 2000). Further, in hypertensive patients, acute or chronic ascorbic acid treatment had no effect on brachial artery FMD even though it decreased blood pressure (Duffy *et al.*, 2001). These results indicate that conduit vessel endothelial dysfunction secondary to hypertension is not reversed by acute or chronic treatment with oral ascorbic acid. They also raise the question of whether, and if so to what extent, long-term ASC supplements impact on clinical endpoints of CVD.

The beneficial effect of ASC on the preservation of normal endothelial function may be rationalized by multiple mechanisms (May, 2000), including inhibition of LDL oxidation, activation of eNOS, and preservation of compounds that exhibit bioactivity similar to authentic ·NO, such as S-nitrosothiols. Vascular endothelial cells internalize and degrade oxidized LDL via a receptor-mediated pathway (Sawamura *et al.*, 1997), and this can result in heightened oxidative stress in the cells. Therefore, oxidized LDL and its oxidized lipids decrease ·NO bioavailability through increased production of $O_2^{·-}$ that scavenges ·NO with the potentially damaging $ONOO^-$ formed as a by-product (Keaney *et al.*, 1995; Cominacini *et al.*, 2001). Principally, ASC may attenuate this cellular oxidative stress by inhibiting LDL oxidation, and/or by scavenging oxidants produced by cells that have taken up oxidized LDL. ASC itself can compete with $O_2^{·-}$ for reaction with ·NO, although effective competition is observed only at supra-physiological concentrations of vitamin C ($\geq$ 10 mM) (Jackson *et al.*, 1998). The normal extracellular (30–150 μM) and cellular concentrations of ASC (low millimolar values) are thus not likely to prevent ·NO from interacting with $O_2^{·-}$.

Oxidized LDL has the potential to interfere with additional endothelial cell functions. For example, oxidized LDL induces apoptosis (Cominacini *et al.*, 2000) that may contribute to atherosclerosis based on the 'response to injury' model (Dimmeler *et al.*, 1997), particularly as oxidized LDL also attenuates 'repair' by inhibiting vascular endothelial growth factor-induced endothelial cell migration (Chavakis *et al.*, 2001).

Inhibition of LDL oxidation by ASC may therefore diminish a number of deleterious effects, although the importance of this depends on the relevance of *in vitro* oxidized LDL used in the studies cited above. In addition to inhibiting formation of oxidized LDL, ASC increases the barrier function of endothelial cell monolayers *in vitro* (Utoguchi *et al.*, 1995) and this may contribute to the maintenance of a functional endothelium.

ASC has also the potential to maintain the synthesis of ·NO by eNOS that requires cofactors such as NADPH, flavin adenine dinucleotide, flavin mononucleotide and tetrahydrobiopterin. A deficiency in some of these cofactors can change the enzyme into an oxidase that produces $O_2^{·-}$ in addition to, or in place of, ·NO (Vásquez-Vivar *et al.*, 1998). Addition of ASC to aortic endothelial cells *in vitro* enhances eNOS activity and this is due to an increase in cellular tetrahydrobiopterin (Huang *et al.*, 2000). ASC chemically stabilizes tetrahydrobiopterin. While this appears independent of reduction of dihydrobiopterin to tetrahydrobiopterin (Vásquez-Vivar *et al.*, 2001), it effectively increases the half-life and concentration of tetrahydrobiopterin in cells (Heller *et al.*, 2001) and decreases $O_2^{·-}$ production by NOS (Vásquez-Vivar *et al.*, 1999). Chemical stabilization also attenuates autoxidation of tetrahydrobiopterin, a process that itself can cause inactivation of ·NO (Mayer *et al.*, 1995). Furthermore, ASC may preserve ·NO bioavailability indirectly by sparing cellular thiols, such as GSH, that have the potential to stabilize ·NO via formation of *S*-nitrosothiols (Ignarro *et al.*, 1981).

In direct contrast to the above indicated beneficial effect, results from two recent *ex vivo* studies indicate that pre-incubation of rabbit aortic rings (de Saram *et al.*, 2002) or bovine ciliary vascular bed and rat mesentery vessels (McNeish *et al.*, 2002) with physiological concentrations of ASC impaired acetylcholine-induced relaxation. It is not clear whether this impaired response reflects species or vascular bed differences. Notably, in these studies, a possible ASC-mediated redox cycling of transition metals contaminating the buffers used was not considered. Therefore, it is plausible that a spurious *ex vivo* oxidative stress may have accounted for the observed impaired vessel relaxation.

The preservation of ˙NO formation and bioavailability has potential downstream effects on vascular homeostasis independent of vessel relaxation, such as inhibition of leukocyte adhesion to endothelial cell, smooth muscle cell proliferation, and platelet aggregation (Carr *et al.*, 2000c). These will be discussed briefly in the following sections.

Vitamin C and cell adhesion molecule expression

Leukocyte adhesion to the endothelium is an early step in atherogenesis (Davies *et al.*, 1993). Adhesion molecules are expressed in a localized fashion by aortic endothelium that overlies early foam cell lesions (Cybulsky and Gimbrone, 1991), and adhesion molecule deficiency protects mice against atherosclerosis (Bourdillon *et al.*, 2000). Expression of adhesion molecules is induced by a range of agents and conditions such as oxidized LDL (Lehr *et al.*, 1991, 1992), certain types of oxidized lipids (Lehr *et al.*, 1995), lysophosphatidylcholine (Kume *et al.*, 1992), pro-inflammatory cytokines (Panes *et al.*, 1999), and smoking (Weber *et al.*, 1996). ˙NO is known to modulate the early steps in the inflammatory response, such as the expression of adhesion molecules and leukocyte adhesion and migration (Kubes *et al.*, 1991), potentially providing the basis for a protective effect of ASC related to increased bioavailability of ˙NO.

Like ˙NO, supplemental vitamin C, but not vitamin E or probucol, protects leukocytes from smoke-induced adhesion to micro- and macro-vascular endothelium in hamsters (Lehr *et al.*, 1995) and humans (Woollard *et al.*, 2002). Smokers have decreased plasma ASC and their monocytes show enhanced adhesion to endothelial cells *in vitro* (Weber *et al.*, 1996). The underlying mechanism of the inhibitory effect of ASC remain unclear. While the activation of important monocyte adhesion proteins, e.g., CD11b, is under redox regulation and hence a potential site of action of ASC (Blouin *et al.*, 1999), vitamin C does not appear to affect CD11b expression on monocytes (Woollard *et al.*, 2002). Another factor involved in the regulation of leukocyte adhesion is platelet-activating factor (PAF), and PAF-like lipids that can be formed during nonenzymatic oxidation of phospholipids (Heery *et al.*, 1995) are implicated in smoke-induced leukocyte adhesion to the vessel wall (Lehr *et al.*, 1997). Similarly, F_2-isoprostanes increase platelet adhesion and reduce the anti-adhesive effects of ˙NO (Minuz *et al.*, 1998). This raises the possibility of a direct link between lipid oxidation and cell adhesion, and hence a potential target for regulation by ASC.

Vitamin C and vascular smooth muscle cell growth

ASC has been reported to affect the proliferation of vascular smooth muscle cells directly (Ivanov *et al.*, 1997) and indirectly, via regulation of extracellular matrix synthesis (Schwartz *et al.*, 1982). As uncontrolled proliferation of smooth muscle cells is a hallmark of atherosclerosis and related CVDs (Ross, 1999), its inhibition may be beneficial. Such benefit could be in addition to ASC maintaining ·NO and inhibiting oxidation of LDL, as oxidized LDL antagonizes ·NO-dependent activation of soluble guanylate cyclase (Schmidt *et al.*, 1991) that itself can regulate smooth muscle cell proliferation *in vitro* (Garg and Hassid, 1989; Mooradian *et al.*, 1995).

15.3 Ascorbate in atherosclerotic lesions

LDL oxidation that contributes to atherosclerosis is implied to take place in the extracellular space of the arterial wall, and most of the potential proatherogenic properties of oxidized LDL require the modified lipoprotein particle to be present outside cells. For this reason, antioxidant defenses surrounding and associated with LDL in the extracellular fluid are of particular importance. The extracellular antioxidants in the vascular wall have not been evaluated systematically since this material is difficult to obtain. However, it is reasonable to assume that the antioxidant composition of the fluid surrounding cells, extracellular matrix and lipoprotein particles in the intima is comparable to that of human blood plasma (Frei *et al.*, 1988) and suction blister fluid (Dabbagh and Frei, 1995). The antioxidant defenses of human extracellular fluids (Halliwell and Gutteridge, 1990; Stocker and Frei, 1991) and the artery wall (Stocker, 2000) have been reviewed previously, so that the following will focus on ASC.

15.3.1 Availability of ascorbate and in the artery wall

Low-molecular weight, nonproteinaceous antioxidants represent an important line of defense in the extracellular compartment. Unfortunately, only a few reports have addressed their contents in normal human arteries and veins, and the information available is limited to homogenates prepared from the vessels, and focus on vitamin E (Suarna *et al.*, 1995; Carpenter *et al.*, 1995; Killion *et al.*, 1996). Overall, normal arteries contain about one-third the amount of ASC present in plasma (c.f. *Tables 15.1* and *15.3*), although it is not known whether the ASC detected is derived from intra- or extracellular sites.

Atherosclerosis is associated with complex changes to the content, structure, chemical composition and location of lipid/lipoprotein particles within the vessel wall, all of which are expected to affect the mechanism and efficacy of the various antioxidants. Very little is known about the changes in antioxidant defenses and to date no systematic study on the disease stage-dependent changes in ASC in the artery wall has been carried out. The interpretation of available data is restricted further by the fact that almost all results are obtained with homogenates prepared from entire or large parts of vessels that simply do not provide insight into focal areas and/or events.

An early study (Willis and Fishman, 1955) reported the ASC concentration in diseased human aortic tissue to be similar to that in plasma, and this was confirmed in a more recent study employing human endarterectomy specimens (Suarna *et al.*, 1995) (*Table 15.3*). Based on recoveries of cells per wet weight of

Table 15.3: Nonproteinaceous and proteinaceous aortic antioxidants. Data from (Stocker, 2000).

Antioxidants	Normal artery	Plaque
Water-soluble (nmol·mg protein^{-1})		
Ascorbate	0.12 ± 0.07	1.30 ± 0.87
Urate	0.11 ± 0.09	3.32 ± 2.26
Lipid-soluble (mmol·mol cholesterol^{-1})		
α-Tocopherol	4.2 ± 1.7	6.3 ± 4.8
α-Tocopherol[a]	ND	28.6 ± 21.8
γ-Tocopherol	ND	0.24 ± 0.18
Ubiquinone-10	1.5 ± 1.5	1.5 ± 0.15
Lycopene	Trace	0.05 ± 0.06
β-Carotene	Trace	0.05 ± 0.04
Glutathione-related (mU·mg protein^{-1})		
GSH Peroxidase	11.9 ± 3.8	3.55 ± 2.1*
GSH Reductase	0.24 ± 0.18	1.07 ± 0.4*
GST Peroxidase	ND	1.37 ± 0.35*
Total GST	22.5 ± 6	20.9 ± 5.3
Superoxide dismutase related (mU·mg protein^{-1})		
EC-SOD	104 ± 49	77 ± 55
Cu,Zn-SOD	100 ± 50	90 ± 42
Mn-SOD	2.5 ± 1.4	3.6 ± 1.9

[a]Expressed as mmol·mol cholesteryl linoleate^{-1}, the most prevalent oxidizable lipid present in human LDL. *$P < 0.05$ vs. normal artery. Abbreviations are as for *Table 15.1*.

human plaque (Jonasson *et al.*, 1986) and assuming an average intracellular ASC concentration of 10 mM, it can be calculated from the data reported by Suarna *et al.* (1995 that >90% of the detected ASC in aortic plaque is derived from extracellular sites.

Available data from animal models of atherosclerosis indicate comparable aortic concentrations of ASC in healthy and diseased vessels (Witting *et al.*, 1999b, 2000b), although care needs to be taken in the interpretation of these results, as mice are able to synthesize the vitamin. The observations that in human plaque samples and normal arteries only small and comparable amounts of vitamin C are present as dehydroascorbic acid, and that plaque contains several times more ASC than normal artery, all suggest that vitamin C does not generally become deficient during atherogenesis. This situation reflects that of α-TOH (see above), and together, available data indicate a generally intact antioxidant capacity in atherosclerotic compared to normal tissue.

The evidence for relatively large amounts of lipid oxidation products in the presence of intact concentrations of vitamins E and C is intriguing given that ASC effectively inhibits lipoprotein lipid oxidation and TMP *in vitro*. Of note, LDL aggregates and associates with proteoglycans upon entering the vessel wall (Frank and Fogelman, 1989; Nievelstein *et al.*, 1991). This raises the question whether aggregates of lipoproteins also oxidize via TMP and, more importantly, whether access of ASC to α-TO$^{\cdot}$ (to prevent TMP) is restricted in these aggregates, an effect due to charge repulsion (as both ASC and proteoglycans are negatively charged). Recent studies indicate, however, that LDL aggregated *in vitro* with proteoglycans still oxidizes via TMP, as do lipoproteins isolated from human lesions, and that this process is effectively inhibited by ASC (Upston *et al.*, 2002b). It remains possible of course, that the *in vitro* conditions used and the lipoproteins isolated from human lesions in these studies do not truly reflect the *in vivo* situation. Further studies are clearly needed to elucidate the reason(s) for the apparent paradox that ASC is unable to prevent substantial lipoprotein lipid oxidation in the diseased vessel wall, although this may have grave implications with regards to a potential benefit of ASC against CVD.

15.4 Vitamin C and atherosclerosis

The literature on ASC and atherosclerosis is limited. A potentially important study is The Antioxidant Supplementation in Atherosclerosis Prevention (ASAP) study (Salonen *et al.*, 2000) that examined the effect of vitamin C or vitamin E or a combination on the rate of carotid intima-to-media thickness (IMT) progression. IMT, used as a marker of on-going atherosclerotic disease, correlates only weakly with disease measured by arteriography (Adams *et al.*, 1995), although it is useful as a surrogate marker for longitudinal studies. The combination of vitamins C and E significantly decreased IMT progression rates in men at the follow-up of 3 years. However, supplementation with vitamin C (or vitamin E) alone had no beneficial effect, and there was no benefit for the antioxidant combination in women (Salonen *et al.*, 2000). At 6-year follow-up, the combination of vitamins C and E were again protective in men, but not in women (Salonen *et al.*, 2003), and the authors concluded that their findings confirm that the supplementation with the combination of vitamins C and E slows down atherosclerosis progression in hypercholesterolemic persons. Notably, at the 3-year treatment point the study switched subjects from the vitamin C (n = 120) or vitamin E supplementation alone (n = 115) to the combination, and analyzed the 6-year data for all subjects on vitamins C plus E (i.e., subjects on co-supplements for years 0–6 and 3–6) versus control (n = 105). Unfortunately, information on the effect of changing from the noneffective supplement with a single vitamin to the cosupplement was not provided. In apparent contrast to ASAP, in the WAVE (Women's Angiographic Vitamin and Estrogen) trial postmenopausal women with ≥ 15–75% stenosis at baseline, the progression of coronary atherosclerosis worsened nonsignificantly with daily supplements of vitamin C (1 g) plus vitamin E (800 IU) for 2.8 years compared to placebo (Waters *et al.*, 2002). Overall, these studies show that supplementation with vitamin C (or vitamin E) alone does not attenuate the progression of atherosclerosis. They leave open the possibility that cosupplements with vitamins C and E may provide benefit in some yet to be clearly identified subgroup(s).

The effect of ASC on atherosclerosis in animals has been reviewed recently (Lynch and Frei, 1994; Keaney, 2000). Briefly, early animal studies on atherosclerosis utilized the scorbutic guinea pig as a model (Willis, 1953). The complications related to scurvy and associated with this model were overcome by using guinea pigs with marginal vitamin C deficiency that just prevents the development of scurvy. Such animals when fed a diet enriched with 0.3% cholesterol for prolonged periods of time accumulate cholesterol in the aorta and develop atherosclerotic lesions that are inhibited by excess ASC (Ginter *et al.*, 1969). Similarly, intravenous injection of ASC (100 $mg{\cdot}kg^{-1}$) was reported to attenuate intimal thickening in Dutch-belted rabbits fed 0.5% cholesterol over 6 weeks without affecting blood lipids (Verlangieri *et al.*, 1977). Also, feeding ASC (150 $mg{\cdot}day^{-1}$) to albino rabbits on a 0.3% cholesterol diet slightly decreased the extent of atherosclerosis (Beetens *et al.*, 1986). Although a beneficial effect of ASC has not been reported in all studies (Lynch and Frei, 1994), it is clear that the vitamin C status can be an important determinant for atherogenesis in animals.

15.5 Vitamin C and CVD

15.5.1 Antioxidants and CVD

Traditional clinical endpoints of CVD are death, stroke, and myocardial infarct. It is generally accepted that the major factors associated with increased risk of CVD include age, sex, familial coronary artery disease (CAD), hypercholesterolemia, hypertension, smoking, obesity and a sedentary lifestyle. CAD and stroke are the leading cause of premature death in Westernized populations. Evidence for a beneficial effect of antioxidant vitamins on CAD has accumulated from several lines of research, with much of the focus on vitamin E, LDL oxidation and endothelial function. However, a systematic review of the literature including all major human and animal studies suggests that there is no clear consensus on the role of vitamin E (for recent reviews see Stocker, 1999a; Heinecke, 2001; Witztum and Steinberg, 2001; Kritharides and Stocker, 2002; Jialal and Devaraj, 2003). Perhaps the most consistent finding is that antioxidant supplements do not cause harm, although even this statement may be challenged, as two studies reported co-administration of antioxidants to significantly attenuate the beneficial effect of drug treatment (Tardif *et al.*, 1997; Cheung *et al.*, 2001). Such controversy impacts on the perceived usefulness of antioxidant supplements against CVD (Brown *et al.*, 2002).

15.5.2 Vitamin C and in vivo *lipid oxidation*

Concordant with a beneficial role, three of seven studies in smokers reported vitamin C supplements to decrease *in vivo* markers of lipid oxidation. Noteworthy, the one positive study (Reilly *et al.*, 1996) pointed out by Carr and Frei (1999b) that used urinary F_2-isoprostane (8-*epi*-prostaglandin $F_{2\alpha}$) as marker, involved only five subjects and no placebo group. A recent placebo-controlled study involving 11 subjects per group reported a decrease in the same urinary marker although this did not reach statistical significance (Huang *et al.*, 2002). As smoking itself is associated with increased oxidative damage (Morrow *et al.*, 1995), the relevance of decreased lipid oxidation in smokers receiving vitamin C supplements, if confirmed, to CAD remains unclear. At present, only one study examined the effect of vitamin C supplementation on lipid oxidation in CAD patients, and found no benefit (Gokce *et al.*, 1999). Vitamin C supplements have been reported to decrease plasma concentrations of F_2-isoprostane in guinea pigs in the presence of iron overload (Chen *et al.*, 2000).

15.5.3 Serum ASC, mortality and prevalence of CVD

Data for past and current studies assessing the association between ASC and CVD are summarized in *Table 15.4*. The second National Institutes of Health and Nutrition Examination (NHANES II) study investigated the relation of serum ASC to CVD (and cancer) mortality among adults living in the USA (Simon *et al.*, 2001). Serum ASC concentrations were available from 8453 participants aged 30 years or over at baseline. Most participants showed serum ASC ranging within normal levels (> 0.4 mg·dL^{-1} or > 22.7 μM for the US population) with ~17% suffering deficiency (< 0.4 mg·dL^{-1}) and mean vitamin C intake 98 mg·day^{-1}. Subjects with normal to high serum ASC showed a marginal (21–25%) decrease in the risk for CVD ($P < 0.09$ for trend) and significant decrease in risk for all cause mortality ($P < 0.001$), suggesting that low serum ASC is associated weakly with increased CVD risk.

The third NHANES study (Simon and Hudes, 1999) examined the relationship between serum levels of ASC and the prevalence of CVD (defined as self-reported angina, myocardial infarct, or stroke). There was a marginal inverse association between ASC and the prevalence of angina, but not infarct or stroke, among alcohol-consuming participants (multivariate odds ratio 0.48 at 95% confidence; $P < 0.06$ for trend) over the course of the study (5 years). As there is an interaction between ASC and alcohol consumption, the association was tested among participants who reported no alcohol consumption; serum ASC concentrations were not associated with CVD prevalence (Simon and Hudes, 1999).

In a European study, the relationship between vitamin C status (as measured by dietary intake or plasma ASC concentration) and mortality from stroke and coronary heart diseases was examined in 730 men and women aged 65 years or more over a period of 20 years (Gale *et al.*, 1995). Mortality from stroke was highest in those with the lowest vitamin C status, however no association was found between vitamin C and risk of death from coronary heart disease. A similar outcome was reported more recently for stroke in a rural Japanese population (Yokoyama *et al.*, 2000).

Smoking decreases plasma ASC, and this is reversed by cessation of smoking (Lykkesfeldt *et al.*, 1996). While smokers have an increased risk for CVD, such risk decreases with increasing time of cessation and, after ~15 years, approaches that of never-smokers (Enstrom, 1999). Whether this decreased risk results from restoration of normal plasma concentrations of ASC appears questionable however, given that the latter is reached after only a few weeks (Lykkesfeldt *et al.*, 1996).

15.5.4 Vitamin C intake or supplementation and CAD

Similar to the situation with vitamin E, the evidence linking CAD and vitamin C is inconsistent, as reviewed recently (Carr and Frei, 1999b). Several prospective cohort studies reported high dietary vitamin C intake or supplementation to associate with reduced risk of CAD. It is worth noting, however, that such studies tend to favor trends within the group of survivors (i.e., a survivor bias) that potentially skew the relative risk assessment. Nevertheless, different studies reported an association (n = 8) or no significant association (n = 8) of dietary vitamin C with CAD/CVD incidence (Carr and Frei, 1999b).

Table 15.4: Association between vitamin C intake and reduced CVD risk (prospective studies). Adapted from Carr and Frei, 1999b.

Reference	Cohort/ duration (years)/sex	End point (events)	Vitamin C intake (mg·day^{-1} unless indicated otherwise)	Relative risk (% decrease ↓)	Comments
Enstrom *et al.*, 1986	3119 (10) M/F	CVD (127 deaths)	> 250 cf. < 250	NS ↓	No relation between vitamin C intake and mortality. Established risk factors show strong association with CVD
	4479 (10) M	CVD (558 deaths)	> 50	↓ 42%	Strong association with males but not females despite adjusting all data for age, sex and 10 other potentially confounding variables
Enstrom, 1993	6809 (10) F	CVD (371 deaths)	> 50	↓ 25% NS	Weak inverse relation in females
Stampfer *et al.*, 1993	87245 (8) F	CAD (552 cases)	> 359 cf. < 93	NS ↓	No correlation between dietary vitamin C intake (over a broad range) and reduced risk of CAD
Stampfer *et al.*, 1993	87245 (8) F	Stroke (183 cases)	> 359 cf. < 93	NS ↓	Trend toward reduced risk for stroke
	39910 (4) M	CAD (667 deaths)	392 cf. 92 (median)	NS ↓	Supplementation to high (~400 mg·day^{-1}) levels has no effect on CAD
Fehily *et al.*, 1993	2512 (5) M	CVD (148 cases)	> 67 cf. < 35	NS ↓	A weak trend between increasing risk of CVD and decreased vitamin C intake
	2748 (14) M	CAD (186 deaths)	> 85 cf. < 60	NS ↓	Supports idea that vitamin C is beneficial. However, it cannot be excluded that foods containing other micronutrients provided protection. Daily intakes >91 mg/day had no additional beneficial effect
	2385 (14) F	CAD (58 deaths)	> 91 cf. < 61	↓51%	

Continued overleaf

Table 15.4: *Continued*

Reference	Cohort/ duration (years)/sex	End point (events)	Vitamin C intake ($mg \cdot day^{-1}$ unless indicated otherwise)	Relative risk (% decrease ↓)	Comments
	730 (20) M/F	Stroke (125 deaths) CAD (182 deaths)	> 45 cf. < 28 > 45 cf. < 28	↓ 50% NS ↓	In elderly subjects dietary or serum vitamin C strongly relates to stroke but not death from coronary heart disease
	4989 (3) M	Carotid atherosclerosis	> 982 cf. < 56	↓ Intimal thickness	Positive effect of vitamin C on a biochemical measure of CVD. IMT correlates strongly with established risk factors and predicts angiographically assessed restenosis in coronary vessels.
	6318 (3) F		> 728 cf. < 64	↓ Intimal thickness	
	1556 (24 y) M	CAD (231 deaths)	> 113 cf. < 82	↓ 30%	Consumption of foods rich in vitamin C and β-carotene reduce risk of CVD in elderly men
	34486 (7) F	CAD (242 deaths)	> 391 cf. < 112[b] > 196 cf. < 87c Supplement cf. no supplement	NS ↓ NS ↓ NS ↓	Both normal dietary intake and regular supplementation with vitamin C has no beneficial effect
Losonczy *et al.*, 1996b	11178 (6) M/F	CAD (1101 deaths)	Supplement cf. no supplement	NS ↓	Although ineffective alone, vitamin C in combination with vitamin E has a positive beneficial effect (53% ↓, 95% confidence)
Sahyoun *et al.*, 1996	725 (10) M	CVD (101 deaths)	> 388 cf. < 90	NS ↓	Vitamin C shows a 62% decrease though this is not significant
Mark *et al.*, 1998	29584 (5) M	Stroke	180 supplement cf. diet alone	NS ↓	Population of Chinese men and women received vitamin C or not in combination with 30 μg molybdenum per day

	8453 (12–16) M/F with 36 exclusions for CVD analyses	CVD (561 cases)	[a]0.1–2.7 mg·dl^{-1}	NS 21-25% ↓ (P = 0.09)	Limitation of the study is that it is not known whether low serum vitamin C preceded or was a consequence of the onset of CVD.
	7658 (5) M/F	CVD (871 cases)	[a]0.1–3.0 mg·dl^{-1}	NS ↓	Among drinkers serum vitamin C is associated with a decreased prevalence of angina (52% ↓, 95% confidence) but does not correlate with stroke or myocardial infarction
Yokoyama *et al.*, 2000	2121 (20) M/F	Stroke (196 cases)	[a]49.9 μM	25% ↓	Strong association between increasing serum vitamin C and reduced risk of stroke for both men and women
	20536 (5)	CVD (1063 deaths)	250 mg·day^{-1}	NS ↓	Supplement included vitamin E, 600 mg and β-carotene 20 mg daily). No significant reduction in CVD mortality, or incidence of any type of vascular disease
	160 (3)	CVD (38 events)	1000 ·day^{-1}	NS ↓	Vitamin C given in combination with vitamin E (800 IU·day^{-1}), β-carotene (25 mg·day^{-1}) and selenium (100 μg·day^{-1}). No evidence to support a role for antioxidants in the prevention of coronary disease
	40 (2)	Transplant CVD	[a]580 μM	7.2% ↓	Antioxidant cocktail of vitamins C + E yields a significant decrease in the progression of intimal index (defined as plaque area / vessel area)

M, Male; F, female; NS, not significant
[a]Values refer to plasma concentration of ASC

An early epidemiological study (Enstrom *et al.*, 1992) examined the relation between dietary vitamin C intake (based on dietary survey) and total or CVD mortality over a median of 10 years in a cohort of 11 348 adults, aged 25–74 years. Overall, males with high intakes of vitamin C showed a standardized mortality ratio (SMR) of 0.65 (range 0.52–0.80) for all causes and 0.58 (0.41–0.78) for all CVD. In contrast, females showed an SMR of 0.90 (0.74–1.09) for all causes and 0.75 (0.55–0.99) for all CVD. These results are suggestive of a beneficial effect of vitamin C, although it is worth noting that no attempt was made in this study to adjust for vitamin E intake or supplementation.

In the cross-sectional Atherosclerosis Risk in Communities Study (Kritchevsky *et al.*, 1995), the relationship between average carotid artery wall thickness and the intake of dietary and supplemented vitamin C, vitamin E and β-carotene was examined in 6318 female and 4989 male participants aged 45–64 years. There was a significant ($P = 0.019$ for women, $P = 0.035$ for men) inverse relationship between vitamin C intake and artery wall thickness. This provides support for the hypothesis that dietary vitamin C may protect against atherosclerotic disease, although carotid wall thickness is only a proxy for the disease.

In three of the eight studies where an association was reported, vitamin C was co-administered together with β-carotene (Pandey *et al.*, 1995; Knekt *et al.*, 1994) or vitamin E (Losonczy *et al.*, 1996a). Multivariate analyses of these study groups suggest that β-carotene and vitamin E intake may also be associated with decreased risk, so that interpretation of the effect of vitamin C alone is complicated. It is also noteworthy that the studies reporting no association include two large epidemiological studies (Kushi *et al.*, 1996; Rimm *et al.*, 1993).

Prospective studies with vitamin C supplementation and CVD

Prospective studies examining the effect of vitamin C supplementation on CVD endpoints are almost entirely limited to combinations of vitamin C and other antioxidants. The large MRC/BHF Heart Protection Study (HPS) for secondary prevention examined the benefit of the combination of vitamins C and E, and β-carotene. Although the supplementation regimen increased blood antioxidant levels substantially, no significant reductions in the mortality from, or incidence of, any type of vascular disease, or other major outcome was found (Group, 2002a). This was in contrast to the beneficial effect observed with simvastatin (Group, 2002b). Two further studies investigated antioxidant combination supplements together with lipid-lowering therapy. In the secondary prevention High-density Lipoprotein Atherosclerosis Treatment Study (HATS), subjects were randomized to simvastatin (lipid-lowering) plus niacin treatment, or antioxidants (vitamin C, α-TOH, β-carotene and selenium) or a combination of both. Compared to no treatment, only simvastatin/niacin significantly lowered the rate of stenosis progression and favorably altered plasma lipid profiles (Brown *et al.*, 2001). Antioxidant supplementation alone had no significant effect on clinical endpoints but, notably, when used in combination with simvastatin/niacin, antioxidants negated the benefit of the latter on plasma lipid profile and stenosis progression. In contrast, in the HPS this combination did not counteract the favorable effect of simvastatin on mortality and cardiovascular events (Group, 2002a).

In a recent small study involving 40 patients, the effect of vitamins C and E in addition to pravastatin on progression of cardiac transplant-associated arteriosclerosis was studied, with the change in average intimal index measured by intravascular ultrasound as the primary endpoint (Fang *et al.*, 2002). Compared to control, antioxidant supplements reduced plaque growth independent of any change in endothelial function. In addition to the problems associated with the small sample size of this study, it is worth noting that the underlying pathogenesis of transplant-associated arteriosclerosis is a fibro-proliferative response to chronic allogenic immune activation, which differs from that of atherosclerosis.

15.6 Summary and conclusions

High dietary intake of vitamin C is associated with reduced risk of CVD. However, trials in atherosclerosis-related CVD indicate that supplements with vitamin C alone do not provide a general benefit, as can be demonstrated with, for example, statins. This is also the case for the combination of vitamin C with other antioxidants and/or with lipid lowering therapy, that they do not consistently convey cardiovascular benefit, in either primary or secondary prevention. It is possible however that certain subpopulations of patients, such as those with endothelial dysfunction, benefit from supplements with vitamin C.

References

Adams MR, Nakagomi A, Keech A, Robinson J, McCredie R, Bailey BP, Freedman SB, Celermajer DS (1995) Carotid intima-media thickness is only weakly correlated with the extent and severity of coronary artery disease. *Circulation* **92**, 2127–2134.

Alcain FJ, Buron MI, Villalba JM, Navas P (1991) Ascorbate is regenerated by HL-60 cells through the transplasmalemma redox system. *Biochim. Biophys. Acta* **1073**, 380–385.

Barclay LR, Baskin KA, Locke SJ, Vinqvist MR (1989) Absolute rate constants for lipid peroxidation and inhibition in model biomembranes. *Can. J. Chem.* **67**, 1366–1369.

Barclay LRC, Locke SJ, MacNeil JM (1985) Autoxidation in micelles. Synergism of vitamin C with lipid-soluble vitamin E and water-soluble Trolox. *Can. J. Chem.* **63**, 366–374.

Bartlett D, Church DF, Bounds PL, Koppenol WH (1995) The kinetics of the oxidation of L-ascorbic acid by peroxynitrite. *Free Radic. Biol. Med.* **18**, 85–92.

Beckman JS, Ye YZ, Anderson PG, Chen J, Accavitti MA, Tarpey MM, White CR (1994) Extensive nitration of protein tyrosine in human atherosclerosis detected by immunohistochemistry. *Biol. Chem. Hoppe Seyler* **375**, 81–88.

Beetens JR, Coene M-C, Verheyen A, Zonnekeyn L, Herman AG (1986) Vitamin C increases the prostacyclin production and decreases the vascular lesions in experimental atherosclerosis in rabbits. *Prostaglandins* **32**, 335–352.

Berger TM, Polidori MC, Dabbagh A, Evans PJ, Halliwell B, Morrow JD, Roberts LJ, Frei B (1997) Antioxidant activity of vitamin C in iron-overloaded human plasma. *J. Biol. Chem.* **272**, 15656–15660.

Berliner JA, Heinecke JW (1996) The role of oxidized lipoproteins in atherogenesis. *Free Radic. Biol. Med.* **20**, 707–727.

Bisby RH, Parker AW (1991) Reactions of the α-tocopheroxyl radical in micellar solutions studied by nanosecond laser flash photolysis. *FEBS Lett.* **290**, 205–208.

Blouin E, Halbwachs-Mecarelli L, Rieu P (1999) Redox regulation of b2-integrin CD11b/CD18 activation. *Eur. J. Immunol.* **29**, 3419–3431.

Bourdillon MC, Poston RN, Covacho C, Chignier E, Bricca G, McGregor JL (2000) ICAM-1 deficiency reduces atherosclerotic lesions in double-knockout mice (ApoE(–/–)/ICAM-1(–/–)) fed a fat or a chow diet. *Arterioscler. Thromb. Vasc. Biol.* **20**, 2630–2635.

Bowry VW, Ingold KU, Stocker R (1992) Vitamin E in human low-density lipoprotein. When and how this antioxidant becomes a pro-oxidant. *Biochem. J.* **288**, 341–344.

Bowry VW, Mohr D, Cleary J, Stocker R (1995) Prevention of tocopherol-mediated peroxidation of ubiquinol-10-free human low density lipoprotein. *J. Biol. Chem.* **270**, 5756–5763.

Bowry VW, Stocker R (1993) Tocopherol-mediated peroxidation. The pro-oxidant effect of vitamin E on the radical-initiated oxidation of human low-density lipoprotein. *J. Am. Chem. Soc.* **115**, 6029–6044.

Braun S, von Bruchhausen F (1994) Vitamin E or probucol as donors for oxidation of human low-density lipoprotein by peroxidases/H_2O_2. *Pharmacology* **49**, 325–335.

Brown BG, Cheung MC, Lee AC, Zhao XQ, Chait A (2002) Antioxidant vitamins and lipid therapy: end of a long romance? *Arterioscler. Thromb. Vasc. Biol.* **22**, 1535–1546.

Brown BG, Zhao XQ, Chait A, Fisher LD, Cheung MC, Morse JS, *et al.* (2001) Simvastatin and niacin, antioxidant vitamins, or the combination for the prevention of coronary disease. *N. Engl. J. Med.* **345**, 1583–1592.

Buettner GR (1993) The pecking order of free radicals and antioxidants: lipid peroxidation, alpha-tocopherol, and ascorbate. *Arch. Biochem. Biophys.* **300**, 535–543.

Bult H, Herman AG, Matthys KE (1999) Antiatherosclerotic activity of drugs in relation to nitric oxide function. *Eur. J. Pharmacol.* **375**, 157–176.

Burton GW, Wronska U, Stone L, Foster DO, Ingold KU (1990) Biokinetics of dietary RRR-α-tocopherol in the male guinea pig at three dietary levels of vitamin C and two levels of vitamin E. Evidence that vitamin C does not 'spare' vitamin E in vivo. *Lipids* **25**, 199–210.

Carpenter KL, Cheeseman KH, van der Veen C, Taylor SE, Walker MK, Mitchinson MJ (1995) Depletion of alpha-tocopherol in human atherosclerotic lesions. *Free Radic. Res.* **23**, 549–558.

Carr A, Frei B (1999a) Does vitamin C act as a pro-oxidant under physiological conditions? *FASEB J.* **13**, 1007–1024.

Carr AC, Frei B (1999b) Toward a new recommended dietary allowance for vitamin C based on antioxidant and health effects in humans. *Am. J. Clin. Nutr.* **69**, 1086–1107.

Carr AC, Frei B (2002) Human neutrophils oxidize low-density lipoprotein by a hypochlorous acid-dependent mechanism: the role of vitamin C. *Biol. Chem.* **383**, 627–636.

Carr AC, Myzak MC, Stocker R, McCall M R, Frei B (2000a) Myeloperoxidase binds to low-density lipoprotein: potential implications for atherosclerosis. *FEBS Lett.* **487**, 176–180.

Carr AC, Tijerina T, Frei B (2000b) Vitamin C protects against and reverses specific hypochlorous acid- and chloramine-dependent modifications of low-density lipoprotein. *Biochem. J.* **346**, 491–499.

Carr AC, Zhu BZ, Frei B (2000c) Potential antiatherogenic mechanisms of ascorbate (vitamin C) and alpha-tocopherol (vitamin E). *Circ. Res.* **87**, 349–354.

Chavakis E, Dernbach E, Hermann C, Mondorf UF, Zeiher AM, Dimmeler S (2001) Oxidized LDL inhibits vascular endothelial growth factor-induced endothelial cell migration by an inhibitory effect on the Akt/endothelial nitric oxide synthase pathway. *Circulation* **103**, 2102–2107.

Chen K, Suh J, Carr AC, Morrow JD, Zeind J, Frei B (2000) Vitamin C suppresses oxidative lipid damage in vivo, even in the presence of iron overload. *Am. J. Physiol. Endocrinol. Metab* **279**, E1406–E1412.

Cheung MC, Zhao XQ, Chait A, Albers JJ, Brown BG (2001) Antioxidant supplements block the response of HDL to simvastatin-niacin therapy in patients with coronary artery disease and low HDL. *Arterioscler. Thromb. Vasc. Biol.* **21**, 1320–1326.

Chisolm GM, Steinberg D (2000) The oxidative modification hypothesis of atherogenesis: an overview. *Free Radic. Biol. Med.* **28**, 1815–1826.

Cominacini L, Pasini AF, Garbin U, Davoli A, Tosetti ML, Campagnola M, Rigoni A, Pastorino AM, Lo Cascio V, Sawamura T (2000) Oxidized low density lipoprotein (ox-LDL) binding to ox-LDL receptor-1 in endothelial cells induces the activation of NF-kB through an increased production of intracellular reactive oxygen species. *J. Biol. Chem.* **275**, 12633–12638.

Cominacini L, Rigoni A, Pasini AF, Garbin U, Davoli A, Campagnola M, Pastorino AM, Lo Cascio V, Sawamura T (2001) The binding of oxidized low density lipoprotein (ox-LDL) to ox-LDL receptor-1 reduces the intracellular concentration of nitric oxide in endothelial cells through an increased production of superoxide. *J. Biol. Chem.* **276**, 13750–13755.

Cybulsky MI, Gimbrone MA (1991) Endothelial expression of a mononuclear leukocyte adhesion molecule during atherogenesis. *Science* **251**, 788–791.

Cynshi O, Kawabe Y, Suzuki T, Takashima Y, Kaise H, Nakamura M, *et al.* (1998) Antiatherogenic effects of the antioxidant BO-653 in three different animal models. *Proc. Natl Acad. Sci. USA* **95**, 10123–10128.

Dabbagh AJ, Frei B (1995) Human suction blister interstitial fluid prevents metal ion-dependent oxidation of low density lipoprotein by macrophages and in cell-free systems. *J. Clin. Invest.* **96**, 1958–1966.

Daugherty A, Dunn JL, Rateri DL, Heinecke JW (1994) Myeloperoxidase, a catalyst for lipoprotein oxidation, is expressed in human atherosclerotic lesions. *J. Clin. Invest.* **94**, 437–444.

Davies MJ, Gordon JL, Gearing AJ, Pigott R, Woolf N, Katz D, Kyriakopoulos A (1993) The expression of the adhesion molecules ICAM-1, VCAM-1, PECAM, and E-selectin in human atherosclerosis. *J. Pathol.* **171**, 223–229.

de Saram K, McNeill KL, Khokher S, Ritter JM, Chowienczyk PJ (2002) Divergent effects of vitamin C on relaxations of rabbit aortic rings to acetylcholine and NO-donors. *Br. J. Pharmacol.* **135**, 1044–1050.

Deterding LJ, Barr DP, Mason RP, Tomer KB (1998) Characterization of cytochrome c free radical reactions with peptides by mass spectrometry. *J. Biol. Chem.* **273**, 12863–12869.

Diaz MN, Frei B, Vita JA, Keaney JF Jr (1997) Antioxidants and atherosclerotic heart disease. *N. Engl. J. Med.* **337**, 408–416.

Dimmeler S, Haendeler J, Galle J, Zeiher AM (1997) Oxidized low-density lipoprotein induces apoptosis of human endothelial cells by activation of CPP32-like proteases. A mechanistic clue to the 'response to injury' hypothesis. *Circulation* **95**, 1760–1763.

Duffy SJ, Gokce N, Holbrook M, Huang A, Frei B, Keaney J F Jr, Vita JA (1999) Treatment of hypertension with ascorbic acid. *Lancet* **354**, 2048–2049.

Duffy SJ, Gokce N, Holbrook M, Hunter LM, Biegelsen ES, Huang A, Keaney JF Jr, Vita JA (2001) Effect of ascorbic acid treatment on conduit vessel endothelial dysfunction in patients with hypertension. *Am. J. Physiol. Heart Circ. Physiol.* **280**, H528–H534.

Eling TE, Curtis JF, Harman LS, Mason RP (1986) Oxidation of glutathione to its free radical metabolite by prostaglandin H synthase. A potential endogenous substrate for the hydroperoxidase. *J. Biol. Chem.* **261**, 5023–5028.

Enstrom JE (1993) Counterpoint – Vitamin C and mortality. *Nutr. Today* **28**, 28–32.

Enstrom JE (1999) Smoking cessation and mortality trends among two United States populations. *J. Clin. Epidemiol.* **52**, 813–825.

Enstrom JE, Kanim LE, Breslow L (1986) The relationship between vitamin C intake, general health practices, and mortality in Alameda County, California. *Am. J. Public Health* **76**, 1124–1130.

Enstrom JE, Kanim LE, Klein MA (1992) Vitamin C intake and mortality among a sample of the United States population. *Epidemiology* **3**, 194–202.

Esterbauer H, Gebicki J, Puhl H, Jürgens G (1992) The role of lipid peroxidation and antioxidants in oxidative modification of LDL. *Free Radic. Biol. Med.* **13**, 341–390.

Esterbauer H, Jürgens G, Quehenberger O, Koller E (1987) Autoxidation of human low density lipoprotein: loss of polyunsaturated fatty acids and vitamin E and generation of aldehydes. *J. Lipid Res.* **28**, 495–509.

Esterbauer H, Quehenberger O, Jürgens G (1988) In: *Free Radicals, Methodology and Concepts.* (eds C Rice-Evans, B Halliwell). Richelieu, London, pp. 243–268.

Esterbauer H, Striegl G, Puhl H, Rotheneder M (1989) Continuous monitoring of in vitro oxidation of human low density lipoprotein. *Free Rad. Res. Commun.* **6**, 67–75.

Fang JC, Kinlay S, Beltrame J, Hikiti H, Wainstein M, Behrendt D *et al.* (2002) Effect of vitamins C and E on progression of transplant-associated arteriosclerosis: a randomised trial. *Lancet* **359**, 1108–1113.

Fang JC, Kinlay S, Hikita H, Suh JH, Piana RN, Selwyn AP, Frei B, Morrow JD, Ganz P (2000) Relation of preservation of nitric oxide-dependent coronary vasomotor function to plasma vitamin C concentrations in cardiac transplant recipients. *Am. J. Cardiol.* **86**, 460–462.

Fehily AM, Yarnell JW, Sweetnam PM, Elwood PC (1993) Diet and incident ischaemic heart disease: the Caerphilly Study. *Br. J. Nutr.* **69**, 303–314.

Ferns GAA, Forster L, Steward-Lee A, Konneh M, Nourooz-Zadeh J, Änggård EE (1992) Probucol inhibits neointimal thickening and macrophage accumulation after balloon injury in the cholesterol-fed rabbit. *Proc. Natl Acad. Sci. USA* **89**, 11312–11316.

Frank JS, Fogelman AM (1989) Ultrastructure of the intima in WHHL and cholesterol-fed rabbit aortas prepared by ultra-rapid freezing and freeze-etching. *J. Lipid Res.* **30**, 967–978.

Frei B (1991) Ascorbic acid protects lipids in human plasma and low-density lipoprotein against oxidative damage. *Am. J. Clin. Nutr.* **54**, 1113S–1118S.

Frei B, England L, Ames BN (1989) Ascorbate is an outstanding antioxidant in human blood plasma. *Proc. Natl Acad. Sci. USA* **86**, 6377–6381.

Frei B, Stocker R, Ames BN (1988) Antioxidant defenses and lipid peroxidation in human blood plasma. *Proc. Natl Acad. Sci. USA* **85**, 9748–9752.

Frei B, Stocker R, England L, Ames BN (1990) Ascorbate: the most effective antioxidant in human blood plasma. *Adv. Exp. Med. Biol.* **264**, 155–163.

Freyschuss A, Stiko-Rahm A, Swedenborg J, Henriksson P, Bjorkhem I, Berglund L, Nilsson J (1993) Antioxidant treatment inhibits the development of intimal thickening after balloon injury of the aorta in hypercholesterolemic rabbits. *J. Clin. Invest.* **91**, 1282–1288.

Fruebis J, Steinberg D, Dresel HA, Carew TA (1994) A comparison of the antiatherogenic effects of probucol and a structural analogue of probucol in low density lipoprotein receptor-deficient rabbits. *J. Clin. Invest.* **94**, 392–398.

Gale CR, Martyn CN, Winter PD, Cooper C (1995) Vitamin C and risk of death from stroke and coronary heart disease in cohort of elderly people. *Br. Med. J.* **310**, 1563–1566.

Garg UC, Hassid A (1989) Nitric oxide-generating vasodilators and 8-bromo-cyclic guanosine monophosphate inhibit mitogenesis and proliferation of cultured rat vascular smooth muscle cells. *J. Clin. Invest.* **83**, 1774–1777.

Ginter E, Babala J, Cerven J (1969) The effect of chronic hypovitaminosis C on the metabolism of cholesterol and atherogenesis in guinea pigs. *J. Atheroscler. Res.* **10**, 341–352.

Gokce N, Keaney JF Jr, Frei B, Holbrook M, Olesiak M, Zachariah BJ, Leeuwenburgh C, Heinecke JW, Vita JA (1999) Long-term ascorbic acid administration reverses endothelial vasomotor dysfunction in patients with coronary artery disease. *Circulation* **99**, 3234–3240.

Group HPSC (2002a) MRC/BHF Heart Protection Study of antioxidant vitamin supplementation in 20,536 high-risk individuals: a randomised placebo-controlled trial. *Lancet* **360**, 23–33.

Group HPSC (2002b) MRC/BHF Heart Protection Study of cholesterol lowering with simvastatin in 20,536 high-risk individuals: a randomised placebo-controlled trial. *Lancet* **360**, 7–22.

Halliwell B (1996) Vitamin C: antioxidant or pro-oxidant in vivo? *Free. Radic. Res.* **25**, 439–454.

Halliwell B, Gutteridge JMC (1990) The antioxidants of human extracellular fluids. *Arch. Biochem. Biophys.* **280**, 1–8.

Hawkins CL, Davies MJ (1999) Hypochlorite-induced oxidation of proteins in plasma: formation of chloramines and nitrogen-centred radicals and their role in protein fragmentation. *Biochem. J.* **340**, 539–548.

Hazell LJ, Arnold L, Flowers D, Waeg G, Malle E, Stocker R (1996) Presence of hypochlorite-modified proteins in human atherosclerotic lesions. *J. Clin. Invest.* **97**, 1535–1544.

Hazell LJ, Stocker R (1993) Oxidation of low-density lipoprotein with hypochlorite causes transformation of the lipoprotein into a high-uptake form for macrophages. *Biochem. J.* **290**, 165–172.

Hazell LJ, van den Berg JJM, Stocker R (1994) Oxidation of low-density lipoprotein by hypochlorite causes aggregation that is mediated by modification of lysine residues rather than lipid oxidation. *Biochem. J.* **302**, 297–304.

Hazen SL, Gaut JP, Crowley JR, Hsu FF, Heinecke JW (2000) Elevated levels of protein-bound p-hydroxyphenylacetaldehyde, an amino acid-derived aldehyde generated by myeloperoxidase, are present in human fatty streaks, intermediate lesions and advanced atherosclerotic lesions. *Biochem. J.* **352**, 693–699.

Hazen SL, Heinecke JW (1997) 3-Chlorotyrosine, a specific marker of myeloperoxidase-catalyzed oxidation, is markedly elevated in low density lipoprotein isolated from human atherosclerotic intima. *J. Clin. Invest.* **99**, 2075–2081.

Heery JM, Kozak M, Stafforini DM, Jones DA, Zimmerman GA, McIntyre TM, Prescott SM (1995) Oxidatively modified LDL contains phospholipids with platelet-activating factor-like activity and stimulates the growth of smooth muscle cells. *J. Clin. Invest.* **96**, 2322–2330.

Heinecke JW (1999) Mechanisms of oxidative damage by myeloperoxidase in atherosclerosis and other inflammatory disorders. *J. Lab. Clin. Med.* **133**, 321–325.

Heinecke JW (2001) Is the emperor wearing clothes? Clinical trials of vitamin E and the LDL oxidation hypothesis. *Arterioscler. Thromb. Vasc. Biol.* **21**, 1261–1264.

Heller R, Unbehaun A, Schellenberg B, Mayer B, Werner-Felmayer G, Werner ER (2001) L-ascorbic acid potentiates endothelial nitric oxide synthesis via a chemical stabilization of tetrahydrobiopterin. *J. Biol. Chem.* **276**, 40–47.

Heydeck D, Upston JM, Vita H, Yla-Herttuala S, Stocker R (2001) Oxidation of low-density lipoprotein by rabbit and human 15-lipoxygenase: Prevalence of non-enzymatic reactions. *J. Lipid Res.* **42**, 1082–1088.

Huang A, Vita JA, Venema RC, Keaney JF Jr (2000) Ascorbic acid enhances endothelial nitric-oxide synthase activity by increasing intracellular tetrahydrobiopterin. *J. Biol. Chem.* **275**, 17399–17406.

Huang HY, Appel LJ, Croft KD, Miller ER 3rd, Mori TA, Puddey IB (2002) Effects of vitamin C and vitamin E on in vivo lipid peroxidation: results of a randomized controlled trial. *Am. J. Clin. Nutr.* **76**, 549–555.

Ignarro LJ, Lippton H, Edwards JC, Baricos WH, Hyman AL, Kadowitz PJ, Gruetter CA (1981) Mechanism of vascular smooth muscle relaxation by organic nitrates, nitrites, nitroprusside and nitric oxide: evidence for the involvement of S-nitrosothiols as active intermediates. *J. Pharmacol. Exp. Ther.* **218**, 739–749.

Ischiropoulos H, Zhu L, Beckman JS (1992) Peroxynitrite formation from macrophage-derived nitric oxide. *Arch. Biochem. Biophys.* **298**, 446–451.

Ivanov VO, Ivanova SV, Niedzwiecki A (1997) Ascorbate affects proliferation of guinea-pig vascular smooth muscle cells by direct and extracellular matrix-mediated effects. *J. Mol. Cell. Cardiol.* **29**, 3293–3303.

Jackson TS, Xu A, Vita JA, Keaney JF Jr (1998) Ascorbate prevents the interaction of superoxide and nitric oxide only at very high physiological concentrations. *Circ. Res.* **83**, 916–922.

Jessup W, Mohr D, Gieseg SP, Dean RT, Stocker R (1992) The participation of nitric oxide in cell free- and its restriction of macrophage-mediated oxidation of low-density lipoprotein. *Biochim. Biophys. Acta* **1180**, 73–82.

Jialal I, Devaraj S (2003) Antioxidants and atherosclerosis: don't throw out the baby with the bath water. *Circulation* **107**, 926–928.

Jonasson L, Holm J, Skall O, Bondjers G, Hansson GK (1986) Regional accumulation of T cells, macrophages, and smooth muscle cells in the human atherosclerotic plaque. *Arteriosclerosis* **6**, 131–138.

Kagan VE, Serbinova EA, Packer L (1990) Recycling and antioxidant activity of tocopherol homologs of differing hydrocarbon chain lengths in liver microsomes. *Arch. Biochem. Biophys.* **282**, 221–225.

Keaney JF Jr (2000) In: *Oxidative Stress and Vascular Disease* (ed. JF Keaney Jr). Kluwer Academic, Boston, pp. 195–211.

Keaney JF Jr, Xu A, Cunningham D, Jackson T, Frei B, Vita JA (1995) Dietary probucol preserves endothelial function in cholesterol-fed rabbits by limiting vascular oxidative stress and superoxide generation. *J. Clin. Invest.* **95**, 2520–2529.

Kenar JA, Havrilla CM, Porter NA, Guyton JR, Brown SA, Klemp KF, Selinger E (1996) Identification and quantification of the regioisomeric cholesteryl linoleate hydroperoxides in oxidized human low density lipoprotein and high density lipoprotein. *Chem. Res. Toxicol.* **9**, 737–744.

Killion SL, Hunter GC, Eskelson CD, Dubick MA, Putnam CW, Hall KA, Luedke CA, Misiorowski RL, Schilling JD, McIntyre KE (1996) Vitamin E levels in human atherosclerotic plaque: the influence of risk factors. *Atherosclerosis* **126**, 289–297.

Knekt P, Reunanen A, Jarvinen R, Seppanen R, Heliovaara M, Aromaa A (1994) Antioxidant vitamin intake and coronary mortality in a longitudinal population study. *Am. J. Epidemiol.* **139**, 1180–1189.

Kritchevsky SB, Shimakawa T, Tell GS, Dennis B, Carpenter M, Eckfeldt JH, Peacher-Ryan H, Heiss G (1995) Dietary antioxidants and carotid artery wall thickness. The ARIC Study. Atherosclerosis Risk in Communities Study. *Circulation* **92**, 2142–2150.

Kritharides L, Stocker R (2002) The use of antioxidant supplements in coronary heart disease. *Atherosclerosis* **164**, 211–219.

Kubes P, Suzuki M, Granger DN (1991) Nitric oxide: an endogenous modulator of leukocyte adhesion. *Proc. Natl Acad. Sci. USA* **88**, 4651–4655.

Kuhn H, Belkner J, Zaiss S, Fahrenklemper T, Wohlfeil S (1994) Involvement of 15-lipoxygenase in early stages of atherogenesis. *J. Exp. Med.* **179**, 1903–1911.

Kume N, Cybulsky MI, Gimbrone MA Jr (1992) Lysophosphatidylcholine, a component of atherogenic lipoproteins, induces mononuclear leukocyte adhesion molecules in cultured human and rabbit arterial endothelial cells. *J. Clin. Invest.* **90**, 1138–1144.

Kushi LH, Folsom AR, Prineas RJ, Mink PJ, Wu Y, Bostick RM (1996) Dietary antioxidant vitamins and death from coronary heart disease in postmenopausal women. *N. Engl. J. Med.* **334**, 1156–1162.

Leeuwenburgh C, Hardy MM, Hazen SL, Wagner P, Oh-ishi S, Steinbrecher UP, Heinecke JW (1997a) Reactive nitrogen intermediates promote low density lipoprotein oxidation in human atherosclerotic intima. *J. Biol. Chem.* **272**, 1433–1436.

Leeuwenburgh C, Rasmussen JE, Hsu FF, Mueller DM, Pennathur S, Heinecke JW (1997b) Mass spectrometric quantification of markers for protein oxidation by tyrosyl radical, copper, and hydroxyl radical in low density lipoprotein isolated from human atherosclerotic plaques. *J. Biol. Chem.* **272**, 3520–3526.

Lehr HA, Becker M, Marklund SL, Hubner C, Arfors KE, Kohlschütter A, Messmer K (1992) Superoxide-dependent stimulation of leukocyte adhesion by oxidatively modified LDL in vivo. *Arterioscler. Thromb.* **12**, 824–829.

Lehr HA, Frei B, Olofsson AM, Carew TE, Arfors KE (1995) Protection from oxidized LDL-induced leukocyte adhesion to microvascular and macrovascular endothelium in vivo by vitamin C but not by vitamin E. *Circulation* **91**, 1525–1532.

Lehr HA, Hubner C, Nolte D, Finckh B, Beisiegel U, Kohlschutter A, Messmer K (1991) Oxidatively modified human low-density lipoprotein stimulates leukocyte adherence to the microvascular endothelium in vivo. *Res. Exp. Med. (Berlin)* **191**, 85–90.

Lehr HA, Weyrich AS, Saetzler RK, Jurek A, Arfors KE, Zimmerman GA, Prescott SM, McIntyre TM (1997) Vitamin C blocks inflammatory platelet-activating factor mimetics created by cigarette smoking. *J. Clin. Invest.* **99**, 2358–2364.

Levine GN, Frei B, Koulouris SN, Gerhard MD, Keaney JF Jr, Vita JA (1996) Ascorbic acid reverses endothelial vasomotor dysfunction in patients with coronary artery disease. *Circulation* **93**, 1107–1113.

Losonczy KG, Harris TB, Havlik RJ (1996a) Vitamin E and vitamin C supplement use and risk of all-cause and coronary heart disease mortality in older persons: the Established Populations for Epidemiologic Studies of the Elderly. *Am. J. Clin. Nutr.* **64**, 190–196.

Losonczy KG, Harris TB, Havlik RJ (1996b) Vitamin E and vitamin C supplement use and risk of all-cause and coronary heart disease mortality in older persons: the Established Populations for Epidemiologic Studies of the Elderly. *Am. J. Clin. Nutr.* **64**, 190–196.

Lykkesfeldt J, Prieme H, Loft S, Poulsen HE (1996) Effect of smoking cessation on plasma ascorbic acid concentration. *BMJ* **313**, 91.

Lynch SM, Frei B (1994) In: *Natural Antioxidants in Human Health and Disease* (ed. B Frei). Academic Press, New York, pp. 353–385.

Mark SD, Wang W, Fraumeni JF Jr, Li JY, Taylor PR, Wang GQ, Dawsey SM, Li B, Blot WJ (1998) Nutritional supplements lower the risk of stroke or hypertension? *Epidemiology* **9**, 9–15.

Martin A, Frei B (1997) Both intracellular and extracellular vitamin C inhibit atherogenic modification of LDL by human vascular endothelial cells. *Arterioscler. Thromb. Vasc. Biol.* **17**, 1583–1590.

Mashima R, Witting PK, Stocker R (2001) Oxidants and antioxidants in atherosclerosis. *Curr. Opin. Lipidol.* **12**, 411–418.

May JM (2000) How does ascorbic acid prevent endothelial dysfunction? *Free Radic. Biol. Med.* **28**, 1421–1429.

May JM, Cobb CE, Mendiratta S, Hill KE, Burk RF (1998a) Reduction of the ascorbyl free radical to ascorbate by thioredoxin reductase. *J. Biol. Chem.* **273**, 23039–23045.

May JM, Qu ZC, Mendiratta S (1998b) Protection and recycling of α-tocopherol in human erythrocytes by intracellular ascorbic acid. *Arch. Biochem. Biophys.* **349**, 281–289.

Mayer B, Klatt P, Werner ER, Schmidt K (1995) Kinetics and mechanism of tetrahydrobiopterin-induced oxidation of nitric oxide. *J. Biol. Chem.* **270**, 655–659.

McNeish AJ, Wilson WS, Martin W (2002) Ascorbate blocks endothelium-derived hyperpolarizing factor (EDHF)-mediated vasodilatation in the bovine ciliary vascular bed and rat mesentery. *Br. J. Pharmacol.* **135**, 1801–1809.

Mendiratta S, Qu ZC, May JM (1998) Enzyme-dependent ascorbate recycling in human erythrocytes: role of thioredoxin reductase. *Free Radic. Biol. Med.* **25**, 221–228.

Minuz P, Andrioli G, Degan M, Gaino S, Ortolani R, Tommasoli R, Zuliani V, Lechi A, Lechi C (1998) The F2-isoprostane 8-epiprostaglandin F2a increases platelet adhesion and reduces the antiadhesive and antiaggregatory effects of NO. *Arterioscler. Thromb. Vasc. Biol.* **18**, 1248–1256.

Mooradian DL, Hutsell TC, Keefer LK (1995) Nitric oxide (NO) donor molecules: effect of NO release rate on vascular smooth muscle cell proliferation in vitro. *J. Cardiovasc. Pharmacol.* 25, 674–678.

Morrow JD, Frei B, Longmire AW, Gaziano JM, Lynch SM, Shyr Y, Strauss WE, Oates JA, Roberts LJ 2nd (1995) Increase in circulating products of lipid peroxidation (F2-isoprostanes) in smokers. Smoking as a cause of oxidative damage. *N. Engl. J. Med.* **332**, 1198–1203.

Nardai G, Braun L, Csala M, Mile V, Csermely P, Benedetti A, Mandl J, Bánhegyi G (2001) Protein-disulfide isomerase- and protein thiol-dependent dehydroascorbate reduction and ascorbate accumulation in the lumen of the endoplasmic reticulum. *J. Biol. Chem.* **276**, 8825–8828.

Nievelstein PFEM, Fogelman AM, Mottino G, Frank JS (1991) Lipid accumulation in rabbit aortic intima 2 hours after bolus infusion of low density lipoprotein. A deep-etch and immunolocalization study of ultrarapidly frozen tissue. *Arteriosc1. Thromb.* **11**, 1795–1805.

Niu X, Zammit V, Upston JM, Dean RT, Stocker R (1999) Co-existence of oxidized lipids and α-tocopherol in all lipoprotein fractions isolated from advanced human atherosclerotic plaques. *Arterioscler. Thromb. Vasc. Biol.* **19**, 1708–1718.

Packer JE, Slater TF, Willson RL (1979) Direct observation of a free radical interaction between vitamin E and vitamin C. *Nature* **278**, 737–738.

Pandey DK, Shekelle R, Selwyn BJ, Tangney C, Stamler J (1995) Dietary vitamin C and beta-carotene and risk of death in middle-aged men. The Western Electric Study. *Am. J. Epidemiol.* **142**, 1269–1278.

Panes J, Perry M, Granger DN (1999) Leukocyte–endothelial cell adhesion: avenues for therapeutic intervention. *Br. J. Pharmacol.* **126**, 537–550.

Parthasarathy S, Wieland E, Steinberg D (1989) A role for endothelial cell lipoxygenase in the oxidative modification of low-density lipoprotein. *Proc. Natl Acad. Sci. USA* **86**, 1046–1050.

Porter NA, Weber BA, Weenen H, Khan JA (1980) Autoxidation of polyunsaturated lipids. Factors controlling the stereochemistry of product hydroperoxides. *J. Am. Chem. Soc.* **102**, 5597–5601.

Porter NA, Wujek DG (1984) Autoxidation of polyunsaturated fatty acids, an expanded mechanistic study. *J. Am. Chem. Soc.* **106**, 2626–2629.

Pratico D, Tangirala RK, Radar D, Rokach J, FitzGerald GA (1998) Vitamin E suppresses isoprostane generation in vivo and reduces atherosclerosis in apoE-deficient mice. *Nature Med.* **4**, 1189–1192.

Quinn MT, Parthasarathy S, Fong LG, Steinberg D (1987) Oxidatively modified low density lipoproteins: a potential role in recruitment and retention of monocyte/macrophages during atherogenesis. *Proc. Natl. Acad. Sci.* USA **84**, 2995–2998.

Radi R, Beckman JS, Bush KM, Freeman BA (1991) Peroxynitrite oxidation of sulfhydryls. The cytotoxic potential of superoxide and nitric oxide. *J. Biol. Chem.* **266**, 4244–4250.

Rae TD, Schmidt PJ, Pufahl RA, Culotta VC, O'Halloran TV (1999) Undetectable intracellular free copper: the requirement of a copper chaperone for superoxide dismutase. *Science* **284**, 805–808.

Raitakari OT, Adams MR, McCredie RJ, Griffiths KA, Stocker R, Celermajer DS (2000) Oral vitamin C and endothelial function in smokers: short-term improvement, but no sustained beneficial effect. *J. Am. Coll. Cardiol.* **35**, 1616–1621.

Reilly M, Delanty N, Lawson JA, FitzGerald GA (1996) Modulation of oxidant stress in vivo in chronic cigarette smokers. *Circulation* **94**, 19–25.

Rimm EB, Stampfer MJ, Ascherio A, Giovannucci E, Colditz GA, Willett WC (1993) Vitamin E consumption and the risk of coronary heart disease in men. *N. Engl. J. Med.* 328, 1450–1456.

Ross R (1999) Atherosclerosis – an inflammatory disease. *N. Engl. J. Med.* **340**, 115–126.

Sahyoun NR, Jacques PF, Russell RM (1996) Carotenoids, vitamins C and E, and mortality in an elderly population. *Am. J. Epidemiol.* **144**, 501–511.

Salonen JT, Nyyssönen K, Salonen R, Lakka HM, Kaikkonen J, Porkkala-Sarataho E, *et al.* (2000) Antioxidant Supplementation in Atherosclerosis Prevention (ASAP) study: a randomized trial of the effect of vitamins E and C on 3-year progression of carotid atherosclerosis. *J. Intern. Med.* **248**, 377–386.

Salonen RM, Nyyssönen K, Kaikkonen J, Porkkala-Sarataho E, Voutilainen S, Rissanen TH, *et al.* (2003) Six-year effect of combined vitamin C and E supplementation on atherosclerotic progression: the Antioxidant Supplementation in Atherosclerosis Prevention (ASAP) Study. *Circulation* **107**, 947–953.

Sato K, Niki E, Shimasaki H (1990) Free radical-mediated chain oxidation of low density lipoprotein and its synergistic inhibition by vitamin E and vitamin C. *Arch. Biochem. Biophys.* **279**, 402–405.

Sawamura T, Kume N, Aoyama T, Moriwaki H, Hoshikawa H, Aiba Y, *et al.* (1997) An endothelial receptor for oxidized low-density lipoprotein. *Nature* **385**, 73–77.

Schmidt K, Graier WF, Kostner GM, Mayer B, Bohme E, Kukovetz WR (1991) Oxidized low-density lipoprotein antagonizes the activation of purified soluble guanylate cyclase by endothelium-derived relaxing factor but does not interfere with its biosynthesis. *Cell. Signal* **3**, 361–367.

Schwartz E, Bienkowski RS, Coltoff-Schiller B, Goldfischer S, Blumenfeld OO (1982) Changes in the components of extracellular matrix and in growth properties of cultured aortic smooth muscle cells upon ascorbate feeding. *J. Cell Biol.* **92**, 462–470.

Sharma MK, Buettner GR (1993) Interaction of vitamin C and vitamin E during free radical stress in plasma: an ESR study. *Free Rad. Biol. Med.* **14**, 649–653.

Simon JA, Hudes ES (1999) Serum ascorbic acid and cardiovascular disease prevalence in U.S. adults: the Third National Health and Nutrition Examination Survey (NHANES III). *Ann. Epidemiol.* **9**, 358–365.

Simon JA, Hudes ES, Tice JA (2001) Relation of serum ascorbic acid to mortality among US adults. *J. Am. Coll. Nutr.* **20**, 255–263.

Stampfer MJ, Hennekens CH, Manson JE, Colditz GA, Rosner B, Willett WC (1993) Vitamin E consumption and the risk of coronary disease in women. *N. Engl. J. Med.* **328**, 1444–1449.

Steinberg D (2002) Atherogenesis in perspective: hypercholesterolemia and inflammation as partners in crime. *Nature Med.* **8**, 1211–1217.

Steinberg D, Parthasarathy S, Carew TE, Khoo JC, Witztum JL (1989) Beyond cholesterol: Modifications of low-density lipoprotein that increase its atherogenicity. *N. Engl. J. Med.* **320**, 915–924.

Stocker R (1999) The ambivalence of vitamin E in atherogenesis. *TiBS* **24**, 219–223.

Stocker R (2000) In: *Oxidative Stress and Vascular Disease* (ed. Keaney JF Jr.) Kluwer Academic, Boston, pp. 27–47.

Stocker R, Bowry VW, Frei B (1991) Ubiquinol-10 protects human low density lipoprotein more efficiently against lipid peroxidation than does α-tocopherol. *Proc. Natl Acad. Sci. USA* **88**, 1646–1650.

Stocker R, Frei B (1991) In: *Oxidative Stress: Oxidants and Antioxidants* (ed. H Sies). Academic Press, London, pp. 213–243.

Sturgeon BE, Sipe HJ Jr, Barr DP, Corbett JT, Martinez JG, Mason RP (1998) The fate of the oxidizing tyrosyl radical in the presence of glutathione and ascorbate. Implications for the radical sink hypothesis. *J. Biol. Chem.* **273**, 30116–30121.

Suarna C, Dean RT, May J, Stocker R (1995) Human atherosclerotic plaque contains both oxidized lipids and relatively large amounts of α-tocopherol and ascorbate. *Arterioscler. Thromb. Vasc. Biol.* **15**, 1616–1624.

Suarna C, Dean RT, Southwell-Keely P, Moore DE, Stocker R (1997) Separation and characterization of cholesteryl oxo- and hydroxy-linoleate in human atheroclerotic plaque. *Free Radic. Res.* **27**, 397–408.

Tamba M, O'Neill P (1991) Redox reactions of thiol free radicals with the anti-oxidants ascorbate and chlorpromazine: role in radioprotection. *J. Chem. Soc. Perkin Trans.* **2**, 1681–1685.

Tangirala RK, Casanada F, Miller E, Witztum JL, Steinberg D, Palinski W (1995) Effect of the antioxidant N,N′-diphenyl 1,4-phenylenediamine (DPPD) on atherosclerosis in apoE-deficient mice. *J. Lipid Res.* **15**, 1625–1630.

Tardif J-C, Côté, G, Lespérance J, Bourassa M, Lambert J, Doucet S, Bilodeau L, Nattel S, de Guise P (1997) Probucol and multivitamins in the prevention of restenosis after coronary angioplasty. *N. Engl. J. Med.* **337**, 365–372.

Thomas SR, Davies MJ, Stocker R (1998) Oxidation and antioxidation of human low-density lipoprotein and plasma exposed to 3-morpholinosydnonimine and reagent peroxynitrite. *Chem. Res. Toxicol.* **11**, 484–494.

Thomas SR, Neuzil J, Mohr D, Stocker R (1995) Co-antioxidants make α-tocopherol an efficient antioxidant for LDL. *Am. J. Clin. Nutr.* **62**, 1357S–1364S.

Upston JM, Neuzil J, Witting PK, Alleva R, Stocker R (1997) 15-Lipoxygenase-induced enzymic oxidation of low density lipoprotein associated free fatty acids stimulates nonenzymic, α-tocopherol-mediated peroxidation of cholesteryl esters. *J. Biol. Chem.* **272**, 30067–30074.

Upston JM, Niu X, Brown AJ, Mashima R, Wang H, Senthilmohan R, Kettle AJ, Dean RT, Stocker R (2002a) Disease stage-dependent accumulation of lipid and protein oxidation products in human atherosclerosis. *Am. J. Pathol.* **160**, 701–710.

Upston JM, Terentis AC, Morris K, Keaney JF Jr, Stocker R (2002b) Oxidized lipid accumulates in the presence of α-tocopherol in atherosclerosis. *Biochem. J.* **363**, 753–760.

Upston JM, Terentis AC, Stocker R (1999) Tocopherol-mediated peroxidation (TMP) of lipoproteins: implications for vitamin E as a potential antiatherogenic supplement. *FASEB J.* **13**, 977–994.

Upston JM, Witting PK, Brown AJ, Stocker R, Keaney JF Jr (2001) Effect of vitamin E on aortic lipid oxidation and intimal proliferation after vascular injury in cholesterol-fed rabbits. *Free Radic. Biol. Med.* **31**, 1245–1253.

Utoguchi N, Ikeda K, Saeki K, Oka N, Mizuguchi H, Kubo K, Nakagawa S, Mayumi T (1995) Ascorbic acid stimulates barrier function of cultured endothelial cell monolayer. *J. Cell. Physiol.* **163**, 393–399.

Vásquez-Vivar J, Hogg N, Martásek P, Karoui H, Pritchard KA Jr, Kalyanaraman B (1999) Tetrahydrobiopterin-dependent inhibition of superoxide generation from neuronal nitric oxide synthase. *J. Biol. Chem.* **274**, 26736–26742.

Vásquez-Vivar J, Kalyanaraman B, Martásek P, Hogg N, Masters BS, Karoui H, Tordo P, Pritchard KA Jr (1998) Superoxide generation by endothelial nitric oxide synthase: the influence of cofactors. *Proc. Natl Acad. Sci. USA* **95**, 9220–9225.

Vásquez-Vivar J, Whitsett J, Martásek P, Hogg N, Kalyanaraman B (2001) Reaction of tetrahydrobiopterin with superoxide: EPR-kinetic analysis and characterization of the pteridine radical. *Free Radic. Biol. Med.* **31**, 975–985.

Verlangieri AJ, Hollis TM, Mumma RO (1977) Effects of ascorbic acid and its 2-sulfate on rabbit aortic intimal thickening. *Blood Vessels* **14**, 157–174.

Vita JA, Keaney JF Jr, Raby KE, Morrow JD, Freedman JE, Lynch S, Koulouris SN, Hankin BR, Frei B (1998) Low plasma ascorbic acid independently predicts the presence of an unstable coronary syndrome. *J. Am. Coll. Cardiol.* **31**, 980–986.

Waddington EI, Croft KD, Sienuarine K, Latham B, Puddey IB (2003) Fatty acid oxidation products in human atherosclerotic plaque: an analysis of clinical and histopathological correlates. *Atherosclerosis* **167**, 111–120.

Wardman P (1995) In: *Biothiols in Health and Disease* (eds L Packer, E Cadenas). Marcel Dekker, New York, pp. 1–19.

Waters DD, Alderman EL, Hsia J, Howard BV, Cobb FR, Rogers WJ, OuYang P, *et al.* (2002) Effects of hormone replacement therapy and antioxidant vitamin supplements on coronary atherosclerosis in postmenopausal women: a randomized controlled trial. *JAMA* **288**, 2432–2440.

Wayner DDM, Burton GM, Ingold KU, Barclay LRC, Locke SJ (1987) The relative contributions of vitamin E, urate, ascorbate and proteins to the total peroxyl radical-trapping antioxidant activity of human blood plasma. *Biochim. Biophys. Acta* **924**, 408–419.

Weber C, Erl W, Weber K, Weber PC (1996) Increased adhesiveness of isolated monocytes to endothelium is prevented by vitamin C intake in smokers. *Circulation* **93**, 1488–1492.

Wieland E, Parthasarathy S, Steinberg D (1993) Peroxidase-dependent metal-independent oxidation of low density lipoprotein in vitro: a model for in vivo oxidation? *Proc. Natl Acad. Sci. USA* **90**, 5929–5933.

Willis GC (1953) An experimental study of the intimal ground substance in atherosclerosis. *CMAJ* **69**, 17–22.

Willis GC, Fishman S (1955) Ascorbic acid content of human arterial tissue. *CMAJ* **72**, 500–503.

Winterbourn CC, Munday R (1990) Concerted action of reduced glutathione and superoxide dismutase in preventing redox cycling of dihydroxypyrimidines, and their role in antioxidant defence. *Free Rad. Res. Commun.* **8**, 287–293.

Witting PK, Douglas DJ, Mauk AG (2000a) Reaction of human myoglobin and H_2O_2. Involvement of a thiyl radical produced at cysteine 110. *J. Biol. Chem.* **275**, 20391–20398.

Witting PK, Pettersson K, Letters J, Stocker R (2000b) Anti-atherogenic effect of coenzyme Q10 in apolipoprotein E gene knockout mice. *Free Radic. Biol. Med.* **29**, 295–305.

Witting PK, Pettersson K, Letters J, Stocker R (2000c) Site-specific anti-atherogenic effect of probucol in apolipoprotein E deficient mice. *Arterioscler. Thromb. Vasc. Biol.* **20**, e26–e33.

Witting PK, Pettersson K, Östlund-Lindqvist A-M, Westerlund C, Wågberg M, Stocker R (1999a) Dissociation of atherogenesis from aortic accumulation of lipid hydro(pero)xides in Watanabe heritable hyperlipidemic rabbits. *J. Clin. Invest.* **104**, 213–220.

Witting PK, Pettersson K, Östlund-Lindqvist A-M, Westerlund C, Westin Eriksson A, Stocker R (1999b) Inhibition by a co-antioxidant of aortic lipoprotein lipid peroxidation and atherosclerosis in apolipoprotein E and low density lipoprotein receptor gene double knockout mice. *FASEB J.* **13**, 667–675.

Witting PK, Upston JM, Stocker R (1997) The role of α-tocopheroxyl radical in the initiation of lipid peroxidation in human low density lipoprotein exposed to horse radish peroxidase. *Biochemistry* **36**, 1251–1258.

Witting PK, Westerlund C, Stocker R (1996) A rapid and simple screening test for potential inhibitors of tocopherol-mediated peroxidation of LDL lipids. *J. Lipid Res.* **37**, 853–867.

Witting PK, Willhite CA, Davies MJ, Stocker R (1999c) Lipid oxidation in human low density lipoprotein induced by metmyoglobin/H_2O_2, involvement of α-tocopheroxyl and phosphatidylcholine alkoxyl radicals. *Chem. Res. Toxicol.* **12**, 1173–1181.

Witztum JL, Steinberg D (2001) The oxidative modification hypothesis of atherosclerosis: does it hold for humans? *Trends Cardiovasc. Med.* **11**, 93–102.

Woollard KJ, Loryman CJ, Meredith E, Bevan R, Shaw JA, Lunec J, Griffiths HR (2002) Effects of oral vitamin C on monocyte:endothelial cell adhesion in healthy subjects. *Biochem. Biophys. Res. Commun.* **294**, 1161–1168.

Yokoyama T, Date C, Kokubo Y, Yoshiike N, Matsumura Y, Tanaka H (2000) Serum vitamin C concentration was inversely associated with subsequent 20-year incidence of stroke in a Japanese rural community. The Shibata study. *Stroke* **31**, 2287–2294.

Zhang R, Brennan ML, Fu X, Aviles RJ, Pearce GL, Penn MS, Topol EJ, Sprecher DL, Hazen SL (2001) Association between myeloperoxidase levels and risk of coronary artery disease. *JAMA* **286**, 2136–2142.

Zou MH, Shi C, Cohen RA (2002) Oxidation of the zinc–thiolate complex and uncoupling of endothelial nitric oxide synthase by peroxynitrite. *J. Clin. Invest.* **109**, 817–826.

16

Dietary allowances for vitamin C: Recommended Dietary Allowances and optimal nutrient ingestion

Mark Levine, Sebastian Padayatty, Arie Katz, Oran Kwon, Peter Eck, Christopher Corpe, Je-Hyuk Lee and Yaohui Wang

16.1 Background

16.1.1 Original purpose of Recommended Dietary Allowances

In 1940 a committee of the Food and Nutrition Board of the United States National Academy of Sciences began work to develop recommendations on the amounts of nutrients to be provided to the armed forces and the general public. The recommendations gradually came to be known as Recommended Dietary Allowances (RDAs), as formulated by the RDA committee (National Research Council, 1941). The first and all subsequent recommendations concerned essential nutrients, defined as those required for human life and tissue growth and/or repair. Without an essential nutrient, deficiency syndromes and disease would occur. Model systems used animals to quantify growth and gain knowledge about deficiency syndromes. Using this information, RDAs were designed to plan what was termed adequate nutrition, meaning nutrition not just to prevent deficiency, but also to provide an additional margin of safety (Food and Nutrition Board, 1994c).

The 1953 RDA edition clarified and expanded these concepts:

> *The allowances are designed for the maintenance of good nutrition of healthy persons in the United States under present conditions. They are not necessarily applicable to situations of stringency or limited food supply. The recommendations are not requirements, since they represent not merely minimal needs of average persons, but nutrient*

Vitamin C, edited by Han Asard, James M May and Nicholas Smirnoff

> *levels selected to cover individual variations in a substantial majority of the population. In addition, the values for each nutrient above the minimal level that will prevent deficiency are considered to provide for increased needs in times of stress and to permit other potential benefits. Although the optimal intake of essential dietary constituents remains largely speculative, there is considerable evidence that improvement in growth and function occurs when the intake of certain nutrients is increased above the level just sufficient to prevent signs of deficiency disease.*
>
> Food and Nutrition Board (1953)

16.1.2 Basis of RDAs from 1974 to 1989

The 1974 RDA edition further defined RDAs as "...the levels of intake of essential nutrients considered, in the judgment of the Food and Nutrition Board on the basis of available scientific knowledge, to be adequate to meet the known nutritional needs of practically all healthy persons" (Food and Nutrition Board, 1974). Although not explicitly stated in RDA publications before 1974, this definition is consistent across all 10 RDA editions. RDAs were set for essential nutrients, were to account for individual variability, and provided a margin of safety above minimal requirements to prevent deficiency.

16.1.3 RDAs as part of Dietary Reference Intakes

After the 10th RDA edition (Food and Nutrition Board, 1989), the Food and Nutrition Board recognized the need to re-evaluate the RDA concepts. In 1994 the Food and Nutrition Board summarized the type of evidence that should be used for calculating RDAs (Food and Nutrition Board, 1994a). This information is:

- nutrient intakes observed in healthy people;
- epidemiological observations of populations in which the clinical consequences of nutrient deficiencies are corrected by dietary improvement;
- balance studies measuring nutrient status in relation to intake;
- nutrient depletion/repletion studies in which subjects are maintained on diets containing marginally or low or deficient levels of a nutrient, followed by correction of the deficit with measured amounts of that nutrient (studies undertaken only when risk is minimal);
- extrapolation from animal experiments;
- biochemical measurements that assess the degree of tissue saturation or adequacy of molecular function in relation to nutrient intake;
- evidence, when available, that nutrient intake can decrease the risk of chronic disease.

The last two concepts are new additions to RDA criteria. The first new addition provides a clearly needed framework to account for the relationship between nutrient function and nutrient concentration. The latter implies that the basis of nutrient recommendations can shift from preventing deficiency with safety margins to preventing disease, or optimizing health. We were the first to recognize that ideal nutrient recommendations could be based on concentration–function relationships rather than on prevention of deficiency with a margin of safety (Levine, 1986; Levine and Hartzell, 1987; Dhariwal *et al.*,

1989; Levine *et al.*, 1991, 1993). Based on work from our laboratory and our concerns about how RDAs were determined and interpreted, we called for and made a strong case for these modifications. Both additions are now utilized in deriving RDAs.

Taken together, the concepts above provide a clear framework as to the types of data considered in deriving RDAs. However, the Food and Nutrition Board members felt strongly that it was necessary for RDA documents to provide more detail on how recommendations were calculated. Specifically, it was believed that several reference values should be derived (Food and Nutrition Board, 1994b). The hopes were that these values could be used for providing a systematic way of organizing the scientific literature; identifying the strengths and weaknesses of existing data; identifying where data were insufficient and so that recommendations could be made to fill research gaps; and deriving endpoints that could help assess the influence of nutrients on risk of chronic disease. The reference values collectively were termed Dietary Reference Intakes (DRIs) (Yates *et al.*, 1998), defined by the Food and Nutrition Board as follows: (Food and Nutrition Board and Panel on Dietary Antioxidants and Related Compounds, 2000b).

- 'Estimated Average Requirement (EAR): a nutrient intake value that is estimated to meet the requirement of half the healthy individuals in a life stage and gender group.... Before setting the EAR, a specific criterion of adequacy is selected, based on a careful review of the literature.
- Recommended Dietary Allowance (RDA): the dietary intake level that is sufficient to meet the nutrient requirement of nearly all (97 to 98 percent) healthy individuals in a particular life stage and gender group.... If the standard deviation (SD) of the EAR is available and the requirement for the nutrient is symmetrically distributed, the RDA is set at 2 SDs above the EAR:
  $RDA = EAR + 2SD_{EAR}$
  If data about variability in requirements are insufficient to calculate an SD, a coefficient of variation (CV) for the EAR of 10 percent is ordinarily assumed, unless available data indicate a greater variation is probable. If 10 percent is assumed to be the CV, then twice that amount added to the EAR is defined as equal to the RDA. The resulting equation for the RDA is then
  $RDA = 1.2 \times EAR$.
- Adequate Intake (AI): a recommended intake value based on observed or experimentally determined approximations or estimates of nutrient intake by a group (or groups) of healthy people that are assumed to be adequate – used when an RDA cannot be determined.
- Tolerable Upper Intake Level (UL): the highest level of nutrient intake that is likely to pose no risk of adverse health effects for almost all individuals in the general population. As intake increases above the UL, the risk of adverse effects increases.'

These reference points were adapted with modifications from a report by the Committee on Medical Aspects of Food Policy (Committee on Medical Aspects of Food Policy, 1991). The calculations in part assume that a population has a normal distribution requirement for a nutrient (Food and Nutrition Board, 1994b).

16.2 Recommended Dietary Allowances and Dietary Reference Intakes for Vitamin C, 2000: rationales and values

16.2.1 Estimated average requirement for vitamin C in adults aged 19–50 years

As just described, the RDA is based on an EAR if available. The EAR selected by the Food and Nutrition Board for vitamin C is intake that maintains near maximal neutrophil vitamin C concentrations with negligible urinary loss. The EAR is based on the concentration of vitamin C in neutrophils that is believed to provide antioxidant protection, and the ingested amount of vitamin C needed to obtain this concentration while minimizing urinary excretion of the vitamin (Food and Nutrition Board and Panel on Dietary Antioxidants and Related Compounds, 2000a).

16.2.2 Basic data for EAR determination: vitamin C as an antioxidant in neutrophils

The Food and Nutrition Board noted in descriptive detail that vitamin C functions as an antioxidant in neutrophils, lung and gastric mucosa, and in protection against lipid peroxidation (Food and Nutrition Board and Panel on Dietary Antioxidants and Related Compounds, 2000a). The EAR is based in part on that amount of vitamin C providing antioxidant protection in leukocytes, inclusive of neutrophils, macrophages, and lymphocytes. Ascorbate concentrations within the range of plasma concentrations (5–80 μM, described below) were protective against inactivation of alpha-1-antiprotease in a cell free system (Halliwell *et al.*, 1987). Extracellular ascorbate concentrations from 25 to 100 μM inhibited superoxide production by activated neutrophils (Anderson and Lukey, 1987) (*Figure 16.1*). The effect in neutrophils was reported to occur without affecting neutrophils' ability to kill bacteria.

16.2.3 Clinical data for EAR determination: vitamin C depletion–repletion in healthy men

Depletion–repletion data

Data from this laboratory were used to provide plasma and neutrophil concentrations in humans as a function of a wide range of vitamin C doses. These data were obtained as part of a depletion–repletion clinical study for vitamin C in healthy men (Levine *et al.*, 1996). Seven healthy men aged 20–26 years were each hospitalized for 5–6 months on a metabolic ward. For the entire hospitalization subjects consumed a diet containing < 5 mg of vitamin C daily, corrected for other vitamins and minerals so that only vitamin C intake was restricted. For the depletion phase subjects received no vitamin C and were depleted to plasma concentrations of approximately 7 μM without clinical scurvy. For the repletion phase, subjects received pure vitamin C in water twice daily in the fasted state until steady state was achieved for the first dose. At steady state bioavailability sampling for the dose was performed; pure circulating monocytes and lymphocytes were isolated by apheresis; neutrophils

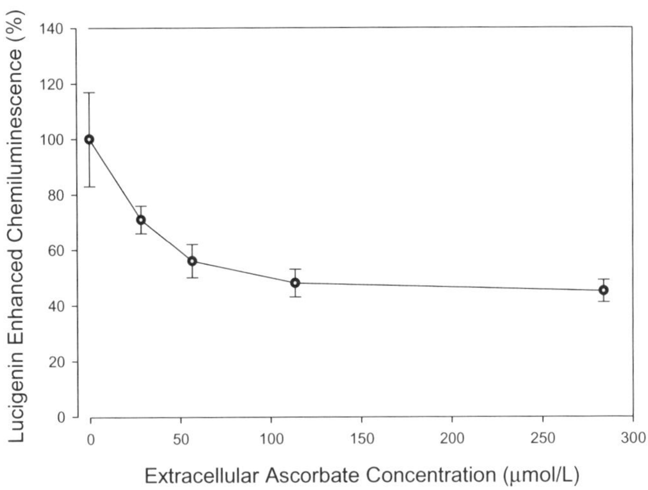

Figure 16.1: Effects of varying concentrations of ascorbic acid (28–280 μM) on superoxide-mediated lucigenin-enhanced chemiluminescence in neutrophils activated with cytochalasin B (1 μg·ml^{-1}) and N-formyl-methionyl-leucyl-phenylalanine (1 μM) added together. Results are data obtained 1 min after activation and are expressed as the mean percent of control (no ascorbic acid) ± standard error of the mean of three experiments. Modified from Anderson and Lukey (1987) where details of the study can be found.

were isolated; and multiple 24-h samples were obtained for creatinine, vitamin C, oxalate, urate, and other metabolites. After these measurements were completed, the dose was escalated and the sequence repeated at the new dose. Subjects achieved steady state in succession for each of the following daily doses in mg: 30, 60, 100, 200, 400, 1000, and 2500. Bioavailability sampling was performed for the following doses in mg: 15, 30, 50, 100, 200, 500, and 1250. A study using the same design with 15 healthy women subjects was published (Levine *et al.*, 2001) subsequent to the DRI recommendations for vitamin C, and these results are described in a later section.

Plasma concentrations in men at steady state as a function of dose are shown in *Figure 16.2*. These data indicate that there is a steep sigmoid relationship between dose and steady state plasma concentration at doses ≤ 100 mg daily, and that there is little change in plasma concentrations at doses > 200 mg daily. Tissue concentrations at steady state for these doses are shown in *Figure 16.3*. These data show that tissues saturate at lower doses than those needed for plasma saturation, and that neutrophils are saturated between 100 and 200 mg daily. Urine excretion at steady state for each dose is shown in *Figure 16.4*. The data show that the threshold of urine excretion is between 60 and 100 mg daily doses, corresponding to plasma concentrations between approximately 25 and 57 μM. At 60 mg daily, there is virtually no excretion of vitamin C in urine, while at 100 mg daily approximately 25% of the dose is excreted.

Bioavailability

It was concluded that bioavailability was 70–90% for usual dietary intakes of vitamin C (Food and Nutrition Board and Panel on Dietary Antioxidants and Related Compounds, 2000a). Based on the available evidence (Mangels *et al.*, 1993; Johnston

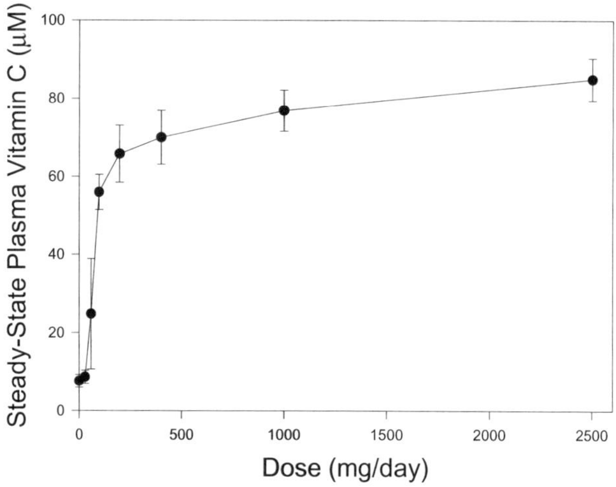

Figure 16.2: Steady state plasma vitamin C concentrations (mean ± SD) as a function of dose for all doses for seven men.
Subjects consumed a vitamin C deficient diet, resulting in plasma and tissue vitamin C depletion. Vitamin C in solution was then administered by mouth at the doses shown until steady state was reached for each dose. Doses through 400 mg daily were received by seven subjects, through 1000 mg daily by six subjects, and through 2500 mg daily by three subjects. Reproduced from Levine *et al.* (1996) where details of the study can be found.

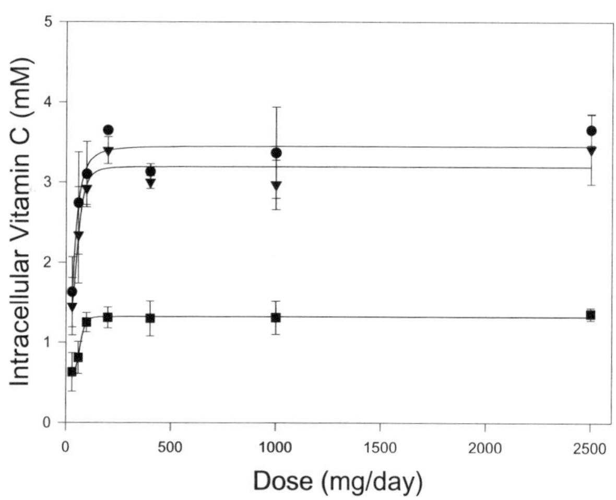

Figure 16.3: Intracellular vitamin C concentrations (mean ± SD) in circulating cells as a function of dose in men.
Cells were isolated when steady state was achieved for each dose. For neutrophils samples were available from seven men at doses of 30–100 mg daily; six men at doses of 200 and 1000 mg daily; five men at 400 mg daily; and two men at 2500 mg daily. For lymphocytes and monocytes samples were available from six men at 30 and 60 mg daily; from four men at 100 and 1000 mg daily; from two men at 2500 mg daily; and from one man at 200 and 400 mg daily. Reproduced from Levine *et al.* (1996) where details of the study can be found.

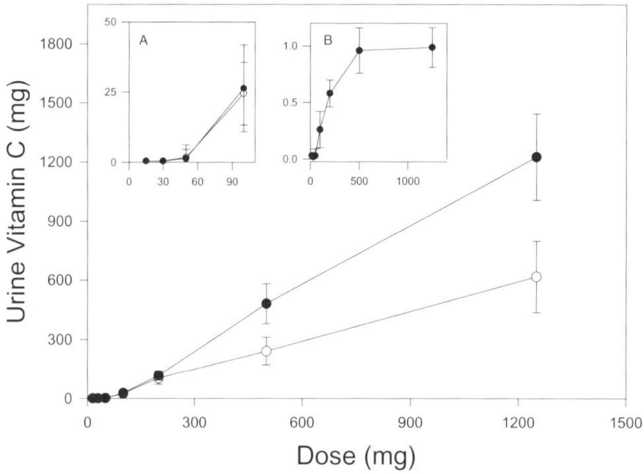

Figure 16.4: Urinary vitamin C excretion as a function of single vitamin C doses at steady state. Vitamin C excretion over 24 h was determined after administration of single doses given either orally (○) or intravenously (●). Data indicate mean values ± S.D. Inset A, vitamin C excretion for single oral (○) or intravenous (●) doses of 15–100 mg. X-axis indicates dose, Y-axis indicates amount (mg) excreted in urine. Inset B, fractional excretion (the fraction of the dose excreted) after intravenous administration of single doses of vitamin C. X-axis indicates dose, Y-axis indicates fractional excretion (vitamin C excreted in urine in mg divided by the vitamin C dose in mg). Data from oral and intravenous administration were available from seven subjects at doses 15, 30, 50, and 100 mg; from six subjects at 500 mg; from five subjects at 200 mg; from three subjects at 1250 mg. Reproduced from Levine *et al.* (1996) where details of the study can be found.

and Luo, 1994; Levine *et al.*, 1996; Graumlich *et al.*, 1997), bioavailability was judged to be the same for vitamin C pure, vitamin C in supplements, and vitamin C in foods. Bioavailability was not believed to impact dietary recommendations.

16.2.4 Calculation of EAR and RDA values for vitamin C in adults aged 19–50 years

Extracellular vitamin C protects neutrophils from oxidants, and protection was assumed to be dependent on vitamin C accumulated intracellularly in relation to its extracellular concentration. An EAR was selected to achieve near maximal neutrophil vitamin C concentration, for oxidant protection, but with minimal urinary loss from the dose (Halliwell *et al.*, 1987; Anderson and Lukey, 1987; Levine *et al.*, 1996) (*Figures 16.1*, *16.3* and *16.4*). At the 100-mg dose, with neutrophils nearly 100% saturated, there is 25% urinary loss. A dose that would achieve 80% neutrophil saturation was arbitrarily chosen so as to provide anti-oxidant protection but with less urinary loss. Data were not available for doses that would produce 80% neutrophil saturation, but it was clear from the available data that this would occur between doses of 60 and 100 mg daily. The data were readily modeled by regression analysis to yield the dose corresponding to 80% saturation, which was approximately 75 mg. This value was selected as the EAR for adults aged 19–50 years (Food and Nutrition Board and Panel on Dietary Antioxidants and Related Compounds, 2000a).

The RDA for men was calculated by assuming that the standard deviation for the EAR was unknown, because neutrophil data were not available at the 75-mg daily dose. A standard deviation of 10% was assumed. The RDA was calculated to be 90 mg daily, using the definition that the RDA was 120% of the EAR when the standard deviation for the EAR was unavailable (see above).

At the time DRIs for vitamin C were released, clinical depletion–repletion data for women described below were not yet published. Compared to men, it was assumed that women would have a lower EAR because of their smaller body size, total body mass, and lean body mass. Based on body weight differences from men, the EAR for women was calculated as 60 mg daily, with a corresponding RDA of 75 mg daily (Food and Nutrition Board and Panel on Dietary Antioxidants and Related Compounds, 2000a).

16.2.5 Calculation of EAR and RDA values for vitamin C in other healthy groups

Using the data above for men and the described rationale for extrapolation to women, EARs and RDAs were calculated in many groups based on differences in relative body weights. The following age groups (years) had recommendations calculated in this manner: children 1–3; children 4–8; boys 9–13; boys 14–18; girls 9–13; girls 14–18; men > 51; women > 51 (see *Table 16.1*).

For pregnant women the EAR was calculated based on data that 7 mg daily will protect infants from scurvy (Goldsmith, 1961; Rajalakshmi *et al.*, 1965; Van Eckelen, 1953). By rounding up these data, the EAR was estimated to increase 10 mg daily in pregnant compared to nonpregnant women. RDAs for pregnant women were then calculated as 120% of the EAR, as described above. Both EARs and RDAs are the same in pregnant women aged 19–30 and 31–50 years.

Infants fed human milk during the first 6 months of life were estimated to consume an average of 40 mg of vitamin C daily (Allen *et al.*, 1991; Butte *et al.*, 1984; Heinig *et al.*, 1993; Salmenpera, 1984; Sneed *et al.*, 1981). For lactating women this 40 mg was added to the EAR for each age group, and the RDA was modified accordingly as 120% of the EAR (see *Table 16.1*).

Smokers have lower vitamin C concentrations than nonsmokers, even when intake is believed to be comparable (Marangon *et al.*, 1998; Lykkesfeldt *et al.*, 2000). Radiolabeled tracer vitamin C was administered to healthy nonsmokers (Kallner *et al.*, 1979) and smokers (Kallner *et al.*, 1981). When the findings were compared, smokers were estimated to metabolize approximately 35 mg of vitamin C more than nonsmokers (Kallner *et al.*, 1981). The EAR and RDA for smokers are adjusted upwards accordingly (see *Table 16.1*).

In summary, the depletion–repletion data for healthy young men (Levine *et al.*, 1996) were used to calculate EARs and RDAs in men and women in multiple age groups across the life span by adjusting for weight. In calculations for pregnant women and smokers, data specific for each group were added to data from men with adjustments as necessary for body weight. Despite these specific modifications, the depletion–repletion data for healthy young men are key parts of the calculations for vitamin C RDAs across almost all groups.

Table 16.1: Dietary reference intake values for vitamin C in milligrams, by life stage and sex.

Life stage	Sex	Age (years)	EAR	RDA	AI	UL
Infants (months)		0–6			40	[a]
		7–12			50	
Children	Male and female	1–3	13	15		400
		4–8	22	25		650
	Male	9–13	39	45		1200
		14–18	63	75		1800
	Female	9–13	39	45		1200
		14–18	56	65		1800
Adults	Male	19–30	75	90		2000
		31–50	75	90		
		51–70	75	90		
		> 70	75	90		
	Female	19–30	60	75		
		31–50	60	75		
		51–70	60	75		
		> 70	60	75		
Pregnancy		14–18	66	80		1800
		19–30	70	85		2000
		31–50	70	85		
Lactation		14–18	96	115		1800
		19–30	100	120		2000
		31–50	100	120		
Smokers	Male	> 19	110	130[b]		
	Female	> 19	95	115[b]		

[a]It is not possible to establish UL for infants and children, for whom the source of vitamin C intake should be infant formula and food only. [b]While EARs were stated for smokers, RDAs for smokers were not explicitly documented. We calculated RDAs for smokers based on stated EAR × 1.2. Adapted from Food and Nutrition Board and Panel on Dietary Antioxidants and Related Compounds (2000a, 2000c).

16.2.6 Vitamin C in infants aged 0–12 months: adequate intakes

Data were judged insufficient to reflect an adequate functional response to dietary vitamin C in infants. Adequate intakes (AIs) were calculated for infants aged 0–6 months and 7–12 months (*Table 16.1*). For the former group the AI was based on average human milk consumption and human milk vitamin C content. Average vitamin C concentration in human milk was 50 $mg \cdot l^{-1}$ (Salmenpera, 1984; Sneed *et al.*, 1981), and average milk consumption was 0.78 l (Allen *et al.*, 1991; Butte *et al.*, 1984; Heinig *et al.*, 1993). The AI was therefore 39 mg daily, rounded to 40 mg daily for infants aged 0–6 months. Infants aged 7–12 months were presumed to obtain some vitamin C from human milk and some from foods. Human milk intake was estimated at 0.6 l daily (Dewey *et al.*, 1984) and human milk from nursing mothers of this group was estimated to contain 45 mg vitamin $C \cdot l^{-1}$ (Salmenpera, 1984), so that milk contributes 276 mg daily. Foods are estimated to contribute 22 mg daily (Montalto *et al.*, 1985). The AI is 49 mg, rounded to 50 mg daily.

16.2.7 Vitamin C requirements and disease

Data were believed to be insufficient to estimate vitamin C requirements for the following: cardiovascular disease, cancers (all types), cataract, asthma, pulmonary diseases, the common cold, periodontal health, and cognition. Vitamin C can reduce oxidative damage in many experimental systems. However, an *in vivo* vitamin C requirement could not be determined because *in vitro* and *in vivo* data were judged to be insufficient. Examples of data considered were vitamin C mediated decreased oxidative DNA and chromosomal damage, vitamin C mediated decreased LDL or lipid oxidation, and vitamin C mediated enhanced endothelium dependent vasodilatation (Food and Nutrition Board and Panel on Dietary Antioxidants and Related Compounds, 2000a).

16.2.8 Tolerable upper intake levels (UL)

Osmotic diarrhea and gastrointestinal disturbances were selected as the endpoints on which to base the UL. Calculation of UL required data for two parameters: the Lowest Observed Adverse Effect Level (LOAEL), and an uncertainty factor. UL was defined as:

$$\text{UL} = \frac{\text{LOAEL}}{\text{uncertainty factor}}$$

The LOAEL, was identified as 3 g daily, based ironically on data of Cameron and Campbell investigating effects of high dose vitamin C in cancer treatment (Cameron and Campbell, 1974). This dose was generally consistent with other case reports (Hoffer, 1971; Hoyt, 1980) and small studies (Stein *et al.*, 1976; Wandzilak *et al.*, 1994). The uncertainty factor was set low, at 1.5, because there was not much uncertainty about the dose of vitamin C that induced osmotic diarrhea. Using the formula above, UL for adults 19 years and older was calculated to be 2 g daily (Food and Nutrition Board and Panel on Dietary Antioxidants and Related Compounds, 2000a).

Limited data are available to calculate UL in those younger than 19 years. The effect of 1 g of vitamin C on children 8–9 years old was not different from that of placebo (Ludvigsson *et al.*, 1977). This is consistent with UL of 1 g daily in this age group and also with adult data on a body weight basis. Therefore, UL values for other age groups < 19 years are extrapolated based on body weight differences (*Table 16.1*). Data for pregnant and lactating women are insufficient to recommend a different UL compared to women ages > 19 years.

16.3 Critical evaluation of RDAs and DRIs for vitamin C

16.3.1 Overview

The 2000 RDA for vitamin C represents a novel attempt by the Food and Nutrition Board to base vitamin C recommendations on vitamin C function in relation to its concentration in humans. We believe that the Food and Nutrition Board is to be congratulated for changing its conceptual approach to dietary recommendations. We proposed this strategy nearly 15 years before (Levine, 1986; Levine and Hartzell, 1987), and believed then and now that this strategy should be applied to many water-soluble vitamins. For success, certain types of data are absolutely essential (Levine *et al.*, 1991, 1993, 1997a, 1997b). In broad terms, these data are biochemical or molecular function in relation to concentration in the relevant tissue, and concentration achieved as a function of dose. These concepts are addressed in detail in a subsequent section and are outlined in *Table 16.2*. Using the approach in this table, our evaluation of DRIs is divided into biochemical and clinical comments.

16.3.2 Biochemical basis of DRIs for vitamin C

The Food and Nutrition Board clearly recognized that very little data were available describing vitamin C concentration–function relationships in cells. Although substantial data were available utilizing cell-free systems, these data might not accurately reflect true intracellular events and would have uncertain clinical meaning. Examples are ascorbate inhibition of LDL oxidation *in vitro* vs. *in vivo* (Jialal *et al.*, 1990), and norepinephrine biosynthesis from dopamine by isolated enzyme vs. in secretory vesicles (Dhariwal *et al.*, 1989). We agree with the Food and Nutrition Board's decision not to utilize data from cell-free systems.

Table 16.2: *In situ* kinetics.

Biochemical component: vitamin biochemical and molecular functions in relation to vitamin concentrations

- Assay for vitamin
- Ability to have different vitamin concentrations *in situ*: transport, depletion, repletion
- Distribution of vitamin *in situ*
- Assay for vitamin function
- Determination of vitamin function *in vitro* and *in situ*
- Localization of vitamin function *in situ* and relationship to vitamin distribution
- Specificity of vitamin function *in vitro* and *in situ*
- Vitamin function in relation to different vitamin concentrations *in situ*

Clinical component: achieving vitamin concentrations clinically and their potential effects

- Availability of the vitamin in the diet
- Steady state vitamin concentration in plasma as a function of dose
- Steady state vitamin concentration in cells/tissues as a function of dose
- Vitamin bioavailability (absorption)
- Vitamin excretion
- Beneficial effects in relation to dose: direct effects and epidemiologic observations
- Vitamin safety and adverse effects

Pharmacologic component: achieving vitamin concentrations with pharmacologic (non-nutritive) benefit

- Concentrations in plasma and cells as a function of oral dose
- Concentrations in plasma and cells as a function of parenteral (intravenous/intramuscular) dose
- Vitamin metabolism/excretion as a function of dose
- Potential pharmacologic benefit in relation to dose
- Safety and adverse effects of pharmacologic doses

Unfortunately, there were few experiments, if any, from which to choose besides the neutrophil experiments selected. The general experimental design was that isolated neutrophils were incubated with cytochalasin B, the chemotactic peptide formylmethionyl-leucyl-phenylalanine as an activator, and a chemiluminescence indicator to measure superoxide production (Bender and Van Epps, 1983; Briheim *et al.*, 1984; Dahlgren *et al.*, 1985; Anderson and Lukey, 1987). Cytochalasin B was added in all experiments because it enhanced chemiluminescence (Bender and Van Epps, 1983). The first chemiluminescence indicator studied was luminol. Luminol gave one peak during the first 2–3 min after activation, and a second peak for the next 15–20 min. The initial peak was believed to represent extracellular superoxide, and the second peak was believed to represent intracellular superoxide. Using luminol as an indicator, formylmethionyl-leucyl-phenylalanine as an activator, and added cytochalasin B, extracellular vitamin C reduced chemiluminescence for the first 5 min after activation, and then had no effect or actually increased chemiluminescence thereafter (Anderson and Lukey, 1987). The interpretation of these data was that vitamin C decreased extracellular but not intracellular superoxide. The second chemiluminescence indicator studied was lucigenin, which was believed to be more specific in measuring extracellular superoxide only, although other explanations were possible (Dahlgren *et al.*, 1985). With lucigenin as an indicator, formylmethionyl-leucyl-phenylalanine as an activator, and added cytochalasin B, vitamin C reduced chemiluminescence for the first 3 min after activation (Anderson and Lukey, 1987). These data were interpreted to indicate that vitamin C decreased extracellular superoxide. Only the latter lucigenin data, and just those from the first minute after activation, were utilized by the Food and Nutrition Board for calculating an EAR for vitamin C (Anderson and Lukey, 1987; Food and Nutrition Board and Panel on Dietary Antioxidants and Related Compounds, 2000a) (*Figure 16.1*).

Regrettably, there are numerous flaws in using these experiments to justify recommended intake. First, these experiments were conducted under nonphysiologic conditions with cytochalasin B. Second, it is uncertain whether an extracellular effect on superoxide concentrations over only the first minute of neutrophil activation has physiologic meaning. Third, the measurements with and without vitamin C were variable and had relatively large standard deviations (SEMs were used by the investigators, see *Figure 16.1*). Fourth, changes in internal neutrophil superoxide in relation to internal vitamin C concentration were ignored. It is necessary to consider this relationship because neutrophil vitamin C concentrations increase as much as 10-fold over minutes when neutrophils are activated by bacteria, via ascorbate recycling (Washko *et al.*, 1993; Welch *et al.*, 1995; Wang *et al.*, 1997). Fifth, it can be debated whether use of the chemotactic peptide formylmethionyl-leucyl-phenylalanine is equivalent to a true physiologic bacterial stimulus. Sixth, although vitamin C had an extracellular effect in these experiments, the data were cited to justify intracellular neutrophil vitamin C concentrations found clinically in relation to dose, as discussed next.

16.3.3 Clinical basis of DRIs for vitamin C

Compared to the last data evaluation for RDAs in 1989, the Food and Nutrition Board was able to consider new data describing dose–concentration relationships for vitamin C in healthy young adults. The Food and Nutrition Board is to be commended for utilizing these data to decide the current RDA for vitamin C.

Despite inclusion of the dose–concentration data, their interpretation and application by the Food and Nutrition Board is open to question. The data justifying biochemical action of vitamin C in neutrophils, described in the previous section, are based on action of extracellular vitamin C as an antioxidant (Anderson and Lukey, 1987) (*Figure 16.1*). Plasma values are the best available indicator of extracellular vitamin C concentrations. We believe that plasma values were the appropriate ones to justify an EAR based on the only available neutrophil experiments (Anderson and Lukey, 1987). However, intracellular neutrophil concentrations were selected instead by the Food and Nutrition Board. Use of plasma values would have resulted in higher EARs and RDAs than those published. There is an additional concern regarding how intracellular neutrophil vitamin C concentrations were used. Neutrophil intracellular concentrations at doses between those actually used were calculated by regression analyses based on the available data. The standard deviation of these calculations was assumed to be 10%, and the EAR was calculated based on this assumption. However, similar assumptions used to calculate the intracellular concentration at an intermediate dose could also have been used to calculate the standard deviation, rather than assuming it was 10% (Levine *et al.*, 2001). A calculated standard deviation would have been 19.4% rather than 10%. Utilizing the EAR of 75 mg, a standard deviation of 19.4% results in an RDA of 104 mg, higher than the published DRI value (Levine *et al.*, 2001).

In contrast to infrequent revisions in prior RDAs, the DRI system was designed to be revised in an ongoing manner as new data became available. Vitamin C data for women were not available when the DRIs/RDAs were published in 2000, and data extrapolated from men were used instead. Data for women were published in 2001 (Levine *et al.*, 2001). Hopefully the data previously not available will be considered and incorporated in a timely way.

The data for women show that there was a sigmoid relationship between dose and plasma concentration, and plasma concentrations saturated at 400 mg daily doses and above. The highest steady state plasma concentration was approximately 80 μM, similar to values in men (*Figure 16.5*). Circulating neutrophils, monocytes, lymphocytes, and platelets were close to saturation by 200 mg daily (*Figure 16.6*). Intracellular vitamin C concentrations at saturation were approximately 1.3–4.0 mM. Urine excretion at steady state for each dose is shown in *Figure 16.7*. The data show that the threshold of urine excretion is close to the 60 mg daily dose, while at 100 mg daily approximately 50% of the dose is excreted.

The same formulas and criteria used to determine EAR and RDA for men as published in the DRI recommendations were applied to these new data (Levine *et al.*, 2001). It was shown that an RDA value based on the actual new data from women was higher when compared to the DRI estimated values that were determined by correcting male data for lean body mass (Food and Nutrition Board and Panel on Dietary Antioxidants and Related Compounds, 2000a). Based on actual rather than estimated data, for adult women the EAR was calculated as 60 mg and the RDA as 90 mg (Levine *et al.*, 2001).

EAR and RDA calculations were based on assumptions that bioavailability was the same for pure vitamin C, vitamin C in supplements, and vitamin C in foods

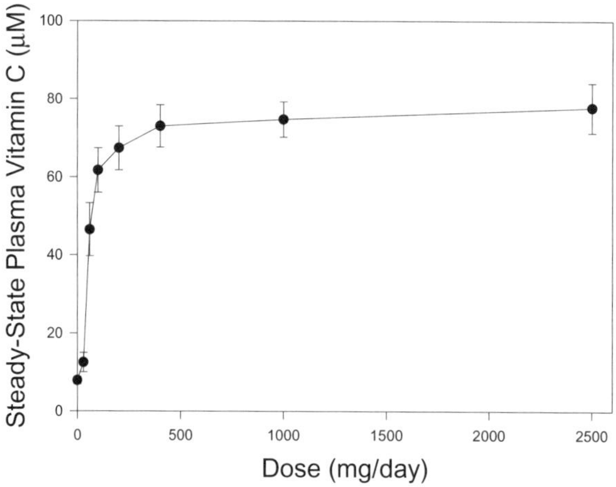

Figure 16.5: Steady state plasma vitamin C concentrations (mean ± SD) as a function of dose for all doses for 15 women.
Subjects consumed a vitamin C deficient diet, resulting in plasma and tissue vitamin C depletion. Vitamin C in solution was then administered by mouth at the doses shown until steady state was reached for each dose. Doses through 200 mg daily were received by 15 subjects, through 1000 mg daily by 13 subjects, and through 2500 mg daily by 10 subjects. Reproduced from Levine *et al.* (2001) where details of the study can be found.

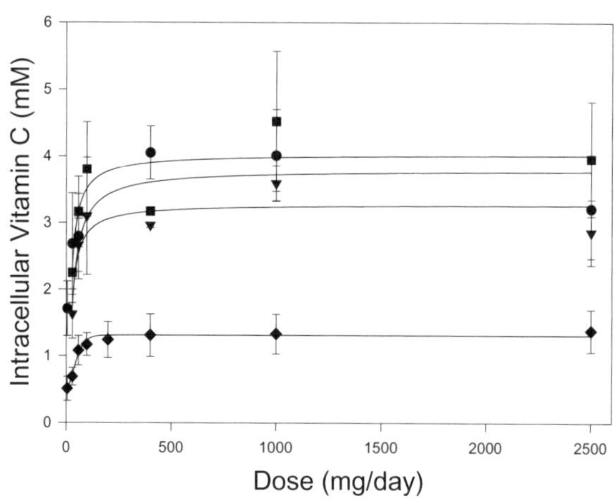

Figure 16.6: Intracellular vitamin C concentrations (mean ± SD) in circulating cells as a function of dose in women.
Cells were isolated when steady state was achieved for each dose. For neutrophils samples were available from 13 women at doses 0–200 mg daily; from 11 women at doses of 400 and 100 mg daily; and from 10 women at 2500 mg daily. For lymphocytes, monocytes, and platelets samples were available from 13 women at 30 mg daily; from 12 women at 60 mg daily; from six women at 100 mg daily; from two women at 400 and 1000 mg daily; from nine women at 2500 mg daily. Reproduced from Levine *et al.* (2001) where details of the study can be found.

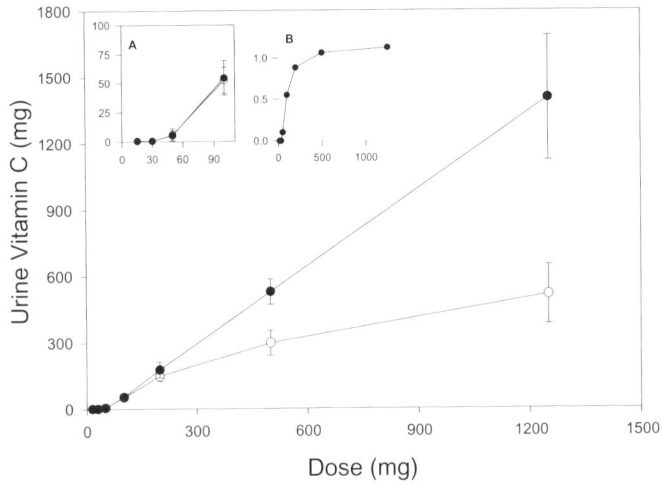

Figure 16.7: Urinary vitamin C excretion as a function of single vitamin C doses at steady state.
Vitamin C excretion over 24 h was determined after administration of single doses given either orally (○) or intravenously (●). Data indicate mean values ± S.D. Inset A, vitamin C excretion for single oral (○) or intravenous (●) doses of 15–100 mg. X-axis indicates dose, Y-axis indicates amount (mg) excreted in urine. Inset B, fractional excretion (the fraction of the dose excreted) after intravenous administration of single doses of vitamin C. X axis indicates dose, Y axis indicates fractional excretion (vitamin C excreted in urine in mg divided by the vitamin C dose in mg). Data from oral and intravenous administration were available from 11 subjects at doses 15, 30, 50, and 200 mg; from 10 subjects at 100 mg; from eight subjects at 500 mg; from nine subjects at 1250 mg. Reproduced from Levine *et al.* (2001) where details of the study can be found.

(Food and Nutrition Board and Panel on Dietary Antioxidants and Related Compounds, 2000a). Based on both the existing data and emerging evidence, these assumptions may not be valid. True vitamin C bioavailability was determined in the NIH depletion–repletion studies for men and women, but findings have been published only for men to date (Levine *et al.*, 1996; Graumlich *et al.*, 1997). These findings were for pure vitamin C only, and no comparisons were made to vitamin C bioavailability in supplements or foods. Only two other studies were cited to support the bioavailability claims (Mangels *et al.*, 1993; Johnston and Luo, 1994). Both of these studies were of relative, not absolute bioavailability, meaning that bioavailability of one vitamin C form was compared to another. One study was a simple comparison of relative bioavailability of vitamin C in three different commercial vitamin tablets (Johnston and Luo, 1994). There was no testing of pure vitamin C without binders or other tablet ingredients. The second study measured repletion rates after depletion as a surrogate for relative bioavailability, with comparison of several different vitamin C forms in tablets or foods (Mangels *et al.*, 1993). This design might obscure differences in true bioavailability, because repletion rates were measured over weeks rather than over hours. Even so, vitamin C in raw broccoli had a significantly lower relative bioavailability compared to the other forms of vitamin C tested. The study design used successive repletion cycles, and the repletion responses from the second depletion were attenuated. Because raw broccoli was given to correct a first depletion, and cooked broccoli to correct a second depletion, the decrease of relative bioavailability from raw broccoli might have been underestimated. Again, pure vitamin C was not tested as a standard. We interpret these data as showing that vitamin C in some foods could be less bioavailable compared to vitamin C in other foods and supplements. New biochemical and animal data also indicate that vitamin C intestinal transport is inhibited by flavonoids, commonly found in fruits and vegetables (Park and Levine, 2000; Song *et al.*, 2002). Based on the evidence considered together, we conclude that it is uncertain whether bioavailability of pure vitamin C is equivalent to bioavailability of vitamin C in foods.

RDAs are designed for intake of vitamin C in foods, not for vitamin C as a pure substance. Recommendations, however, are calculated based on experiments in which pure vitamin C was ingested. If bioavailability of vitamin C from foods is less than for pure vitamin C, EAR and RDA values will have to be revised upwards.

16.3.4 Use of DRIs

The Food and Nutrition Board recognized prior inconsistencies of standardization in formulating nutrition recommendations (Food and Nutrition Board, 1994b). It could be argued that previous nutrition recommendations were determined in part by arbitrary judgments without standardized criteria. The new terms were a result of careful design by the Food and Nutrition Board to begin solving these problems (Yates *et al.*, 1998). However, we are concerned that these new terms have the potential to be double-edged swords. The terms are not easily understood, except by nutrition professionals. Use of the terms by scientists in other disciplines and the general public is uncertain. As nutrition professionals, we are concerned that there is risk of alienating the general public and our supporters if the new terms are not easily understood and readily applied. We fear that other scientists will regard the new terms as camouflage of lack of data, and conclude that nutrition is not at the forefront of scientific advances. New terms will not solve the problem of absence of key data on which to base decisions. In addition, we believe there still remains potential for arbitrary judgments that simply can be fit to the new terms. It is essential that world experts for each nutrient are recruited to participate in analyses of data for RDAs and DRIs.

While well intentioned, the application of EAR to RDAs may be too stringent as designed. If data are not available to reflect variance of an EAR, ten percent standard deviation is used. Unless large populations are studied, ten percent standard deviation is too stringent a standard for most clinical experiments. Because the RDA is EAR plus two times its standard deviation, use of the ten percent figure will result in underestimate of an RDA. Calculation of RDAs is based on the assumption that the target population has a normal distribution for the nutrient, but it is unknown whether this assumption is correct for one or many nutrients and across different populations, life stages, and between the sexes.

16.4 Optimal nutrient ingestion

16.4.1 In situ *kinetics*

Background

We proposed that vitamin C is a model nutrient to determine optimal nutrient ingestion. Optimal nutrient ingestion is taken to mean broadly the amount(s) of nutrient needed to optimize health and prevent disease. For more than half a century, the problem nutrition scientists have faced has been to define and measure 'optimal' (Perla and Marmortston, 1941). Part of our proposal is that 'optimal' can be quantified using specific concentration–function relationships, particularly reaction kinetics. For these measurements to have true clinical application, they must reflect not just measurements of isolated reactions, but rather reactions 'in position', or *in situ*. This means in living cells, in tissue, and eventually in humans. We called this proposal *in situ* kinetics (Levine, 1986; Levine *et al.*, 1991, 1997a, 1997b). Although we have modeled *in situ* kinetics based on vitamin C and discuss its application to this vitamin, the principles should apply to all water-soluble vitamins. Principles of *in situ* kinetics are listed in *Table 16.2* and described next.

Biochemical component

The goal of the biochemical component of *in situ* kinetics is to determine how vitamin C concentrations regulate vitamin C specific reactions in living cells. Both enzymatic and nonenzymatic reactions can be considered for reaction kinetics. We prefer enzymatic reactions because of potential specificity for vitamin C. Nonenzymatic or chemical reactions have the possibility of being affected nonspecifically by other electron donors, or antioxidants, present in cells. The building blocks of biochemical *in situ* kinetics are summarized in *Table 16.2*. More information about *in situ* kinetics principles is found elsewhere (Levine *et al.*, 1997a, 1997b). Utilizing these principles will allow determination of vitamin concentrations that achieve maximal function without toxicity. Because vitamin C is involved in many reactions in cells, each will have to be characterized as function of vitamin C concentration. The combined curves will define optimal biochemical concentrations (Levine *et al.*, 1997a, 1997b).

Clinical component

Information about effective biochemical concentrations is then utililized clinically. The overall goals of clinical *in situ* kinetics are to determine what vitamin concentrations are achieved clinically in relation to dose, and what doses achieve concentrations that effect biochemical reactions. Clinical *in situ* kinetics consists of seven parts, described in *Table 16.2*.

Examples of clinical *in situ* kinetics are the studies describing vitamin C concentrations in relation to dose (Levine *et al.*, 1996; Levine *et al.*, 2001). These studies, described earlier, show that vitamin C concentrations are very tightly controlled in healthy humans. With increasing oral ingestion, vitamin C concentrations achieve saturation in cells and plasma and then do not rise higher (*Figures 16.2, 16.3, 16.5* and *16.6*). The controlling factors and underlying responsible transporters are described in *Table 16.3* and are discussed further in a subsequent section.

Table 16.3: Mediation of tight control of vitamin C concentrations and responsible transporters.

Process	Transporters responsible
Ingestion	–
Bioavailability	SVCT1 ?? others ??
Tissue transport/distribution	SVCT2 ?? others ??
Utilization	??
Renal reabsorption/excretion	SVCT1 ?? others??
Measurement technique	–

Pharmacologic application

A third corollary component of *in situ* kinetics is that vitamin concentrations achieved under some conditions may have pharmacologic rather than physiologic functions. If vitamin C is administered intravenously rather than orally, tight control mechanisms are bypassed until the administered dose is excreted by the kidney and a new steady state is established This means that doses given intravenously can achieve transient high concentrations, over hours, that could never occur with oral ingestion. Whether these pharmacologic concentrations have consequences remains to be explored (Padayatty and Levine, 2000, 2001). While these concentrations may not have nutritional consequences, they may have pharmacologic ones. There are other examples of vitamins having potential pharmacologic roles independent of nutritional ones. Niacin given in far higher doses than needed to prevent deficiency has therapeutic effects in hyperlipidemia management (Xydakis and Ballantyne, 2002; Gotto, 2002), and riboflavin given in the same manner may be useful in migraine prevention (Schoenen *et al.*, 1994, 1998). The pharmacologic components of *in situ* kinetics are listed in *Table 16.2*.

General considerations

In situ kinetics provides a novel framework for achieving optimal nutrient concentrations. While *in situ* kinetics may not be perfect, to our knowledge it is the only comprehensive framework for achieving optimal nutrition, and using nutrition to prevent disease. *In situ* kinetics represents one means of fulfilling the true power of nutrition in health maintenance and disease prevention.

The Food and Nutrition Board is to be commended for recognizing some of the principles of *in situ* kinetics, for deciding to base nutritional recommendations on preventing disease when data are available, and moving beyond prevention of deficiency as the benchmark. There has also been inherent recognition that concentration–function relationship data are key for deriving nutrient recommendations. However, intake amount and resulting concentration can be confused as equivalent, when they may not be (Levine *et al.*, 1999). As described below, as a consequence of disease and/or genes it is possible that nutrient intake may not produce the expected nutrient concentration. In addition, intake of vitamin C may not always be a good index of achieved concentration because there is a steep dose–concentration relationship across a narrow dose range (*Figures 16.2* and *16.5*). For these reasons, whenever possible both clinical and biochemical outcome should be correlated with measured nutrient concentration, rather than intake (Levine *et al.*, 1999).

Experiments are sorely needed that describe concentration–function relationships for many different vitamins in cells and that can then be extended to clinical testing and relevance. The Food and Nutrition Board recognized these problems with respect to vitamin C, as well as other vitamins, and called for experiments to address them (Food and Nutrition Board and Panel on Dietary Antioxidants and Related Compounds, 2000a). These experiments are very difficult to conduct, but are key to solving the issue of concentration–function relationships.

16.4.2 New directions

Biochemical and molecular functions

Concentration dependent enzymatic functions of vitamin C are probably the most specific reactions that can be studied for *in situ* kinetics. These reactions are not easily studied in intact cells especially because of assay limitations and problems inherent in determining kinetics in living cells. Nevertheless, we firmly believe that characterization of vitamin C dependent enzyme kinetics in cells is essential for determining optimal nutrient recommendations. Discovery of new enzymatic functions of vitamin C will broaden these studies. New techniques are needed to characterize concentration dependent enzyme function *in vitro* and *in vivo*.

Antioxidant, or chemical, nonenzymatic vitamin C functions offer a second broad class of reactions for *in situ* kinetics. These reactions may not be specific for vitamin C as an electron donor; this is a potential disadvantage. However, there are many unknowns concerning how cell function *in vivo* is mediated by these reactions, and their depth and breadth remain to be discovered (Padayatty *et al.*, 2003). It is quite possible that some will have specificity for vitamin C. Measurement limitations of oxidants and their products must be overcome for these possibilities to be explored fully (Padayatty *et al.*, 2003). Once target reactions are identified, concentration function principles will have to be applied.

Enzymatic and many antioxidant effects of vitamin C are post-translational events, at the protein or protein lipid level. It is possible that vitamin C, particularly as an antioxidant, may have effects on maintaining genomic integrity, preventing or repairing DNA damage, or modulating gene transcription or message translation. These areas remain to be explored fully.

For enzymatic and chemical vitamin C dependent reactions, whether genomic, transcriptional, translational, or post-translational, the first goal is to identify target reactions. Vitamin C dependence, however, is not sufficient. Concentration–function relationships will have to be characterized, and these should have clinical relevance. For example, an intracellular vitamin C dependent reaction that achieves V_{max} at 10 μM vitamin C has little clinical application. This is because intracellular vitamin C concentrations are always far higher clinically except perhaps in red cells from patients who have severe vitamin C deficiency.

Effects of vitamin C on enzymatic and antioxidant reactions may vary as a consequence of genomic diversity. It is possible that a given enzyme may be variably affected by vitamin C in different individuals due to single nucleotide polymorphisms, deletions, rearrangements, or other genomic differences. Roger Williams originally described this concept as 'biochemical individuality' more than 30 years ago (Williams and Pelton, 1966; Williams and Deason, 1967). More recent examples of the potential effects of genetic diversity on enzyme kinetics have been reviewed (Ames *et al.*, 2002). These examples reinforce Williams' observations in animals that there are substantial variations with respect to nutrient need, that these variations have a genomic basis, and that such variations may substantially impact nutrient recommendations. Variations in vitamin C transport and reducing proteins could also have similar impact, by affecting the amount of vitamin available intracellularly at any given extracellular concentration.

Vitamin C is transported by sodium dependent transporters SVCT1 and SVCT2 (*Table 16.3*; see Chapter 6 by Wilson). SVCT1 is an absorptive/reabsorptive transporter responsible for intestinal absorption and renal reabsorption. SVCT2 is responsible for vitamin C accumulation in many tissues. SVCT1 has a K_m of ~200 μM and is a high capacity transporter. SVCT2 has a K_m of ~20 μM, with an apparent V_{max} similar to the concentration at which plasma achieves vitamin C saturation (Daruwala *et al.*, 1999; Levine *et al.*, 1999). The fact that the apparent V_{max} of SVCT2 is similar to plasma concentrations of approximately 70 μM is evidence, although indirect, for an optimal plasma vitamin C concentration (Levine *et al.*, 1999). Other vitamin C transporters may exist, either sodium dependent or independent, and these should be searched for and characterized.

Vitamin C recycling may also play a role in vitamin C economy. In vitamin C recycling, vitamin C outside cells is oxidized to dehydroascorbic acid, which is transported by glucose transporters and immediately reduced intracellularly (Washko *et al.*, 1993; Welch *et al.*, 1995; Rumsey *et al.*, 1997) (see chapter by May and Asard). The contribution of vitamin C recycling to overall vitamin C accumulation *in vivo* is uncertain, and probably varies as a function of local dehydroascorbic acid concentrations. Tools to specifically measure vitamin C transport and/or vitamin C recycling are needed. Also needed are assays for dehydroascorbic acid that are not dependent on sample reduction and on subtraction of background vitamin C also present in the same sample. The role of glucose transporters in dehydroascorbic acid transport has not been fully characterized, and this is necessary to determine what tissues have the ability to recycle vitamin C.

Clinical research

Mechanisms of tight control. Clinical data for men and women show that vitamin C concentrations are tightly controlled in plasma and cells when vitamin C is taken by mouth (*Figures 16.2* and *16.5*). If vitamin C ingestion is not rate limiting, healthy young adult humans strive for steady state plasma concentrations of approximately 70–80 μM, with saturation at millimolar concentrations of many cellular stores. These concentrations are achieved with vitamin C ingestion of 200–300 mg daily, generally available from foods. Once these concentrations are achieved, ingestion of doses as much as sixfold higher does not change steady state values in plasma and tissues. The factors responsible for tight control and the underlying known vitamin C transporters are shown in *Table 16.3*.

Modifications in tight control. We predict there will be direct clinical consequences to vitamin C concentrations if mechanisms of tight control are altered. A prime example is based on the role of renal reabsorption and renal excretion in maintaining vitamin C concentrations. If vitamin C renal reabsorption is incomplete or aberrant, then we predict there will be a 'renal leak' of vitamin C. If this occurred, ingested doses would have to be increased to achieve steady state values found in healthy subjects. If the leak is substantial, it may not be possible to reach steady state plasma values found in healthy subjects who ingest 200–300 mg daily. Before this hypothesis can be tested, the threshold of vitamin C excretion must first be characterized in subjects with normal renal function. Once this threshold is known, variations can be sought in subjects with subtle or stark abnormalities renal function.

Similar to patients with aberrant renal function, tight control in healthy people could be altered by variations in the renal threshold for vitamin C excretion. Such variations could be a consequence of genomic polymorphisms (single nucleotide polymorphisms, SNPs) or mutations in vitamin C transporters. Changes in the coding region of SVCT1 have the potential to affect gastrointestinal absorption and renal reabsorption, while similar changes in SVCT2 could affect cellular accumulation in many tissues. In addition, it is possible that there are other vitamin C transporters that have not been identified, and changes in these transporters also have the potential to affect either absorptive or cellular vitamin C transport.

Another example of modifying tight control with consequences for healthy people is altered gastrointestinal absorption. Fruits and vegetables contain flavonoids that dampen vitamin C absorption, as shown in animals and for the expressed intestinal transporter SVCT1. Clinical recommendations are based in part on our bioavailability data. The data, obtained using pure vitamin C studied like a drug, showed that vitamin C bioavailability was nearly complete for doses up to 100 mg (Levine *et al.*, 1996; Graumlich *et al.*, 1997). However, RDAs are specifically for food intake. If food substances such as flavonoids decrease vitamin C absorption, then EARs and RDAs for vitamin C may be too low. Experiments to test the effects of food components on vitamin C bioavailability are clearly necessary, and should be possible despite obstacles of food source uniformity. Altered vitamin C absorption could also occur in intestinal diseases associated with aberrant nutrient absorption, as a consequence of diseases of the intestinal mucosa, such as in patients with Crohn disease or gluten enteropathy (celiac sprue).

Tight control could also be modified due to excessive vitamin C utilization, which could be caused by increased oxidants, increased enzyme activity, or altered metabolism of vitamin C dependent reactions. Increased oxidant concentrations that accelerate utilization of vitamin C have been proposed to occur in a number of diseases, including diabetes, myocardial infarction, acute pancreatitis, sepsis, and critical illnesses (Padayatty *et al.*, 2003) Increased utilization of vitamin C has been difficult to test in part because sensitive and specific assays to measure oxidants *in vivo* are not available. Oxidants by their nature are unstable, and reliable *in vivo* assays have remained elusive. Assays that measure putative consequences of *in vivo* oxidant activity rather than oxidants themselves are also problematic. Ideal assays would measure circulating oxidants *in vivo*, consequences of *in vivo* oxidant activity, or specific vitamin C dependent function *in vivo*.

Potential purposes of tight control. Tight control of vitamin C concentrations assures that concentrations are maintained within a narrow range at steady state, unless there is diminished ingestion. It is unknown why vitamin C concentrations in plasma and cells are highly regulated, but there are two broad possibilities. First, vitamin C concentrations may be kept within a narrow range in the general circulation because local, or paracrine, concentrations are much higher and have regulatory function (Levine and Morita, 1985). If vitamin C concentrations approached local values, putative regulatory function could be lost. Second, vitamin C concentrations may be kept below some value, either intracellularly or extracellularly, because of potential harm, either from vitamin C itself or a metabolic product. The most likely possibility for harm is that when plasma vitamin C concentrations rise above a certain value, the vitamin becomes an inappropriate electron donor, with formation of free radicals that damage cells. Data are limited in support of either hypothesis, although each is attractive, testable, and has clinical consequences.

As an example of the second hypothesis, tight control can be bypassed at least transiently by intravenous administration of vitamin C, to purposely generate oxidants for possible therapeutic purposes (Padayatty and Levine, 2000, 2001). Renal excretion will restore plasma concentrations to a steady state baseline over hours, dependent on dose. The concentrations obtained from intravenous administration could have pharmacologic but not physiologic consequences, because the relevant concentrations would be achieved only by intravenous administration. Exploring this area will not be pertinent to optimal recommendations, but instead is a logical consequence of the available clinical data about tight control. If concentrations achieved only intravenously have potential benefit, the area should be explored because of little expected adverse effects.

16.4.3 Interim strategies

Only partial data are available for optimal vitamin C intake. Data are far from complete for many reasons as noted above. In summary, these include difficulty of performing kinetics *in situ*, incomplete information about vitamin C functions at many cellular levels, and lack of assays to detect vitamin C functional consequences *in vitro* and *in vivo*. We conclude that studies of vitamin C optimal intake are only in their infancy, with many obstacles to be overcome.

Despite the complexities, interim strategies are possible. We believe a promising one is based on the role of the kidney in maintaining tight control. Until data are available for concentration dependent functions, an interim hypothesis is that recommended dietary allowances can be calculated from doses that produce vitamin C plasma concentrations at the threshold of renal excretion. The basis of this hypothesis is observational. Because vitamin C is not excreted below a threshold, concentrations up to the threshold are proposed to be physiologically desirable, although for unknown reasons. This hypothesis is not new, and was the basis of previous recommended dietary allowances for vitamin C until DRIs were released in 2000 (Food and Nutrition Board, 1980, 1989). This hypothesis is also the basis in part of other RDAs (Food and Nutrition Board, 1998a, 1998b).

Prior problems concerning this notion were not with the hypothesis *per se*, but with the data to support it. Data were based on nonspecific vitamin C assays subject to interference, on experimental design without control of vitamin C ingestion, narrow dose ranges of vitamin C, unknown subject compliance, few synchronized measurements of vitamin C in plasma and urine, and limited ability to compare accurate dose–concentration data and urine threshold data.

These problems were addressed and solved in the depletion–repletion studies in men and women (Levine *et al.*, 1996, 2001). Subjects had vitamin C intake strictly controlled, a wide dose range was used, and vitamin C was measured by a specific HPLC electrochemical assay. Concurrent urine and plasma vitamin measurements will soon be available from each bioavailability sampling in all subjects at every dose. Taken together, these data should allow accurate determination of the threshold of vitamin C excretion as a function of plasma concentration for each subject. This plasma concentration can then be fit to the dose–concentration curve for each subject to determine the dose that produced the threshold concentration. This dose represents a true EAR, and an RDA can be calculated based on the measured, rather than calculated, standard deviation of this measurement.

16.4.4 Long-term goals

The long-term goals of ideal nutrient recommendations are to determine optimal concentrations based on clinical functional consequences, in healthy people and in people with disease. For healthy people, goals may be to attain concentrations that maximize biochemical or molecular function with clinical benefit or potential to prevent disease. For patients with disease, goals may be to restore or maintain aberrant vitamin C dependent functions, and show clinical benefit.

For vitamin C, these studies are in their infancy. We have dose–concentration relationships only in young healthy people, without clear functional consequences. We need to expand these studies to other ages and develop models to do so without prolonged hospitalizations as used in the depletion–repletion studies. Exploration of functional consequences is essential at all levels, from molecular to clinical. We also must expand our visions to investigate patients with renal disease, diabetes, and conditions where increased vitamin C utilization might occur. Dose–concentration relationships in smokers should be characterized, as smokers account for approximately 20% of the US adult population and have increased utilization of vitamin C (Food and Nutrition Board and Panel on Dietary Antioxidants and Related Compounds, 2000a). Assays are clearly needed to measure *in vivo* oxidants that are specific, sensitive, free from interferences, and reproducible. We must account for genomic variations in transporters and functional outcomes. Corollary data that implicate vitamin C as a potential pharmacologic agent should also be explored, because of little anticipated toxicity with potential for benefit.

The principles of *in situ* kinetics can be modified and applied to many other water-soluble vitamins. We have little understanding of the potential synergy between nutrients. There is possibility to improve functional consequences if many reactions in a pathway are modified together, rather than one at a time. The concept of combined therapy rather than a single agent is used now in treating tuberculosis, bacterial infection, and HIV. This concept of nutritional synergy has tremendous potential for optimal nutrition, if we can measure functional consequences. Using the concepts presented here, many nutrient recommendations can have a functional and concentration dependent basis. The ideal is to maximize cellular milieu for optimal function for all nutrients. These strategies have tremendous potential to use the full but unappreciated power of nutrition to optimize health and prevent disease.

References

Allen JC, Keller RP, Archer P, Neville MC (1991) Studies in human lactation: milk composition and daily secretion rates of macronutrients in the first year of lactation. *Am. J. Clin. Nutr.* **54**, 69–80.

Ames BN, Elson-Schwab I, Silver EA (2002) High-dose vitamin therapy stimulates variant enzymes with decreased coenzyme binding affinity (increased K(m)): relevance to genetic disease and polymorphisms. *Am. J. Clin. Nutr.* **75**, 616–658.

Anderson R, Lukey PT (1987) A biological role for ascorbate in the selective neutralization of extracellular phagocyte-derived oxidants. *Ann. N.Y. Acad. Sci.* **498**, 229–247.

Bender JG, Van Epps DE (1983) Analysis of the bimodal chemiluminescence pattern stimulated in human neutrophils by chemotactic factors. *Infect. Immun.* **41**, 1062–1070.

Briheim G, Stendahl O, Dahlgren C (1984) Intra and extracellular events in luminol dependent chemiluminescence of polymorphonuclear leukocytes. *Infect. Immun.* **45**, 1–5.

Butte NF, Garza C, Smith EO, Nichols BL (1984) Human milk intake and growth in exclusively breast-fed infants. *J. Pediatr.* **104**, 187–195.

Cameron E, Campbell A (1974) The orthomolecular treatment of cancer. II. Clinical trial of high-dose ascorbic acid supplements in advanced human cancer. *Chem. Biol. Interact.* **9**, 285–315.

Committee on Medical Aspects of Food Policy (1991). Dietary Reference Values for Food Energy and Nutrients for the United Kingdom: Report of the Panel on Dietary Reference Values of the Committe on Medical Aspects of Food Policy. Report on Health and Social Subjects No. 41. Department of Health, London.

Dahlgren C, Aniansson H, Magnusson K-E (1985) Pattern of formylmethionyl-leucyl-phenylalanine-induced luminol and lucigenin dependent chemiluminescence in human neutrophils. *Infect. Immun.* **47**, 326–328.

Daruwala R, Song J, Koh WS, Rumsey SC, Levine M (1999) Cloning and functional characterization of the human sodium-dependent vitamin C transporters hSVCT1 and hSVCT2. *FEBS Lett.* **460**, 480–484.

Dewey KG, Finley DA, Lonnerdal B (1984) Breast milk volume and composition during late lactation (7–20 months). *J. Pediatr. Gastroenterol. Nutr.* **3**, 713–720.

Dhariwal KR, Washko P, Hartzell WO, Levine M (1989) Ascorbic acid within chromaffin granules. In situ kinetics of norepinephrine biosynthesis. *J. Biol. Chem.* **264**, 15404–15409.

Food and Nutrition Board (1953) *Recommended Dietary Allowances.* 1–2. National Research Council, Washington, D.C., pp. 1–18.

Food and Nutrition Board (USRC) (1974) *Recommended Dietary Allowances.* National Academy of Sciences, Washington, D.C., p. 2.

Food and Nutrition Board (USRC) (1980) *Recommended Dietary Allowances.* National Academy Press, Washington, D.C., pp 12–17.

Food and Nutrition Board (USRC) (1989) *Recommended Dietary Allowances.* National Academy Press, Washington, D.C., pp. 8–10.

Food and Nutrition Board (USRC) (1994a) Concepts Underlying the Recommended Dietary Allowances. In: *How Should the Recommended Dietary Allowances be Revised?* National Academy Press, Washington, D.C., pp. 10–11.

Food and Nutrition Board (USRC) (1994b) Concepts Underlying the Recommended Dietary Allowances. In: *How Should the Recommended Dietary Allowances be Revised?* National Academy Press, Washington, D.C., pp. 17–24.

Food and Nutrition Board (USRC) (1994c) In: *How Should the Recommended Dietary Allowances be Revised?* National Academy Press, Washington, D.C., pp. 7–15.

Food and Nutrition Board (1998a) Riboflavin. In: *Dietary Reference Intakes: Thiamin, Riboflavin, Niacin, Vitamin B6, Folate, Vitamin B12, Pantothenic Acid, Biotin, and Choline* (ed. Food and Nutrition Board). National Academy Press, Washington, D.C., pp. 87–122.

Food and Nutrition Board (1998b) Thiamin. In: *Dietary Reference Intakes: Thiamin, Riboflavin, Niacin, Vitamin B6, Folate, Vitamin B12, Pantothenic Acid, Biotin, and Choline* (ed. Food and Nutrition Board). National Academy Press, Washington, D.C., pp. 58–86.

Food and Nutrition Board, Panel on Dietary Antioxidants, Related Compounds (2000a) Vitamin C. In: *Dietary Reference Intakes for Vitamin C, Vitamin E, Selenium, and Carotenoids.* National Academy Press, Washington, D.C., pp. 95–185.

Food and Nutrition Board, Panel on Dietary Antioxidants, Related Compounds (2000b) Summary. In: *Dietary Reference Intakes for Vitamin C, Vitamin E, Selenium, and Carotenoids.* National Academy Press, Washington, D.C., pp. 2–4.

Food and Nutrition Board, Panel on Dietary Antioxidants, Related Compounds (2000c) Uses of Dietary Reference Intakes. In: *Dietary Reference Intakes for Vitamin C, Vitamin E, Selenium, and Carotenoids.* National Academy Press, Washington, D.C., pp. 383–400.

GoldSmith GA (1961) Human requirements for vitamin C and its use in clinical medicine. *Ann. N.Y. Acad. Sci.* **92**, 230–245.

Gotto AMJ (2002) Management of dyslipidemia. *Am. J. Med.* **112** Suppl 8A: 10S–18S.

Graumlich JF, Ludden TM, Conry-Cantilena C, Cantilena LR Jr, Wang Y, Levine M (1997) Pharmacokinetic model of ascorbic acid in healthy male volunteers during depletion and repletion. *Pharm. Res.* **14**, 1133–1139.

Halliwell B, Wasil M, Grootveld M (1987) Biologically significant scavenging of the myeloperoxidase-derived oxidant hypochlorous acid by ascorbic acid. Implications for antioxidant protection in the inflamed rheumatoid joint. *FEBS Lett.* **213**, 15–17.

Heinig MJ, Nommsen LA, Peerson JM, Lonnerdal B, Dewey KG (1993) Energy and protein intakes of breast-fed and formula-fed infants during the first year of life and their association with growth velocity: the DARLING Study. *Am. J. Clin. Nutr.* **58**, 152–161.

Hoffer A (1971) Ascorbic acid and toxicity. *N. Engl. J. Med.* **285**, 635–636.

Hoyt CJ (1980) Diarrhea from vitamin C. *JAMA* **244**, 1674–1674.

Jialal I, Vega GL, Grundy SM (1990) Physiologic levels of ascorbate inhibit the oxidative modification of low density lipoprotein. *Atherosclerosis* **82**, 185–191.

Johnston CS, Luo B (1994) Comparison of the absorption and excretion of three commercially available sources of vitamin C. *J. Am. Diet. Assoc.* **94**, 779–781.

Kallner A, Hartmann D, Hornig D (1979) Steady-state turnover and body pool of ascorbic acid in man. *Am. J. Clin. Nutr.* **32**, 530–539.

Kallner AB, Hartmann D, Horni, DH (1981) On the requirements of ascorbic acid in man: steady-state turnover and body pool in smokers. *Am. J. Clin. Nutr.* **34**, 1347–1355.

Levine M (1986) New concepts in the biology and biochemistry of ascorbic acid. *N. Engl. J. Med.* **314**, 892–902.

Levine M, Cantilena CC, Dhariwal KR (1993) In situ kinetics and ascorbic acid requirements. *World Rev. Nutr. Diet.* **72**, 114–127.

Levine M, Conry-Cantilena C, Wang Y, Welch RW, Washko PW, Dhariwal KR, *et al.* (1996) Vitamin C pharmacokinetics in healthy volunteers: evidence for a Recommended Dietary Allowance. *Proc. Natl Acad. Sci. USA* **93**: 3704–3709.

Levine M, Dhariwal KR, Washko PW, Butler JD, Welch RW, Wang YH, Bergsten P (1991) Ascorbic acid and in situ kinetics: a new approach to vitamin requirements. *Am. J. Clin. Nutr.* **54**, 1157S–1162S.

Levine M, Hartzell W (1987) Ascorbic acid: the concept of optimum requirements. *Ann. N.Y. Acad. Sci.* **498**, 424–444.

Levine M, Morita K (1985) Ascorbic acid in endocrine systems. *Vitam. Horm.* **42**, 1–64.

Levine M, Rumsey S, Wang Y (1997a) Principles involved in formulating recommendations for vitamin C intake: a paradigm for water–soluble vitamins. *Methods Enzymol.* **279**, 43–54.

Levine M, Rumsey SC, Wang Y, Park J, Kwon O, Amano N (1997b) In situ kinetics: an approach to recommended intake of vitamin C. *Methods Enzymol.* **281**: 425–437.

Levine M, Rumsey SC, Daruwala RC, Park JB, Wang Y (1999) Criteria and recommendations for vitamin C intake. *JAMA* **281**: 1415–1423.

Levine M, Wang Y, Padayatty SJ, Morrow J (2001) A new recommended dietary allowance of vitamin C for healthy young women. *Proc. Natl Acad. Sci. USA* **98**, 9842–9846.

Ludvigsson J, Hansson LO, Tibbling G (1977) Vitamin C as a preventive medicine against common colds in children. *Scand. J. Infect. Dis.* **9**, 91–98.

Lykkesfeldt J, Christen S, Wallock LM, Chang HH, Jacob RA, Ames BN (2000) Ascorbate is depleted by smoking and repleted by moderate supplementation: a study in male smokers and nonsmokers with matched dietary antioxidant intakes. *Am. J Clin. Nutr.* **71**, 530–536.

Mangels AR, Block G, Frey CM, Patterson BH, Taylor PR, Norkus EP, Levander OA (1993) The bioavailability to humans of ascorbic acid from oranges, orange juice and cooked broccoli is similar to that of synthetic ascorbic acid. *J. Nutr.* **123**: 1054–1061.

Marangon K, Herbeth B, Lecomte E, Paul-Dauphin A, Grolier P, Chancerelle Y, Artur Y, Siest G. (1998) Diet, antioxidant status, and smoking habits in French men. *Am. J. Clin. Nutr.* 67, 231–239.

Montalto MB, Benson JD, Martinez GA (1985) Nutrient intakes of formula-fed infants and infants fed cow's milk. *Pediatrics* **75**, 343–351.

National Research Council (1941) *Recommended Dietary Allowances*. National Research Council, Washington, D.C.

Padayatty SJ, Katz A, Wang Y, Eck P, Kwon O, Lee JH, *et al.* (2003) Vitamin C as an antioxidant: evaluation of its role in disease prevention. *J. Am. Coll. Nutr.* **22**, 18–35.

Padayatty SJ, Levine M (2000) Reevaluation of ascorbate in cancer treatment: emerging evidence, open minds and serendipity. *J. Am. Coll. Nutr.* **19**, 423–425.

Padayatty SJ, Levine M (2001) New insights into the physiology and pharmacology of vitamin C. *CMAJ.* **164**, 353–355.

Park JB, Levine M (2000) Intracellular accumulation of ascorbic acid is inhibited by flavonoids via blocking of dehydroascorbic acid and ascorbic acid uptakes in HL-60, U937 and Jurkat cells. *J. Nutr.* **130**, 1297–1302.

Perla D, Marmortston J (1941) The effect of vitamin C on resistence. In: *Natural Resistance and Clinical Medicine*. Little Brown & Co, Boston, pp. 1038–1091.

Rajalakshmi R, Deodhar AD, Ramarkrishnan CV (1965) Vitamin C secretion during lactation. *Acta Paediatr. Scand.* **54**, 375–382.

Rumsey SC, Kwon O, Xu GW, Burant CF, Simpson I, Levine M (1997) Glucose transporter isoforms GLUT1 and GLUT3 transport dehydroascorbic acid. *J. Biol. Chem.* **272**, 18982–18989.

Salmenpera L (1984) Vitamin C nutrition during prolonged lactation: optimal in infants while marginal in some mothers. *Am. J Clin. Nutr.* **40**, 1050–1056.

Schoenen J, Jacquy J, Lenaerts M (1998) Effectiveness of high-dose riboflavin in migraine prophylaxis. A randomized controlled trial. *Neurology* **50**, 466–470.

Schoenen J, Lenaerts M, Bastings E (1994) High-dose riboflavin as a prophylactic treatment of migraine: results of an open pilot study. *Cephalalgia* **14**, 328–329.

Sneed SM, Zane C, Thomas MR (1981) The effects of ascorbic acid, vitamin B6, vitamin B12, and folic acid supplementation on the breast milk and maternal nutritional status of low socioeconomic lactating women. *Am. J. Clin. Nutr.* **34**, 1338–1346.

Song J, Kwon O, Chen S, Daruwala R, Eck P, Park JB, Levine M (2002) Flavonoid inhibition of SVCT1 and GLUT2, intestinal transporters for vitamin C and glucose. *J. Biol. Chem.* **277**, 15252–15260.

Stein HB, Hasan A, Fox IH (1976) Ascorbic acid-induced uricosuria. A consequency of megavitamin therapy. *Ann. Intern. Med.* **84**, 385–388.

Van Eckelen M (1953) Occurrence of vitamin C in foods. *Proc. Nutr. Soc.* **12**, 228–232

Wandzilak TR, D'Andre SD, Davis PA, Williams HE (1994) Effect of high dose vitamin C on urinary oxalate levels. *J. Urol.* **151**, 834–837.

Wang Y, Russo TA, Kwon O, Chanock S, Rumsey SC, Levine M (1997) Ascorbate recycling in human neutrophils: Induction by bacteria. *Proc. Natl Acad. Sci.USA* **94**, 13816–13819.

Washko PW, Wang Y, Levine M (1993) Ascorbic acid recycling in human neutrophils. *J. Biol. Chem.* **268**, 15531–15535.

Welch RW, Wang Y, Crossman A Jr, Park JB, Kirk KL, Levine M (1995) Accumulation of vitamin C (ascorbate) and its oxidized metabolite dehydroascorbic acid occurs by separate mechanisms. *J. Biol. Chem.* **270**, 12584–12592.

Williams RJ, Deason G (1967) Individuality in vitamin C needs. *Proc. Natl Acad. Sci. USA* **57**, 1638–1641.

Williams RJ, Pelton RB (1966) Individuality in nutrition: effects of vitamin A-deficient and other deficient diets on experimental animals. *Proc. Natl Acad. Sci. USA* **55**, 126–134.

Xydakis AM, Ballantyne CM (2002) Combination therapy for combined dyslipidemia. *Am. J. Cardiol.* **90**, 21K–29K.

Yates AA, Schlicker SA, Suitor, CW (1998) Dietary reference intakes: the new basis for recommendations for calcium and related nutrients, B vitamins, and choline. *J. Am. Diet. Assoc.* **98**, 699–706.

Index